Digital Television

THE WILEY BICENTENNIAL–KNOWLEDGE FOR GENERATIONS

Each generation has its unique needs and aspirations. When Charles Wiley first opened his small printing shop in lower Manhattan in 1807, it was a generation of boundless potential searching for an identity. And we were there, helping to define a new American literary tradition. Over half a century later, in the midst of the Second Industrial Revolution, it was a generation focused on building the future. Once again, we were there, supplying the critical scientific, technical, and engineering knowledge that helped frame the world. Throughout the 20th Century, and into the new millennium, nations began to reach out beyond their own borders and a new international community was born. Wiley was there, expanding its operations around the world to enable a global exchange of ideas, opinions, and know-how.

For 200 years, Wiley has been an integral part of each generation's journey, enabling the flow of information and understanding necessary to meet their needs and fulfill their aspirations. Today, bold new technologies are changing the way we live and learn. Wiley will be there, providing you the must-have knowledge you need to imagine new worlds, new possibilities, and new opportunities.

Generations come and go, but you can always count on Wiley to provide you the knowledge you need, when and where you need it!

WILLIAM J. PESCE
PRESIDENT AND CHIEF EXECUTIVE OFFICER

PETER BOOTH WILEY
CHAIRMAN OF THE BOARD

Digital Television
Technology and Standards

John Arnold

Michael Frater

Mark Pickering

The University of New South Wales, ADFA
Canberra, ACT, Australia

WILEY-INTERSCIENCE

A JOHN WILEY & SONS, INC., PUBLICATION

Published by John Wiley & Sons, Inc., Hoboken, New Jersey
Published simultaneously in Canada

Wiley Bicentennial Logo: Richard J. Pacifico

Library of Congress Cataloging-in-Publication Data:

Arnold, John, 1954-
Digital television : technology and standards / by John Arnold, Michael Frater, Mark Pickering.
p. cm.
ISBN 978-0-470-14783-2
1. Digital television. I. Frater, Michael II. Pickering, Mark 1966- III. Title.
TK6678.A77 2007
621.388'07–dc22 2007007077

Printed in the United States of America

10 9 8 7 6 5 4 3 2 1

This book is dedicated to our wives

Gemma Arnold
Emma Frater
Kim Pickering

whose support made possible both this book and our numerous absences at international standards meetings which preceded it.

Contents

Preface

In the last 50 years, television has arguably become the dominant source of entertainment and information in many countries. In the western world, most households own at least one television. Many have two or more. Over this half century, television technology has proved to be very adaptable, able to accommodate upgrades taking advantage of new technology without requiring existing receivers to be replaced. Indeed, a 50-year-old television receiver could still be used in most countries. The maintenance of compatibility with the existing receivers has been achieved through incremental improvements in service quality. The transitions from black-and-white to color television and from mono to stereo audio are both examples of this.

Digital television offers a number of potential advantages over the older, analog technology. High-definition services, providing much greater resolution than the conventional standard-definition television, are possible, as is the packing of several standard definition programs into the same bandwidth as a single analog television channel. The current international move to digital television is more revolutionary in nature than the previous changes, requiring the phasing-in of digital receivers and the subsequent phasing-out of the existing analog receivers. Consumers will therefore be required to purchase new equipment, in the form of a digital television or a decoder, to convert digital signals into a form that can be passed to their existing analog receiver.

This book describes the technology and standards behind digital television. It introduces the basic techniques used in video coding, audio coding, and systems, which provide for the multiplexing of these services and other ancillary data into a single bit stream. The description of standards covers the north-American Advanced Television System Committee (ATSC) and the European Digital Video Broadcasting (DVB). Aspects relating to these standards are described independently, allowing the reader to cover only those parts relevant to one system if desired.

The first chapter provides an introduction to analog and digital television, setting up the basic division of functionality into the representation of video, the representation of audio, and the underlying systems that provide services such as multiplexing of video and audio onto a single channel and modulation. The remaining chapters are grouped into three parts, covering the video, audio, and systems aspects of digital television, respectively. Each of these parts is written so that it can be read independent of the other parts. The first part deals with the coding of digital video signals using the MPEG-2 standard to produce a compressed digital video bit stream.

Chapter 2 describes the characteristics of video material. Chapters 3 and 4 describe the signal processing used to reduce the spatial and temporal redundancy of digital video signals, with Chapter 3 describing predictive coding and Chapter 4 transform coding. Chapter 5 describes the principles behind the syntax used to represent the various data elements carried in a compressed video bit stream, whereas Chapter 6 introduces the specific features of the MPEG-2 video standard.

The second part covers the coding and compression of digital audio, using a similar structure to the first part. Chapter 7 introduces the aspects of the human ear that are critical in determining subjective audio quality, followed in Chapter 8 by a description of the signal processing used for digital audio compression, including the use of subband filter banks in audio coding. Chapters 9 describes the specific methods used by the MPEG-1 and MPEG-2 standards, respectively, with Chapter 10 describing the Dolby AC-3 system used primarily by ATSC.

The third part describes the modulation of digital television services for transmission, the system protocols used for multiplexing, timing, and control, and the other components of a digital television service that provide a range of data services, including closed captioning (also known as subtitling) and teletext. In this part, separate descriptions are provided for the different techniques provided by DVB and ATSC. Chapter 11 describes MPEG-2 systems, which provide the multiplexing, timing, control data for digital television. MPEG-2 systems also carry a collection of data, known as program-specific information, which describes the contents of a systems' bit stream. ATSC and DVB each provide its own extensions to the program-specific information, which are described separately in Chapter 12. Chapter 13 describes the terrestrial broadcast modulation schemes of ATSC and DVB, including the use of channel coding to protect bit streams from errors introduced in transmission. Finally, the closed-captioning and teletext systems are described in Chapter 14.

This book might be used as a textbook supporting a variety of different types of courses. An undergraduate digital television course might be based on a selection of material from Chapters 1–3, 5, 6, and 11. A postgraduate course in digital television, for which background in digital signal processing and digital communications theory is assumed, could extend this to include all material from Chapters 1–3 on video, 5 and 6 on audio, and 10 and 11 on systems, incorporating selections from other chapters. A more specialized course on video coding could be based on Chapters 1–5.

John Arnold
Michael Frater
Mark Pickering
Canberra, Australia
June 2007

Chapter 1

Introduction to Analog and Digital Television

1.1. INTRODUCTION

From small beginnings less than 100 years ago, the television industry has grown to be a significant part of the lives of most people in the developed world, providing arguably the largest single source of information to its viewers.

The first true television system was demonstrated by John Logie Baird in the 1920s. Further experiments were conducted in the following decade, leading to trial broadcasts in Europe and the United States, and eventually to the regular television service we know today. Originally, only monochrome pictures were supported. Color television was introduced in the United States in 1954 and in Europe in 1967.

Television systems have evolved as *simplex* transmission systems, as shown in Figure 1.1. The term simplex means that information flows only in one direction across the channel. A transmitter, whose antenna is usually mounted on a tall tower, broadcasts a signal to a large number of receivers. Each receiver decodes the transmission and passes it on to a display device. Sometimes, the receiver and display are integrated into a single device, such as in a standard television that incorporates a means for the user to select the channel to be viewed. Sometimes, the receiver and display are separate devices, such as when a signal is received through a video cassette recorder (VCR) and passed to an external display. This system is known as *terrestrial broadcast* television.

Satellite and cable television systems operate on similar models. Figure 1.2 shows the outline structure of a cable television system.

1.2. ANALOG TELEVISION

Traditional television services make use of analog technology to provide an audio-visual, broadcast service. The basic structure of an analog television transmitter is shown in Figure 1.3. Video and audio signals, which may be derived from live sources such as cameras and microphones or from storage devices such as video

Digital Television, by John Arnold, Michael Frater and Mark Pickering.

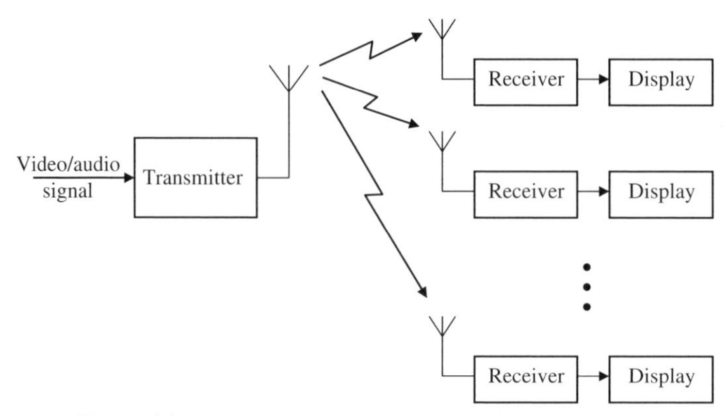

Figure 1.1 Simplex structure of terrestrial broadcast television.

recorders, are fed into separate modulators, whose output is multiplexed and upconverted to form the broadcast signal.

Various methods of modulating, multiplexing, and upconverting the signals to specific broadcast frequencies (as shown in Figure 1.3) are defined in the various analog television standards. Three of the major standards used for analog television are National Television System Committee (NTSC) [1], used primarily in North, Central, and South America, Systeme Electronique (pour) Couleur avec Memoire (SECAM), used in France and countries in eastern Europe such as Poland and Russia, and Phase Alternating Line (PAL) [2], used in many other countries including western Europe and Australia.

In this chapter, we discuss the operation of analog television with reference to three areas: the representation of video, the representation of audio, and the systems that provide the multiplexing of video and audio services into a single channel.

1.2.1. Video

An analog video signal is created by a time sequence of pictures, with 25 or 30 of these pictures displayed every second. Each picture consists of a number of lines,

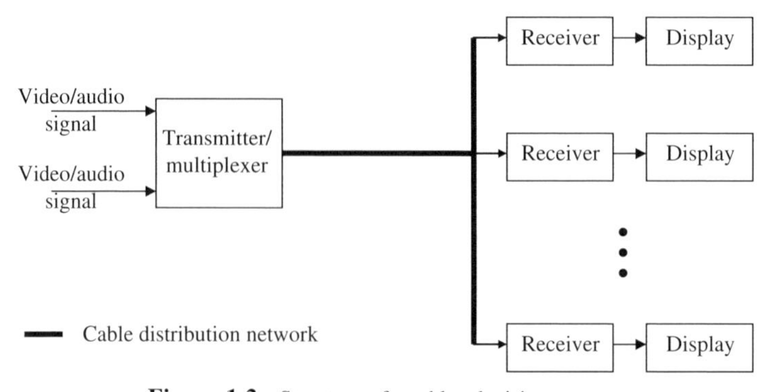

Figure 1.2 Structure of a cable television system.

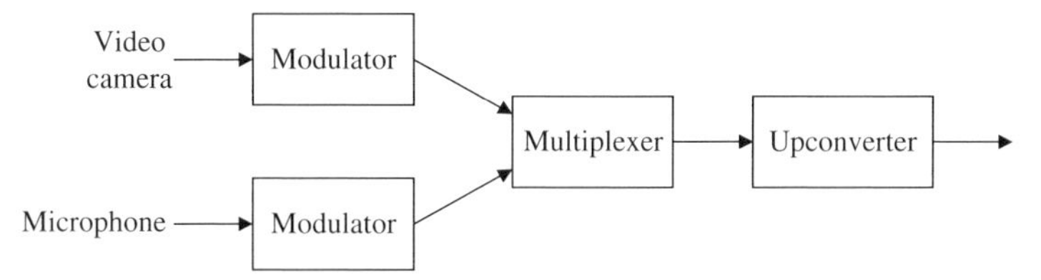

Figure 1.3 Basic structure of an analog television system.

each of which is scanned left to right, as illustrated in Figure 1.4. The vertical resolution is usually 576 lines for 25 Hz systems and 480 lines for 30 Hz systems.

In addition to the displayed lines, a number of other lines of data are transmitted. These are intended to provide time for the scan in a cathode ray tube to return from the bottom right of the display at the end of one picture to the top left of the display at the beginning of the next picture. The inclusion of these nondisplayed lines brings the total number of lines per picture to 625 for 25 Hz systems and 525 for 30 Hz systems. The time in which these nondisplayed lines are transmitted is known as the *vertical blanking interval* (*VBI*).

1.2.1.1. Horizontal Synchronization

In an analog television signal, a synchronization pulse is provided at the start of every line in the picture as shown in Figure 1.5, which shows the waveform for a single line where the brightness decreases in steps from left to right. This means that the display begins its horizontal scan at the same place in the signal as the camera that captured the video signal. In addition, a longer synchronization pulse is used to indicate that the scan should restart at the top left of the display. These synchronization pulses allow the receiver to achieve synchronism with the incoming signal.

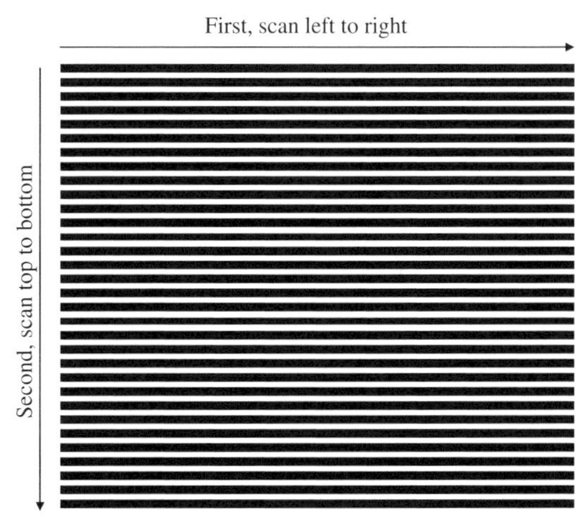

Figure 1.4 Simple left-to-right, top-to-bottom scan.

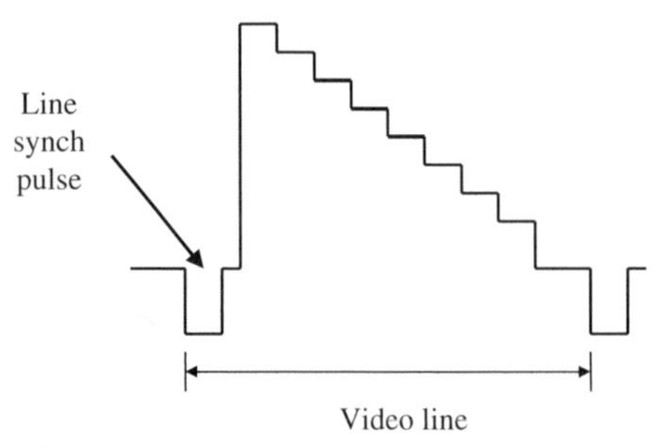

Figure 1.5 Waveform of a single picture line of analog video.

The interval allocated for transmission of the line synchronization pulse and the immediately surrounding regions (known as the front and back porches) is known as the *line blanking interval* or the *horizontal blanking interval*. Its length is 11 μs in NTSC and 12 μs in PAL and SECAM systems.

1.2.1.2. Horizontal Resolution

The horizontal resolution of an analog television system depends on the bandwidth of the video signal. Roughly speaking, the resolution of the system is 2 pixels per Hertz of video bandwidth. These pixels are shared equally between the transmitted lines. The number of useful pixels in each line is reduced by the length of the line blanking interval. The horizontal resolution r_h of an analog video system with bandwidth B is therefore

$$r_h = 2Bt_{ULI}$$

where t_{ULI} is the useful line interval. The horizontal resolutions for a number of in-service analog television systems are shown in Table 1.1. In the case of PAL and SECAM, there are a number of different implementations, each denoted by a

Table 1.1 Approximate horizontal resolution for selected analog television systems.

System	Lines per second (KHz)	Line period (μs)	Useful line interval (line period – line blanking interval) (μs)	Video bandwidth (*B*) (MHz)	Approximate horizontal resolution (pixels)
NTSC	15.750	63.5	52.5	4.2	441
PAL (B, G, H)/ SECAM (B, G)	15.625	64.0	52	5.0	520
PAL (I)	15.625	64.0	52	5.5	572
PAL (D)/SECAM (D, K, K1, L)	15.625	64.0	52	6.0	624

single letter. The video bandwidth varies between implementations, and a number of options are shown.

1.2.1.3. Interlaced Video

When analog television was designed, an important design trade-off was between the service picture rate and the service bandwidth. The picture rate chosen needs to be sufficiently fast to ensure that a human viewer perceives an apparently continuous service (as opposed to a rapid series of individual pictures—called flicker—which would be subjectively most unpleasant). Once the appropriate horizontal and vertical resolution of a television picture had been decided, the desired bandwidth meant that a relatively low picture rate (25 or 30 Hz) was all that could be achieved. These picture rates are insufficient to avoid flicker in all circumstances. However, simply increasing the picture rate would lead to an increase in the required service bandwidth. This was an unacceptable outcome. The developers of analog television overcame this problem using a technique called interlacing.

Interlacing divides each picture into two fields, as shown in Figure 1.6. One field contains the odd lines from the picture (i.e., lines 1, 3, 5, ...) and is called the odd field, whereas the other field contains the even lines from the picture (i.e., lines 2, 4, 6,...) and is called the even field (Figure 6(a)). The odd lines are scanned from the camera system and then half a picture time later (i.e., 1/50th or 1/60th of a second) the even lines are scanned (Figure 6(b)). This approach improves the rendition of moving objects and also completely removes the flicker problem discussed earlier. The trade-off is some loss in vertical resolution of the picture.

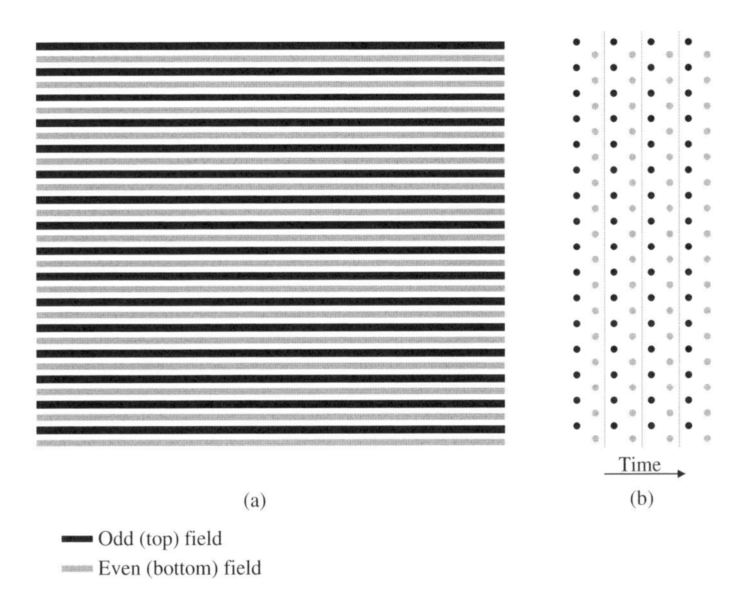

Figure 1.6 Interlace structure showing location of odd and even fields, (a) as seen on the display, and (b) the formation of pictures from two consecutive fields.

Table 1.2 Numbers of video lines per field and picture.

System	Notional picture frequency Hz	Field frequency Hz	Displayed lines per picture	Displayed lines per field	Total lines per picture	Total lines per field
PAL, SECAM	25	50	576	288	625	313 (odd)/ 312 (even)
NTSC	30	60	480	240	525	263 (odd)/ 262 (even)

Table 1.2 shows the number of lines per picture and field for 25 and 30 Hz analog television systems.

EXAMPLE 1.1—MATLAB

The aim of this example is to demonstrate the impact of combining two fields containing a moving object into a single picture.

SOLUTION Figure 1.7 shows two fields of 128 × 128 pixels consisting of a black background and a white square of size 32 × 32 pixels that has moved four pixels to the right between fields.

```
A = zeros(128);                % black background for odd field
B = zeros(128);                % black background for even field
A(49:80, 49:80) = 255 × ones(32);  % white square in odd field
B(49:80, 57:88) = 255 × ones(32);  % white square in even field
(moved 8 pixels right)
```

The individual fields can be displayed using the MATLAB function image.m:

```
image(A)                       % display odd field
image(B)                       % display even field
```

The images obtained by displaying A and B are shown in Figure 1.7.

The two fields can be merged into a single picture, which is then displayed, using the commands below.

```
C = zeros(256,128);
C(1:2:255,:) = A;
C(2:2:256,:) = B;
image(C);
```

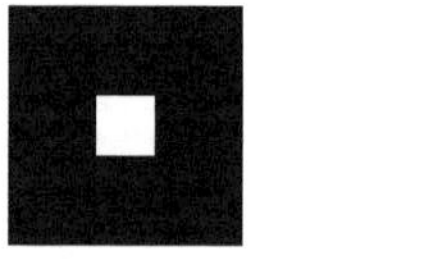

Odd field Even field

Figure 1.7 Odd and even fields produced in Example 1.

Figure 1.8 Merged fields to form a picture.

The resulting image is shown in Figure 1.8. The jagged edges are caused by the movement of the white block between the odd and even fields. In some circumstances, these jagged edges can lead to localized flickering. ■

When fields containing moving objects are merged to form a single picture, straight edges that are moving horizontally are turned into jagged edges. An example from the "Mobile and Calendar" sequence is shown in Figure 1.9, in which the jagged edges of moving objects such as the spots on the ball and the numbers on the calendar are clearly apparent.

1.2.1.4. Color Television

Television was initially a monochrome (black-and-white) service. When color television was to be introduced, the color information needed to be introduced in a way that did not affect substantially the quality of service received by consumers who still had a black-and-white television receiver. As is well known, color receivers display only three colors (red (R), green (G), and blue (B)). The mixing of these colors at the human eye provides the range of colors that we are used to with color television.

Transmitting separate signals for red, green, and blue would triple the bandwidth requirement for color television compared to monochrome television. Because a monochrome signal is not present in this set, the only way to provide a good quality monochrome picture for existing receivers would be to send yet another signal just for this purpose; this would be a very wasteful use of valuable spectrum. The quality of reception at monochrome receivers would have been significantly compromised.

Figure 1.9 Picture from the "Mobile and Calendar" sequence.

The approach taken was to transmit not the color signals *R*, *G*, and *B* but the monochrome signal (known as the luminance *Y*) accompanied by two color difference, or chrominance, signals (*U* and *V*) from which the three colors *R*, *G*, and *B* can be reconstructed. The values of the luminance signal and two chrominance signals can be calculated from *R*, *G*, and *B* according to

$$Y = 0.299R + 0.587G + 0.114B$$

$$U = \frac{B - Y}{2.03}$$

$$V = \frac{R - Y}{1.14}$$

Slightly different versions of these equations are used in different television systems.

The three color signals *R*, *G*, and *B* are reconstructed at the receiver and displayed. Because the luminance signal is still transmitted, it is still available to monochrome receivers and so there is a minimal impact on existing viewers. The color difference signals can also be transmitted with a significantly smaller bandwidth than the luminance signal. This is acceptable because the resolution of the human eye is lower for chrominance than it is for luminance. The use of color difference signals was therefore an early attempt at bandwidth compression.

1.2.2. Audio

The audio accompanying video in an analog television system usually has a bandwidth of approximately 15 kHz. The audio system in analog television originally supported only a single (monophonic) channel. It has been extended with the same philosophy of backward compatibility used for adding color information in the video to provide a range of services, including options for stereo audio, and two independent audio channels. In all cases, the original monophonic audio is still transmitted to support older receivers, with other signals added to provide higher levels of functionality.

1.2.3. Systems

Specification of the representation of audio and video is not sufficient to define a television service. A means is required to multiplex the video signals (luminance and chrominance) and the audio (mono or stereo) onto a single channel. We refer to this capability as the "systems" part of the television service.

Each country specifies a channel bandwidth for broadcast television systems. In North and Central America, 6 MHz is used, whereas 7 or 8 MHz is commonly used in the rest of the world. Approximately 70% of the bandwidth of the channel is allocated to video, with the remaining capacity available for audio and guard bands between channels.

Figure 1.10 shows the spectrum of a typical, monochrome, analog television channel with a single audio channel. Most of the capacity of the channel is allocated to the video, with a small amount available for audio. The video signal is usually modulated using vestigial sideband amplitude modulation with the upper sideband dominant, whereas the audio signal is frequency modulated with a maximum deviation of approximately 50 kHz (giving an audio bandwidth of 100 kHz). The audio carrier is located within the channel, but outside that part of the channel specified for the transmission of video. Each of the analog television standards specifies the locations of the video and audio carriers.

Extension to support color television can be achieved by the multiplexing of the chrominance signals onto the channel, as shown in Figure 1.11. This is done by using

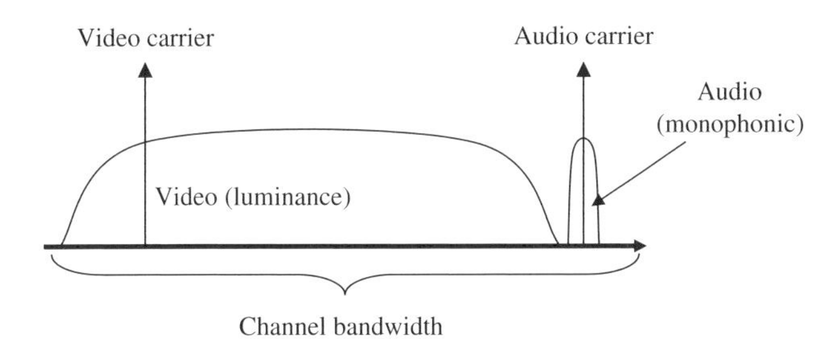

Figure 1.10 Spectrum of a typical monochrome analog television channel.

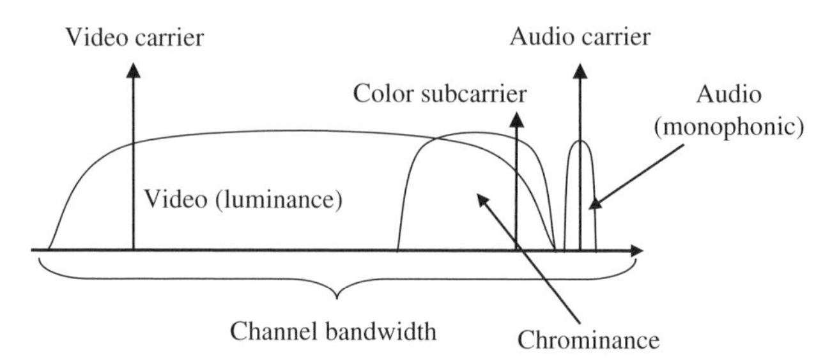

Figure 1.11 Spectrum of a color analog television channel.

vestigial sideband modulation for the chrominance signal, with the lower sideband dominant. Each of the three standards specifies the location for the color subcarrier, which is the carrier frequency associated with the modulation of the chrominance signal. The carriers for the two chrominance signals have the same frequency, but differ in phase by 90°. This "phase multiplexing" allows separation of the signals at the receiver. Noting that most of the energy in video signals occurs at low frequencies, the chrominance information is transmitted toward the upper end of the video spectrum. This does have the effect that high-frequency luminance information can sometimes be mistakenly decoded as color information. It is for this reason that herringbone tweed jackets sometime flair purple on color television receivers. The high-frequency monochrome information from the tweed is incorrectly decoded as color information. The problem has been addressed by television producers becoming aware of the problem and making sure that presenters do not wear inappropriate clothing.

A second audio channel can be incorporated simply by specifying the location of its carrier. Frequency modulation is usually also used for the second audio channel. Backward compatibility is maintained by ensuring that a valid monophonic audio signal for the program is transmitted on the original audio carrier. This is illustrated in Figure 1.12.

Each of the various standards for analog television (NTSC, PAL, and SECAM) specifies frequencies for the video carrier, color subcarrier, and audio carriers. Each standard also specifies maximum bandwidths for the video and each of the audio channels.

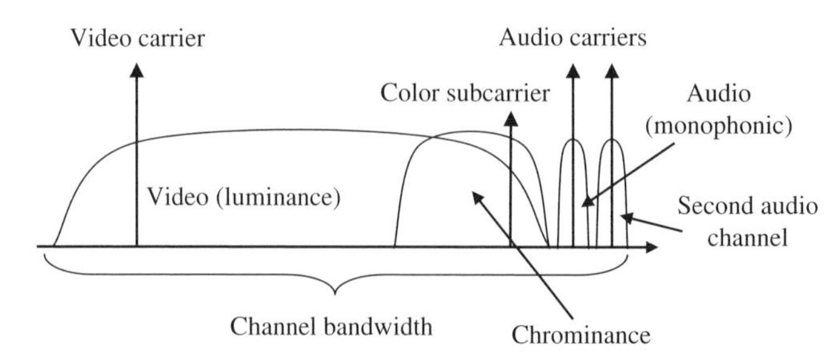

Figure 1.12 Spectrum of a typical color analog television channel with stereo audio.

1.2.3.1. Ancillary Services

Analog television systems have evolved to carry not only audio and video signals, but also a range of ancillary data services. These ancillary services make use of the nondisplayed lines of the video vertical blanking interval to provide low-rate data services such as closed captioning (also known as subtitling) and teletext. Because these services are carried in the vertical blanking interval, they have no impact on receivers that are not equipped to decode them.

1.3. THE MOTIVATION FOR DIGITAL TELEVISION

The initial impetus for moving to a digital signal was standards conversion (e.g., from 525 line NTSC at 30 pictures/s to 625 line PAL at 25 pictures/s). This is an extremely difficult process in the analog domain. Significant signal processing is still required in the digital domain. However, appropriate high-speed hardware can be built to allow the task to be successfully carried out. Other motivations for the change from analog to digital television include carrying multiple digital television channels within the existing bandwidth allocated to a single analog television service, the ability to carry higher resolution services (such as high-definition television) in a single channel, and the integration of a range of interactive services into the television broadcast.

From a communication point of view, digital transmission has many advantages. In particular, it offers considerable noise immunity. Consider the analog signal shown in Figure 1.13. The original analog signal is perturbed by noise. If the noise is

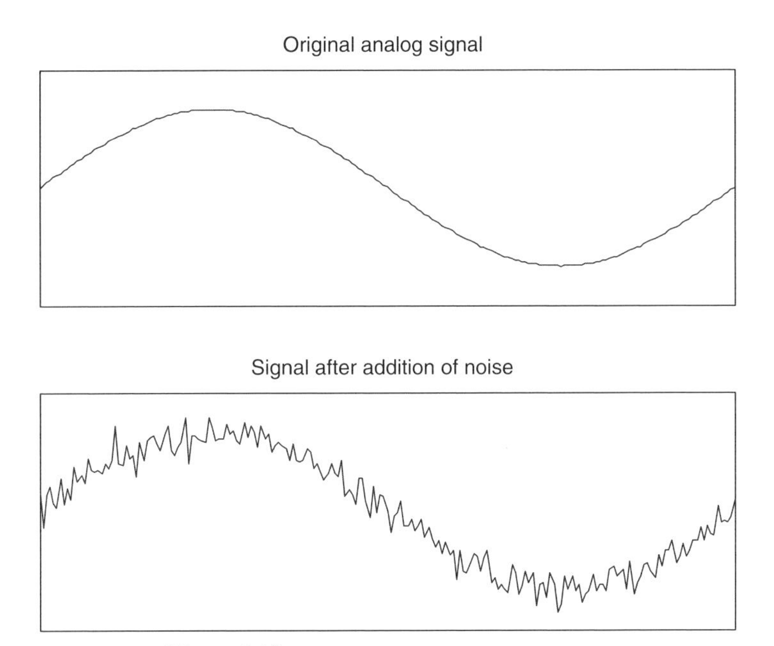

Figure 1.13 Impact of noise on an analog signal.

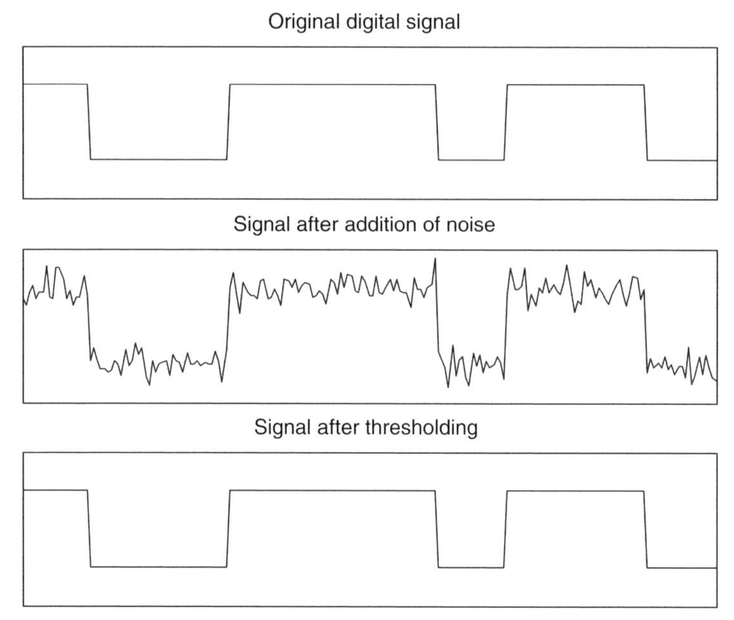

Figure 1.14 Impact of noise on a digital signal.

in the same area of the spectrum as the signal (so-called in-band noise), then there is little that can de done to remove it.

The impact of noise on a digital signal is quite different, as illustrated in Figure 1.14. In this case, a simple thresholding operation allows the original signal to be perfectly reconstructed. Even when the noise is large enough to cross the threshold, enhanced signal processing techniques such as matched filtering [3] can be employed to achieve good performance (which can be improved still further using error correction techniques such as those described in a later chapter). The ability of digital signals to reject noise makes digital systems ideal for long-distance transmission because quality can be maintained through many repeaters.

Other advantages of digital systems include the fact that digital components are of low cost and are very stable. In addition, many digital networks are now emerging for the transmission of audiovisual material at a range of transmission rates.

1.4. THE NEED FOR COMPRESSION

If digital systems offer so many advantages, why have we not moved to digital television long ago? The answer lies in the very high data rates required for transmitting raw, uncompressed digital video and the complexity of the digital systems required to provide real-time processing for compression and decompression.

Table 1.3 Resolution of digital television.

Picture rate (Hz)	25	30
Lines per picture	576	480
Luminance samples per line	720	720
Fields per second	50	60
Interlace	Two fields per picture (2:1)	Two fields per picture (2:1)

The resolution defined for digital television by the ITU-R Recommendation BT.601 [4] is given in Table 1.3. The number of lines per picture is the same as the number of displayed lines for the analog services. When an analog television signal is converted to digital, the nondisplayed lines in the vertical blanking interval are removed. For both 25 and 30 Hz transmission, 14,400 lines per second are transmitted, which means that 10,368,000 pixels (or luminance samples) must be transmitted each second.

For distribution of digital television, each chrominance signal is sampled at half the rate of the luminance signal, that is, at 360 samples per line. Thus, there is one sample of each of the chrominance components (U and V) for every two luminance components (Y). If each of the Y, U, and V is represented to 8-bit accuracy, then an average of 16 bits is required for each luminance sample.

The raw bit rate is therefore 10,368,000 luminance samples per second multiplied by 16 bits per sample, giving a data rate of 165.89 Mbit/s. Even in the highest capacity, modern, communications networks, this is an extremely high capacity to be allocated to a single service.

The corresponding bandwidth requirement for various digital modulation schemes is shown in Table 1.4, each of which is much greater than the 6, 7, or 8 MHz allocated for the transmission of an analog television service. If digital television is to compete effectively with analog television, it needs to be able to utilize a bandwidth not more than (and preferably significantly less than) an equivalent analog service. Of course, the raw data rate could be reduced to achieve this

Table 1.4 Bandwidth requirement for uncompressed digital video using various digital modulation schemes.

Modulation scheme	Bits/second/Hertz of bandwidth	Required bandwidth (MHz)
Binary phase-shift keying (BPSK)	1	165.89
Quadrature phase-shift keying (QPSK)	2	82.94
8-ary phase-shift keying (8-ary PSK)	3	55.30
256-ary quadrature amplitude modulation (256-ary QAM)	8	20.74

goal. This could be done by reducing either the number of samples per line, the number of lines per picture, or the number of pictures per second. Such an approach would seriously affect the quality of the received service and so is not a viable solution.

A similar method can be used to calculate the rate required for transmitting uncompressed digital audio. If each channel of audio is sampled at 44.1 kHz with a resolution of 16 bits per sample, 705.6 kbit/s is required per channel. For five-channel audio (such as that used for surround-sound systems), a total of 3.5 Mbit/s is required. Although this is much less than the rate required for raw digital video, it still represents a significant expansion of the bandwidth requirement for the audio service compared to analog television.

Fortunately, the characteristics of the video and audio signals are such that significant savings are possible in the amount of data that needs to be transmitted in order to adequately represent the original signals. The digital signal processing techniques that allow this aim to be achieved will be a major focus of the first two parts of this text. It turns out that 5–10 Mbit/s is a reasonable target bit rate for a digital television service, with approximately 10% of the available data rate taken by transmission overheads, 10% allocated to audio, and the remaining 80% to video. Under these circumstances, compression factors of approximately 40 are required for digital video (meaning that the compressed digital video should require one fortieth of the rate required by the uncompressed video) and 10 for digital audio.

1.5. STANDARDS FOR DIGITAL TELEVISION

The use of standards in television broadcast systems is critical to their success. It is necessary that a consumer be able to purchase a receiver from any manufacturer and be confident of being able to watch television transmissions from any television broadcaster. Standards have always played a major part in providing this interoperability. For analog television, these were NTSC, PAL, and SECAM.

Modern digital television systems are based on one of the two standards, both named after the groups that developed them. The US *Advanced Television Systems Committee* (*ATSC*) [5] family of standards is used in North America, whereas the *Digital Video Broadcast* (*DVB*) [6] family of standards is used in much of the rest of the world, including Europe, much of Asia, and Australia.

DVB uses the MPEG-2 video standard [7] to provide video compression, the MPEG-2 audio standard [8] for audio compression, and the MPEG-2 systems standard [9] to multiplex the compressed video and audio with other data for transmission. Additional DVB standards extend the functionality of the MPEG-2 systems specification and specify how additional data (including subtitling and teletext) are carried in the bit stream.

ATSC also uses the MPEG-2 standards for video compression and multiplexing. Instead of using the MPEG-2 audio standard, ATSC specifies its own standard for

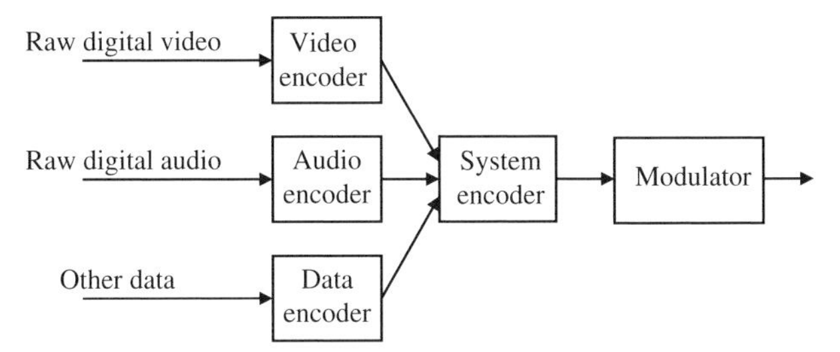

Figure 1.15 Outline structure of a digital television encoding and transmission system.

audio compression, which uses the Dolby AC-3 compression system [10]. Like DVB, ATSC specifies additional standards for carrying data (including closed captioning) in the bit stream.

The DVB and ATSC standards are available free of charge, at the time of writing. MPEG standards are available for purchase through national bodies affiliated to the International Standards Organization. In all cases, sufficient information is provided in this text for the reader to understand how the technology embedded in each relevant standard works. Access to the standard would be required, however, for a complete implementation to be developed.

The notional structure of a digital video transmitter is shown in Figure 1.15, consisting of separate encoders for each type of signal to be included in the transmitted program, a system encoder that multiplexes the outputs of these encoders and a modulator that converts the multiplexed bit stream into a form suitable for transmission in the same channel as that used for analog television. Part 1 of this book is concerned with the characteristics of the video encoder, its output bit stream, and the corresponding decoder. Part 2 is concerned with the audio encoder, its output bit stream, and decoder. Part 3 of the book covers the system encoder, encoders for other types of data, and modulation.

REFERENCES

1. See, for example,
   (a) D.G. Fink (Ed.), *Color Television Standards—NTSC*, New York: McGraw-Hill, 1955.
   (b) D.H. Pritchard, US color television fundamentals—a review, *IEEE Trans. Consumer Electron.*, **CE-23**, 1977.
2. Television systems; Enhanced 625-line Phased Alternate Line (PAL) television; PALplus, Sofia Antipolis: ETSI, 1997.
3. Further details on matched filtering can be found in many textbooks on digital communications, including B. Sklar, *Digital Communications: Fundamentals and Applications*, Englewood Cliffs, NJ: Prentice Hall, 2001.
4. Recommendation BT.601, Studio encoding parameters of digital television for standard 4:3 and wide-screen 16:9 aspect ratios, Geneva: ITU-R, 1995.
5. See, for example, http://www.atsc.org.
6. See, for example, http://www.dvb.org.

7. ISO/IEC 13818-2, Information technology—Generic coding of moving pictures and associated audio information—Part 2: Video, 1996.
8. ISO/IEC 13818-2, Information technology—Generic coding of moving pictures and associated audio information—Part 3: Audio, 1996.
9. ISO/IEC 13818-2, Information technology—Generic coding of moving pictures and associated audio information—Part 1: Systems, 1996.
10. ATSC Standard A/58, Digital audio compression standard (AC-3), Advanced Television Systems Committee, 1995.

Chapter 2

Characteristics of Video Material

We saw in the previous chapter that simply converting an analog video service into digital form results in an unacceptable increase in the bandwidth required to transmit the service even when sophisticated modulation schemes that transmit several bits/s/Hz are employed. For digital video to be a practical reality, it is essential that the number of bits required to represent each picture in a video sequence be significantly reduced. Fortunately, the characteristics of video material are such that substantial reductions in the number of bits can be achieved without noticeably affecting the subjective quality of the video service. In this chapter, we describe the characteristics that allow these savings to be made.

Figure 2.1 shows the first picture of the "Mobile and Calendar" video sequence that will be used to illustrate concepts as we consider the various signal processing techniques employed to compress video material. Longer versions of this sequence (among others) were used during the development of the MPEG digital standards that are at the heart of digital television. The picture contains 576 rows of pixels with each row containing 704 pixels. Although slightly less than CCIR Recommendation 601,[1] most of the development work for international standards was performed at this resolution.

2.1. PICTURE CORRELATION

Consider the picture shown in Figure 2.1, which is taken from the "Mobile and Calendar" sequence. Although this is an extremely "busy" picture, there are still large areas that are of a similar gray level. This includes the white background in the calendar, the light gray of the goat, the light background of the wallpaper, and the black of the body of the train. This "sameness" within a picture can be exploited to reduce the amount of data that needs to be transmitted to accurately represent the

[1] The missing pixels and rows of pixels are taken up by the horizontal and vertical blanking intervals.

Digital Television, by John Arnold, Michael Frater and Mark Pickering.

Figure 2.1 First frame of the video sequence "Mobile and Calendar."

picture. Let us take an extreme example. Consider a picture in which every pixel is the same shade of gray. In order to completely represent the picture all that would be needed would be the gray level of the first (top left) pixel together with the statement that every other pixel is the same shade of gray. The information about this one pixel is sufficient to allow the values of all the other pixels to be correctly determined.

Going to the other extreme, consider a picture made up of white noise. In this case, the value of every pixel needs to be individually specified because knowing the value of a particular pixel tells nothing about the value of any other pixel in the picture.

Mathematically, the "sameness" of a picture is measured by the autocorrelation function. This function measures how pixel "sameness" varies as a function of the distance between the pixels. The correlation coefficient r between two blocks of pixels $A(i,j)$ and $B(i,j)$ where i and j are the pixel positions within each block is defined as

$$r = \frac{\sum_i \sum_j \left(A(i,j) - \mu_A\right)\left(B(i,j) - \mu_B\right)}{\sqrt{\sum_i \sum_j \left(A(i,j) - \mu_A\right)^2 \sum_i \sum_j \left(B(i,j) - \mu_B\right)^2}} \tag{2.1}$$

where μ_A and μ_B are the mean values of $A(i,j)$ and $B(i,j)$, respectively.

For two blocks that are identical (e.g., any two blocks extracted from the picture where every pixel is identical), the correlation coefficient is one. For blocks that are completely uncorrelated (e.g., any two blocks extracted from the white noise picture),

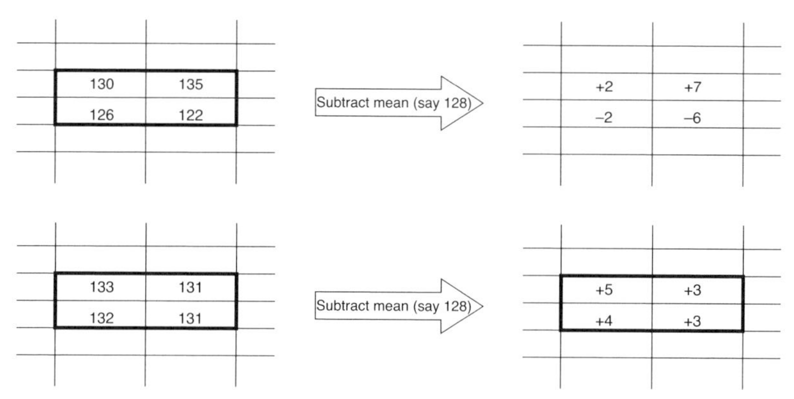

Figure 2.2 Calculation of correlation coefficient.

the correlation will be zero. In fact the correlation coefficient can take values in the range from −1 to 1 corresponding to the cases where $A(i,j) = -B(i,j)$ and $A(i,j) = B(i,j)$, respectively.

EXAMPLE 2.1

Consider the two 2 × 2 block of pixels shown in Figure 2.2. The mean of each block is first subtracted from each pixel in that block. Failing to do this leads to a correlation coefficient that is always positive and close to one irrespective of the pixel values because the mean value dominates the calculation.

$$r = \frac{(+2)(+5)+(+7)(+3)+(-2)(+4)+(-6)(+3)}{\sqrt{((+2)(+2)+(+7)(+7)+(-2)(-2)+(-6)(-6))((+5)(+5)+(+3)(+3)+(+4)(+4)+(+3)(+3))}}$$

$$= 0.0328$$

This corresponds to a pair of (admittedly small) blocks that are almost uncorrelated. Note that if the mean was not subtracted the correlation coefficient would be 0.9993. This clearly demonstrates how the mean can mask the underlying behavior within the block. ■

This idea can be easily extended to whole pictures using MATLAB.

EXAMPLE 2.2—MATLAB

Calculate the correlation coefficient for the luminance component of the first picture of the "Mobile and Calendar" sequence for horizontal shifts in the range from −10 to +10 pixels.

SOLUTION In order to obtain a reasonably global value for the correlation coefficient, it is necessary to use a large block of pixels. We use a 576 × 684 pixel block within the 576 × 704 pixel picture. This allows the block to be moved in the range from −10 to +10 pixels without running off the edge of the picture. This is shown in Figure 2.3.

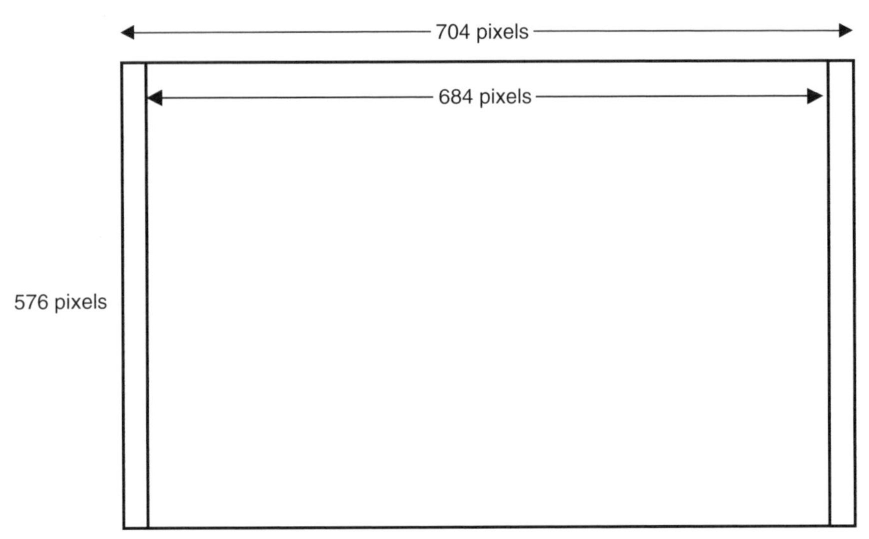

Figure 2.3 Block area for horizontal correlation coefficient calculation.

Assume that the array calendar_1 contains the first picture of the "Mobile and Calendar" sequence. The appropriate MATLAB script is

```
horiz_corr = zeros(1,21);
[row,col] = size(calendar_1);
block = calendar_1(:,11:col-10);             %define block to be moved over picture
block = block — mean(mean(block));           %subtract block mean
for position = −10:10
  compare_block = calendar_1(:,11+position:col-10+position);          %define block
  compare_block = compare_block − mean(mean(compare_block));          %subtract mean
  horiz_corr(1,position+11) = corr2(block,compare_block);             % calculate coefficient
end
x=−10:10;
plot(x,horiz_corr);
grid
```

After appropriate labeling of the axes, the result is shown in Figure 2.4.

As expected, at a displacement of 0 the correlation coefficient is 1. Note that the correlation coefficient at a displacement of 1 pixel is, however, in excess of 0.9. Even at a pixel shift of 10 pixels, the correlation coefficient is still greater than 0.65. This clearly demonstrates that even for a complex picture such as this one, there is still a large amount of correlation that can be exploited to reduce the required data rate. ■

This idea can simply be extended to calculate a two-dimensional plot of the correlation coefficient. In this case the block of interest is moved over a range of −5 to +5 pixels in both the horizontal and the vertical directions with the result being plotted as a three-dimensional surface plot. After appropriate interpolation, the result is shown in Figure 2.5.

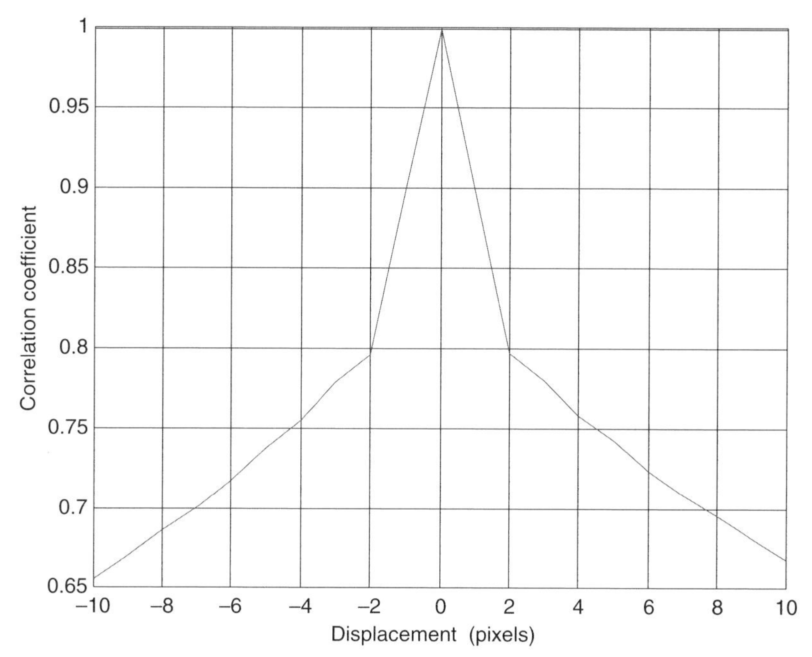

Figure 2.4 Horizontal correlation coefficient for luminance component of the first picture of sequence "Mobile and Calendar."

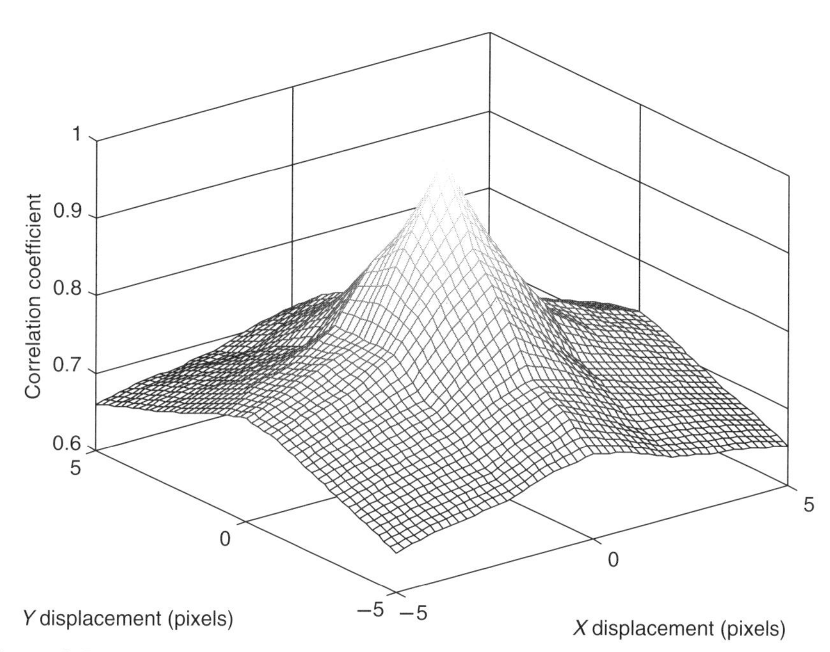

Figure 2.5 The three-dimensional surface plot of the correlation coefficient for the luminance component of the first picture of sequence "Mobile and Calendar."

It is clear that correlation in all directions remains high for several pixel shifts from the pixel of interest. Even at a shift of 5 pixels left and 5 pixels down, the correlation is still as high as 0.65.

The conclusion from these examples is that if we know the value of a particular pixel then we also have a considerable amount of information about the values of nearby pixels. It turns out that correlation between adjacent pictures in a video sequence is as large, and often even larger, than is the correlation between pixels within a picture. How this information can be exploited to reduce the amount of data required to represent a picture is discussed in later chapters.

2.2. INFORMATION CONTENT

It is important to be able to quantify in some way the amount of information produced by a source. Consider the emission of a symbol "a" from a source with probability p_a. If $p_a = 1$ then symbol a is produced continuously by the source. There is therefore no surprise when it is produced and hence there is no information provided by its appearance.

On the contrary, if the various symbols produced by the source have a range of possibilities and, if the probability p_a is low, there is more surprise—and hence more information—in its occurrence. The smaller the value of p_a, the greater is the information content. The information content is related to the reciprocal of the probability of occurrence.

The amount of information I_a is defined as

$$I_a = \log_2\left(\frac{1}{P_a}\right) = -\log_2(P_a) \tag{2.2}$$

The base for the logarithmic function is in fact quite arbitrary. However, it is usual to use the base 2 as shown. In this case, the unit of information is the bit (a contraction of *bi*nary digi*t*).

When p_a is equal to ½, the information content is 1 bit. Thus 1 bit of information is sufficient to distinguish one of two equally likely events.

This definition of information content displays the following pleasing characteristics:

- $I_a = 0$ for $p_a = 1$. As stated earlier, if we are sure of the symbol prior to its occurrence then no information is gained.
- $I_a \geq 0$ for $0 \leq p_a \leq 1$. The occurrence of a symbol provides at worst no information but never brings about a loss of information.
- $I_a \geq I_b$ if $p_a < p_b$. Less likely events provide more information than more likely events.

- $I_{ab} = I_a \cap I_b = I_a + I_b$. If a and b are statistically independent events (i.e., if $p_{ab} = p_a p_b$) then the information content of the two events is the sum of the information content of each event.

If we consider a source producing m symbols $s_0, s_1, \ldots, s_{m-1}$ with probabilities $p_0, p_1, \ldots, p_{m-1}$, then the mean information content is given by

$$
\begin{aligned}
H &= E\left(I_{S_k}\right) \\
H &= \sum_{k=0}^{m-1} p_k I_{S_k} \\
H &= \sum_{k=0}^{m-1} p_k \log_2\left(\frac{1}{p_k}\right) \\
H &= -\sum_{k=0}^{m-1} p_k \log_2 p_k
\end{aligned}
\tag{2.3}
$$

H is called the entropy of the source and measures the average information content per source symbol. The entropy provides a lower bound on the average number of bits required to transmit a sequence of independent symbols, given the probability of each symbol.

EXAMPLE 2.3

A source generates four symbols a, b, c, and d that have probability of occurrence of 0.4, 0.3, 0.2, and 0.1, respectively. What is the entropy of the source?

SOLUTION The entropy calculation is shown in Table 2.1.

Table 2.1 Entropy calculation for Example 2.3.

Symbol	p_i	$I = -\log_2(p_i)$	$-p_i \log_2(p_i)$
a	0.4	1.32	0.53
b	0.3	1.74	0.52
c	0.2	2.32	0.46
d	0.1	3.32	0.33
		Total source entropy	1.84 bits/ symbol

■

This process can be extended to measure the information content of a picture. In an 8-bit gray scale picture, each pixel takes one of 256 possible values (0–255). By calculating the probability of each of these 256 values, it is possible to calculate the entropy of the picture. Figure 2.6 shows the histogram of pixel intensities for the first luminance picture of the sequence "Mobile and Calendar."

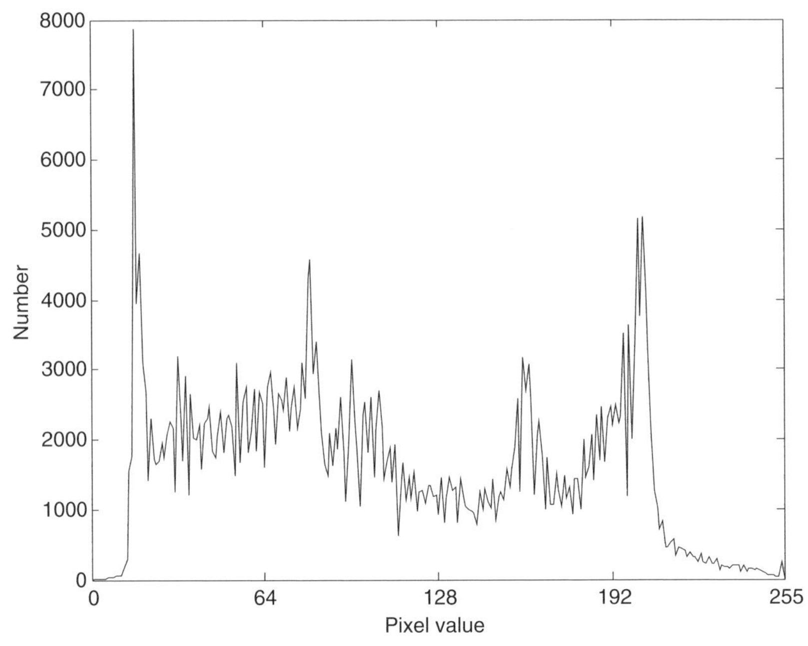

Figure 2.6 The histogram of the first luminance picture of the sequence "Mobile and Calendar."

From Figure 2.6, it is clear that not all pixels are equally likely. When the picture entropy is calculated, it turns out that the entropy is 7.61 bits/pixel. This implies that the coding efficiency, defined as the entropy divided by the actual number of bits per symbol in the original picture (7.61/8 in this case), is already more than 95% efficient. Future chapters show that significantly more complex signal processing is needed to achieve the level of compression required for the successful introduction of digital television. If all 256 values were equally likely, the entropy would be 8 bits/pixel.

The key question now is how do we attempt to exploit the fact that the entropy is less than the number of bits per pixel currently required to represent the picture. This question can be answered by considering an old form of coding symbols (in this case letters of the alphabet)—Morse code. The Morse code representation of the 26 letters of the alphabet is given in Table 2.2.

The amount of time required to transmit a message is minimized by representing commonly occurring letters with a small number of keystrokes (e.g., the letter E) whereas less commonly occurring letters are represented by a larger number of keystrokes (e.g., the letters Q, Y, and Z). In Morse code, a dot (.) requires only one third of the transmission time of a dash (−). This implies that the number of keystrokes does not exactly correlate with transmission time. Nonetheless, the implications are clear. If we want to transmit symbols at the rate suggested by the entropy, we need to represent each symbol by a binary number with the same number of bits as the information content of the symbol. This is not always possible because the information content is not necessarily exactly an integer number of bits.

Table 2.2 Morse code for letters of the alphabet.

Letter	Morse code	Letter	Morse code	Letter	Morse code
A	.-	J	.---	S	...
B	-...	K	-.-	T	-
C	-.-.	L	.-..	U	..-
D	-..	M	–	V	...-
E	.	N	-.	W	.--
F	..-.	O	---	X	-..-
G	--.	P	.--.	Y	-.--
H	....	Q	--.-	Z	--..
I	..	R	.-.		

EXAMPLE 2.4

A message consists of four symbols (a, b, c and d) with probabilities 0.75, 0.125, 0.0625, and 0.0625, respectively.

(a) Calculate the entropy of the message.

(b) Calculate the required number of bits per symbol and the coding efficiency if fixed length coding is employed.

(c) Calculate the required number of bits per symbol and the coding efficiency if variable length coding is employed.

SOLUTION (a) The entropy of the message can be calculated as shown in Table 2.3.

Table 2.3 Calculation of message entropy.

Symbol	p_i	$I = -\log_2(p_i)$	$-p_i \log_2(p_i)$
a	0.7500	0.415	0.311
b	0.1250	3.000	0.375
c	0.0625	4.000	0.250
d	0.0625	4.000	0.250
		Total source entropy	1.186 bits/ symbol

(b) As there are four symbols, representing each symbol by a fixed length code would require 2 bits/symbol as shown in Table 2.4.

Table 2.4 Representation with fixed length code words.

Symbol	p_i	Code word	Code word length
a	0.7500	00	2
b	0.1250	01	2
c	0.0625	10	2
d	0.0625	11	2

The average code word length (L) can then be calculated according to

$$\begin{aligned} L &= \sum_i (p_i)(\text{code word length}_i) \\ &= 0.75 \times 2 + 0.125 \times 2 + 0.0625 \times 2 + 0.0625 \times 2 \\ &= 2 \text{ bits/symbol} \end{aligned} \quad (2.4)$$

Given that every symbol is represented by 2 bits, this is hardly a remarkable result. The coding efficiency is defined below.

$$\text{Coding efficiency} = \frac{\text{Entropy}}{\text{Number of bits per symbol}} \quad (2.5)$$

In this case the coding efficiency is

$$\frac{1.186}{2.0} = 59.3\% \quad (2.6)$$

(c) Now consider the variable word length coding shown in Table 2.5. The most likely symbol is represented by a short code word whereas less likely symbols are represented by longer code words.

Table 2.5 Representation with variable length code words.

Symbol	p_i	Code word	Code word length
a	0.7500	0	1
b	0.1250	10	2
c	0.0625	110	3
d	0.0625	111	3

In this case, the average code word length (L) can then be calculated according to

$$\begin{aligned} L &= \sum_i (p_i)(\text{code word length}_i) \\ &= 0.75 \times 1 + 0.125 \times 2 + 0.0625 \times 3 + 0.0625 \times 3 \\ &= 1.375 \text{ bits/symbol} \end{aligned} \quad (2.7)$$

The coding efficiency has been increased to 1.186/1.375 = 86.2% a substantial improvement. ■

Details of techniques used to design variable length coding schemes will be considered in the next chapter.

2.3. THE HUMAN VISUAL SYSTEM

The previous two sections have discussed the mathematical properties of picture data that can be exploited to reduce the amount of data required to represent a picture. In considering the data compression problem, it needs to be firmly borne in mind that the aim of the process is to produce a reconstructed picture that, when viewed by a human viewer, is of sufficient quality to meet the needs of the service. In particular, the aim is not to produce a reconstructed picture that is identical with the original picture that was imaged at the source. It follows that any parts of the picture that are not apparent to a viewer need not be preserved in the reconstructed picture. This allows huge savings to be made in the amount of data that needs to be transmitted. For these savings to be made, we need to have a reasonable understanding of the characteristics, and in particular the limitations, of the human visual system. These characteristics are discussed briefly in this section.

2.3.1. Perception of Changes in Brightness

According to Weber's law, the just detectable change in luminance (ΔY) is proportional to the luminance (Y). In fact

$$\frac{\Delta Y}{Y} \approx 0.02 \tag{2.8}$$

The response of the human visual system is therefore logarithmic. It also implies that a given change in luminance value due to a coding artifact is more visible in darker rather than light areas of the picture.

There is, however, another effect that needs to be considered. The luminance from a cathode ray tube (CRT) is related to the applied voltage (V) according to the nonlinear relationship.

$$Y = V^{\gamma}, \qquad 2 \leq \gamma \leq 3 \tag{2.9}$$

When taken together, Weber's law and the nonlinear CRT relationship mean that the luminance perceived by a viewer is in fact approximately linearly related to the voltage applied to the CRT. Figure 2.7 shows a complete video compression system starting with a scene and camera followed by video coding (and decoding), and finally display on a CRT for the viewer. We have already seen that the combined effect of the CRT and the viewer's human visual system (as shown in the smaller rectangle in Fig. 2.7) is approximately linear. For the displayed image to look the

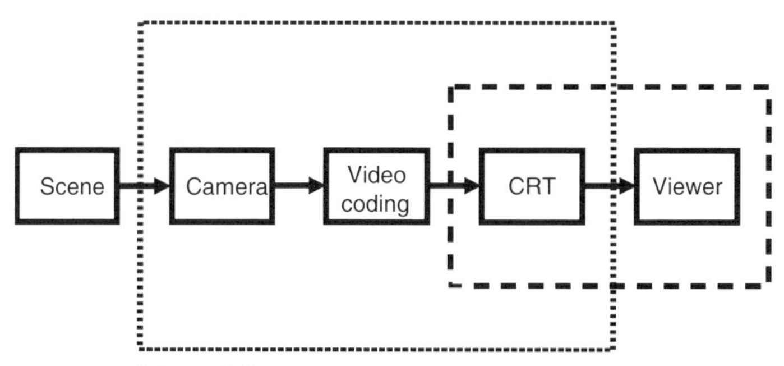

Figure 2.7 A complete video compression system.

same as the scene when viewed by the camera, the system contained in the larger rectangle from camera to CRT should also be linear. We want to perform video coding in the linear domain, and so video cameras invariably contain gamma correction circuitry.

$$V_{\text{out}} = V_{\text{camera}}^{\frac{1}{\gamma}} \tag{2.10}$$

2.3.2. Spatial Masking

It has been demonstrated experimentally that coding artifacts[2] in active regions of a picture (i.e., areas where there are sharp edges or other fast intensity changes) are less subjectively noticeable than those in inactive (i.e., flat) regions of a picture. This means that active areas of a picture can be coded more coarsely than inactive areas. Spatial masking is extensively exploited in modern video coders.

2.3.3. Temporal Masking

It has also been demonstrated that near scene changes in a video sequence there is a significant reduction in the subjective impact of coding artifacts. This means that pictures immediately after a scene cut (and to a lesser extent prior to the scene cut as well) can be coded more harshly than other pictures in the sequence without having any significant impact on subjective service quality.

2.3.4. Frequency Sensitivity

Figure 2.8 shows the approximate spatial frequency sensitivity of the human visual system for luminance information. It is clear that the frequency sensitivity is highest at low and medium frequencies and decreases rapidly at high frequencies. This does not mean that all high-frequency information can be deleted because edges introduce significant amounts of such high-frequency data that, if removed, would lead to a general blurring of the reconstructed image. Rather it means that high-frequency information can be represented less accurately than lower frequency information. This aspect of the human visual system is also widely exploited in modern video coders.

It is also worth noting that the contrast sensitivity for chrominance (color) information falls off at significantly lower frequencies than is the case for luminance

[2] The representation of parts of a picture can be made more and more approximate, or "coarse," if fewer bits are available to represent it. In such situations, artifacts of the coding occur as either visible or invisible distortion.

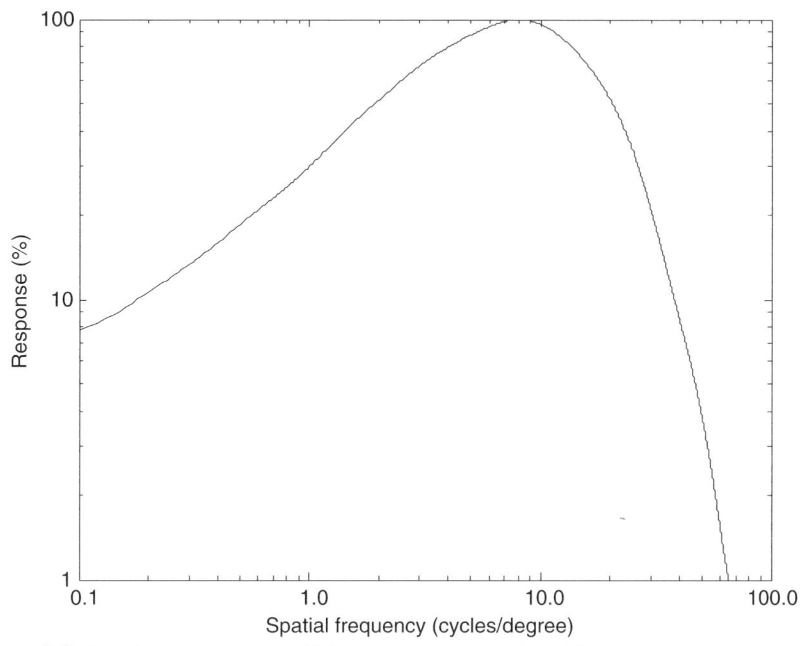

Figure 2.8 Luminance contrast sensitivity response as a function of frequency for the human visual system.

information. For this reason, chrominance information is invariably sampled in the spatial domain at a lower sampling rate than the luminance information.

2.3.5. Tracking of Motion

When an object in a video sequence first begins to move, there is a short period of time before the eye starts to track the motion. The eye's ability to resolve fine spatial detail in the object is reduced during this period. This is a similar effect to that described earlier at scene cuts. As an object continues to move, the eye's ability to track this detail improves providing that the motion is sufficiently slow that the eye can track it. Coding artifacts near objects that have just begun to move are therefore less noticeable than in other areas of the picture.

2.3.6. Conclusion

A thorough understanding of the human visual system is essential in the design of effective video encoders. Much of the compression achieved relies upon the introduction of coding artifacts in areas of a picture where they will not be subjectively noticeable. Although some conclusions have been drawn about the characteristics of the human visual system in this section, this is still an area of active research.

2.4. SUMMARY

In this chapter, we have introduced the characteristics of images and the human visual system that can be exploited to allow the efficient representation of video material. In the next two chapters, we introduce the signal processing techniques that have been developed to facilitate this exploitation.

PROBLEMS

2.1 A progressive television service called 480p is to be introduced with each picture being made up of 720 pixels/line and 480 lines/picture. The picture rate is 60 pictures/s. Each pixel requires 12 bits of data to represent luminance and chrominance information. Calculate the efficiency required from the modulation scheme (in bits/s/Hz) so that the digitized video service can be carried in a channel with bandwidth 6 MHz.

2.2 Repeat Problem 2.1 for a 1080i service that is made up of 2:1 interlaced pictures transmitted at a picture rate of 30 pictures/s. Each picture (i.e., a pair of fields) contains 1440 pixels/line and 1080 lines/picture.

2.3 Calculate the correlation coefficient between the 2×2 pixel blocks shown in Figure 2.9. Assume that the mean value of all pixels in the picture is 128.

146	123		157	117
130	105		140	91

Figure 2.9 Pixel data for Problem 2.3.

2.4 Calculate the correlation coefficient between the 3×3 pixel blocks shown in Figure 2.10. Assume that the mean value of all pixels in the picture is 128.

205	130	247		217	172	174
163	224	188		134	215	97
153	185	182		52	5	213

Figure 2.10 Pixel data for Problem 2.4.

2.5 What areas of a picture would you expect to display high correlation values and which areas would you expect to display low correlation? Consider the first picture of the "Mobile and Calendar" sequence. What areas of the picture will display local high and low values of correlation?

2.6 Consider the first picture of other video sequences. What areas of each picture would you expect to display local high and low values of correlation?

(a) Explain how this could be calculated.

(b) Under what circumstances would you expect high and low values of temporal correlation?

2.7 As indicated in the chapter, the correlation coefficient can take values in the range from -1 to $+1$. Consider the 4×4 pixel block shown in Figure 2.11 that has been drawn for a set of pixel values with mean value 128.

150	145	161	168
144	153	170	101
141	155	148	84
145	160	160	92

Figure 2.11 Pixel data for Problem 2.7.

(a) Calculate the 4×4 pixel block that has a correlation coefficient of $+1$ when compared to this block.

(b) Calculate a 4×4 pixel block that has a correlation coefficient of 0 when compared to this block.

(c) Calculate a 4×4 pixel block that has a correlation coefficient of -1 when compared to this block.

2.8 A source generates four symbols a, b, c, and d with probabilities 0.50, 0.20, 0.20, and 0.10, respectively. What is the entropy of the source? If each symbol is represented by a fixed length binary code word, what is the coding efficiency?

2.9 A source generates six symbols a, b, c, d, e, and f with probabilities 0.30, 0.20, 0.20, 0.15, 0.10, and 0.05 respectively. What is the entropy of the source? If each symbol is represented by a fixed length binary code word, what is the coding efficiency?

2.10 A source produces three symbols with probabilities 0.90, 0.05, and 0.05, respectively. What is the entropy of the source? If each symbol is represented by a fixed length binary code word, what is the coding efficiency?

2.11 The source described in Problem 2.10 has symbols grouped in pairs prior to encoding. Each symbol can be assumed to be independent of any other symbol. What is the entropy of the source now? If each pair of symbols is represented by a fixed length binary code word, what is the coding efficiency?

2.12 Repeat Problem 2.11 for the case where the symbols are grouped in threes.

2.13 A source generates three symbols. One symbol has probability p whereas both of the other two symbols have the same probability. Plot source entropy as a function of the probability p.

2.14 The coding efficiency using variable length code words is usually less than 100%. Under what circumstances will this coding efficiency be exactly 100%?

2.15 When variable word length code words are used to represent symbols of differing probabilities, it is usual to ensure that no short code word is a prefix for a longer code word (i.e., if one code word is 00 then 001 could not be another code word as it begins with 00). Explain briefly why this might be an advantage. For Problems 2.8 and 2.9 suggest appropriate variable length code words that fulfill this requirement.

2.16 The Morse code is not a code where short code words cannot be a prefix of longer code words (e.g., the letter E is represented by a single dot whereas the letter I is represented by two dots). When transmitting Morse code, the operator needs to pause between letters to allow correct decoding at the receiver. What would be the impact of using such a code in a digital television system?

2.17 The length of variable word length codes is designed to match the long-term probabilities of each of the symbols generated by the source. Although this should lead to a saving overall, it is possible in some instances that the number of bits generated to represent a string of symbols can be more if the symbols were encoded using fixed length code word. Explain the circumstances where this is possible.

2.18 Four symbols are represented by the variable length code words as shown in Table 2.6. A string of transmitted symbols consists of the symbols A,B,A,B,C,D,A,A, espectively.

Table 2.6 Variable length code words for Problem 2.18.

Symbol	Code word
A	0
B	10
C	110
D	111

(a) Calculate the generated bit stream corresponding to this run of symbols.

(b) The second bit of this bit stream is received in error. Decode the received bit stream and comment on the result.

(c) Suggest a method that might be used to overcome the problem of transmission errors when variable length coding is used.

2.19 The characteristics of the human visual system are very complex. Over the years, researchers have developed tests that attempt to quantify these characteristics. Design an experiment that might allow you to verify Weber's law.

2.20 Repeat Problem 2.19 but this time design an experiment that measures the frequency response of the human visual system.

MATLAB EXERCISE 2.1: CORRELATION COEFFICIENT WITHIN A PICTURE

The text has discussed in some detail the idea of correlation within a picture. In this exercise, you use MATLAB to measure the correlation coefficient of pictures from different video sequences. You also study the variation in the local value of the correlation coefficient within an image.

Section 1 Correlation coefficient of pictures from different sequences

(a) For the first picture of a video sequence, calculate the two-dimensional correlation coefficient over the range of ± 5 pixels. Plot the result as a three-dimensional plot. Your final result should look very similar to the plot shown in Figure 2.5. Note that the readability of the plot can be increased by interpolating between integer points us the two-dimensional interpolation function available in MATLAB.

(b) Repeat this exercise for the first picture of other video sequences. Comment on the results. Which pictures would you expect to be easy to compress? Which pictures would you expect to be more difficult to compress? Why?

Section 2 Correlation coefficient within a picture

(a) Extend the MATLAB program developed in the previous section so that you can select individual 16 $\times$ 16 pixel blocks within a picture. Choose different blocks within the first picture of your sequence. Calculate the two-dimensional correlation coefficient for each of these blocks again in the range of ± 5 pixels. You might find it helpful to highlight the selected block in some way (such as surrounding it with a white boarder). Use you program to determine what picture characteristics give rise to high and low values for the correlation coefficient.

(b) Use your program to perform a similar study on the first picture from other video sequences. Comment on whether you observations for these pictures correspond to those for the first sequence.

MATLAB EXERCISE 2.2: CORRELATION COEFFICIENT BETWEEN PICTURES IN A SEQUENCE

As well as being able to calculate the correlation coefficient within a picture, it is also possible to calculate the correlation coefficient between pictures in a sequence. This is studied in this exercise.

(a) Load the third picture from a video sequence into MATLAB. Now calculate the temporal correlation coefficient with the first five pictures in the sequence. Plot the correlation coefficient as a function of temporal displacement over the range of ± 2 frame times.

(b) Repeat this calculation for other video sequences. Plot these on the same graph as the one used for the first sequence.

(c) What characteristics of a sequence lead to high and low values of temporal correlation?

MATLAB EXERCISE 2.3: ENTROPY OF A PICTURE

In this exercise we plot the histogram of a picture. We then calculate the entropy of the picture and the coding efficiency when represented by fixed length code words.

Section 1 Histogram of a picture

(a) For the first picture of a video sequence, calculate the number of pixels at each of the possible 256 gray scale values (range 0–255). Plot the histogram.

(b) Calculate the histogram for the first picture of other sequences.

(c) Comment on any differences between the histograms. Which picture would you expect to have the highest entropy? Which picture would you expect to have the lowest entropy?

Section 2 Calculation of entropy

(a) Extend the program developed in the previous section to calculate the entropy (H) of each picture using the formula

$$H = -\sum_{k=0}^{m-1} p_k \log_2 p_k \tag{2.11}$$

where p_k is the probability of a pixel having intensity k.
Note that there may be some gray scale values that have a probability of zero. Your program needs to be able to deal with this situation.

(b) How do the actual entropy values compare with the estimate you made in the previous section?

Chapter 3

Predictive Encoding

In the previous chapter we saw that most pictures can be characterized by high spatial and temporal picture correlation. All pictures exhibit nonuniform pixel distribution to a certain degree and signal processing can greatly increase this nonuniformity. If we can exploit these characteristics, we can reduce the amount of information that needs to be transmitted to adequately represent a picture. In the next two chapters, the video coding tools that exploit these characteristics are discussed. This chapter covers predictive encoding techniques including motion-compensated prediction. Chapter 4 looks at transform encoding. Together, predictive encoding and transform encoding form the heart of the MPEG-2 video encoder.

3.1. ENTROPY CODING

The idea of using variable length coding to efficiently represent messages is known as entropy coding. If we are to be able to make regular use of it, we need some automatic method for calculating the appropriate variable word length codes dictated by a particular set of symbol probabilities. The best and most widely used technique is called Huffman coding. The current video and audio compression standards used for digital television include variable length code words generated as Huffman codes. We will now study how to generate these useful code word sets.

3.1.1. Huffman Coding

The steps involved in generating a set of Huffman code words are summarized below.

Step 1: List the symbols to be transmitted in decreasing order of probability.

Step 2: Combine the two symbols with the smallest probabilities and reorder the symbols in decreasing order of probability.

Step 3: Repeat Step 2 until only two combined symbols remain.

Digital Television, by John Arnold, Michael Frater and Mark Pickering.

Step 4: Mark each combination consistently with either

]$^{1}_{0}$ or]$^{0}_{1}$

Step 5: Read off the Huffman code words from right to left.

The entire process is best illustrated by way of an example.

EXAMPLE 3.1

The symbols a, b, c, and d have probabilities 0.75, 0.125, 0.0625, and 0.0625, respectively. Calculate the Huffman code words appropriate for these symbols.

SOLUTION We begin by listing the code words in decreasing order of probability (Step 1) as shown in Figure 3.1.

a	0.75
b	0.125
c	0.0625
d	0.0625

Figure 3.1 Initial ordering of symbol probabilities.

Now we combine the two symbols with the smallest probabilities and reorder in decreasing order of probability (Step 2). In this case after combining, the two smallest probabilities are equal. It is not important to the final outcome that one is placed higher in the reordered list. In this case we have placed the newly merged symbols last as shown in Figure 3.2. We will return at the end of the example to look at the impact of the other choice.

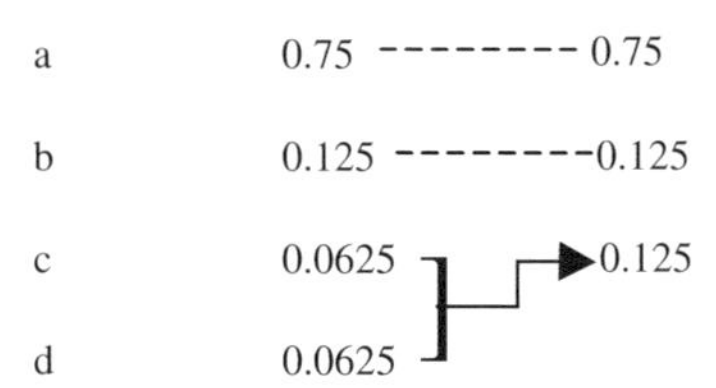

Figure 3.2 Probabilities after first combining and reordering.

We now again combine the two smallest probabilities and reorder (Step 2). The result is shown in Figure 3.3. As this leaves us with only two probabilities, we can stop the combining process (Step 3).

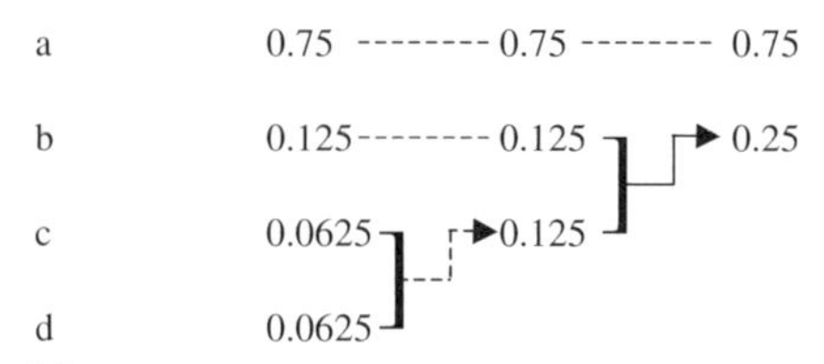

Figure 3.3 Probabilities after second combining and reordering.

This is called the Huffman tree. We now proceed to mark each combination operation (Step 4). This is shown in Figure 3.4.

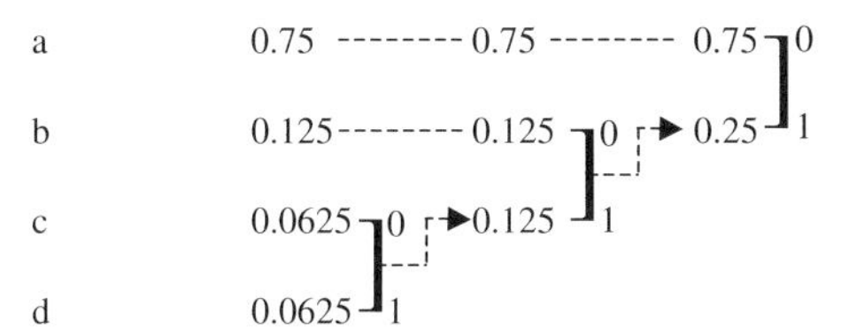

Figure 3.4 The marking of each combination operation.

The Huffman code word for each symbol can be read back from right to left in the Huffman tree as shown in Figures 3.5–3.8.

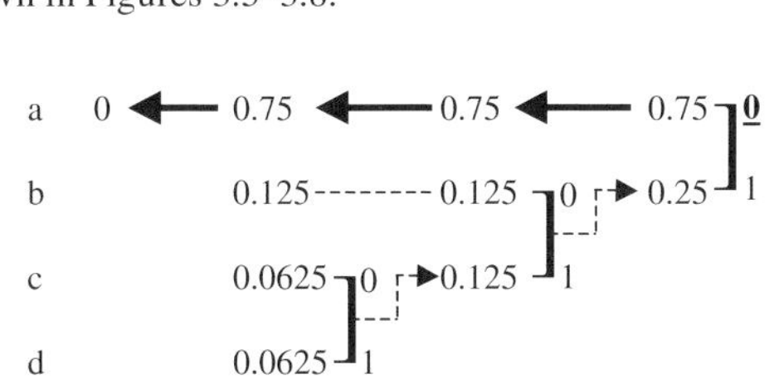

Figure 3.5 Reading off the Huffman code word for symbol a.

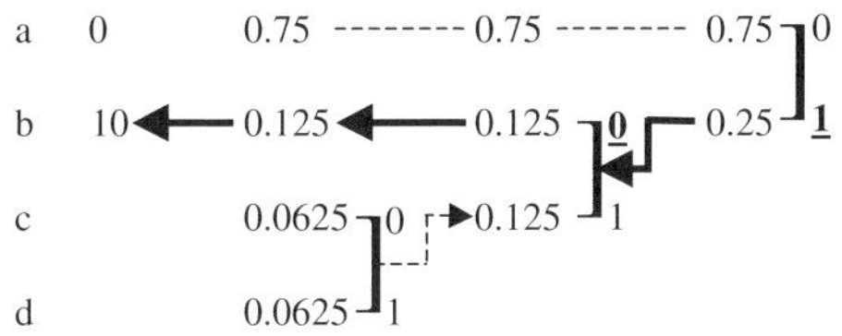

a 0 0.75 0.75 0.75 0
b 10 0.125 0.125 0 0.25 1
c 0.0625 0 0.125 1
d 0.0625 1

Figure 3.6 Reading off the Huffman code word for symbol b.

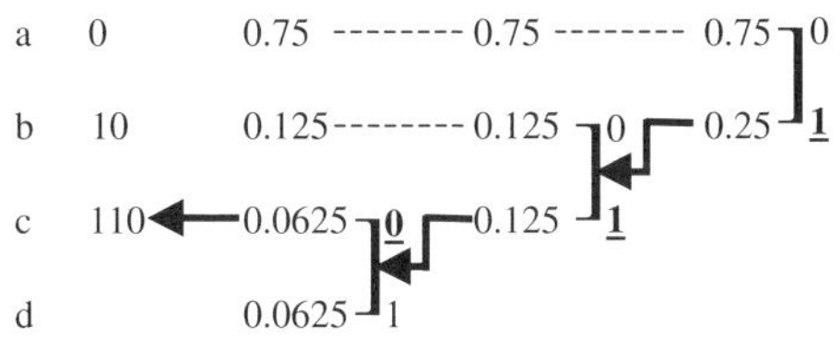

a 0 0.75 0.75 0.75 0
b 10 0.125 0.125 0 0.25 1
c 110 0.0625 0 0.125 1
d 0.0625 1

Figure 3.7 Reading off the Huffman code word for symbol c.

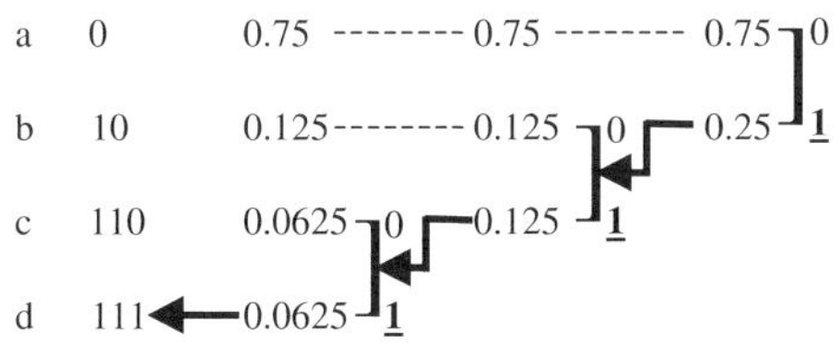

a 0 0.75 0.75 0.75 0
b 10 0.125 0.125 0 0.25 1
c 110 0.0625 0 0.125 1
d 111 0.0625 1

Figure 3.8 Reading off the Huffman code word for symbol d.

Table 3.1 Huffman code words for Example 3.1.

Symbol	Huffman code word
a	0
b	10
c	110
d	111

Thus the Huffman code words for this example are as shown in Table 3.1.

No short code word ever forms a prefix for a longer code word in a Huffman code. This means that there is no need for any form of space between code words, and so any bit stream is uniquely decipherable.

Had we reordered the other way after the first combination operation, the final Huffman tree would be as shown in Figure 3.9.

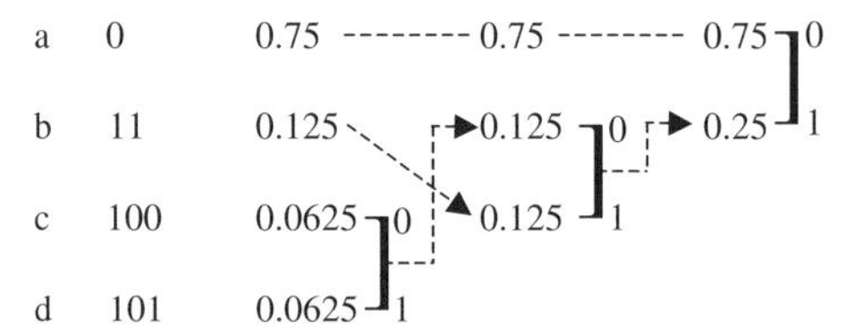

Figure 3.9 Huffman tree if alternate reordering after first symbol combination.

In this case, the Huffman code words are as shown in Table 3.2.

Table 3.2 Huffman code words for alternate ordering in Example 3.1

Symbol	Huffman code word
a	0
b	11
c	100
d	101

The code words for symbols b, c, and d have changed. However, each code word is the same length as in the previous Huffman code calculation in this example. The average code word length in each case would therefore be identical.

We now look at a simple example of coding and decoding symbols using Huffman codes. ■

EXAMPLE 3.2

Use the Huffman code defined in Table 3.1 to encode the stream of symbols shown in Figure 3.10.

a b c d a a a b

Figure 3.10 Original stream of symbols for Example 3.2.

After the encoding process, we have the code words shown in Figure 3.11.

0 10 110 111 0 0 0 10

Figure 3.11 Symbols from Figure 3.10 encoded using Huffman code from Table 3.1.

This can be concatenated to form the final transmitted bit stream shown in Figure 3.12.

01011011100010

Figure 3.12 Final transmitted bit stream.

The decoding process consists of reading values from the bit stream until a valid code word is detected. Because no Huffman code word ever forms the prefix of a longer Huffman code word, this always results in correct decoding in an error-free environment. The decoding process is shown in Figure 3.13.

0	1011011100010							
a								
0	1	011011100010						
a	?							
0	10	11011100010						
a	b							
0	10	1	1011100010					
a	b	?						
0	10	11	011100010					
a	b	?						
0	10	110	11100010					
a	b	c						
0	10	110	1	1100010				
a	b	c	?					
0	10	110	11	100010				
a	b	c	?					
0	10	110	111	00010				
a	b	c	d					
0	10	110	111	0	0010			
a	b	c	d	a				
0	10	110	111	0	0	010		
a	b	c	d	a	a			
0	10	110	111	0	0	0	10	
a	b	c	d	a	a	a		
0	10	110	111	0	0	0	1	0
a	b	c	d	a	a	a	?	
0	10	110	111	0	0	0	10	
a	b	c	d	a	a	a	b	

Figure 3.13 Huffman decoding process. ■

The introduction of transmission errors can have a significant effect on the decoding process. This is illustrated in Example 3.3.

EXAMPLE 3.3

The bit stream generated in Example 3.2 is received with a single bit error in the second bit of the bit stream. Determine the decoded sequence in this case.

SOLUTION The received bit stream is given in Figure 3.14.

$0\underline{0}011011100010$

Figure 3.14 Bit stream generated in Example 3.2 with a single bit error.

The decoding process is shown in Figure 3.15.

0	0011011100010								
a									
0	0	011011100010							
a	a								
0	0	0	11011100010						
a	a	a							
0	0	0	1	1011100010					
a	a	a	?						
0	0	0	11	011100010					
a	a	a	?						
0	0	0	110	11100010					
a	a	a	c						
0	0	0	110	1	1100010				
a	a	a	c	?					
0	0	0	110	11	100010				
a	a	a	c	?					
0	0	0	110	111	00010				
a	a	a	c	d					
0	0	0	110	111	0	0010			
a	a	a	c	d	a				
0	0	0	110	111	0	0	010		
a	a	a	c	d	a	a			
0	0	0	110	111	0	0	0	10	
a	a	a	c	d	a	a	a		
0	0	0	110	111	0	0	0	1	0
a	a	a	c	d	a	a	a	?	
0	0	0	110	111	0	0	0	10	
a	a	a	c	d	a	a	a	b	

Figure 3.15 Huffman decoding process for the errored bit stream of Figure 3.14.

Comparing the decoded sequence for the errored bit stream with the correctly decoded sequence, we see that the second and third symbols (b and c) have been incorrectly received (as a and a) and an additional symbol (c) is also decoded before correct decoding resumes with the next symbol (d). The compression achieved using Huffman coding is achieved at the expense of greater vulnerability of the generated bit stream to errors. Modern video coders are designed to minimize the effect of this vulnerability. However, it cannot be completely removed. ■

Another limitation of Huffman code words is the requirement that each code word is an integer number of bits in length. If the probability of a symbol is 1/3, the optimum number of bits to assign is 1.6. The Huffman code assigns either one or two bits—either choice leads to a longer compressed message than is theoretically necessary. The problem is particularly serious when the probability of one symbol is very high. For example, if the probability of one symbol is 0.9 then the optimal code word length would be 0.15 bits. A Huffman code would assign a one bit code word. This is the reason that Huffman codes cannot exactly achieve the entropy of the symbol stream. This problem can be addressed by using a different scheme known as arithmetic coding. Although arithmetic coding features in standards such as MPEG-4 and JPEG-2000, it is not a part of the MPEG-2 standards that form the basis of digital television broadcasting services. It is therefore not discussed further here.

3.1.2. Run Length Coding

In run length coding, a run of consecutive identical symbols is combined together and represented by a single variable length (e.g., Huffman) code word. A simple example is the transmission of a black-and-white facsimile. Consider the line segment shown in Figure 3.16.

The message to be transmitted would be

(3 white pixels) (2 black pixels) (5 white pixels) (1 black pixel) (2 white pixels) (6 black pixels) (3 white pixels) (3 black pixels) (3 white pixels) ...

If the statistics of the various run lengths are calculated, appropriate Huffman code words can be designed. In the case of facsimile transmission, separate Huffman code word sets have been designed for runs of white pixels and runs of black pixels. This design strategy has been used because the statistics for the different types of runs are not the same. In typical documents, long runs of white pixels are more common than long runs of black pixels.

3.2 PREDICTIVE CODING

In Chapter 2, we have seen that knowing the value of a particular pixel provides a considerable amount of information about the pixels that surround it due to the

Figure 3.16 Segment of a black-and-white facsimile.

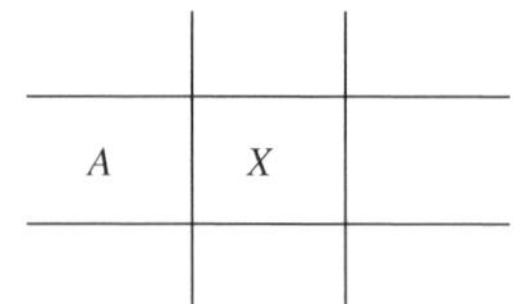

Figure 3.17 Simple one-dimensional prediction.

high correlation that typically exists between nearby pixels. What is needed is some way to exploit this correlation so that the amount of information that needs to be transmitted can be reduced. One way of achieving this is by a technique known as predictive encoding.

The aim of predictive encoding is to use the values of already transmitted pixels to predict the value of the current pixel to be transmitted. It is then only necessary to transmit the difference between the prediction and the actual pixel value. If the prediction is accurate, this is usually a small value and so less bits of data will need to be transmitted in order to represent it.

The simplest form of predictive encoding is to predict the value of the pixel to be transmitted using the value of the pixel immediately to its left. This is illustrated in Figure 3.17. This approach is often called one-dimensional *differential pulse code modulation* (DPCM). The difference between the prediction and the actual pixel value is then transmitted.

In Figure 3.17, X is the pixel to be transmitted and A is the value of the pixel immediately to its left. We estimate the value of X as A that is

$$\hat{X} = A$$

The value transmitted to the receiver is then

$$\text{Difference} = X - \hat{X}$$
$$\text{Difference} = X - A$$

EXAMPLE 3.4

Consider the set of pixel values given in Figure 3.18. Calculate the values of the prediction difference when one-dimensional DPCM is employed. We assume that the first pixel in a line is predicted by the mid-gray value 128.

130	135	141	129	151
124	140	165	200	199
119	132	175	205	203

Figure 3.18 Pixel values for Example 3.4.

The pixel values are repeated in Figure 3.19 with the prediction value for each pixel in italics directly below it. Thus the first pixel in the first row (130) is predicted by mid-gray value (128), the second value in the first row (135) is predicted by the first value in the first row (130), and so on.

Pixel value	130	135	141	129	151
Prediction	*128*	*130*	*135*	*141*	*129*

Pixel value	124	140	165	200	199
Prediction	*128*	*124*	*140*	*165*	*200*

Pixel value	119	132	175	205	203
Prediction	*128*	*119*	*132*	*175*	*205*

Figure 3.19 Pixel values with appropriate predictor immediately below the pixel value.

The prediction difference is then simply calculated by subtracting the prediction from the true pixel value giving the result shown in Figure 3.20.

+2	+5	+6	−12	+22
−4	+16	+25	+35	−1
−9	+13	+43	+30	−2

Figure 3.20 Prediction difference for Example 3.4. ■

The following MATLAB example applies one-dimensional DPCM to the first picture in the "Mobile and Calendar" sequence.

EXAMPLE 3.5—MATLAB

Calculate the one-dimensional DPCM prediction difference picture for the first picture in the "Mobile and Calendar" sequence. Also calculate the entropy of this prediction difference picture.

SOLUTION The process is simply that the prediction difference for column i is the pixel value in column i minus the pixel value in column $(i-1)$. The exception is the first column where there is no pixel to the left from which to form a prediction. This column is predicted with the mid-gray value 128. If the prediction difference is stored in an array DIFF then appropriate MATLAB code is given below. It is assumed that the an array calendar_1 contains the luminance pixel values of the first frame of the sequence "Mobile and Calendar."

```
[row,col] = size(calendar_1);
DIFF = zeros(row,col);
DIFF(:,1) = calendar_1(:,1) – 128*ones(row,1);
for col_number = 2:col
DIFF(:,col_number) = calendar_1(:,col_number) – calendar_1(:,col_number-1);
end
```

The histogram for the prediction difference contained in an array DIFF is shown in Figure 3.21.

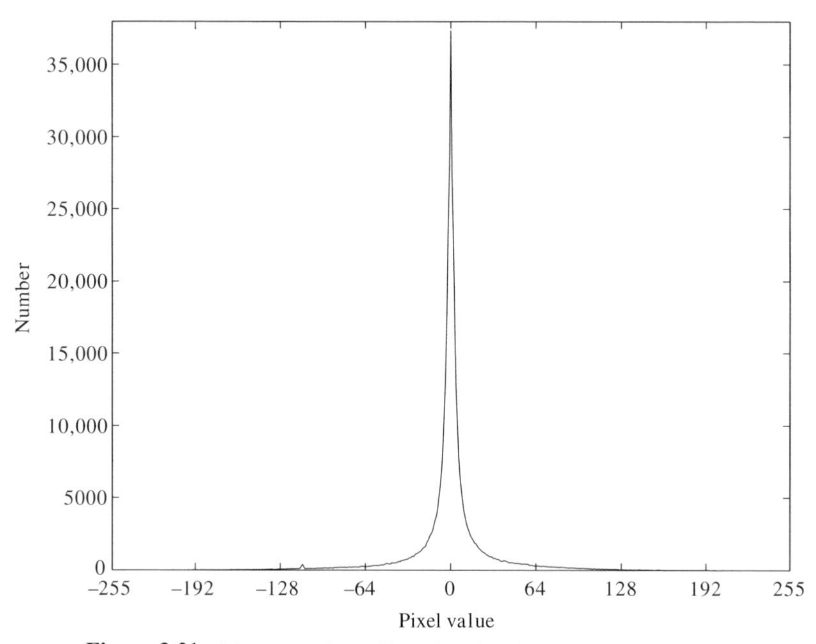

Figure 3.21 Histogram of one-dimensional DPCM prediction difference.

As we had hoped, Figure 3.21 shows that the prediction difference is usually small with zero, the most common prediction difference and almost all prediction differences in the range of ±64. The possible range of pixel values has increased from 0 to 255 in the original picture to −255 to +255 in the prediction difference. If the prediction was not reasonably accurate, we could actually have an increase in the average number of bits needed to represent each pixel. The entropy for the prediction difference is 5.95 bits/pixel. This represents a small saving (just over 20%) on the original picture entropy of 7.61 bits/pixel. Although this is far less than is needed for the introduction of digital television services, it has been achieved in a lossless manner.

Figure 3.22 shows the prediction difference picture for one-dimensional DPCM (after shifting by 128 so that mid-gray value represents 0 and scaling by a factor of 2 to make the prediction difference clearer).

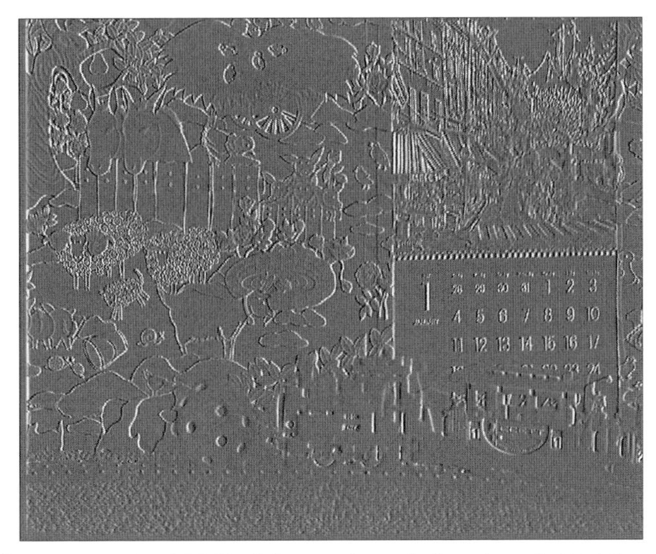

Figure 3.22 One-dimensional DPCM difference picture for first picture in "Mobile and Calendar" sequence (shifted by 128 and scaled by 2).

It is apparent that most of the large prediction differences occur at the edges of objects within the picture. ■

In the previous example, the pixel to the left of the pixel to be transmitted was used as the predictor. There is, of course, no reason why this pixel needs to be the one used. Any pixel whose value has already been transmitted to the receiver (so that the receiver can form an identical prediction) can be used. For example, the pixel immediately above (i.e., on the previous line to) the pixel to be transmitted could be used to form the prediction. Indeed, there is no need for the prediction to be based on just a single pixel. Figure 3.23 shows a generalized two-dimensional predictor.

In this case, the prediction of X ($\hat{X}$) can be formed according to

$$\hat{X} = k_1 A + k_2 B + k_3 C + k_4 D$$

The values of the weighting factors (k_i) are picture-dependant. Using pixels both to the left and above the pixel to be predicted means that both the horizontal and vertical correlations within the picture are exploited. Some improvement in performance would therefore be expected.

B	*C*	*D*
A	*X*	

Figure 3.23 Two-dimensional prediction.

EXAMPLE 3.6—MATLAB

Implement a two-dimensional predictor of the form given below.

$$\hat{X} = 0.5A + 0.5C$$

Calculate the entropy of the prediction difference using MATLAB for the first picture in the "Mobile and Calendar" sequence.

SOLUTION First we need to form the prediction. For pixels in the first row of the picture, we assume that the pixel above the pixel to be predicted has value 128 (mid-gray). For pixels in the first column of the picture, we assume that the pixel to the left of the pixel to be predicted has value 128. The prediction and the prediction difference can then be formed according to the MATLAB code given below. As before, the array calendar_1 contains the luminance pixel values of the first frame of the sequence "Mobile and Calendar."

```
[row,col] = size(calendar_1);
Guess = zeros(row,col);

for irow = 1:row
  if(irow == 1)
    Guess(irow,:) = Guess(irow,:) 1 0.5*128*ones(1,col);
  else
    Guess(irow,:) = Guess(irow,:) + 0.5* calendar_1(irow-1,:);
  end
end

for icol = 1:col
  if(icol == 1)
    Guess(:,icol) = Guess(:,icol) + 0.5*128*ones(row,1);
  else
    Guess(:,icol) = Guess(:,icol) + 0.5* calendar_1(:,icol-1);
  end
end

PRED = A - fix(Guess);
```

The picture entropy is 5.79 bits/pixel in this case, a reduction of 0.16 bits/pixel (2.5%) compared to one-dimensional prediction. This is a modest improvement given the additional computational requirements, but has still been achieved in a lossless manner. ■

Researchers have shown that once pixels A and C in Figure 3.23 have been used, there is little to be gained by using additional pixels in the same picture. Predictive encoders where all of the pixels used to form the prediction are in the same picture as the pixel to be predicted are called intrapicture predictive encoders.

As well as predicting from other pixels within the current picture, it is also possible to predict from a previous picture or pictures. If there has been little motion between the pictures then the prediction can be efficient. This is called interpicture prediction.

EXAMPLE 3.7—MATLAB

Use the first picture of the sequence "Mobile and Calendar" to predict the second picture in the sequence. Display the prediction difference picture and calculate its entropy.

SOLUTION The appropriate MATLAB code is given below. It is assumed that calendar_1 and calendar_2 contains the first and second frames, respectively, of the sequence "Mobile and Calendar."

```
DIFFERENCE = calendar_2 – calendar_1;
```

The resulting prediction difference picture (after shifting by 128 so that mid-gray represents 0 and scaling by a factor of 2 to make the prediction difference clearer scaled by a factor of 2) is shown in Figure 3.24.

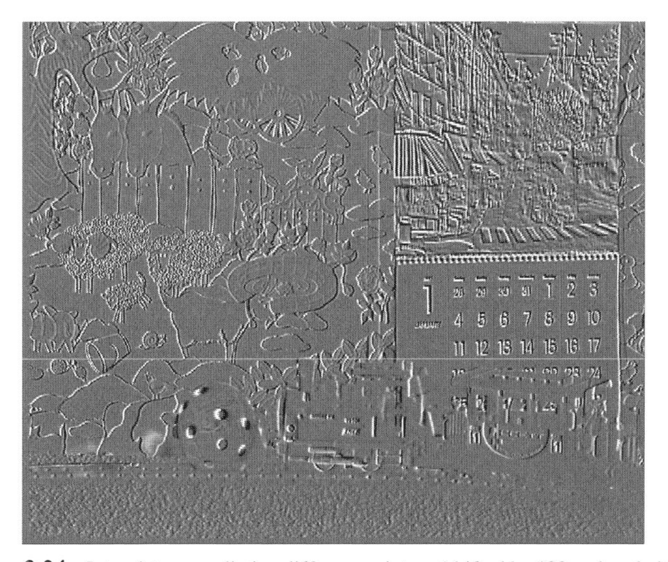

Figure 3.24 Interpicture prediction difference picture (shifted by 128 and scaled by 2).

The entropy of the difference picture is 6.18 bits/pixel. This is higher than either the one-dimensional or the two-dimensional intrapicture predictors, meaning that the prediction in this case requires more information to be transmitted when compared to the intrapicture predictors. ■

It might initially seem surprising that the interpicture prediction performs worse than the intrapicture prediction. The first thing to note is that the sequence "Mobile and Calendar" from which these pictures are extracted contains a large amount of motion. MATLAB provides a convenient means of viewing this motion using the movie function, and this is illustrated in Example 3.8.

EXAMPLE 3.8—MATLAB

Display as a sequence the first five pictures of the sequence "Mobile and Calendar." Assume that the first five pictures are held in arrays calendar_1 to calendar_5, respectively.

SOLUTION The MATLAB code to achieve this is given below.

```
display_picture(calendar_1);
M(:,1) = getpicture;
display_picture(calendar_2);
M(:,2) = getpicture;
display_picture(calendar_3);
M(:,3) = getpicture;
display_picture(calendar_4);
M(:,4) = getpicture;
display_picture(calendar_5);
M(:,5) = getpicture;
movie(M,-50,25);
```

Each picture needs to be displayed prior to being stored into the movie array *M*. This is achieved by the function display_picture. The MATLAB code for this function is given below.

```
function display_picture(CURR)
image(CURR);
colormap(gray(256))
set(gca,'XTick',[])
set(gca,'YTick',[])
imHandle=gco;
imageWidth = size(CURR,2);
imageHeight = size(CURR,1);
set(gca,'Units','pixels');
set(gcf,'Units','pixels');
figPos=[10,10,imageWidth,imageHeight];
axPos=[1,1,imageWidth,imageHeight];
set(gcf,'Position',figPos);
set(gca,'Position',axPos);
pause(1);
```

The final movie command causes the pictures stored in the array *M* to be played forward and backward 50 times at 25 pictures/s. Using +50 for the second variable in the function call causes the picture to be played in the forward direction only. In this case, the jerkiness between picture 5 at the end of a forward play and picture 1 at the start of the next forward play is subjectively distracting.

Looking at the sequence we see that the camera is panning continuously to the left. At the same time, the train is moving from right to left pushing the ball while the calendar is moving up. Almost every part of the picture is in motion, which explains why interpicture prediction is not especially successful in this case. ■

The impact of motion can be clearly seen if we compare a pair of pictures separated by more than one picture time. Figure 3.25 shows the prediction difference picture (shifted and scaled by 2) when the first picture of the sequence "Mobile and Calendar" is used to predict the fourth picture of the sequence.

The impact of the motion is clearly apparent, as many objects appear more than once in the prediction difference picture (e.g., spots on the ball, numbers on the calendar). This is because we need to put the object in its new position in the current picture as well as to remove it from its former position in the prediction

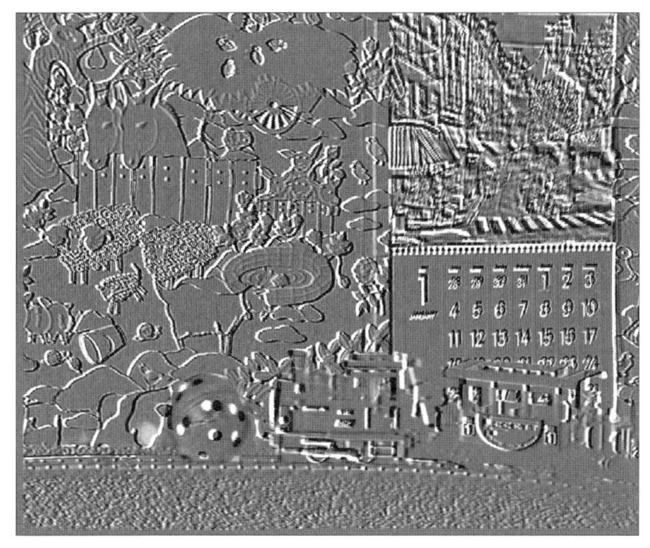

Figure 3.25 Interpicture prediction difference when pictures are separated by three picture times (shifted by 128 and scaled by 2).

picture. These objects thus contribute twice to the entropy of the prediction difference picture.

The simple interpicture predictor described above uses the pixel in the same location in the previous picture to predict the pixel in the current picture. This could be easily extended to include pixels in both the current and the prediction picture. Such a predictor is known as a three-dimensional predictor and is illustrated in Figure 3.26.

In this case the predictor would be

$$\hat{X} = k_1 A + k_2 B + k_3 C$$

Such predictors are rarely used in practice as temporal prediction is invariably combined with motion compensation. Motion-compensated prediction is covered in the next section of this chapter. Another technique, known as transform coding, is used to code the prediction difference after this motion-compensated temporal prediction. Transform coding is the topic of Chapter 4.

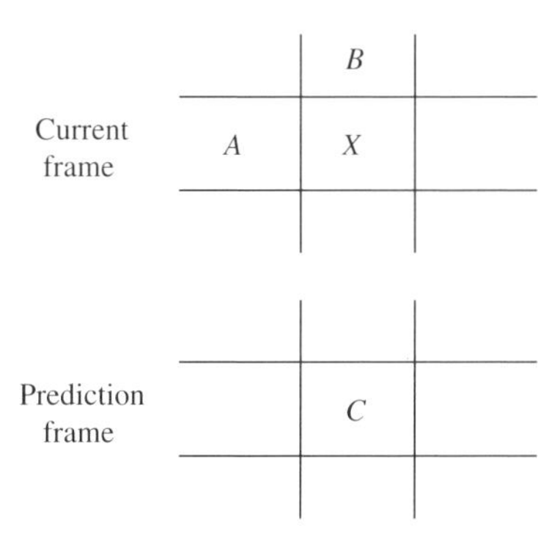

Figure 3.26 Three-dimensional prediction.

3.3 MOTION-COMPENSATED PREDICTION

The performance of interpicture prediction would be greatly improved if we were somehow able to take account of the motion that occurs between pictures. Success would mean that instead of moving objects needing to be coded twice in a picture, we may be able to remove the requirement to code them at all. Instead, information about how the object had moved between pictures would be transmitted.

Although this sounds fine in theory, the estimation of motion in a sequence of pictures is complex. Even if we can assume that the object is rigid (i.e., does not change shape), motion can be made up of several components such as

- zoom (i.e., change of the camera focal length)
- pan (i.e., rotation around an axis normal to the camera axis)
- rotation around the camera axis
- translation along the camera axis
- translation in the plane normal to the camera axis.

Each of these motion types affects the temporal appearance of the resulting video sequence in a particular way. Indeed, complex motion models have been proposed in an attempt to accurately take account of all of these various types of motion. Attempting to implement these models in real time hardware would be impossibly complex. Fortunately approximating all motion as translational motion in a plane normal to the camera axis is a satisfactory approach provided that this is done on sufficiently small objects.

The first part of the process is to attempt to estimate the motion of the object. Motion estimation is performed by searching the prediction picture for the best match to the object of interest in the current picture to be coded. Let us consider an example where rotational motion is estimated using only translation.

EXAMPLE 3.9—MATLAB

The picture in Figure 3.27 is a rotated version of the picture in Figure 3.28. Each picture has size 256 × 256 pixels. The picture is broken into 16 × 16 pixel blocks for motion estimation purposes. Calculate the motion-compensated prediction of Figure 3.27 using the picture in Figure 3.28 as the basis for the prediction.

Figure 3.27 Current picture.

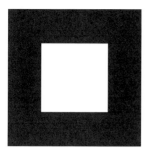

Figure 3.28 Prediction picture.

SOLUTION The motion-compensated prediction picture is shown in Figure 3.29. Using only translational motion to estimate an object that has undergone rotational motion leads to

Figure 3.29 Motion-compensated prediction.

a reasonable approximation of the real motion in this case. The motion compensated prediction difference and the interpicture difference are shown in Figure 3.30. The improvement

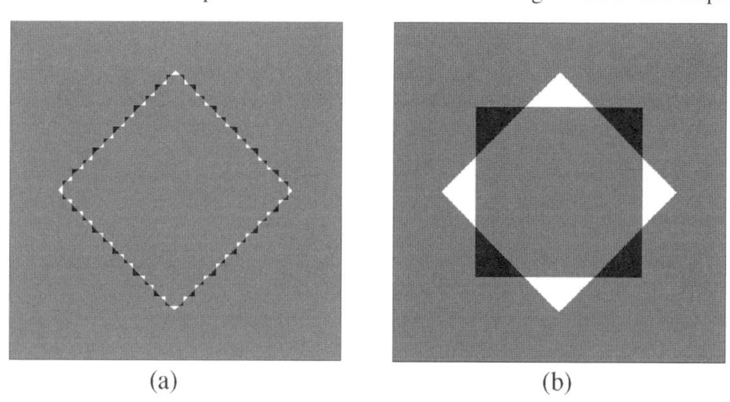

(a) (b)

Figure 3.30 (a) Motion-compensated prediction difference. (b)Interpicture difference.

as a result of the motion compensation process is clearly apparent. In fact, all of the types of motion listed earlier can usually be approximated by translational motion providing that the block size is sufficiently small. ■

3.3.1. Motion Estimation

The previous example demonstrates that translational motion can be used to form a reasonable estimate to rotational motion. However, we still need to develop a technique by which this translational motion can be estimated.

One approach is to partition each picture in the video sequence into its individual objects. A previous prediction picture can then be searched for the same object and the motion of the object between pictures can be estimated. Although this may provide excellent performance, there are a number of difficulties associated with this approach. The first of these is difficulty of finding individual objects within a picture. This process, known as segmentation, has been studied for a long time. Despite this, reliable segmentation techniques have not been found. The second problem is that objects do not move in a purely translational manner. It is therefore highly likely that there would be differences between the object as it appears in the current picture and how it appears in the prediction picture making the matching process far more difficult.

The usual approach to motion estimation is to divide the current picture to be predicted into a number of blocks. The prediction picture, which has already been coded, is then searched for the best matching block to each of these blocks in the current picture. Example 3.10 demonstrates the process.

EXAMPLE 3.10—MATLAB

Use picture 1 of the "Mobile and Calendar"sequence (the prediction picture) to predict picture 5 of the same sequence (the current picture) using motion-compensated prediction. Use a block size of 144 × 176 pixels.

SOLUTION The two pictures are shown in Figure 3.31. The spacing between the pictures has been chosen to allow the motion between pictures to be more easily seen.

(a)

Figure 3.31 (a) Prediction picture: first picture in "Mobile and Calendar" sequence. (b) Current picture: fifth picture in "Mobile and Calendar" sequence.

(b)

Figure 3.31 *(Continued)*

The picture difference is displayed in Figure 3.32 after shifting by 128 and scaling by 2.

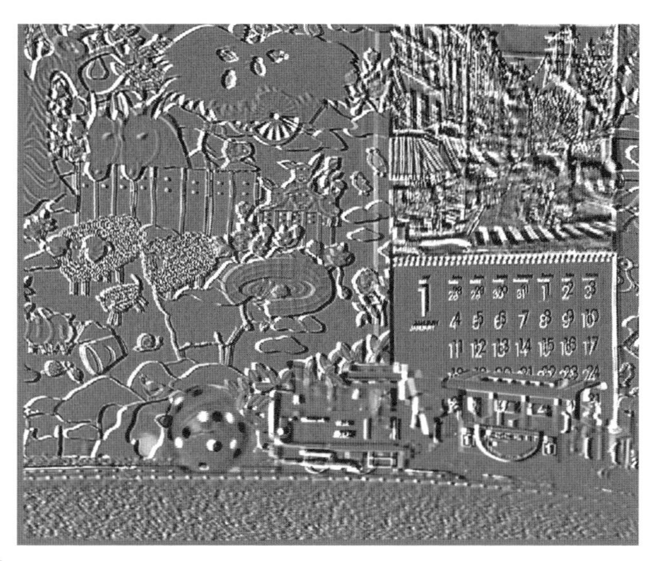

Figure 3.32 Picture difference for the pictures shown in Figure 3.31 (shifted by 128 and scaled by 2).

The motion that has occurred between pictures is clearly evident in the difference picture especially in the spots on the rolling ball and the numbers on the calendar. The aim of motion estimation is to find a block in the prediction picture (picture 1) that best matches each block in the current picture (picture 5). Figure 3.33(a) shows the current picture divided

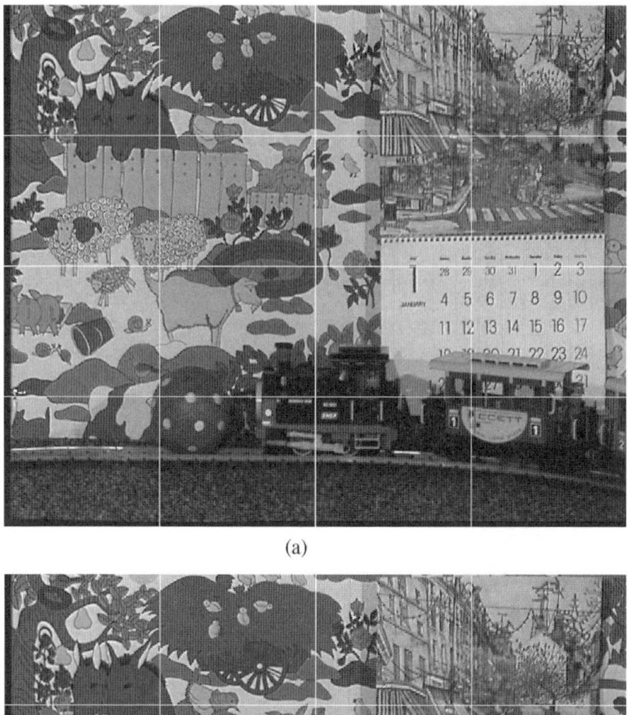

(a)

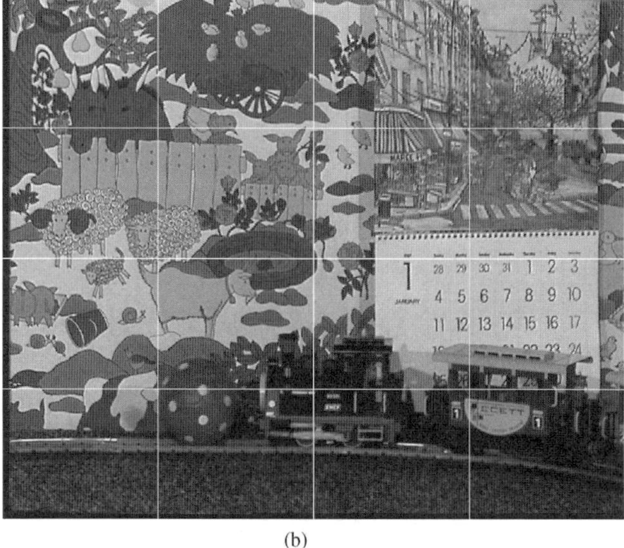

(b)

Figure 3.33 (a) Current picture (picture 5) divided into blocks. (b) Prediction picture (picture 1) divided into blocks.

into 144 × 176 pixel blocks whereas Figure 3.33(b) shows the prediction picture divided into identical blocks. The grid overlay highlights the motion that has occurred between pictures.

We now need to find the block from the prediction picture that best matches each block in the current picture. For simplicity, we consider just one of the blocks so formed—the third block in the third row. This block contains a section of the background wallpaper, part of the train, and a section of the calendar including the large number one. Comparing this block with the block in the same position in the prediction picture, the prediction difference is as shown in Figure 3.34. The motion between the pictures is again clearly evident.

Figure 3.34 Interpicture difference for block of interest (shifted by 128 and scaled by 2).

The motion estimation process consists of finding the same sized block in the prediction picture that best matches the block in the current picture. One simple method of calculation would indicate that the best matching block would be found four pixels to the left and one pixels up from the colocated block in the prediction picture. This block is highlighted in the prediction picture in Figure 3.35.

Figure 3.35 Best matching block in the prediction picture.

The absolute difference between the block in the current picture and the displaced block from the prediction picture is shown in Figure 3.36. This is called the motion-compensated difference.

Figure 3.36 Motion-compensated prediction difference (shifted and scaled by 2).

The improvement when the result shown in Figure 3.36 is compared with the inter-picture difference shown in Figure 3.34 is clearly evident. The mean absolute difference is reduced from 25.5 to 18.4 (a reduction of 30%). Note that though the background wallpaper is quite well predicted, the numbers on the calendar and the train are much less well predicted. This is a result of the fact that the motion of the each of these objects is different. The motion estimation approach employed does not allow for two or more different motions in the same block to be estimated. ■

Motion-compensated prediction consists then of two parts: motion estimation (which is performed at the encoder only) and motion compensation (which needs to be performed both at the encoder and at the decoder).

The motion estimation process consists of dividing the current picture to be encoded into a number of blocks. The prediction picture is then searched for the best match to each of the blocks in the current picture. The process can be best thought of as involving three pictures—the current picture, the prediction picture, and the motion-compensated prediction, as illustrated in Figure 3.37. Both the current and the prediction picture are composed of two round objects. The position of each object has moved between pictures. The motion-compensated prediction is initially blank.

Each block in the current picture is considered in turn. For convenience, we will start with the block that contains the left-hand round object. We define a search area in the prediction picture and then look for the block of pixels that best matches the pixels in the selected block in the current picture. This is illustrated in Figure 3.38 with the search area indicated by the highlighted square. The size of the search area is user selectable. The larger the search area, the larger the motion that can be tracked and hence the more likely a good match. However, this comes at the expense of higher computation requirements. We will consider this topic in more detail a little later.

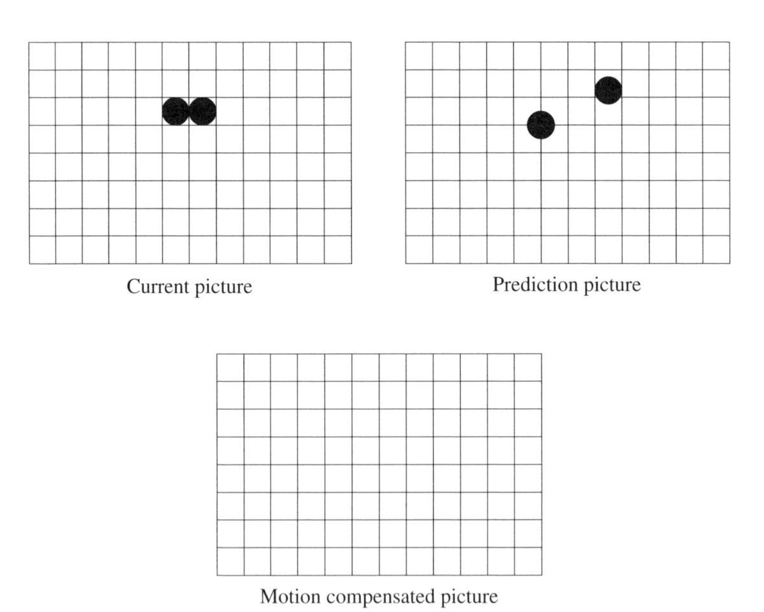

Figure 3.37 Current picture, prediction picture, and motion-compensated prediction.

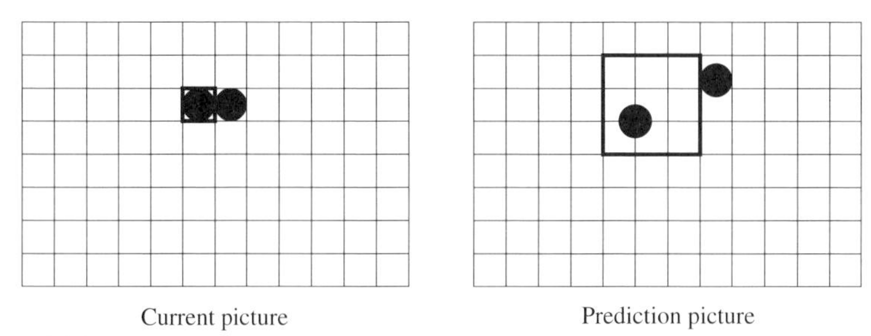

Figure 3.38 Search region in the prediction picture for the selected block in the current picture.

The best matching block[1] in the prediction picture is the one indicated in Figure 3.39. This block is then copied into the same position in the motion-compensated prediction as the block to be predicted in the current picture. This is also illustrated in Figure 3.39.

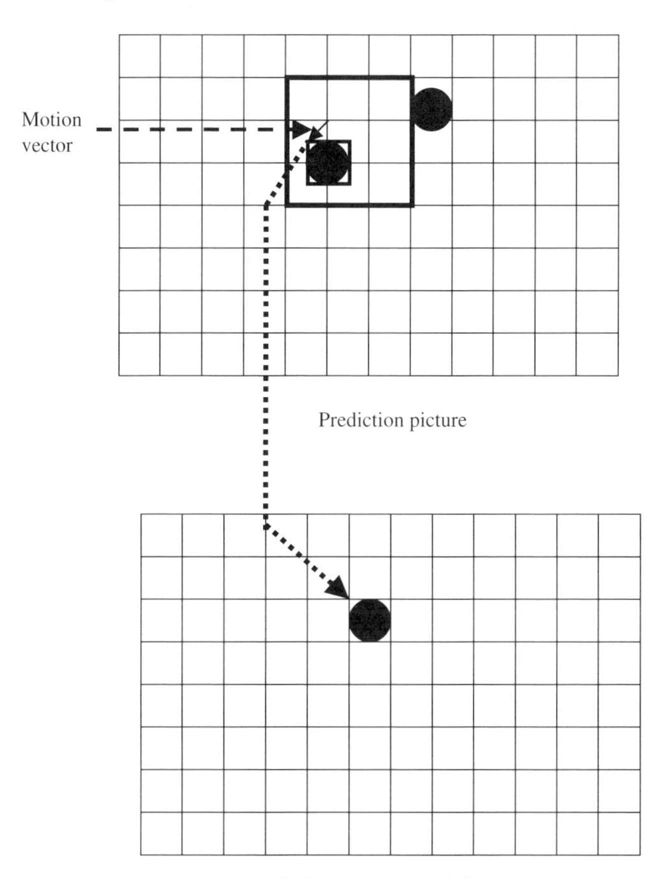

Figure 3.39 Generation of block in motion-compensated prediction.

[1]The best matching block in this case is obvious. For a more general pair of pictures, we need some matching criteria that can be used to select the best match. This is covered a little later.

The location of the block in the prediction picture used to predict the selected block in the current picture is transmitted to the decoder because the decoder does not have access to the same information as the encoder (e.g., original versions of the current and prediction pictures) and so cannot calculate the location itself. The information transmitted to the decoder is called a motion vector. The motion vector in this case is shown in the prediction picture in Figure 3.39 by a solid arrow. It marks the displacement from the location of the top left-hand corner of the block in the current picture that is being predicted to the location of the top left-hand corner of the chosen prediction block in the prediction picture. The arrow in Figure 3.39(a) indicates that the block to be used is several pixels to the left and below the location of the block in the same position as the block being predicted in the current picture.

This process is repeated for the next block in the current picture, and this is shown in Figure 3.40.

In this highly idealized case, the motion compensated prediction is identical to the original picture. Thus the prediction difference when the current picture is predicted using the motion-compensated prediction would be a picture consisting entirely of zeros. The only information that would need to be transmitted to the decoder is the motion vectors.

The search area could of course be made larger. An example of this is shown in Figure 3.41 for the case of the block containing the left-hand object. There are now two possible best match blocks. Each is shown with its corresponding motion vector

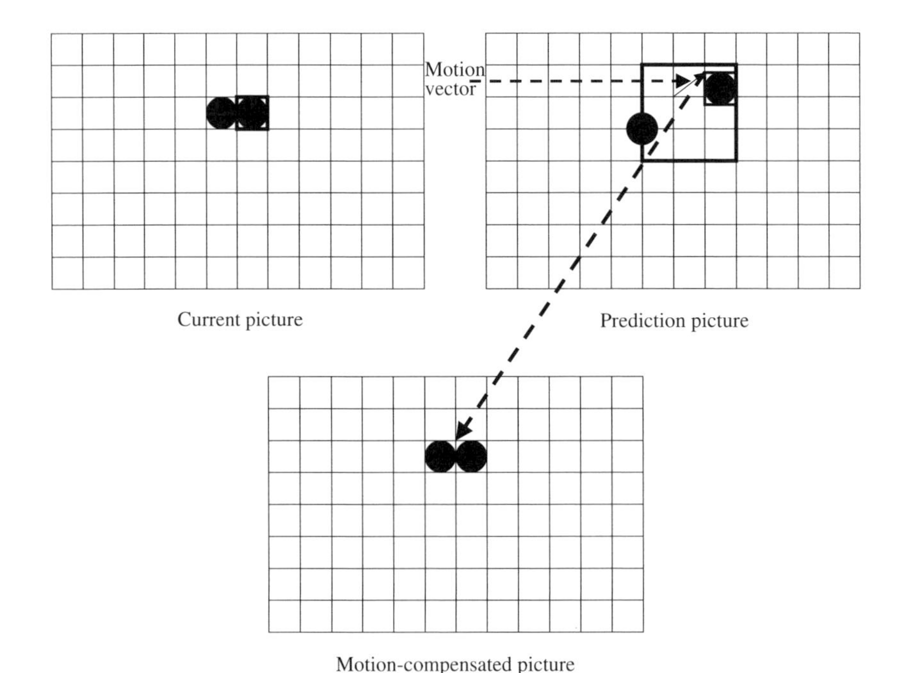

Figure 3.40 Motion estimation process for the next block in the current picture.

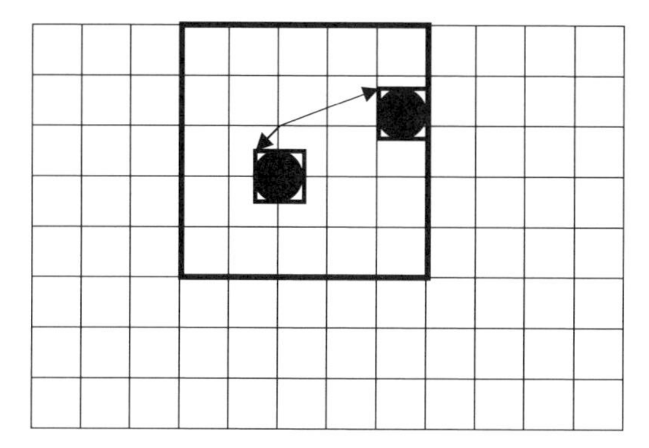

Prediction picture

Figure 3.41 Prediction picture from Figure 3.39 with larger search area.

in Figure 3.41. As the aim of motion-compensated prediction is to find a good match for the block to be predicted in the current picture, it does not really matter which one of these is selected. In video compression, the aim is to find a good match, not necessarily to estimate the correct motion.[2]

In the example just discussed, the best matching block was easily identified. In general, the task is not quite so easy as the best matching block is usually not identical to the block in the current picture. We need a measure of the similarity between the block in the current picture and the various possible prediction blocks in the prediction picture. The average difference between the blocks is of little use because large positive and negative differences might well cancel over the block. Various matching criteria have been proposed, of which the two most common are the summed absolute difference between current and candidate prediction blocks and the summed squared difference. Of these, summed absolute difference is most commonly chosen as there is no need for the relatively slow multiplication operation that is needed for squared difference, even though the use of squared difference often produces a slightly superior result. After the search operation, the block with the smallest summed absolute difference is selected as the best matching block.

Finally, we need to determine how the search is to be performed. The optimum approach (in a minimum absolute error sense assuming that total absolute error is the matching metric being used) is to compare the block we are trying to match with every block of the same size within the search area.

Consider the two 2×2 pixel block from the current picture and the search area (± 2 pixels horizontally and vertically) from the prediction picture shown in Figure 3.42. The colocated pixels in the search area are shown shaded.

There are a total of 25 possible search positions, all of which are shown in Figure 3.43. The order of the search is of no particular significance provided that all of the 25 search positions in this case are considered. Figure 3.43 shows the search

[2]There are other applications such as standards conversion or slow motion replay where correct motion estimation is the prime requirement.

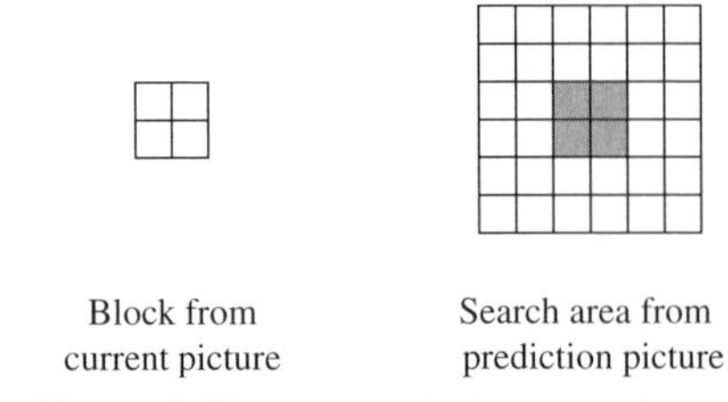

Figure 3.42 Current block and search area.

starting in the top left-hand corner of the search area with the block being moved one pixel to the right for each succeeding search until the top right-hand corner is reached. The search then reverts to the left-hand side one pixel from the top and again moves to the right one pixel at a time. This process continues until the block reaches the bottom row and concludes when the bottom right-hand corner is reached. At each search position, the summed absolute difference is calculated between the block at the search position and the block in the current picture. The block with the smallest summed absolute difference would be chosen as the motion-compensated prediction block.

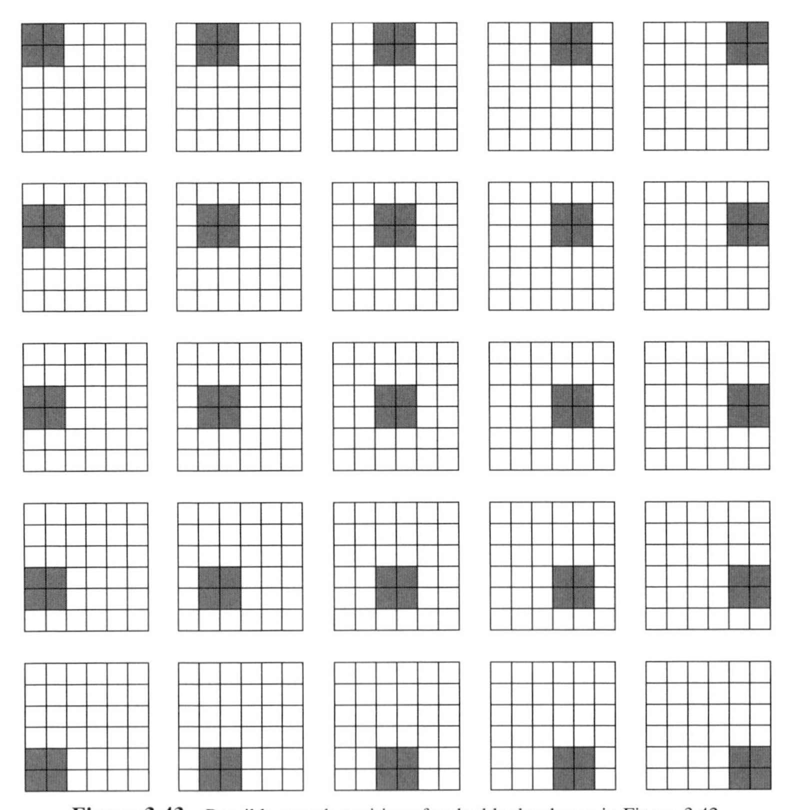

Figure 3.43 Possible search positions for the blocks shown in Figure 3.42.

When every possible search position is considered, the process is referred to as full search. For a search area of $\pm N$ pixels both horizontally and vertically, the total number of search positions to be considered is $(2N + 1) \times (2N + 1)$. For a large search area, this can be a very large number of search positions.

EXAMPLE 3.11

Block-based full search motion estimation with a search area of ± 16 pixels is used to form the prediction of the current picture from a previously coded prediction picture. Block size is 16×16 pixels. Each picture is of resolution 704×480 pixels at a picture rate of 30 pictures/s. Calculate

- the number of search positions for each block
- the number of pixel comparisons per block
- the number of pixel comparisons per picture
- the number of pixel comparisons per second.

SOLUTION The number of searches per block is $(2 \times 16 + 1)(2 \times 16 + 1) = 33 \times 33 = 1089$ searches.

For each search, two blocks of size 16×16 pixels are compared. This requires 256 pixel comparisons per search. The total number of pixel comparisons is then $1089 \times 256 = 278{,}874$ per block.
Each picture consists of

$$\frac{704 \times 480}{16 \times 16} = 1320 \text{ blocks.}$$

The total number of pixel operations per picture is
$278{,}874 \times 1320 = 3.68 \times 10^8$.

The picture rate is 30 pictures/s. The total number of pixel operations per second is therefore $3.68 \times 10^8 \times 30 = 11.0 \times 10^9$. ■

The computational requirements of motion estimation are large indeed. Further, in digital television applications search areas considerably greater than ±16 pixels are regularly employed. Fortunately, special purpose digital signal processing hardware is available that is capable of working at these rates.

Reconsider the pair of pictures shown in Figure 3.31 and the blocked current picture shown in Figure 3.33. If we perform motion compensated prediction for these blocks using a search area of ±8 pixels, then the motion-compensated prediction picture is as shown in Figure 3.44. The motion-compensated prediction difference is shown in Figure 3.45. The discontinuities in the ball, the engine of the train, and the edge of the calendar occur because within each of these blocks there is more than one form of motion—the moving foreground object (ball, engine, or calendar) and the background wallpaper. The final block chosen as the motion-compensated prediction is therefore a compromise between the two different motions. In the case of the first two blocks in the second column particularly, successful matching has

Figure 3.44 Motion-compensated prediction of picture shown in Figure 3.33(a).

been achieved because there is only one moving object (the wallpaper) in each of these blocks.

This raises the question as to what is the most appropriate block size to use. As we have already seen, the block size should be sufficiently small that the likelihood that a block contains two or more different motions is small. On the contrary, the block should be sufficiently large that the overhead associated with the motion vector is not excessive. Let us attempt to quantify this overhead. Assuming a search range

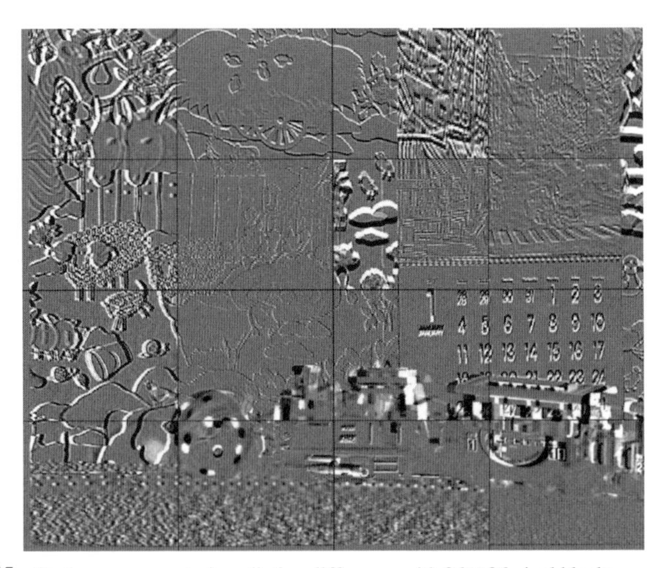

Figure 3.45 Motion-compensated prediction difference with 96×96 pixel blocks (shifted by 128 and scaled by 2).

Table 3.3 Motion vector overhead as a function of block size.

Blocks size (pixels)	Motion vector overhead (bits/pixel)
1×1	8.0000
2×2	2.0000
4×4	0.5000
8×8	0.1250
16×16	0.0313
32×32	0.0078

of ± 7 pixels, we need 4 bits to represent the horizontal component of the motion vector and 4 bits to represent the vertical component of the motion vector—a total of 8 bits/block. Table 3.3 shows the motion vector overhead for various block sizes.

As expected, the motion vector overhead is large for very small block sizes. However, by the time we reach a block size of 16×16 pixels the motion vector overhead reaches a negligibly small level. Most video coding standards use a block size of 16×16 pixels for motion-compensated prediction.

Figure 3.46(a) shows motion-compensated prediction for the pictures of Figure 3.31 using a block size of 16×16 pixels. The obvious improvement of this prediction compared to Figure 3.44 comes directly from the use of smaller blocks. There are still some minor distortions in the numbers on the calendar, and two of the spots on the ball are missing (as a result of the search area being not large enough). Of course, motion-compensated prediction can only rarely provide perfect prediction. The aim is to improve the prediction quality so that the amount of prediction difference information that needs to be transmitted is minimized. Figure 3.46(b) shows the motion vectors calculated during the motion compensation process overlayed on the motion-compensated prediction picture. This is easily done using the MATLAB *quiver* command. It can be seen that the motion estimation process has determined the true motion for much of the picture. However, there are a number of exceptions including

- The left-hand edge of the picture. The panning left of the camera causes new information to be introduced at the left-hand edge that cannot be found in the previous picture. Motion-compensated prediction simply cannot work in this circumstance.
- The white area below the large number one in the calendar. As this area is flat white, almost any motion vector would provide a good prediction and the actual vector chosen will be determined by noise within the picture. Motion-compensated prediction is still effective. A similar effect can be seen in the flat area of hay on the cart (top left of picture) and in the fence behind the sheep (middle left of picture).
- The middle of the ball. The motion estimation search area is not large enough for the correct prediction block containing the white dot to be found. Fast moving objects will invariably prove a problem for motion estimation if the

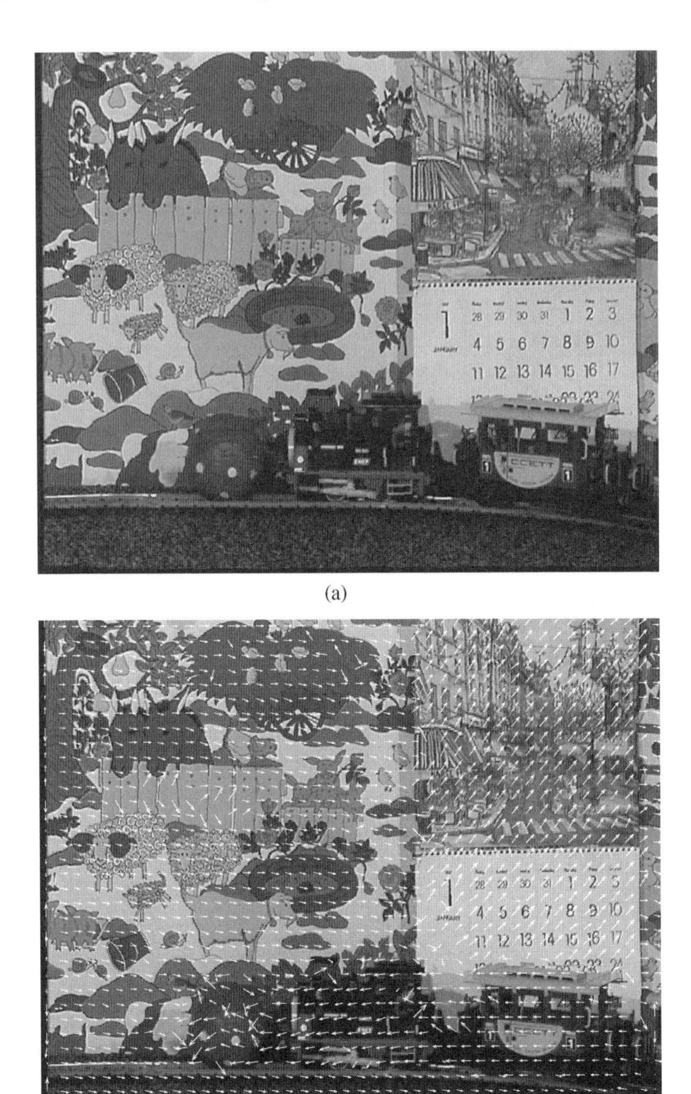

Figure 3.46 (a) Motion-compensated prediction using a block size of 16 × 16 pixels. (b) Motion vector overlay on motion-compensated prediction (vectors scaled by 2).

search area is not sufficiently large. Increasing the search area will improve the performance of motion estimation at the expense of an increase in computational requirements.

The motion-compensated prediction difference in this case is shown in Figure 3.47.

Figure 3.48 shows the histogram for the motion-compensated difference picture shown in Figure 3.47.

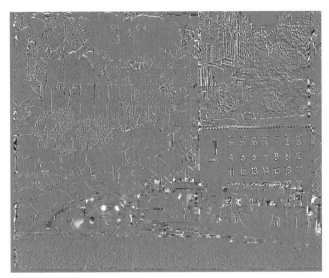

Figure 3.47 Motion-compensated prediction difference with 16×16 pixel blocks (shifted by 128 and scaled by 2).

The entropy of the motion-compensated difference picture is 5.46 bits/pixel. By comparison, the interpicture difference has an entropy of 6.18 bits/pixel and the two-dimensional intrapicture difference has an entropy of 5.79 bits/pixel. So motion compensation has reduced the entropy compared to intrapicture prediction but only by a relatively small amount.

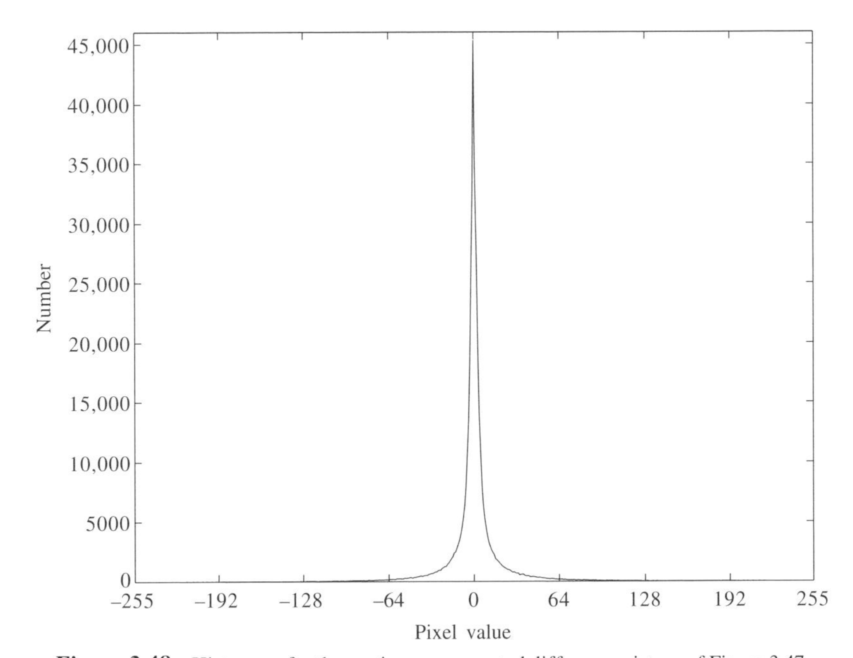

Figure 3.48 Histogram for the motion-compensated difference picture of Figure 3.47.

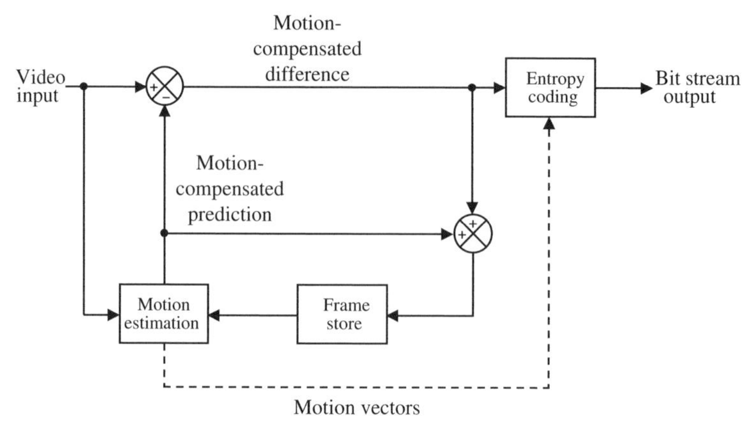

Figure 3.49 Motion-compensated encoder.

A block diagram of the motion-compensated encoder is shown in Figure 3.49 whereas the matching decoder is shown in Figure 3.50.

The decoder is simply the feedback loop of the encoder except for the entropy encoding in the coder and entropy decoding in the decoder. The motion-compensated difference signal is added to the motion-compensated prediction with the result stored in a picture store (a block of memory that holds the data for a digitized video picture).

3.3.2. Motion-Compensated Prediction to Subpixel Accuracy

Up to this point, we have assumed that motion estimation is performed to pixel accuracy. By this, we mean that each pixel in the current picture is predicted using a displaced pixel from the prediction picture. This assumes that motion occurs in multiples of single pixel displacements between pictures. This is obviously not the case in reality. Motion-compensated prediction to subpixel accuracy is possible and is supported in many video encoding standards including those used for digital television services.

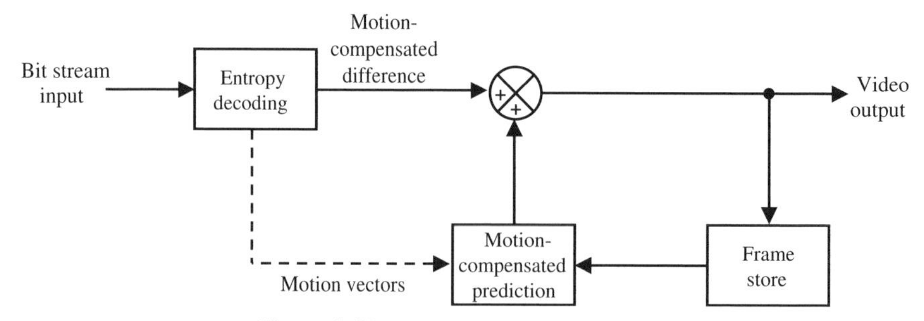

Figure 3.50 Motion-compensated decoder.

Motion-compensated prediction to subpixel accuracy can be achieved by estimating the values of pixels at subpixel displacements using bilinear interpolation from the known pixel values. This is illustrated in Example 3.12.

EXAMPLE 3.12

A 6×6 pixel block of picture data is shown in Figure 3.51. The center 4×4 pixel block is highlighted.

(a) Calculate the value of the 4×4 pixel block half a pixel to the right of the highlighted block.
(b) Calculate the value of the 4×4 pixel block half a pixel below the highlighted block.
(c) Calculate the value of the 4×4 pixel block half a pixel to the left and half a pixel above the highlighted block.

99	93	70	71	96	96
100	82	70	88	97	87
71	96	78	77	98	92
54	67	71	93	98	84
109	55	50	69	92	92
164	110	50	46	73	83

Figure 3.51 Picture data for Example 3.12.

SOLUTION (a) The block half a pixel to the right can be calculated by averaging each pixel in the selected region with the pixel immediately to its right. Rounding is employed to produce integer results. Half values round away from zero that is 1.5 round to 2. The final result is shown in Figure 3.52.

76	79	93	92
87	78	88	95
69	82	96	91
53	60	81	92

Figure 3.52 Half pixel shift to the right.

(b) The block half a pixel below the highlighted block can be calculated by averaging each pixel in the selected region with the pixel immediately below. Rounding is employed to produce integer results. Again, half values round away from zero. The final result is shown in Figure 3.53.

89	74	83	98
82	75	85	98
61	61	81	95
83	50	58	83

Figure 3.53 Half pixel shift down.

(c) The task of calculating the block half a pixel to the left and half a pixel up from the highlighted block is slightly more difficult. Consider the 2 × 2 pixel block at the top right of Figure 3.51 that is shown in Figure 3.54. The pixel half a pixel to the left and half a pixel above (X in the diagram) is calculated by averaging these four pixel values. Rounding is employed to produce integer results. Again, half values round away from zero. The final result is a pixel value of 94.

99		93
	X	
100		82

Figure 3.54 Top 2 × 2 pixels of Figure 3.51.

When applied to the entire 4 × 4 pixel block, the result obtained is shown in Figure 3.55.

94	79	75	88
87	82	78	90
72	78	80	92
71	61	71	88

Figure 3.55 Half pixel shift up and left. ■

It is possible to perform motion-compensated prediction to less than half pixel accuracy. For example, it would be possible to interpolate between half pixel points to calculate quarter pixel points. However, for digital television services based on MPEG-2, motion-compensated prediction is limited to a maximum of half pixel accuracy.

3.4 QUANTIZATION

All of the techniques discussed up to this point have been lossless. This means that the decoded video signal is identical in all respects to the signal that enters the encoder. Although this ensures that the received signal is of the highest possible quality, it also ensures that the amount of compression that can be achieved is

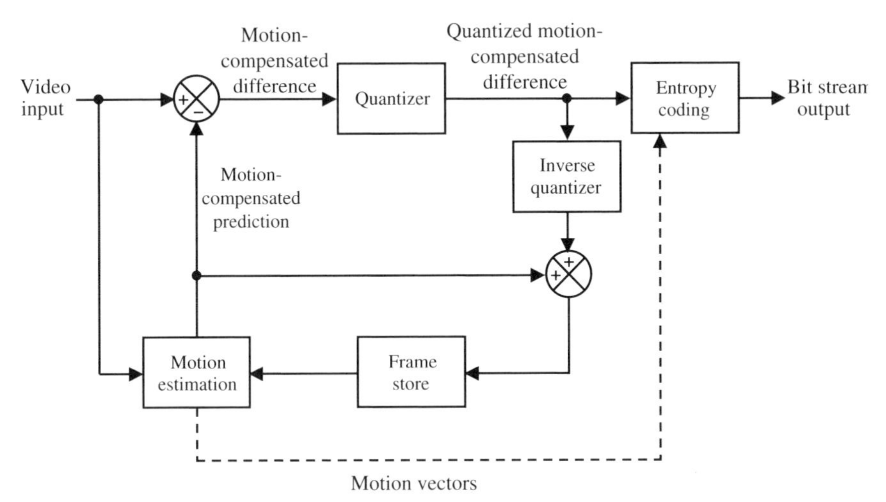

Figure 3.56 Lossy motion-compensated encoder.

limited. Fortunately, it is possible to relax this constraint and thus achieve higher compression while still maintaining more than adequate video service quality. This is achieved by removing information that is not important to the subjective quality of the video material.

While determining which information is subjectively important and which is not is quite a difficult task, the technique for removing information is more easily understood. If we revisit the lossless motion-compensated encoder shown in Figure 3.49, a lossy coder can be produced simply by introducing a quantizer to the motion-compensated difference signal as shown in Figure 3.56.

The motion-compensated difference is quantized prior to entropy coding and transmission to the decoder. As shown in Figure 3.48, the motion-compensated difference can take values in the range of ±255 (511 possible values) although most values will lie close to zero if the motion estimation process has been effective.

Quantizer design has been extensively studied over the years. A quantizer is defined by the position of its decision and reconstruction levels. This is illustrated in the quantizer transfer function shown in Figure 3.57 for a linear quantizer where the spacing between decision levels is constant. Such a quantizer is most useful for an input signal where all values are equally likely.

The role of the quantizer is summarized in Table 3.4. Each row should be read as Lower decision level $<$ Input $<$ Upper decision level $\Rightarrow$ Output $=$ Reconstruction level

For a linear quantizer, the reconstruction level is usually the average of the two decision levels that surround it (i.e., for a level defined by decision levels of $+2$ and $+6$, the reconstruction level would be $+4$) as this is optimum in a mean square error sense.

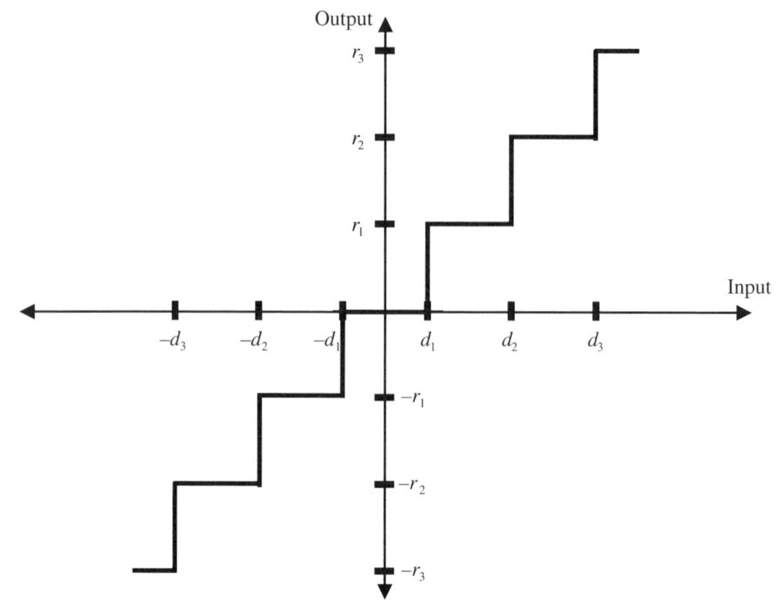

Figure 3.57 Quantizer transfer function.

Table 3.4 Decision and reconstruction levels for the quantizer of Figure 3.57.

Lower decision level	Upper decision level	Reconstruction level
$-\infty$	$-d_3$	$-r_3$
$-d_3$	$-d_2$	$-r_2$
$-d_2$	$-d_1$	$-r_1$
$-d_1$	$+d_1$	0
$+d_1$	$+d_2$	$+r_1$
$+d_2$	$+d_3$	$+r_2$
$+d_3$	$+\infty$	$+r_3$

EXAMPLE 3.13

The data shown in Figure 3.58 has been extracted from the top left-hand corner of a picture. The picture consists of gray scale values in the range 0–255.

145 162 199 155 112

149 151 153 154 162

92 85 88 187 212

Figure 3.58 Picture data for Example 3.13.

(a) Calculate the prediction difference when one-dimensional prediction using the pixel immediately to the left of the pixel to be predicted is employed. Assume that any predictor that falls outside the picture has value 128.
(b) The prediction difference is quantized using a linear quantizer with step size 8. Calculate the reconstructed data after prediction and quantization.

SOLUTION (a) The prediction difference is calculated by subtracting the prediction pixel from the current pixel. Remembering that the first pixel in each line is predicted by the value 128, the prediction difference is as shown in Figure 3.59.

+17	+17	+37	–44	–43
+21	+ 2	+ 2	+1	+ 8
–36	–7	+ 3	+99	+25

Figure 3.59 Unquantized prediction difference for Example 3.13.

(b) Quantization cannot be performed by simply quantizing the prediction difference shown in Figure 3.59. This is because the prediction differences shown in Figure 3.59 were formed using original pixel values. Although these values are available at the encoder, only reconstructed values are available at the decoder. For this reason, reconstructed values need to be used for prediction in the encoder as well. This means that we need to determine the reconstructed value of the current pixel before we can predict the next pixel. The solution is shown in Figure 3.60.

Current pixel	145	162	199	155	112
Prediction	128	144	160	200	152
Prediction difference	+17	+18	39	–45	–40
Quantized prediction difference	+2	+2	+5	–6	–5
Reconstructed prediction difference	+16	+16	+40	–48	–40
Reconstructed pixel	144	160	200	152	112
Current pixel	149	151	153	154	162
Prediction	128	152	152	152	152
Prediction difference	+21	–1	+1	+2	+10
Quantized prediction difference	+3	0	0	0	+1
Reconstructed prediction difference	+24	0	0	0	+8
Reconstructed pixel	152	152	152	152	160
Current pixel	92	85	88	187	212
Prediction	128	96	88	88	184
Prediction difference	–36	–11	0	+99	+28
Quantized prediction difference	–4	–1	0	+12	+3
Reconstructed prediction difference	–32	–8	0	+96	+24
Reconstructed pixel	96	88	88	184	208

Figure 3.60 Determination of reconstruction after one-dimensional prediction and quantization.

The calculations used in Figure 3.60 can be summarized as follows.

- Prediction is the reconstructed value of the previous pixel in the line.
- Prediction difference is the difference between the value of the current pixel and the prediction.
- The quantizer prediction difference is the prediction difference divided by the quantizer step size (eight in this case) rounded to the nearest integer (with a fractional part of 0.5 rounded toward zero thus 14.5 rounds to +14 and −14.5 rounds to −14). It is the quantized prediction difference that would be transmitted to the decoder.
- The reconstructed prediction difference is the quantized prediction difference multiplied by the quantizer step size (eight in this case).
- Finally the reconstructed pixel is the prediction plus the reconstructed prediction difference. This reconstructed value is used as the prediction for the next pixel. ■

The briefest perusal of Figure 3.48 indicates that the values are anything but equally likely. Quantizers can be designed to minimize the mean square error on the basis of the statistical distribution $p(x)$ of the signal to be quantized. The mean square error of a quantizer with decision levels $d_0 < d_1 < d_2 < \ldots < d_L$ and reconstruction levels $r_0 < r_1 < r_2 < \ldots < r_L$ is given by

$$\mathrm{MSE} = \sum_{k=1}^{L} \int_{d_{k-1}}^{d_k} (x - r_k)^2 p(x)\,\mathrm{d}x$$

Differentiating with respect to d_k and r_k for a fixed number of levels L leads to the following conditions for the decision and reconstruction levels.

$$d_k = \frac{r_k + r_{k+1}}{2}$$

$$r_k = \frac{\int_{d_{k-1}}^{d_k} xp(x)\,\mathrm{d}x}{\int_{d_{k-q}}^{d_k} p(x)\,\mathrm{d}x}$$

For the predictive encoders considered up to this point, $p(x)$ can be approximated by a Laplacian distribution with mean zero and variance $2/\lambda^2$.

$$p(x) = \frac{\lambda}{2} \mathrm{e}^{-\lambda|x|}$$

The prediction difference is highly peaked around zero. This results in the quantizer steps being closely spaced near zero prediction difference and getting larger as the prediction difference gets larger. Such a nonlinear quantizer is shown in Figure 3.61.

Although quantization is an important aspect of video compression, in modern coding standard quantization is not performed on prediction differences. Rather, it is transform coefficients that are quantized. We therefore leave any further

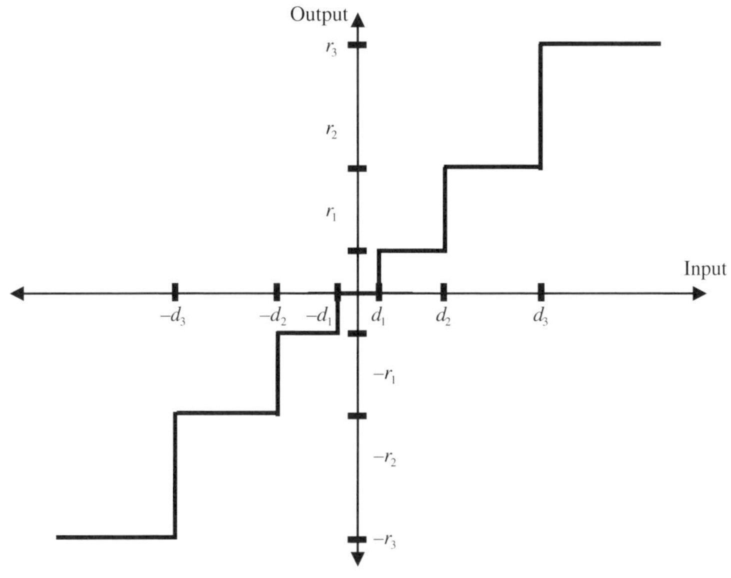

Figure 3.61 Nonlinear quantizer.

consideration of quantization until after transform coding has been introduced in Chapter 4.

3.5. RATE-DISTORTION CURVES

The introduction of quantization allows the data rate produced by the encoder to be traded against the quality of the reconstructed service. A small value of the quantizer leads to a high-quality reconstruction at a high data rate. A large value of the quantizer greatly reduces the data rate, but produces a much lower quality reconstruction.

A common method of measuring the quality of a reconstructed video service is the peak signal-to-noise ratio (PSNR). The mean square error between an original and a reconstructed picture of size $M \times N$ pixels is defined as

$$\text{MSE} = \frac{1}{MN} \sum_{i=1}^{M} \sum_{j=1}^{N} \left(x_{i,j} - \hat{x}_{i,j} \right)^2$$

where $x_{i,j}$ is the value of the original pixel and $\hat{x}_{i,j}$ is the value of the reconsructed pixel.

The PSNR is then defined as

$$\text{PSNR} = 10 \log_{10} \frac{(\text{Peak-to-peak signal})^2}{\text{MSE}}$$

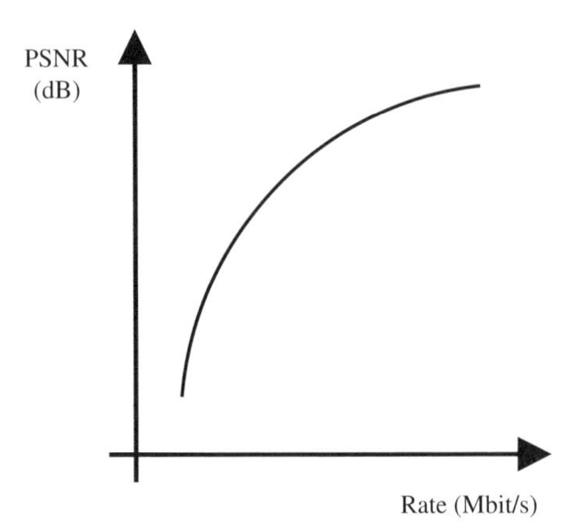

Figure 3.62 Typical rate-distortion curve.

For a 256 level gray scale image the peak to peak signals is 255 and so

$$\text{PSNR} = 10\log_{10}\frac{(255)^2}{\text{MSE}}$$

Plotting PSNR versus the generated data rate produces a curve called the rate-distortion curve. Rate-distortion curves can be useful in the comparison of the performance of various coding techniques on the same video material. An example of a rate-distortion curve, which has been drawn based on the distortion at a number of data rates, is shown in Figure 3.62.

3.6. SUMMARY

In this chapter we have introduced the idea of predictive encoding to reduce the amount of data required to represent a picture. We have seen that prediction can be performed in the spatial domain (intrapicture prediction) or between pictures separated in time (interpicture prediction). The prediction difference is then further coded using a Huffman code to exploit the fact that not all values of prediction difference are equally likely.

The chapter has also introduced the concept of motion-compensated prediction that attempts to compensate for the movement of objects between pictures prior to prediction. This was shown to significantly enhance the performance of interpicture prediction.

Finally, the concept of quantization was introduced. This lossy technique allows the prediction difference to be represented more coarsely. Distortion is introduced while at the same time a major saving in the required data rate is achieved. Distortion and rate can be traded off by varying the value of the quantizer step size.

Motion-compensated prediction is an important technique in the coding of digital television services. However, rather than transmitting the quantized prediction difference, another technique known as transform coding is employed prior to quantization. Transform coding is the topic of the next chapter.

PROBLEMS

3.1 Explain briefly the advantages and disadvantages of variable length codes.

3.2 An encoder produces an output consisting of four different symbols (a to d) having the probabilities of appearance given in Table 3.5.

Table 3.5 Probabilities for Problem 3.2.

Symbol	a	b	c	d
Probability	0.20	0.30	0.40	0.10

(a) What is the entropy of encoder output?

(b) If fixed length code words are used, how many bits would be required to represent each symbol? What is the coding efficiency?

(c) The output symbols are encoded with a variable length Huffman code. What is the average number of bits per symbol needed to transmit the symbols? What is the coding efficiency?

3.3 An encoder produces an output consisting of 10 different symbols (a to j) having the probabilities of occurrence given in Table 3.6.

Table 3.6 Probabilities for Problem 3.3.

Symbol	a	b	c	d	e	f	g	h	i	j
Probability	0.02	0.05	0.08	0.10	0.25	0.20	0.18	0.07	0.04	0.01

(a) What is the entropy of encoder output?

(b) If fixed length code word are used, how many bits would be required to represent each symbol? What is the coding efficiency?

(c) The output symbols are encoded with a variable length Huffman code. What is the average number of bits per symbol needed to transmit the symbols? What is the coding efficiency?

3.4 An encoder has been designed that produces three output symbols namely go up (U), no change (N), or go down (D). The relative probabilities of each symbol are given in Table 3.7.

Table 3.7 Probabilities for Problem 3.4.

Symbol	U	N	D
Probability	0.05	0.90	0.05

(a) Determine the coding efficiency if these symbol are transmitted individually using

(i) a fixed word length coding scheme

(ii) a variable word length Huffman code.

(b) Repeat part (a) after the three symbols have been grouped in pairs (i.e., UU, UN, UD, NU, NN, ND, DU, DN, DD) prior to encoding. You may assume the symbols are independent, that is $P(AB) = P(A)P(B)$.

(c) Repeat part (a) after the three symbols have been grouped in triplets.

(d) Comment on the results obtained in terms of the suitability of Huffman codes in cases where one symbol is much more likely than any of the others.

3.5 An encoder produces a binary output (i.e., a 0 or a 1). The probability of a 0, P_0, is given by p whereas the probability of a 1, P_1, is given by $(1-p)$.

(a) Plot the information content of the encoder output in bits per symbol as a function of p.

(b) If each output symbol is represented by a fixed length code word, plot the encoder output in bits per symbol as a function of p.

(c) If output symbols are grouped in pairs and then variable length coded, plot the encoder output in bits per symbol as a function of p.

(d) If output symbols are grouped in triplets and then variable length coded, plot the encoder output in bits per symbol as a function of p.

(e) Comment on the range of values of p where each of these approaches is appropriate. Give reasons for your decision.

3.6 Five symbols (a to e) are encoded using the variable length codes shown in Table 3.8.

Table 3.8 Variable length code words for Problem 3.7.

Symbol	code word
a	0
b	10
c	110
d	1110
e	1111

(a) Determine the bit stream when the transmitted string of symbols is as shown in Figure 3.63.

b a a c a d a e a a

Figure 3.63 Symbol stream for Problem 3.7.

(b) A single transmission results in an error in the reception of the first bit of the bit stream. Determine the symbol stream that will be output by the decoder.

(c) Repeat (b) for the case where the transmission error occurs in the second bit of the bit stream.

(d) Comment briefly on the impact of transmission errors on information coded with variable length code words.

3.7 A source encoder produces strings of zero bits of various length followed by the value one. The output of the source encoder is to be transmitted as a series of variable length code words that define the length of the string of zeros that occur before each one. The variable length code words are defined in Table 3.9.

Table 3.9 Variable length code words for Problem 3.7.

Length of string of zeros	Source encoder output	code word
0	1	1100
1	01	100
2	001	01
3	0001	00
4	00001	101
5	000001	1101
6	0000001	1110

Use this scheme to encode the sample encoder output shown in Figure 3.64.

0011001000000010000101100001000100000100100100010010001

Figure 3.64 Sample encoder output for Problem 3.7.

3.8 The run length coding scheme described in Problem 3.7 is to be extended so that it can handle a string of zeros up to a length of 63. Rather than define a new Huffman code word for each case, the scheme defined in Table 3.9 is modified by including an additional code word that covers any string of zeros greater than six.

The new code word consists of the prefix 1111 (which is not an existing valid code word) followed by the number of zeros as a 6-bit number. Thus a string of ten zeros followed by a one would be represented by the code word 1111001010. Use the modified coding scheme to encode the sample encoder output shown in Figure 3.65.

00110010000000001000010110000000000000010001000001001001000100100000001

Figure 3.65 Sample encoder output for Problem 3.8.

3.9 Figure 3.66 shows a block of data from a picture containing pixels with values in the range 0–255.

62	61	45	25
28	32	30	27
21	20	22	20
17	16	18	18

Figure 3.66 Picture data for Problem 3.9.

(a) Calculate the prediction difference after one-dimensional predictive encoding using the value of the pixel to the left of the pixel to be predicted as the prediction. Assume that all pixels surrounding the pixels to be predicted have value 128.

(b) Calculate the prediction difference after two-dimensional predictive encoding using the average of the value of the pixel to the left of the pixel to be predicted and the value of the pixel directly above the pixel to be predicted as the prediction. Assume that all pixels surrounding the pixels to be predicted have value 128. Fractional values should be rounded toward zero.

3.10 Consider the picture data given in Figure 3.66. In this case, the prediction difference is to be quantized by a linear quantizer of the type shown in Figure 3.57. The quantizer step size is 4 (i.e., decision levels are at ±2, ±6, ±10, etc.) and the reconstruction level is at the mid-point of the relevant decision levels. Calculate the reconstructed pixel values that would be output from the decoder in these circumstances.

3.11 Repeat Problem 3.10 using a variety of quantizer step sizes in the range 1–256. Plot the mean square error introduced by quantization versus the quantizer step size. Comment on the result.

3.12 Figure 3.67 shows a block of data from a picture containing pixels with values in the range 0–255.

60	121	222	247	97	11	48	66
105	227	240	172	48	16	62	72
74	186	246	206	169	41	25	69
184	254	191	212	131	12	58	68
115	241	225	187	215	96	17	71
236	227	186	210	189	37	35	81
193	242	202	196	220	157	20	42
240	207	206	201	207	79	14	78

Figure 3.67 Picture data for Problem 3.12.

(c) Calculate the prediction difference after one-dimensional predictive encoding using the value of the pixel to the left of the pixel to be predicted as the prediction. Assume that all pixels surrounding the pixels to be predicted have value 128.

(d) Calculate the prediction difference after two-dimensional predictive encoding using the average of the value of the pixel to the left of the pixel to be predicted and the value of the pixel directly above the pixel to be predicted as the prediction. Assume that all pixels surrounding the pixels to be predicted have value 128. Fractional values should be rounded toward zero.

3.13 Consider the picture data given in Figure 3.67. In this case, the prediction difference is to be quantized by a linear quantizer of the type shown in Figure 3.57. The quantizer step size is 8 (i.e., decision levels are at ±4, ±12, ±20, etc.) and the reconstruction level is at the mid-point of the relevant decision levels. Calculate the reconstructed pixel values that would be output from the decoder in these circumstances.

3.14 Repeat Problem 3.13 using

(a) a quantizer step size of 1

(b) a quantizer step size of 128

(c) Comment on the results achieved.

3.15 A video encoder designed for digital television operates on pictures of resolution 720 × 576 pixels at a frame rate of 25 pictures/s. Full search motion estimation is employed using 16×16 pixel blocks with minimum absolute difference as the matching criteria. The search area employed is ±64 pixels.

(a) Calculate the total number of search positions and the total number of absolute pixel difference operations required each second in the encoder.

(b) Calculate the number of search positions and the total number of absolute pixel difference operations required each second in the decoder.

3.16 A video encoder designed for high-definition digital television operates on pictures of resolution 1440 × 480 pixels at a rate of 60 pictures/s. Full search motion estimation is employed using 16×16 pixel blocks with minimum absolute difference as the matching criteria. The search area employed is ±128 pixels.

(a) Calculate the total number of search positions and the total number of absolute pixel difference operations required each second in the encoder.

(b) Calculate the number of search positions and the total number of absolute pixel difference operations required each second in the decoder.

3.17 The enormous number of computations required in the motion estimation process can be achieved in practice, but requires significant use of parallel processing (i.e., several calculations being performed at the same time) to be successfully achieved. What parts of the motion estimation process can occur in parallel. What is the speed-up factor achieved by each type of parallel processing?

3.18 In what circumstances would you expect motion-compensated prediction to perform poorly? Give real world examples in support of your answer.

3.19 Consider the block of pixel data shown in Figure 3.68. The block of interest is surrounded by a black border.

(a) Calculate the block half a pixel to the left block of interest.

50	61	73	75	81	84	84	91	91	82
78	110	106	100	100	102	100	106	100	101
42	102	121	102	98	97	96	98	95	93
52	108	103	96	93	91	94	88	89	92
46	80	111	96	95	89	95	81	71	94
76	104	98	92	93	91	89	48	58	103
79	92	105	97	90	92	93	72	26	72
97	98	99	95	90	96	90	49	59	106
98	93	102	100	91	92	89	77	61	87
91	96	106	90	90	94	85	66	85	101

Figure 3.68 Pixel data for Problem 3.19.

(b) Calculate the block half a pixel up from the block of interest.

(c) Calculate the block half a pixel to the right and half a pixel down from the block of interest.

3.20 How would you go about performing motion estimation to quarter pixel accuracy? Reconsider the pixel data given in Figure 3.68.

(a) Calculate the block one quarter of a pixel to the left block of interest.

(b) Calculate the block three quarters of a pixel up from the block of interest.

(c) Calculate the block three quarters of a pixel to the right and one quarter of a pixel down from the block of interest.

3.21 When a perfect match is found during the motion estimation process, there is no need to perform further searches. However, most hardware implementations of motion estimation algorithms would continue the search in any case. Why do you think this would be the case?

3.22 Most video encoders that incorporate motion-compensated prediction allow poorly predicted blocks to be coded in intrapicture mode (i.e., without any form of temporal prediction). What impact would this have on the bit stream transmitted between the encoder and the decoder?

MATLAB EXERCISE 3.1: HUFFMAN CODING

Huffman encoding is commonly employed in modern digital video encoders to reduce the amount of information needed to represent digital information. In this exercise you will develop a MATLAB program that calculates the Huffman code for an original picture.

Section 1 Design of Huffman code words
Write a MATLAB program to calculate the Huffman code words for a video picture. Hence calculate the average number of bits per pixel required to represent a number of different pictures.

In order to perform this task, you will need to complete the following steps:

- Calculate the probability of each pixel value in the picture.
- Develop the Huffman tree. This requires the development of a representation of the linkages within the computer. One way of doing this is to track where each element in column i of the tree maps to in column $i+1$. It is then possible to read the tree from left to right. Reversing the code word read gives the Huffman code word.
- Calculate the average number of bits per pixel by multiplying the probability of each pixel value by the number of bits in the code word used to represent it and then summing over all possible pixel values. Note that not all 256 possible pixel values may occur in a particular picture.

Section 2 Huffman code efficiency
Write a MATLAB program to calculate the entropy of a picture or a prediction difference picture. Use your routine to calculate the entropy of the various pictures

studied in Section 1. Hence calculate the coding efficiency of the Huffman code for each picture.

Section 3 Huffman code effectiveness
Plot in the same figure the number of bits allocated by your Huffman algorithm to each pixel brightness and the number of bits suggested by the entropy value for that pixel. Comment on the result. Note that not all pixel values are used in some pictures so care will be needed in calculating the individual pixel entropies.

MATLAB Hint: The MATLAB command

hold on

allows more than one graph to be plotted on the same set of axes. The hold is turned off by the

hold off

command

MATLAB EXERCISE 3.2: DIFFERENTIAL PULSE CODE MODULATION

In this exercise you will explore the bit-rate savings that can be achieved using one-dimensional and two-dimensional lossless predictive encoding. Linear quantization is then incorporated in Section 2, thus allowing a study in the trade-offs possible between bit rate and picture quality in differential pulse code modulation.

Section 1 Lossless intrapicture predictive encoding
For at least one picture taken from different video sequences calculate

- the entropy of the horizontal one-dimensional prediction difference (i.e., using the pixel directly to the left of the pixel to be predicted as the prediction).
- the entropy of the vertical one-dimensional prediction difference (i.e., using the pixel directly above the pixel to be predicted as the prediction).
- the entropy of the two-dimensional prediction difference (i.e., using the average of the pixel directly to the left of the pixel to be predicted and the pixel directly above the pixel to be predicted as the prediction).

Comment on these results.

Section 2 Lossy intrapicture predictive encoding
Section 1 involves the lossless encoding of a picture since the original picture can be reconstructed at the decoder. The process is made lossy by the introduction of quantization. Extend the predictive encoders developed in Section 1 to include linear

quantization of the type shown in Figure 3.57. For a quantizer with step size 4, the decision levels would be ±2, ±6, ±10, ... and the reconstruction levels 0, ±4, ±8, ...

Use your simulation to calculate the PSNR in decibels of the reconstructed picture and the entropy of the quantized prediction difference signal for a range of quantizer step sizes. Plot PSNR versus entropy for each of the three prediction techniques. Such a plot is called a rate-distortion curve. Also examine the quality of the reconstructed picture as the quantizer step size is increased. Summarize your results.

Note: A quick check that your program is operating correctly is to use a quantizer step size of 1. In this case, the PSNR should be infinite and the entropy the same as in Section 1.

MATLAB EXERCISE 3.3: TEMPORAL PREDICTION AND MOTION ESTIMATION

In this exercise we study the effectiveness of temporal prediction techniques. This begins with a study of interpicture prediction (i.e., predicting the current pixel using the pixel in the same location in the previous picture). We then move on to look at motion-compensated prediction. Quantization is incorporated into both techniques.

Section 1 Lossless and lossy interpicture prediction

For several video sequences:

- Calculate the entropy of the interpicture prediction difference (i.e., using the pixel in the same location as the current pixel in the previous picture as the predictor). Naturally picture $N+1$ will be predicted by picture N in this experiment.
- Take the fifth picture in each sequence and perform interpicture prediction using each of the pictures that precedes it (i.e., pictures 1–4). Comment on the effectiveness of interpicture prediction as the spacing between the pictures increases.
- The previous two parts deal with the lossless interpicture encoding of a picture because the original picture can be reconstructed at the decoder. The process is made lossy by the introduction of quantization. Extend the predictive encoders developed to include linear quantization of the type shown in Figure 3.57. For a quantizer with step size 4, the decision levels would be ±2, ±6, ±10, ... and the reconstruction levels 0, ±4, ±8, ...
- Use your simulation to calculate the peak signal-to-noise ratio in decibels of the reconstructed picture and the entropy of the quantized prediction difference signal for a range of quantizer step sizes. Plot PSNR versus entropy for interpicture prediction. Such a plot is called a rate-distortion curve. Also examine the quality of the reconstructed picture as the quantizer step size is increased. You should note that the picture quality looks good when a small quantizer is used, gets worse as the quantizer is increased and then starts to improve again when the quantizer step size becomes large. Explain why this

is the case and why this improvement at large quantizer step sizes is of no value in real applications.

Note: A quick check that your program is operating correctly is to use a quantizer step size of 1. In this case, the PSNR should be infinite and the entropy the same as in the lossless case.

Section 2 Lossless motion-compensated prediction
For several video sequences

- Calculate the entropy of the motion-compensated prediction difference using a block size for motion estimation of 16×16 pixels and a search range of ± 8 pixels both vertically and horizontally. Use minimum absolute error as the matching criteria for all of the motion estimation experiments contained in this exercise. Do not forget to include the overhead associated with the motion vectors in this calculation. Naturally picture $N+1$ will be predicted by picture N in this experiment. Make use of the MATLAB quiver command to overlay the motion vectors onto the picture being predicted.
- For a fixed search area of ± 8 pixels both vertically and horizontally, perform experiments to determine the effect of changing the block size used for motion estimation on the total entropy of the motion-compensated prediction difference. Make sure that you include the overhead associated with the motion vectors.
- For a fixed block size of 16×16 pixels, perform experiments to determine the effect of changing the search area to be used for motion estimation on the entropy of the motion-compensated prediction difference. Make sure that you include the overhead associated with the motion vectors.
- Find a pair of pictures from a video sequence that are on either side of a scene cut (or alternatively use two pictures from different video sequences) and calculate the entropy of the motion-compensated prediction difference. This situation occurs at every scene cut in a sequence. Suggest methods by which this limitation of motion-compensated prediction could be overcome.
- Take the fifth picture in each video sequence and perform motion-compensated prediction using each of the pictures that precedes it (i.e., pictures 1–4). Comment on the effectiveness of motion-compensated prediction as the spacing between the pictures increases.

Section 3 Lossy motion-compensated prediction
The previous section dealt with the lossless motion-compensated prediction of a picture because the original picture can be reconstructed at the decoder. The process is made lossy by the introduction of quantization. Extend the motion-compensated encoders developed in Section 2 to include linear quantization of the type shown in Figure 3.57. For a quantizer with step size 4, the decision levels would be ± 2, ± 6, $\pm 10, \ldots$ and the reconstruction levels 0, ± 4, ± 8, ….

Use your simulation to calculate the PSNR in decibels of the reconstructed picture and the entropy of the quantized motion-compensated prediction difference signal (including motion vector overhead) for a range of quantizer step sizes. Plot the rate-distortion curve. Also examine the quality of the reconstructed picture as the quantizer step size is increased. Compare the results obtained with those for lossy interpicture prediction calculated in Section 1.

MATLAB EXERCISE 3.4: FAST SEARCH MOTION ESTIMATION

Full search motion estimation always provides optimum performance (in terms of the matching criteria used at least). However, for a search area of $\pm M$ horizontally and $\pm N$ vertically, the total number of search required is $(2M + 1) \times (2N + 1)$ which can be very large. A number of fast search motion estimation techniques have been proposed by various researchers. Invariably, the aim is to search only a subset of the possible search positions and to choose the search positions checked in an intelligent way that increases the chances of obtaining the best (or at least a good) match. In this exercise we will study one of these techniques and compare its performance to full search motion estimation.

The fast search technique to be studied is three-step search motion estimation that is designed to work with a search area of ±7 pixels both horizontally and vertically. Figure 3.69 shows a 2×2 pixel block shaded surrounded by a search area of ±7 pixels.

The initial nine searches in the prediction picture are at the colocated block and at blocks ±4 pixels above, ±4 pixels below, and a combination of both of these displacements as shown in Figure 3.70. This is a total of nine searches as shown in the figure although the order of search is not important. One of these yields the best match in terms of the measure being used (say absolute error). In this case, we assume that this occurs at search position 1 that is shown in a different color in Figure 3.70.

Position 1 becomes the center of the next search and the new search positions are located at blocks ±2 pixels above, ±2 pixels below, and a combination of both of these displacements as shown in Figure 3.71. This requires eight new searches (searches 10–17 although again the search order is not significant). One of these nine

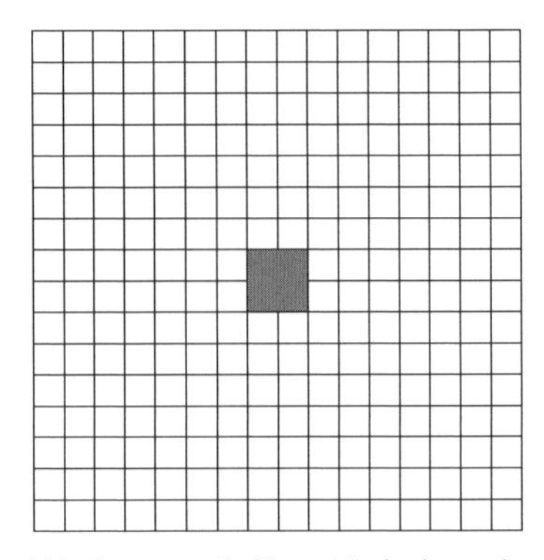

Figure 3.69 2×2 pixel block surrounded by a ±7 pixel search area in the current picture.

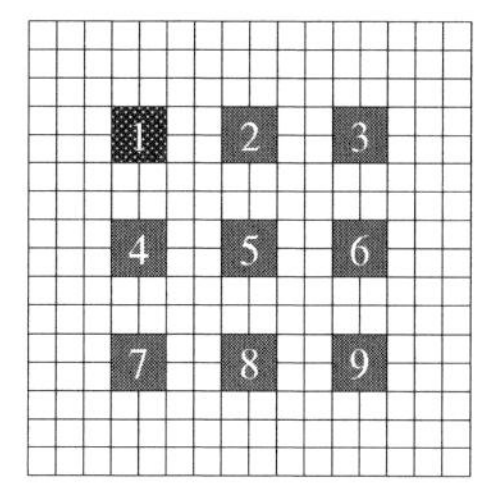

Figure 3.70 First step of three-step search.

positions (the best match could still be at position 1) is the best match in terms of the matching criteria. In this case, we will assume that this occurs at search position 14 that is shown in a different color in Figure 3.71.

Position 14 becomes the center of the next search, and the new search positions are located at blocks ±1 pixels above, ±1 pixels below, and a combination of both of these displacements. Because there is significant overlap between the search positions, no attempt will be made to draw them—hopefully the process is now sufficiently clear that this is not necessary. This again requires 8 new searches (searches 18–25 and again the search order is not significant). One of these nine positions (the best match could still be at position 14) will be the best match in terms of the matching criteria. This match will be taken as the best matching block and its displacement compared to the colocated is the motion vector.

The technique works on the basis that the closer one moves to the best match block, the smaller is the matching error. If this were always the case (which, of course, it is not), the three-step search algorithm would always produce the best matching block. In the remainder of this exercise we will explore the success of this fast search method.

Section 1 Full search motion estimation

- For a range of pairs of adjacent pictures extracted from different video sequences, calculate the entropy of the motion-compensated prediction difference using a block size for motion estimation of 16×16 pixels and a search range of ±7 pixels both vertically and horizontally. Use minimum absolute error as the matching criteria for all of the motion estimation experiments contained in this exercise. Do not forget to include the overhead associated with the motion vectors in this calculation. Naturally picture

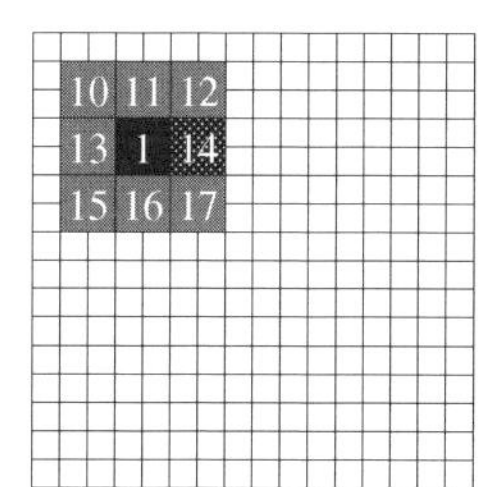

Figure 3.71 Second step in three-step search.

N+1 will be predicted by picture N in this experiment. Make use of the MATLAB quiver command to overlay the motion vectors onto the picture being predicted.

Section 2 Fast search motion estimation

- Repeat the previous calculation using the three-step search fast motion estimation algorithm. Comment on the performance in terms of the entropy of the motion-compensated prediction difference, the accuracy of the motion estimation, and the amount of computation needed.
- Take the fifth picture in each sequence and perform motion-compensated prediction using each of the pictures that precedes it (i.e., pictures 1–4) using both full search and the fast search technique. Comment on the effectiveness of motion-compensated prediction techniques as the spacing between the pictures increases.
- Extend the three-step search to cover a search area of ±15 pixels by including a fourth step in the search process. Again compare the performance of the fast search algorithm with full search in terms of the entropy of the motion-compensated prediction difference, the accuracy of the motion estimation, and the amount of computation needed.

Chapter 4

Transform Coding

4.1. INTRODUCTION TO TRANSFORM CODING

In Chapter 3 we saw that predictive encoding has some success in reducing the amount of information that needs to be transmitted to accurately represent a video picture. The total savings possible are, however, limited. A major drawback of predictive coding is that the number of elements to be transmitted (pixel differences) is the same after prediction as it was before prediction (pixels). The saving results from the reduction in the entropy of the prediction difference compared to the entropy of the original pixel values. For further gains, we need to reduce the number of elements that are transmitted. This can be achieved by transforming the data into a different domain.

In transform coding, the original pixel values are multiplied by a set of basis functions to produce a set of products. These products are added together to produce the coefficient for that basis function. The coefficient indicates how similar the original pixels are to the particular basis function. If the pixels and the basis function are similar then each product is positive and the result is a large positive value. If the pixels and the basis function are dissimilar then some products are positive whereas others are negative. When summed, the result is close to zero. Finally, if the pixels and the basis functions are similar in shape but different in sign then each of the products is negative and the sum is a large negative value. This is illustrated in Example 4.1.

EXAMPLE 4.1

Consider the simple one-dimensional transform[1] whose basis functions are shown in Figure 4.1 that takes four pixels as input and produces four transform coefficients.

Suppose that the pixel vector to be transformed is [21 8 12 21]. This is shown in Figure 4.2.

[1]This is an example of the Hadamard transform.

Digital Television, by John Arnold, Michael Frater and Mark Pickering.

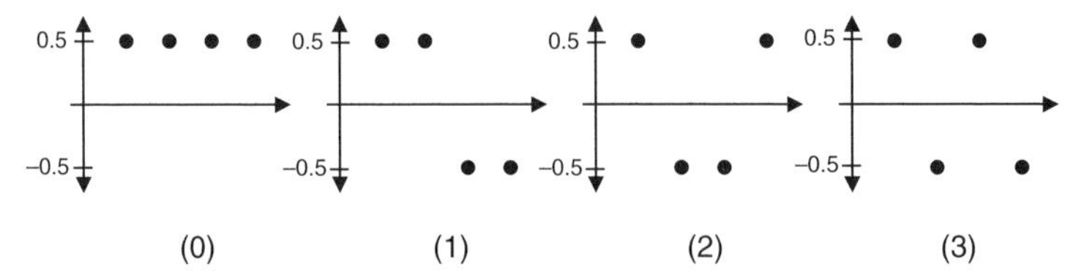

Figure 4.1 Basis functions for a simple four point one-dimensional transform.

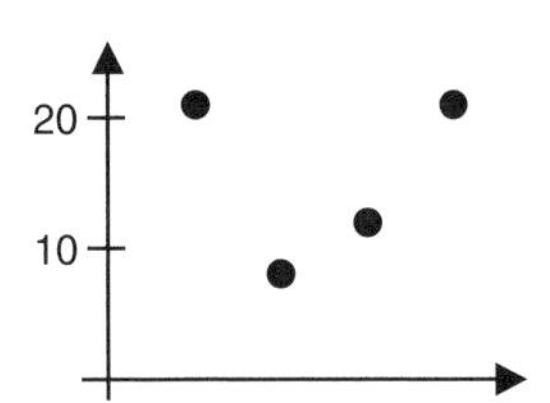

Figure 4.2 Pixel vector to be transformed.

The transform coefficients are then calculated by multiplying the pixel values by the basis functions on a point by point basis and then summing. Thus

$$\text{Coefficient } 0 = 21 \times (+0.5) + 8 \times (+0.5) + 12 \times (+0.5) + 21 \times (+0.5) = 31$$

$$\text{Coefficient } 1 = 21 \times (+0.5) + 8 \times (+0.5) + 12 \times (-0.5) + 21 \times (-0.5) = -2$$

$$\text{Coefficient } 2 = 21 \times (+0.5) + 8 \times (-0.5) + 12 \times (-0.5) + 21 \times (+0.5) = 11$$

$$\text{Coefficient } 3 = 21 \times (+0.5) + 8 \times (-0.5) + 12 \times (+0.5) + 21 \times (-0.5) = 2$$

The resulting coefficients are shown in Figure 4.3.

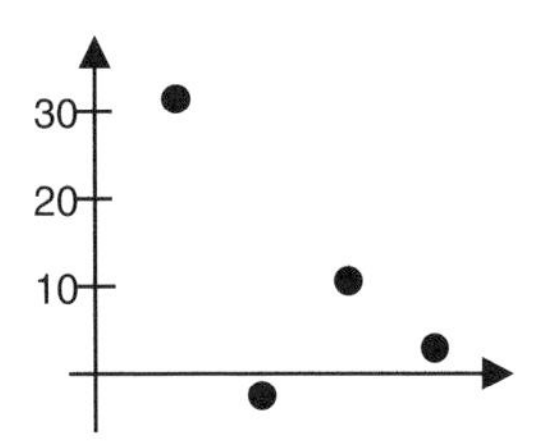

Figure 4.3 Resulting transform coefficients.

The first coefficient gives the average value of the pixels. For this reason it is often called the DC[2] coefficient. The remaining coefficients give information about the variation of the pixel values and are sometimes referred to as AC[3] coefficients. In this case, the second AC coefficient (coefficient 3) is larger than the other two because the general shape of the pixels to be transformed is a close match to the basis function of this coefficient.

The original pixel values can be recovered by the inverse transform process. This consists of weighting each basis function by its coefficient and then summing the four resulting vectors.

[2]Direct current.

[3]Alternating current.

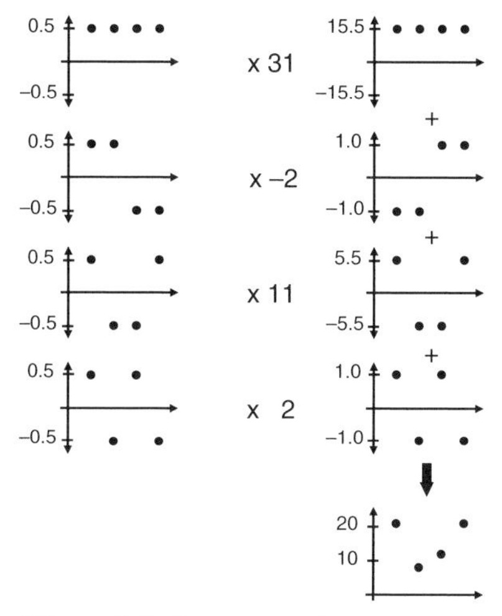

Figure 4.4 Inverse transform process.

$$
\begin{array}{lllll}
[+0.5 & +0.5 & +0.5 & +0.5] \times 31 = & [+15.5 \quad +15.5 \quad +15.5 \quad +15.5] \\
[+0.5 & +0.5 & -0.5 & -0.5] \times -2 = & [-1.0 \quad -1.0 \quad +1.0 \quad +1.0] \\
[+0.5 & -0.5 & -0.5 & +0.5] \times 11 = & [+5.5 \quad -5.5 \quad -5.5 \quad +5.5] \\
[+0.5 & -0.5 & +0.5 & -0.5] \times 2 = & [+1.0 \quad -1.0 \quad +1.0 \quad -1.0]
\end{array}
$$

On summing we obtain [21 8 12 21]

This was the original pixel vector. The inverse transformation process is illustrated in Figure 4.4. ■

4.2. THE FOURIER TRANSFORM

A common transform, one that many engineers and scientists are familiar with, is the Fourier transform. Some explanation of how the Fourier transform works and how it represents signals will help in understanding how transforms are used for compression. The Fourier transform is used to take a signal in the time domain and transform it into the frequency domain. This is possible because any time-domain signal can be represented by a weighted sum of sine and cosine waveforms. Thus the periodic time domain signal $f(t)$ with period T_0 can be represented as

$$f(t) = a_0 + \sum_{k=1}^{\infty} a_k \cos n\omega_0 t + \sum_{k=1}^{\infty} b_k \sin n\omega_0 t \qquad \text{where } \omega_0 = 2\pi / T_0.$$

The sinusoidal and cosinusoidal signals are called basis functions. In transform coding, the original signal is reproduced when these basis functions (appropriately weighted) are summed.

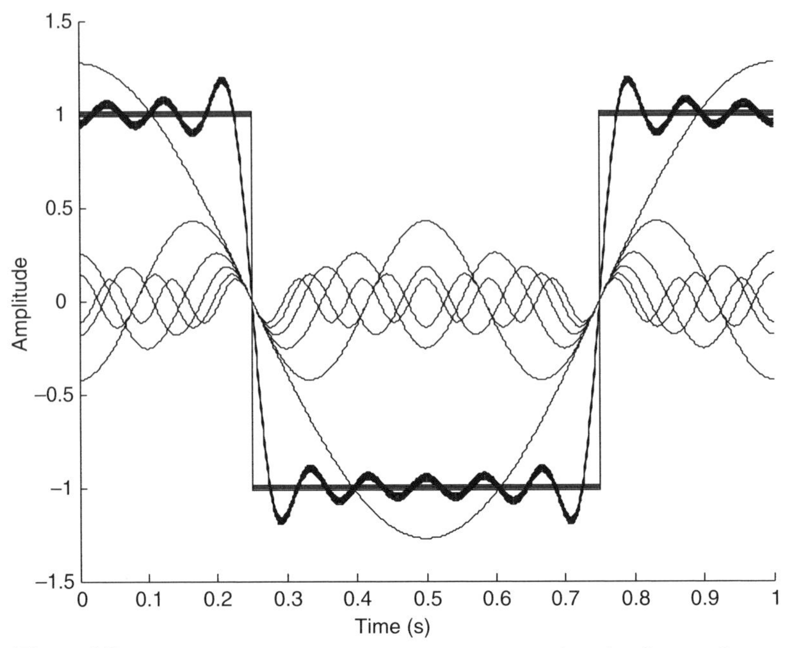

Figure 4.5 Square wave approximated by a weighted summation of cosine waveforms.

Figure 4.5 shows a square wave whose period is 1 s. Fourier analysis shows that the following Fourier series can represent this waveform.

$$f(t) = \frac{4}{\pi}\left(\cos\omega_0 t + \frac{1}{3}\cos 3\omega_0 t + \frac{1}{5}\cos 5\omega_0 t + \frac{1}{7}\cos 7\omega_0 t + \frac{1}{9}\cos 9\omega_0 t + \frac{1}{11}\cos 11\omega_0 t + \ldots\ldots\right)$$

where $\omega_0 = 2\pi$.

Figure 4.5 also shows the first six terms of the Fourier expansion plus the sum of these six terms. The convergence toward a square wave is apparent.

The weighting factor for each basis function is determined by comparing the basis function to the signal to be transformed. If the signal looks very similar to the basis function then a large weighting factor will result. If the signal and the basis function are dissimilar, a small weighting factor will result.

In the case of the Fourier transform of a periodic signal with period T_0, the weighting factors are determined according to

$$a_0 = \frac{1}{T_0}\int_0^{T_0} f(t)\mathrm{d}t$$

$$a_k = \frac{2}{T_0}\int_0^{T_0} f(t)\cos k\omega_0 t \; \mathrm{d}t, \qquad k = 1, 2, \ldots, n$$

$$b_k = \frac{2}{T_0} \int_0^{T_0} f(t) \sin k\omega_0 t \, dt, \qquad k = 1, 2, \ldots, n$$

where $\omega_0 = 2\pi / T_0$

The a_0 term gives the average value of the periodic signal whereas the a_k and b_k values measure the similarity of the waveform to appropriate cosinusoidal and sinusoidal signals, respectively.

The Fourier transform is designed to represent continuous functions. The pictures that we wish to encode are, of course, sampled. In this case, the appropriate transform is the discrete Fourier transform. The method of calculating weighting coefficients of the sampled function *f*(*n*) with a total of *N* sample points is given below.

$$a_k = \sum_{n=0}^{N-1} f(n) \cos\left(\frac{2\pi kn}{N}\right)$$

$$b_k = \sum_{n=0}^{N-1} f(n) \sin\left(\frac{2\pi kn}{N}\right)$$

The sampled values of the basis functions are called basis vectors. Figure 4.6 shows the first cosinusoidal basis vector when $N = 8$. This is a sampled cosine waveform with period eight samples.

The Fourier transform, though an important tool in understanding communication systems, is not the best choice for transform encoding. One reason for this is

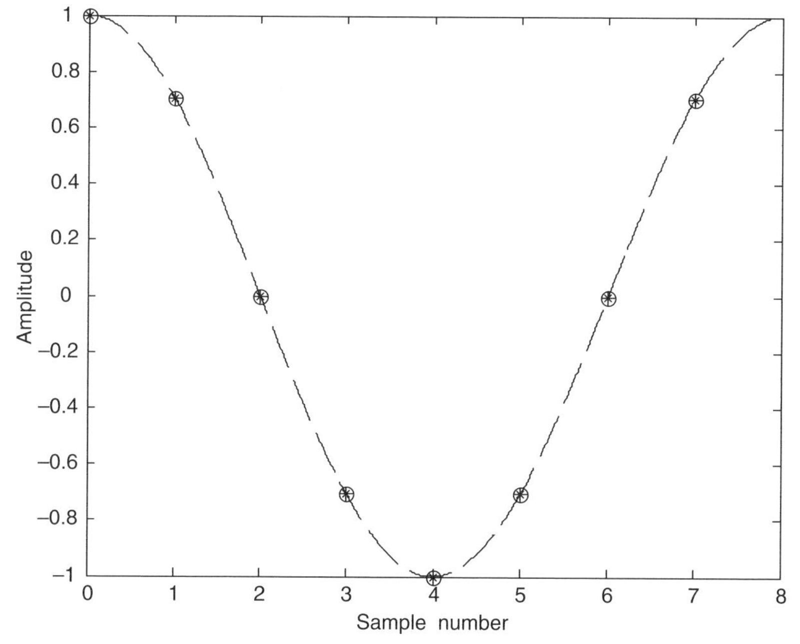

Figure 4.6 First cosine basis vector of discrete Fourier transform.

that there are two weighting factors (a_k and b_k) required for each frequency.[4] This requires a substantial amount of computation.

4.3. THE KARHUNEN–LOEVE TRANSFORM

The aim of a good signal compression transform is to pack the information in the data into the smallest number of transform coefficients. The data to be transformed is collected into a block. For an N point transform, the data would consist of an N element vector.

$$x = \left[x_1 \, x_2 \ldots x_N \right]$$

The transform consists of multiplying the data vector by an $N \times N$ transform matrix.

$$\mathbf{C} = \begin{bmatrix} C_1 \\ C_2 \\ \vdots \\ C_N \end{bmatrix} = \begin{bmatrix} t_{11} & t_{12} & \ldots & t_{1N} \\ t_{21} & t_{22} & \ldots & t_{2N} \\ \vdots & \vdots & \ldots & \vdots \\ t_{N1} & t_{N2} & \ldots & t_{NN} \end{bmatrix} \begin{bmatrix} x_1 \\ x_2 \\ \vdots \\ x_N \end{bmatrix} = \mathbf{Tx}$$

Each row of matrix $\mathbf{T}$ ($\mathbf{t_m}$) represents a basis vector of the transform. An effective transform would produce transform coefficients $\mathbf{C}$ that pack the information contained in the data into as few transform coefficients as possible. This means that it is not necessary to transmit all of the coefficients to obtain a satisfactory reconstruction of the original picture or, if the coefficients are very small, they may not need to be represented very accurately. The lower the number of coefficients that need to be transmitted, the less the amount of data required to transmit them.

Transforms are always orthonormal,[5] that is, when two basis vectors are multiplied together and summed

$$\mathbf{t_m}^{\mathrm{T}} \mathbf{t_n} = \begin{cases} 1, & m = n \\ 0, & m \neq n \end{cases}$$

where the superscript T represents the matrix transpose.

It follows directly that the inverse transform (T^{-1}) is simply the transpose of the forward transform (T^{T}).

$$\mathbf{T}^{-1} = \mathbf{T}^{\mathrm{T}}$$

[4]Many texts on digital signal processing refer to the discrete Fourier transform as generating a complex value for each weighting coefficient of the form $a_k + jb_k$.

[5]This ensures that if we initially approximate a desired function with a limited number of weighted basis functions, the subsequent addition of further weighted basis functions to improve the approximation will not change the values of the coefficients weighting the original basis functions.

We can therefore write

$$\mathbf{x} = \mathbf{T}^{-1}\mathbf{C} = \mathbf{T}^{\mathrm{T}}\mathbf{C} = \begin{bmatrix} t_{11} & t_{21} & \dots & t_{N1} \\ t_{12} & t_{22} & \dots & t_{N2} \\ \vdots & \vdots & \dots & \vdots \\ t_{1N} & t_{2N} & \dots & t_{NN} \end{bmatrix} \begin{bmatrix} C_1 \\ C_2 \\ \vdots \\ C_N \end{bmatrix}$$

and so

$$\mathrm{x} = \sum_{k=1}^{N} C_k \mathrm{t_k} = C_1 \mathrm{t_1} + C_2 \mathrm{t_2} + \dots + C_N \mathrm{t_N}$$

Thus the original signal can be thought of as a weighted sum of the basis vectors. These weights, **C**, are calculated by the forward transform.

If we want to understand the theory behind transform coding of pictures, some statistical measures of pictures need to be defined. The mean of an $N \times M$ pixel picture is given by

$$\overline{x} = \frac{1}{NM} \sum_{n=1}^{N} \sum_{m=1}^{M} x_{n,m}$$

where $x_{n,m}$ is the value of pixel m on line n of the picture. The expected value operator $E(\cdot)$ is often used to define an averaging operation. We can therefore write

$$\overline{x} = E(x)$$

The variance of an $N \times M$ pixel picture is defined as

$$\sigma^2 = \frac{1}{NM} \sum_{n=1}^{N} \sum_{m=1}^{M} (x_{n,m} - \overline{x})^2$$

which can be written using the expected value operator as

$$\sigma^2 = E\left((x - \overline{x})^2\right)$$

If we have two random variables, x_1 and x_2, the covariance of the two variables can be written as

$$\sigma_{12}^2 = E\left((x_1 - \overline{x})(x_2 - \overline{x})\right)$$

Using this terminology, the variance of a single variable would be referred to as σ_{11}. If we have k random variables, the covariance matrix $\mathbf{R_x}$ is defined as

$$\mathbf{R_x} = \begin{bmatrix} \sigma_{11}^2 & \sigma_{12}^2 & \dots & \sigma_{1k}^2 \\ \sigma_{21}^2 & \sigma_{22}^2 & \dots & \sigma_{2k}^2 \\ \vdots & \vdots & \vdots & \vdots \\ \sigma_{k1}^2 & \sigma_{k2}^2 & \dots & \sigma_{kk}^2 \end{bmatrix}$$

Consider a single line of pixels in a picture as defined by

$$x_{1,1} \quad x_{1,2} \quad x_{1,3} \quad \cdots \quad x_{1,m}$$

Let random variable x_1 and x_2 represent the pixel values

$$\begin{aligned} x_1 &= x_{1,1} \quad x_{1,2} \quad x_{1,3} \quad \cdots \quad x_{1,m-1} \\ x_2 &= x_{1,2} \quad x_{1,3} \quad x_{1,4} \quad \cdots \quad x_{1,m} \end{aligned}$$

The covariance for these two random variables is given by

$$\sigma_{12}^2 = E\left((x_1 - \overline{x}_1)(x_2 - \overline{x}_2)\right)$$

and is called the one-step horizontal covariance of the picture line. The k-step horizontal covariance is defined for the random variables x_1 and x_{k+1} that represent the pixel values

$$\begin{aligned} x_1 &= x_{1,1} \quad x_{1,2} \quad x_{1,3} \quad \cdots \quad x_{1,m-k} \\ x_{k+1} &= x_{1,k+1} \quad x_{1,k+2} \quad x_{1,k+3} \quad \cdots \quad x_{1,m} \end{aligned}$$

and is given by

$$\sigma_{1k}^2 = E\left((x_1 - \overline{x}_1)(x_{k+1} - \overline{x}_{k+1})\right)$$

If we want to calculate the covariance matrix for a whole picture, the covariance of each line is calculated using the mean value of the entire picture rather than the mean value for each variable on each line. The k-step horizontal covariance for each line is then given by

$$\sigma_{1k}^2 = E\left((x_1 - \overline{x})(x_{k+1} - \overline{x})\right)$$

where $\overline{x}$ is the mean of the entire image.

The k-step horizontal covariance for the picture is the average of the covariance for each line of the picture. A vertical covariance matrix can also be produced using pixel values down columns as opposed to across rows in the picture. For the first luminance picture in the "Mobile and Calendar" sequence, the 4×4 horizontal covariance matrix (rounded to the nearest integer) is given by

$$\begin{bmatrix} 3914 & 3525 & 3126 & 3050 \\ 3525 & 3907 & 3518 & 3118 \\ 3126 & 3518 & 3901 & 3511 \\ 3050 & 3118 & 3511 & 3894 \end{bmatrix}$$

By similar reasoning, the 4×4 vertical covariance matrix (rounded to the nearest integer) for this picture is given by

$$\begin{bmatrix} 3914 & 3540 & 3312 & 3031 \\ 3540 & 3914 & 3541 & 3313 \\ 3312 & 3541 & 3915 & 3541 \\ 3031 & 3313 & 3541 & 3915 \end{bmatrix}$$

As stated earlier, the aim of a good transform is to pack the information content of the data into the smallest possible number of transform coefficients. Although information content is somewhat difficult to quantify, one approach to achieving this aim would be to pack the maximum amount of energy into the smallest number of transform coefficients. If we represent the original data by only the first M of N transform coefficients with the remaining coefficients replaced by a constant a_j then we obtain

$$\hat{\mathbf{x}}_{\mathbf{M}} = \sum_{k=1}^{M} C_j \mathbf{t_j} + \sum_{k=M+1}^{N} a_j \mathbf{t_j}$$

The resulting error is

$$\begin{aligned} \mathbf{x}_{\mathbf{eM}} &= \mathbf{x} - \hat{\mathbf{x}}_{\mathbf{M}} \\ &= \sum_{k=M+1}^{N} C_k \mathbf{t_k} - \sum_{k=M+1}^{N} a_k \mathbf{t_k} \\ &= \sum_{k=M+1}^{N} (C_k - a_k)\mathbf{t_k} \end{aligned}$$

We now want to calculate e_{M} the average energy in $\mathbf{x}_{\mathrm{eM}}$ where

$$\begin{aligned} e_{\mathrm{M}} &= E\left\{\left[\sum_{k=M+1}^{N} (C_k - a_k)\mathbf{t_k}\right]^2\right\} \\ &= E\left\{\left[(C_{M+1} - a_{M+1})\mathbf{t_{M+1}} + (C_{M+2} - a_{M+2})\mathbf{t_{M+2}} + \ldots + (C_N - a_N)\mathbf{t}_N\right]^2\right\} \\ &= E\left\{(C_{M+1} - a_{M+1})^2 \mathbf{t}_{\mathbf{M+1}}^{\mathrm{T}}\mathbf{t_{M+1}} + (C_{M+1} - a_{M+1})(C_{M+2} - a_{M+2})\mathbf{t}_{\mathbf{M+1}}^{\mathrm{T}}\mathbf{t_{M+2}} + \ldots \right. \\ &\quad \left. + (C_N - a_N)(C_N - a_N)\mathbf{t}_{\mathbf{N}}^{\mathrm{T}}\mathbf{t_N}\right\} \end{aligned}$$

As the transform is orthonormal

$$\mathbf{t}_{\mathbf{i}}^{\mathrm{T}}\mathbf{t_j} = \begin{cases} 1, & i = j \\ 0, & i \neq j \end{cases}$$

Thus the equation for e_{M} simplifies to

$$e_{\mathrm{M}} = E\left(\sum_{k=M+1}^{N} (C_k - a_k)^2\right)$$

We can now choose the value of a_k that minimizes e_{M} by partial differentiation with respect to each a_j where j is in the range $M + 1$ to N.

$$\begin{aligned} \frac{\partial}{\partial a_j} E\left(\sum_{k=M+1}^{N} (C_k - a_k)^2\right) &= E\left(-2(C_j - a_j)\right) \\ &= -2(E(C_j) - \mathrm{E}(a_j)) \\ &= -2(E(C_j) - a_j) \\ &= 0, \text{ at the minimum} \end{aligned}$$

Therefore, at the minimum

$$a_j = E(C_j)$$

Now since

$$C_j = \mathbf{t}_\mathbf{j}^\mathrm{T}\mathbf{x}$$

It follows that

$$\begin{aligned} a_j &= E(\mathbf{t}_\mathbf{j}^\mathrm{T}\mathbf{x}) \\ &= \mathbf{t}_\mathbf{j}^\mathrm{T}E(\mathbf{x}) \end{aligned}$$

$E(\mathbf{x})$ is the mean vector of the data. Setting the mean value to zero results in the optimum value a_j ($j = M + 1$ to N) being zero and so

$$\hat{\mathbf{x}}_\mathbf{M} = \sum_{k=1}^{M} C_j \mathbf{t}_\mathbf{j}$$

From the above, it follows that

$$C_k - a_k = \mathbf{t}_\mathbf{j}^\mathrm{T}(\mathbf{x} - E(\mathbf{x}))$$

This can be substituted into

$$e_\mathrm{M} = E\left(\sum_{k=M+1}^{N} (C_\mathrm{k} - a_\mathrm{k})^2\right)$$

to obtain

$$\begin{aligned} e_\mathrm{M} &= E\left(\sum_{k=M+1}^{N} (\mathbf{t}_\mathbf{k}^\mathrm{T}(\mathbf{x} - E(\mathbf{x}))^2\right) \\ &= E\left(\sum_{k=M+1}^{N} \mathbf{t}_\mathbf{k}^\mathrm{T}(\mathbf{x} - E(\mathbf{x}))(\mathbf{x} - E(\mathbf{x})^\mathrm{T}\mathbf{t}_\mathbf{k}\right) \\ &= \sum_{k=M+1}^{N} \mathbf{t}_\mathbf{k}^\mathrm{T}E\left((\mathbf{x} - E(\mathbf{x}))(\mathbf{x} - E(\mathbf{x})^\mathrm{T}\right)\mathbf{t}_\mathbf{k} \\ &= \sum_{k=M+1}^{N} \mathbf{t}_\mathbf{k}^\mathrm{T}(\mathrm{COV}(\mathbf{x}))\mathbf{t}_\mathbf{k} \end{aligned}$$

where $\mathrm{COV}(\mathbf{x})$ is the covariance matrix of the input data.

We now need to minimize e_M with respect to the basis vectors t_k while maintaining the orthonormal property. Using Lagrange multipliers, we therefore minimize

$$\begin{aligned} e'_\mathrm{M} &= e_\mathrm{M} - \sum_{k=M+1}^{N} \lambda_k(\mathbf{t}_\mathbf{k}^\mathrm{T}\mathbf{t}_\mathbf{k} - 1) \\ &= \sum_{k=M+1}^{N} \mathbf{t}_\mathbf{k}^\mathrm{T}\mathrm{COV}(\mathbf{x})\mathbf{t}_\mathbf{k} - \lambda_k(\mathbf{t}_\mathbf{k}^\mathrm{T}\mathbf{t}_\mathbf{k} - 1) \end{aligned}$$

Because we want to minimize e'_M, we want to find the point where the gradient of e'_M with respect to $\mathbf{t}_j$ is zero. Noting[6] that

$$\text{GRAD}_{\mathbf{t}_j}\left(\mathbf{t}_k^T[A]\mathbf{t}_k\right) = 2[A]\mathbf{t}_k$$

and

$$\text{GRAD}_{t_j}(\mathbf{t}_k^T\mathbf{t}_k) = 2\mathbf{t}_k$$

It follows that

$$\text{GRAD}_{\mathbf{t}_j} e'_M = 2\text{COV}(x)\mathbf{t}_j - \lambda_k(2\mathbf{t}_j) = 0$$

from which it follows that

$$\text{COV}(\mathbf{x})\mathbf{t}_k = \lambda_k \mathbf{t}_j$$

which when solved for $\mathbf{t}_k$ and λ_k will lead to the kth basis vector of the optimum transform. This is exactly the eigenvalue equation, and so $\mathbf{t}_k$ is an eigenvector of the covariance matrix whereas λ_k is the corresponding eigenvalue. Because the covariance matrix is positive and symmetrical about the leading diagonal, it will always have eigenvectors and eigenvalues that will be real.

If the eigenvalues are arranged from largest to smallest this optimum transform, which is called the Karhunen–Loeve transform (KLT),[7] packs the maximum amount of energy into any given number of coefficients. Putting this another way, the mean square error introduced by deleting a given number of coefficients is a minimum for this transform.

EXAMPLE 4.2—MATLAB

Calculate the 8 × 8 horizontal covariance matrix for the first luminance picture in the "Mobile and Calendar" sequence. Calculate the eigenvectors for this matrix. Hence determine the percentage of energy in each coefficient.

SOLUTION The 8 × 8 horizontal covariance matrix can be calculated using the following MATLAB function.

```
function covar = covariance_h(PICTURE)
[row,col] = size(PICTURE);
PICTURE = PICTURE - mean(mean(PICTURE));
Covar = Zeros(8,8);
Sum = 0;

For matrix_row = 1:8
    For matrix_col = 1:8
      delta = matrix_col-matrix_row;
```

[6]See R.J. Clarke, *Transform Coding of Pictures*, Academic Press, 1985 Appendix 4 for further details.

[7]It is also sometimes referred to as the Hotelling transform or the principal components transform.

```
        If delta ≤ 0
          col_max = col;
        else
          col_max = col-delta;
        end
        for row_num = 1:row
          for col_num = matrix_row:col_max
            sum = sum + PICTURE(row_num,col_num)*PICTURE(row_num,col_num+delta);
        end
        sum = sum/(col_max-matrix_row+1);
        covar(matrix_row,matrix_col) = covar(matrix_row,matrix_col) + sum;
        sum = 0;
      end
    end
  end
covar = covar/row;
```

When applied to the first picture of the "Mobile and Calendar" sequence, the covariance matrix obtained (after rounding to the nearest integer) is

$$\begin{bmatrix} 3914 & 3525 & 3126 & 3050 & 2956 & 2883 & 2797 & 2730 \\ 3525 & 3907 & 3518 & 3118 & 3042 & 2948 & 2875 & 2789 \\ 3126 & 3518 & 3901 & 3511 & 3110 & 3035 & 2940 & 2867 \\ 3050 & 3118 & 3511 & 3894 & 3504 & 3102 & 3027 & 2933 \\ 2956 & 3042 & 3110 & 3504 & 3888 & 3496 & 3095 & 3019 \\ 2883 & 2948 & 3035 & 3102 & 3496 & 3881 & 3489 & 3087 \\ 2797 & 2875 & 2940 & 3027 & 3095 & 3489 & 3874 & 3482 \\ 2730 & 2789 & 2867 & 2933 & 3019 & 3087 & 3482 & 3868 \end{bmatrix}$$

The MATLAB command

```
[V,D] = eig(A)
```

produces the eigenvalues in matrix D and the corresponding eigenvectors as the columns of matrix V. MATLAB arranges the eigenvectors vertically down columns. It also arranges the eigenvalues so that each is larger than those that come before it whereas it is usual to list the eigenvalues in decreasing order of magnitude. For consistency with the definition of basis vectors given earlier, it is necessary to employ the MATLAB command

```
V = fliplr(V)';
```

which flips the columns of the matrix from left to right and then calculates the transpose.

When all this is done with the horizontal covariance matrix shown above, the eigenvector matrix, as determined in MATLAB, is

$$\begin{bmatrix} +0.344 & +0.354 & +0.359 & +0.361 & +0.360 & +0.357 & +0.352 & +0.341 \\ -0.426 & -0.448 & -0.317 & -0.115 & +0.115 & +0.318 & +0.452 & +0.430 \\ -0.394 & -0.274 & +0.144 & +0.495 & +0.499 & +0.151 & -0.273 & -0.397 \\ -0.406 & -0.039 & +0.484 & +0.318 & -0.308 & -0.487 & +0.033 & +0.407 \\ -0.397 & +0.253 & +0.434 & -0.299 & -0.306 & +0.429 & +0.258 & -0.396 \\ +0.357 & -0.441 & -0.010 & +0.423 & -0.419 & +0.002 & +0.442 & -0.356 \\ -0.264 & +0.486 & -0.404 & +0.175 & +0.180 & -0.406 & +0.484 & -0.263 \\ -0.157 & +0.318 & -0.405 & +0.460 & -0.459 & +0.404 & -0.316 & +0.156 \end{bmatrix}$$

The eight eigenvectors are plotted in Figure 4.7. We see that the frequency of the eigenvectors increases from top to bottom in the figure.

The energy compaction capability of the transform is simply calculated from the eigenvalues. For the eigenvectors shown in Figure 4.7, the corresponding eigenvalues together with their relative energy are listed in Table 4.1. Note that most of the energy is compacted into the first eigenvalue with about 96% of the total energy compacted into the first four eigenvalues.

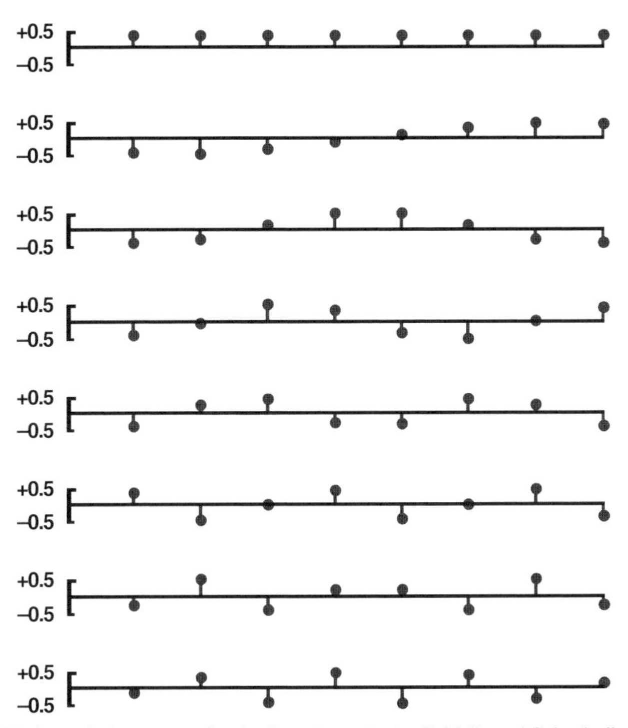

Figure 4.7 Horizontal eigenvectors for the first picture in the "Mobile and Calendar" sequence.

Table 4.1 Horizontal eigenvalues for the first picture in the "Mobile and Calendar" sequence.

Eigenvalue number	Eigenvalue	Percentage of energy
1	25,664	82.4
2	2,104	6.8
3	1,224	3.9
4	877	2.8
5	615	2.0
6	383	1.2
7	203	0.7
8	56	0.2

■

4.4. THE DISCRETE COSINE TRANSFORM

Although the Karhunen–Loeve transform is the optimum transform in terms of energy compaction, it suffers from the significant difficulty that the transform needs to be defined for each picture (or even each 8×8 block within a picture). This requires a significant amount of computation both to calculate the covariance matrix and then the eigenvectors that are used as the basis vectors of the transform. In addition, the transform basis functions (eigenvectors) required for each picture (or each 8×8 block within a picture) need to be transmitted to the decoder so that the picture can be correctly decoded. This represents a significant overhead.

For this reason, a fixed transform known as the discrete cosine transform (DCT) is commonly used in picture and video coding applications. Although suboptimal when compared to the Karhunen–Loeve transform, it has the advantage that it can be used without the need for calculating covariance matrices and eigenvectors. In addition, there is no need to transmit information on the basis vectors used to the receiver.

The basis vectors of an N-point DCT in one-dimension are defined as

$$C(u) = \alpha(u)\sum_{x=0}^{N-1} f(x)\cos\left[\frac{(2x+1)u\pi}{2N}\right] \qquad u = 0,1,2,...,N-1$$

Similarly, the inverse DCT is defined as

$$f(x) = \sum_{u=0}^{N-1}\alpha(u)C(u)\cos\left[\frac{(2x+1)u\pi}{2N}\right] \qquad x = 0,1,2,...,N-1$$

In both of these equations

$$\alpha(u) = \begin{cases} \sqrt{\dfrac{1}{N}} & \text{for } u = 0 \\ \sqrt{\dfrac{2}{\text{N}}} & \text{for } u = 1,2,...N-1 \end{cases}$$

$$
\begin{bmatrix}
+0.354 & +0.354 & +0.354 & +0.354 & +0.354 & +0.354 & +0.354 & +0.354 \\
+0.490 & +0.416 & +0.278 & +0.098 & -0.098 & -0.278 & -0.416 & -0.490 \\
+0.462 & +0.191 & -0.191 & -0.462 & -0.462 & -0.191 & +0.191 & +0.462 \\
+0.416 & -0.098 & -0.490 & -0.278 & +0.278 & +0.490 & +0.098 & -0.416 \\
+0.354 & -0.354 & -0.354 & +0.354 & +0.354 & -0.354 & -0.354 & +0.354 \\
+0.278 & -0.490 & +0.098 & +0.416 & -0.416 & -0.098 & +0.490 & -0.278 \\
+0.191 & -0.462 & +0.462 & -0.191 & -0.191 & +0.462 & -0.462 & +0.191 \\
+0.098 & -0.278 & +0.416 & -0.490 & +0.490 & -0.416 & +0.278 & -0.098
\end{bmatrix}
$$

Figure 4.8 Basis functions for the DCT.

The basis vectors for an eight-point DCT are given in the matrix shown in Figure 4.8. They are also shown in Figure 4.9.

These are just sampled versions of cosine waveforms of increasing frequency ranging from 0 periods per vector (i.e., constant) in the case of the first vector to 3.5 periods per vector in the case of the last vector with each vector containing 0.5 additional periods to the one before it.

Comparing the eigenvectors from the Karhunen–Loeve transform shown in Figure 4.7 with those for the DCT shown in Figure 4.9, we note a remarkable

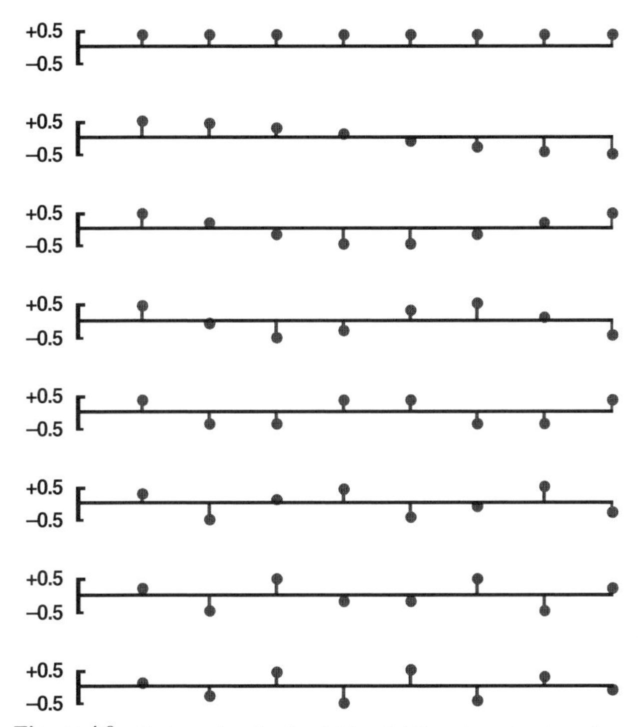

Figure 4.9 Basis vectors for the eight-point discrete cosine transform.

similarity in general shape. The most significant difference is that several of the corresponding eigenvectors in each figure are approximately the negative of each other. However, the effect of changing the sign of an eigenvector is just that the sign of the associated transform coefficients will be changed—energy compaction is identical. The DCT then seems to be a reasonable approximation to the Karhunen–Loeve transform at least for this picture.

For pictures, a two-dimensional DCT is required. The forward and reverse transform are calculated according to the equations

$$C(u,v) = \alpha(u)\alpha(v) \sum_{x=0}^{N-1} \sum_{y=0}^{N-1} f(x,y) \cos\left[\frac{(2x+1)u\pi}{2N}\right]\left[\frac{(2y+1)v\pi}{2N}\right] \quad u,v = 0,1,2,\ldots,N-1$$

and

$$f(x,y) = \sum_{u=0}^{N-1} \sum_{v=0}^{N-1} \alpha(u)\alpha(v)C(u,v) \cos\left[\frac{(2x+1)u\pi}{2N}\right]\left[\frac{(2y+1)v\pi}{2N}\right] \quad x,y = 0,1,2,\ldots,N-1$$

where again

$$\alpha(u) = \begin{cases} \sqrt{\dfrac{1}{N}} & \text{for } u = 0 \\ \sqrt{\dfrac{2}{N}} & \text{for } u = 1, 2, \ldots, N-1 \end{cases}$$

The basis vectors $t(u,v)$ in this case are two-dimensional arrays defined by

$$t(u,v) = \alpha(u)\alpha(v) \cos\left[\frac{(2x+1)u\pi}{2N}\right]\left[\frac{(2y+1)v\pi}{2N}\right] \quad \text{for } x,y = 0,1,\ldots,N-1$$

Pictures of the 64 two-dimensional basis vectors from an 8×8 transform are shown in Figure 4.10 that includes an offset so that negative values appear dark and positive values appear light with mid-gray representing values close to zero. Vertical frequency increases from top to bottom whereas horizontal frequency increases from left to right. Any 8×8 pixel block can be represented by a weighted sum of these two-dimensional basis vectors.

Fortunately, the DCT is a separable transform. This means that the two-dimensional transform can be obtained by first applying a one-dimensional transform across the rows of the data. The result after the horizontal transform has a one-dimensional transform applied vertically to yield the final two-dimensional transform result. This greatly simplifies the transform procedure as well as significantly increasing the speed of the transform.

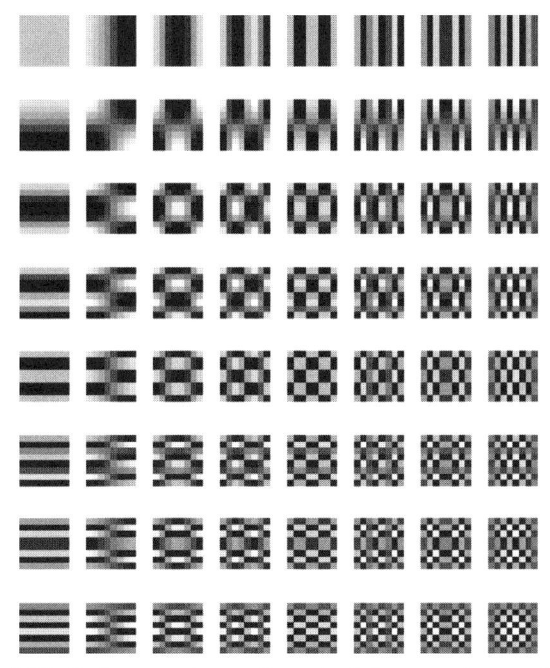

Figure 4.10 Basis vectors for two-dimensional discrete cosine transform.

EXAMPLE 4.3—MATLAB

For the 8×8 pixel block shown in Figure 4.11, calculate the two-dimensional DCT. Hence, show that the picture can be represented by a weighted sum of the two-dimensional basis vectors shown in Figure 4.10.

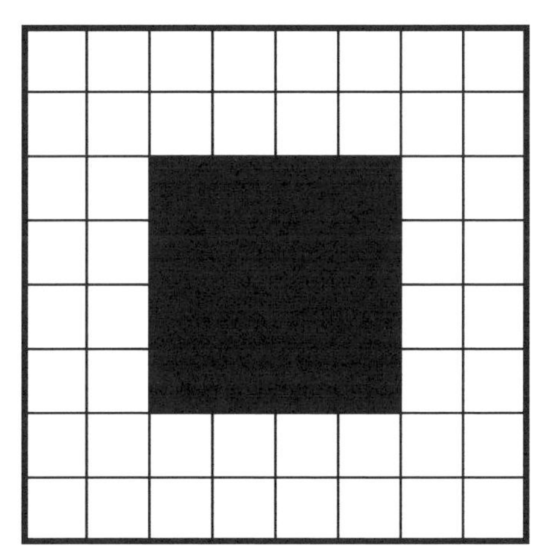

Figure 4.11 Block to be analyzed in Example 4.3.

This picture is represented by the 8×8 pixel array of data shown in Figure 4.12 where white is represented by 255 and black by 0.

255	255	255	255	255	255	255	255
255	255	255	255	255	255	255	255
255	255	0	0	0	0	255	255
255	255	0	0	0	0	255	255
255	255	0	0	0	0	255	255
255	255	0	0	0	0	255	255
255	255	255	255	255	255	255	255
255	255	255	255	255	255	255	255

Figure 4.12 Numerical picture data for Example 4.3.

As explained earlier, the two-dimensional DCT can be calculated by first taking the one-dimensional DCT horizontally and then vertically. After applying the horizontal DCT and rounding to the nearest integer we obtain the result shown in Figure 4.13.

+721	0	0	0	0	0	0	0
+721	0	0	0	0	0	0	0
361	0	333	0	0	0	−138	0
361	0	333	0	0	0	−138	0
361	0	333	0	0	0	−138	0
361	0	333	0	0	0	−138	0
+721	0	0	0	0	0	0	0
+721	0	0	0	0	0	0	0

Figure 4.13 Data of Figure 4.12 after one-dimensional horizontal DCT.

The final two-dimensional DCT is obtained by repeating the one-dimensional discrete cosine transform down the columns of this result. The final result is shown in Figure 4.14.

+1530	0	471	0	0	0	−195	0
0	0	0	0	0	0	0	0
+471	0	−435	0	0	0	+180	0
0	0	0	0	0	0	0	0
0	0	0	0	0	0	0	0
0	0	0	0	0	0	0	0
−195	0	+180	0	0	0	−75	0
0	0	0	0	0	0	0	0

Figure 4.14 Data of Figure 4.12 after two-dimensional DCT.

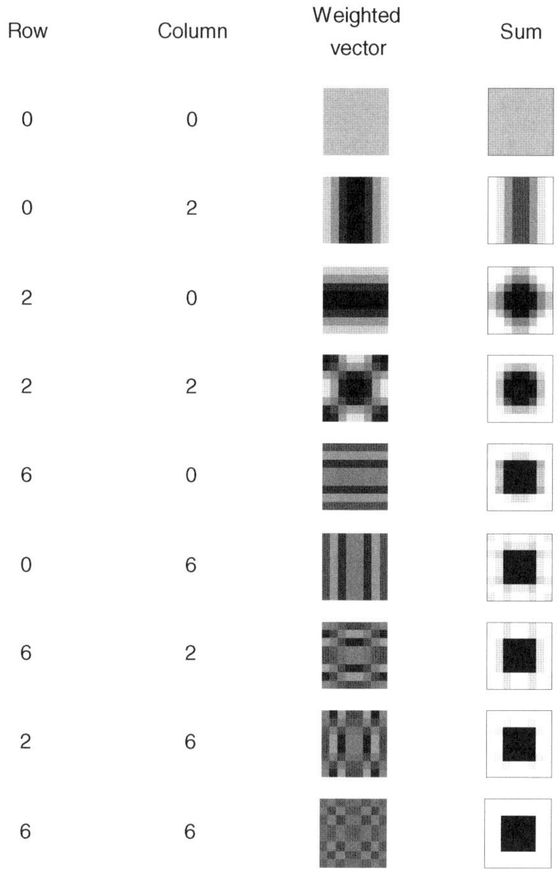

Row	Column	Weighted vector	Sum
0	0		
0	2		
2	0		
2	2		
6	0		
0	6		
6	2		
2	6		
6	6		

Figure 4.15 Weighted two-dimensional basis vectors for the block of Figure 4.11 together with the result from summing the weighted basis vectors to reconstruct the original block.

The weighted two-dimensional basis vectors for each nonzero term in the two-dimensional DCT are shown in Figure 4.15. Also shown is the result when the current weighted basis vector is added to the sum of all the weighted basis vectors appearing above it in Figure 4.15. When all the weighted basis vectors have been summed, we end up with the original block. The last few basis vectors make only a small change to the final block despite the fact that the block chosen has a number of sharp discontinuities that usually imply significant energy at high frequencies. A smoother block would show even less distortion when high frequency basis vectors were omitted. ■

4.4.1. Choice of Transform Block Size

Example 4.3 uses a block size of 8×8 pixels. The appropriate block size is a compromise between the amount of compression achieved (which tends to increase with block size), the correlation within the picture (which tends to decrease with block size), the ability to adapt to local picture statistics (which is better as block size decreases), and computational complexity (which increases with block size). The block size is invariably chosen to be a power of two (i.e., 4×4, 8×8, and 16×16 pixels) as this simplifies computational complexity.

EXAMPLE 4.4—MATLAB

For the first picture of the "Mobile and Calendar" sequence, divide the picture into square blocks of size $N = 2, 4, 8, 16, 32$, and 64 pixels and calculate the DCT of each block. Now retain only the top $N/2 \times N/2$ pixels and calculate the RMS error for each reconstructed picture. Hence, comment on the most appropriate choice of transform block size.

The MATLAB function dct2(A) will calculate the two-dimensional DCT of a block of data. Appropriate MATLAB code to perform this task is given below. The DCT block size is set by the variable tf_size.

```
tf_size = 8;
for irow = 1:tf_size:row
  for icol = 1:tf_size:col
    dct_block = dct2(A(irow:irow1(tf_size-1),icol:icol+(tf_size-1)));
    limit = (tf_size/2)+1;
    dct_block(limit:tf_size,:) = zeros((tf_size/2),tf_size);
    dct_block(:,limit:tf_size) = zeros(tf_size,(tf_size/2));
    rec(irow:irow+(tf_size-1),icol:icol+(tf_size-1)) = round(idct2(dct_block));
    end
end
    rms = sqrt(mean(mean((A - rec). * (A-rec))));
```

The result when applied to the picture is shown in Figure 4.16.

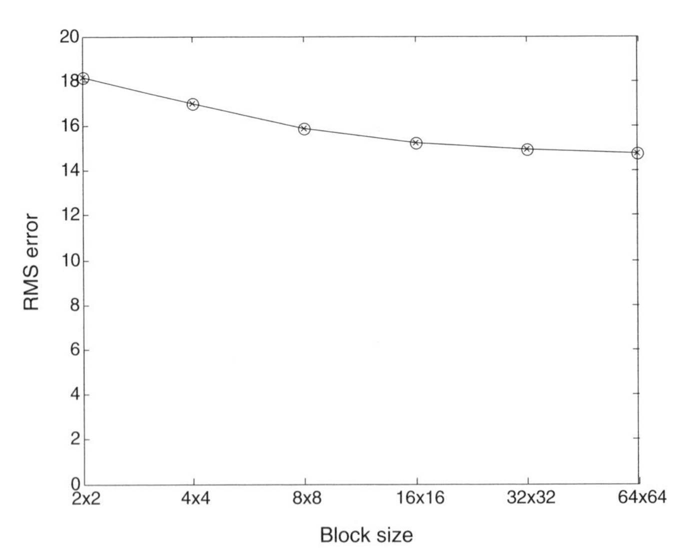

Figure 4.16 Result of deleting all but the top left quarter coefficients for the first picture of the "Mobile and Calendar" sequence.

It is clear that most of the savings are achieved by the time a block size between 8×8 and 16×16 pixels is reached. Hardware complexity considerations lead to the choice of an 8×8 pixel block size. ■

4.4.2. Quantization of DCT Transform Coefficients

We have now succeeded in transforming integer pixel values into real transform coefficients. Transmitting these coefficients without any further processing would probably lead to an increase in the number of bits of information required to represent the picture. However, the transform has packed most of the energy of the picture into a small number of coefficients. Quantizing these coefficients and then transmitting only the significant ones can result in a significant saving. The question remains how best to do this. As the energy is compacted primarily into the first few (low frequency) coefficients, one approach would be to simply not transmit a number of the other (high frequency) coefficients. This is considered in Example 4.5.

EXAMPLE 4.5—MATLAB

For the first luminance picture in the sequence "Mobile and Calendar," calculate the resulting picture when only the top left 4×4, 2×2, and 1×1 DCT coefficients are retained.

The results are shown in Figure 4.17. Even retaining the top 4×4 low-frequency coefficients (Fig. 4.17a) leads to significant blurring in the reconstructed picture. Reducing this

(a)

Figure 4.17 Effect of deleting high-frequency DCT coefficients: (a) top 4×4 coefficients retained; (b) top 2×2 coefficients retained; (c) top 1×1 coefficient retained.

(b)

(c)

Figure 4.17 (*Continued*)

to the top 2×2 low-frequency coefficients (Fig. 4.17b) greatly increases the blurring. In addition, the edges of the individual 8×8 pixel blocks start to become obvious. When only the single DC coefficient is retained (Fig. 4.17c) then the picture becomes a series of blocks. This is hardly surprising as retaining only the DC coefficient means that each pixel in the 8×8 pixel block is replaced by the average value of the block. ■

The previous example has demonstrated that performing the DCT and then simply deleting the higher frequency coefficients is not a satisfactory approach if high-quality reconstructed pictures are required. Although low-frequency information is almost always important, simply removing high-frequency information leads to blurring at sharp edges where high-frequency information is significant.

After quantization, it is desirable that the maximum number of coefficients are zero as this reduces the amount of information that needs to be transmitted. For this reason, a quantizer with a larger than normal "dead zone" (i.e., a quantization region where the coefficient will be set to zero) as shown in Figure 4.18 is commonly employed.

By comparison, a completely linear quantizer would have decision levels at … $-2.5Q$, $-1.5Q$, $-0.5Q$, $+0.5Q$, $+1.5Q$, $+2.5Q$… and reconstruction levels at …$-2Q$, $-Q$, 0, $+Q$, $+2Q$… The larger dead zone ensures that all coefficients in the range $-Q$ to Q are set to zero. The value of the quantizer (Q) is chosen by the user to ensure an adequate representation of the picture. More is said on this topic later.

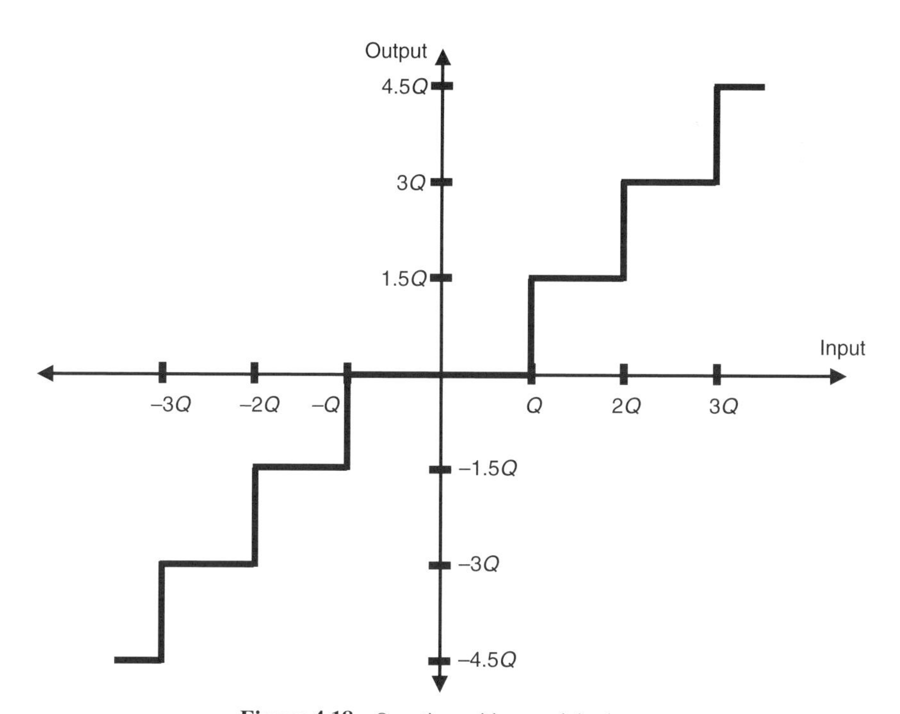

Figure 4.18 Quantizer with central dead zone.

4.4.3. Quantization of DCT Coefficients Based on the Human Visual System

Despite the results shown in Example 4.5, it is well known that the sensitivity of the human visual system does indeed decrease as the spatial frequency (usually measured in cycles per degree of arc) increases. Figure 4.19 shows an indicative plot of the relative spatial frequency response of the eye as a function of spatial frequency measured in cycles per degree of sight.

It is clear that the response peaks at a spatial frequency around 5–10 cycles/degree and falls off sharply at higher frequencies. However, even at these higher frequencies a significant signal will still be observable.

Almost invariably, the transform coefficients are quantized by a linear or near linear quantizer. However, the step size of the quantizer can be varied according to the spatial frequency represented with the step size increasing as the spatial frequency increases. This ensures that high-frequency coefficients will be quantized to zero unless they are sufficiently large that they are likely to be observable to the human visual system.

One way that this can be done is to use a quantization relationship such as

$$\hat{C}_{i,j} = \text{round}\left(\frac{8 \times C_{i,j}}{Q \times W_{i,j}}\right)$$

where $\hat{C}_{i,j}$ is the value of the quantized transform coefficient, $C_{i,j}$ is the value of the original transform coefficient, Q is the quantizer step size for a particular

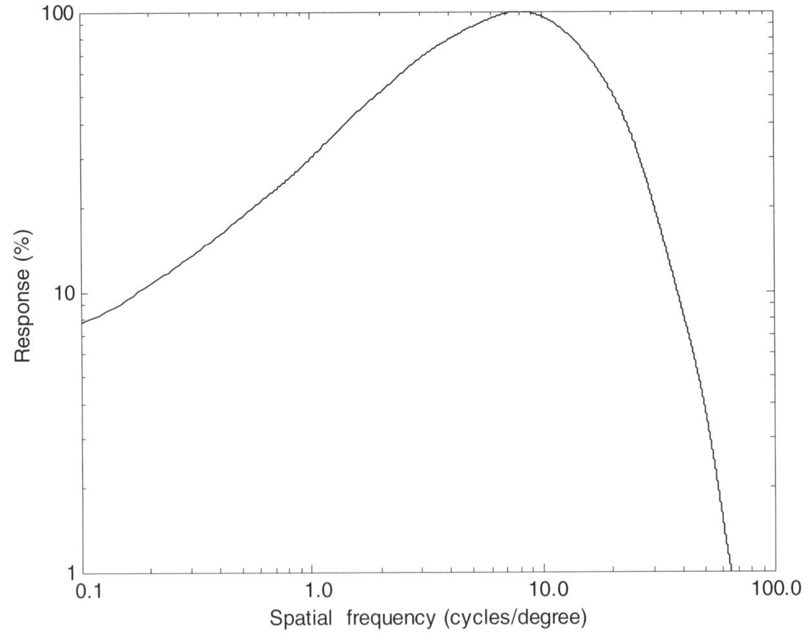

Figure 4.19 Relative spatial frequency response of human visual system.

8	16	19	22	26	27	29	34
16	16	22	24	27	29	34	37
19	22	26	27	29	34	34	38
22	22	26	27	29	34	37	40
22	26	27	29	32	35	40	48
26	27	29	32	35	40	48	58
26	27	29	34	38	46	56	69
27	29	35	38	46	56	69	83

Figure 4.20 Typical values of weighting matrix $W_{i,j}$.

block of data, and $W_{i,j}$ is the weighting value for this particular transform coefficient.

The weighting value $W_{i,j}$ increases as the horizontal and vertical frequencies increase. An example of a matrix of weighting values is shown in Figure 4.20.

Thus if the quantizer step size is 16 and the value of $c(4,4)$[8] for a particular intrablock is 75 then the value of the quantized DCT coefficient, noting that the appropriate quantizer matrix value is 32, would be calculated as shown.

$$c_q(4,4) = \text{round}\left(\frac{8\times 75}{16\times 32}\right) = \text{round}(1.17) = 1$$

If the coefficient $c(1,1)$ had the same value of the original DCT coefficient and the quantizer step size was unchanged then the quantized DCT coefficient value would be calculated as shown.

$$c_q(1,1) = \text{round}\left(\frac{8\times 75}{16\times 16}\right) = \text{round}(2.34) = 2$$

EXAMPLE 4.6

An 8×8 block of data from a picture is shown in Figure 4.21.

(a) Calculate the two-dimensional DCT of the data.
(b) Quantize the data using a quantizer step size of 8.
(c) Quantize the data using a quantizer step size of 8 using the weighting matrix given in Figure 4.20.

91	42	67	72	83	189	245	241
75	74	171	245	240	227	216	221
50	45	75	65	119	228	245	234
72	93	198	246	239	225	214	222
33	58	75	72	155	242	242	229
75	106	215	248	237	223	216	223

Figure 4.21 Picture data for Example 4.6

[8]The DC DCT coefficient would of course be $c(0,0)$.

(a) The result after a two-dimensional DCT performed in MATLAB after rounding to the nearest integer is shown in Figure 4.22.

1294	−495	−104	0	−22	34	48	7
−66	−13	84	13	−1	30	5	−8
−15	7	26	1	21	24	−1	−4
−35	−7	53	12	−17	−4	−5	1
−10	8	19	−9	6	15	2	−2
−60	−24	71	25	−26	−7	−1	1
−7	19	35	−25	−8	24	11	0
−189	−117	173	99	−76	−31	0	−2

Figure 4.22 Picture data of Figure 4.21 after two-dimensional DCT.

(b) Quantizing using a quantizer step size of 8 and rounding to the nearest integer yields the quantized DCT coefficients shown in Figure 4.23.

161	−61	−13	0	−2	4	6	0
−8	−1	10	1	0	3	0	−1
−1	0	3	0	2	3	0	0
−4	0	6	1	−2	0	0	0
−1	1	2	−1	0	1	0	0
−7	−3	8	3	−3	0	0	0
0	2	4	−3	−1	3	1	0
−23	−14	21	12	−9	−3	0	0

Figure 4.23 Quantized DCT coefficients—no weighting matrix.

(c) Quantizing when the weighting factors used in Figure 4.20 are employed yields the results shown in Figure 4.24.

161	−30	−5	0	0	1	1	0
−4	0	3	0	0	1	0	0
0	0	1	0	0	0	0	0
−1	0	2	0	0	0	0	0
0	0	0	0	0	0	0	0
−2	0	2	0	0	0	0	0
0	0	1	0	0	0	0	0
−7	−4	4	2	−1	0	0	0

Figure 4.24 Quantized DCT coefficients—weighting matrix in Figure 4.20 are employed.

Note that the number of DCT coefficients quantized to zero when the weighting matrix is used (45) is considerably greater than when the quantization matrix is not employed (16). ■

4.4.4. Coding of Nonzero DCT Coefficients

We have now produced an 8×8 array of quantized DCT coefficients many of which are zero. We need to be able to entropy code these coefficients and then transmit them to the receiver. The first step of this process is to scan the two-dimensional array of coefficients into a one-dimensional array. This is achieved by scanning the coefficients in the zig-zag scan order shown in Figure 4.25.

This order ensures that the DC coefficient is scanned first followed by the low-frequency AC coefficients. Higher frequency coefficients are scanned toward the end of the scan. Because there is usually less energy at high-frequencies and also because high-frequency coefficients are often quantized more coarsely than low-frequency coefficients to match the characteristics of the human visual system, it is likely that the last nonzero coefficient will be met well before the end of the scan. As we shall see, the scan process can be terminated after the last nonzero coefficient.

After the zig-zag scanning, each nonzero coefficient is grouped with a the run of zero coefficients that proceeds it to form a (run,level) pair. Consider the quantized DCT coefficients shown in Figure 4.26.

After zig-zag scanning in accordance with Figure 4.25, the one-dimensional array is shown in Figure 4.27.

The resulting run-coefficient pairs are as shown in Figure 4.28 with all of the remaining coefficients being zero.

Each run–coefficient pair is not equally likely and so a saving in the number of bits required to transmit the information occurs if the run–coefficient pairs are encoded using a Huffman code. A special Huffman code word is used to indicate that the last nonzero coefficient in a block has been transmitted and is called the end of

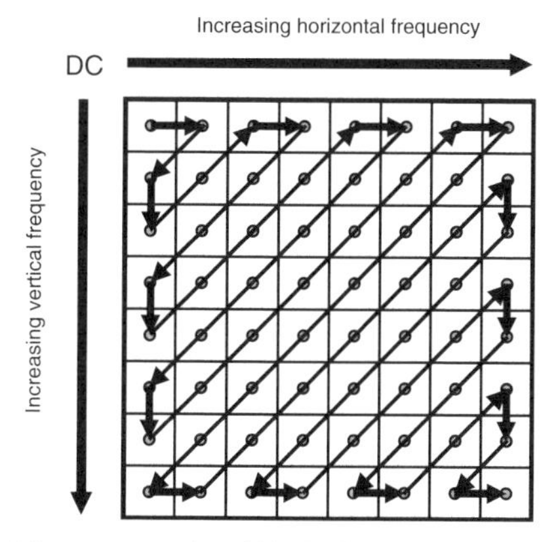

Figure 4.25 Zig-zag scan order of block of quantized transform coefficients.

$$
\begin{bmatrix}
+38 & 0 & +4 & -2 & 0 & 0 & 0 & 0 \\
-5 & +1 & 0 & +1 & 0 & 0 & 0 & 0 \\
+7 & +2 & -2 & +1 & 0 & 0 & 0 & 0 \\
+2 & -1 & +1 & -1 & 0 & 0 & 0 & 0 \\
-2 & 0 & 0 & 0 & 0 & 0 & 0 & 0 \\
0 & 0 & 0 & 0 & 0 & 0 & 0 & 0 \\
0 & 0 & 0 & 0 & 0 & 0 & 0 & 0 \\
0 & 0 & 0 & 0 & 0 & 0 & 0 & 0
\end{bmatrix}
$$

Figure 4.26 Block of quantized DCT coefficients.

+38, 0, –5, +7, +1, +4, –2, 0, +2, +2, –2, –1, –2, +1, 0, 0, 0, +1, +1, 0, 0, 0, 0, 0, –1, 0, 0,

0, 0, 0, 0, 0, 0, 0, 0, 0, 0, 0, 0, 0, 0, 0, 0, 0, 0, 0, 0, 0, 0, 0, 0, 0, 0, 0, 0, 0, 0, 0, 0, 0, 0,

0, 0, 0

Figure 4.27 Coefficients for Figure 4.26 after zig-zag scanning.

(0, +38) (1, –5) (0, +7) (0, +1) (0, +4) (0, –2) (1, +2) (0, +2) (0, –2) (0, –1) (0, –2)

(0, +1) (3, +1) (0, +1) (5, –1)

Figure 4.28 Coefficients of Figure 4.27 after coding into (run,coefficient) pairs.

Code word$_{(0,\ +38)}$, Code word$_{(1,\ -5)}$, Code word$_{(0,\ +7)}$, Code word$_{(0,\ +1)}$, Code word$_{(0,\ +4)}$,

Code word$_{(0,\ -2)}$, Code word$_{(1,\ +2)}$, Code word$_{(0,\ +2)}$, Code word$_{(0,\ -2)}$, Code word$_{(0,\ -1)}$,

Code word$_{(0,\ -2)}$, Code word$_{(0,\ +1)}$, Code word$_{(3,\ +1)}$, Code word$_{(0,\ +1)}$, Code word$_{(5,\ -1)}$,

Code word$_{EOB}$.

Figure 4.29 Huffman code words used to represent the quantized DCT coefficients of Figure 4.26.

block (EOB) code word. Because the EOB code word is sent with every transmitted block, it occurs quite commonly and so is able to be represented by a short Huffman code word. For the (run,level) pairs given above, the transmitted code words would be as given in Figure 4.29.

The Huffman coding tables for the set of possible (run,level) pairs have been developed by standards bodies and are based on the statistics of typical sequences of video material.

4.5. MOTION-COMPENSATED DCT ENCODERS AND DECODERS

While we have so far only considered the application of the DCT to original pictures, it can also be used to code the prediction difference after motion-compensated

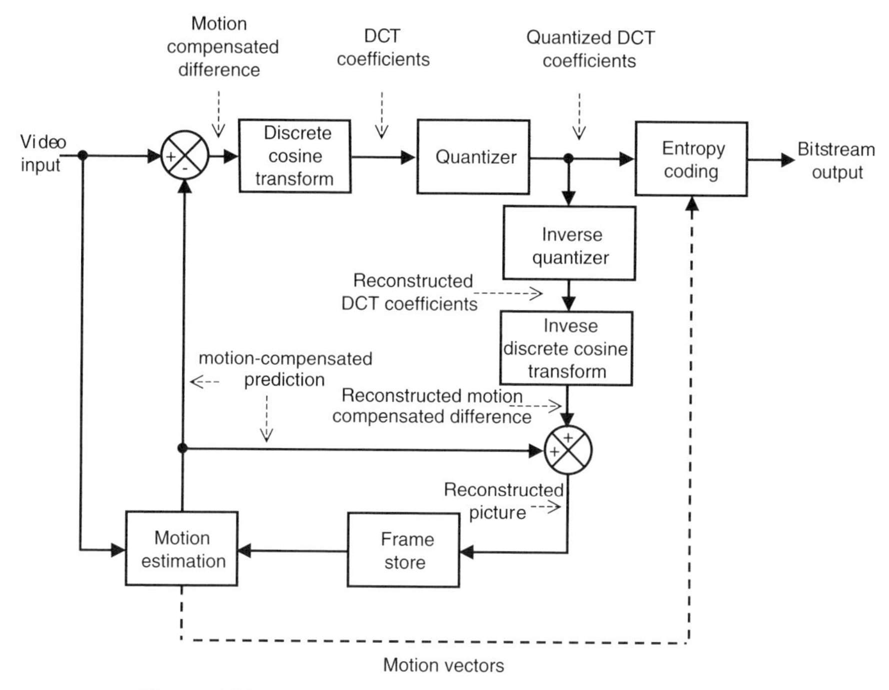

Figure 4.30 Motion-compensated discrete cosine transform encoder.

prediction. Figure 4.30 shows the block diagram of a motion-compensated DCT encoder. The video input has the motion-compensated prediction subtracted from it. The motion-compensated prediction difference is then processed with a two-dimensional DCT prior to quantization. Finally, the quantized DCT coefficients together with the relevant motion vectors are entropy coded and transmitted. The feedback loop of the encoder is equivalent to a decoder and consists of an inverse quantizer[9] followed by an inverse two-dimensional DCT. This produces a reconstruction of the motion-compensated prediction difference. The motion-compensated prediction is then added to the reconstruction of the motion-compensation prediction difference to form the reconstructed picture, which is stored in a frame store for use in the prediction of a subsequent picture.

The corresponding decoder is shown in Figure 4.31. Apart from the initial entropy decoding stage to produce the transform coefficients and motion vectors, this is identical to the feedback loop of the encoder.

Motion-compensated DCT encoders and decoders are the key coding tools of the MPEG-2 video compression standard that is used for digital television broadcasting. Refinements and improvements introduced during the standardization process significantly enhance the performance of the basic architecture. We will consider this topic in considerable detail in Chapter 6.

[9]However, remember that quantization is a lossy process and so cannot be perfectly reversed.

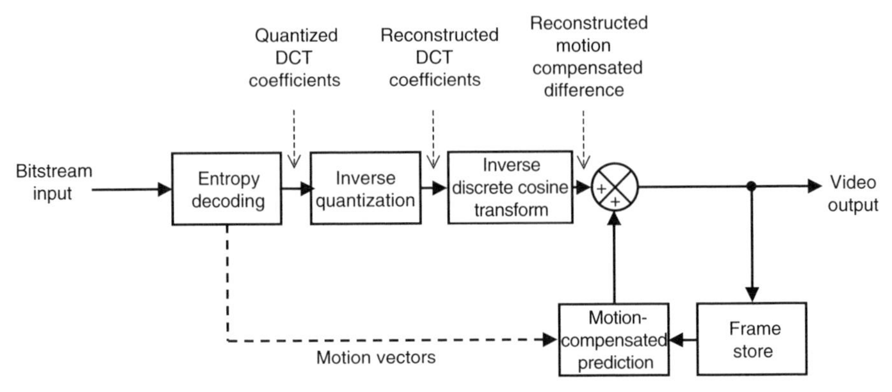

Figure 4.31 Motion-compensated discrete cosine transform decoder.

4.6. RATE CONTROL

The motion-compensated DCT coder shown in Figure 4.32 still has one major problem. Pixels arrive at the encoder at a regular rate and are grouped into 16×16 pixel blocks for motion-compensated prediction. These blocks are subsequently divided into 8×8 pixel blocks for processing with the DCT. Blocks arrive at the motion estimation and the DCT hardware at a regular rate. However, the output of the encoder is inherently a variable rate bit stream. This arises inherently from the fact that

- the number of transform coefficients to be encoded will vary from block to block;
- the number of bits required to encode each transform coefficient is variable and depends on the value of the nonzero coefficients as well as its position within the DCT block (i.e., the number of zeros that precedes each nonzero coefficient).

Most transmission media operate only with constant bit rate (CBR) data streams. It is therefore necessary to turn the variable bit rate (VBR) output of the entropy encoder into a constant bit rate data stream. This process is known as rate control.

The simplest way to achieve rate control is to introduce a buffer between the output of the entropy encoder and the transmission channel. The buffer is simply a block of memory. Data is clocked into the memory at a variable rate and clocked out of the memory at a constant rate in a first in–first out (FIFO) manner. Such a scheme is illustrated in Figure 4.32.

The rate control buffer needs to be sufficiently large to avoid becoming full (called buffer overflow). If the buffer overflows then the data which overflows is lost. This results in serious problems for the decoder in reconstructing the video service.

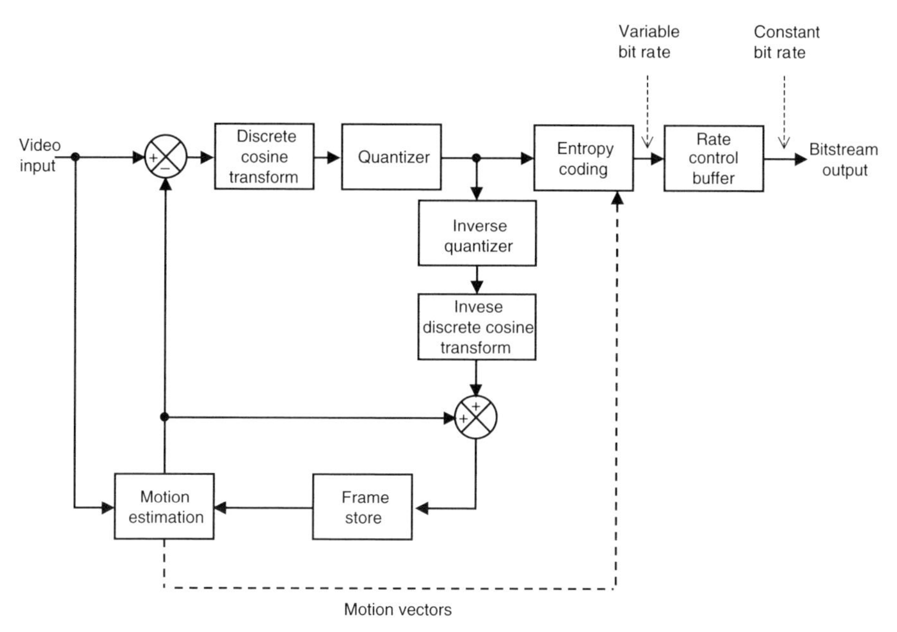

Figure 4.32 Motion-compensated DCT encoder with rate control buffer.

If the rate control buffer was of unlimited size then any variable bit stream could be losslessly transformed to a constant bit rate stream.

In practice, some limit needs to be placed on the size of the rate control buffer. The reason for this becomes clear when it is realized that a second rate control buffer is required at the decoder. In the case of the decoder, the rate control buffer accepts data at a constant rate from the channel but transfers data to the entropy decoder at the variable rate that it requires. Such a decoder is shown in Figure 4.33.

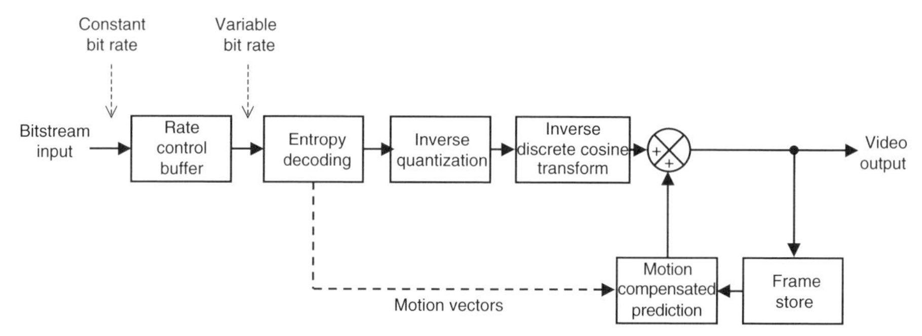

Figure 4.33 Motion-compensated DCT decoder with rate control buffer.

EXAMPLE 4.7

(a) The rate control buffer of an encoder and a decoder are 12 bits deep each. The average data rate output of the encoder (and hence the average data rate input of the decoder) is 2 bits/s. The channel transmission rate is a constant 2 bits/s. A new code word arrives from the encoder every second. Determine the fullness of each buffer if a series of 3-bit code words arrive consecutively from the encoder. Assume that the encoder and the decoder buffers initially contains exactly two 3-bit code words each. Assume that there is synchronization between the encoder and the decoder so that data is clocked in and clocked out of the buffers at the same rate.

(b) Repeat this example for the case where a series of 1 bit code words arrive consecutively from the encoder. In this case, assume that the encoder and the decoder buffers each initially contains exactly six 1-bit code words.

For the first case, the state of the two rate control buffers is shown in Figure 4.34.

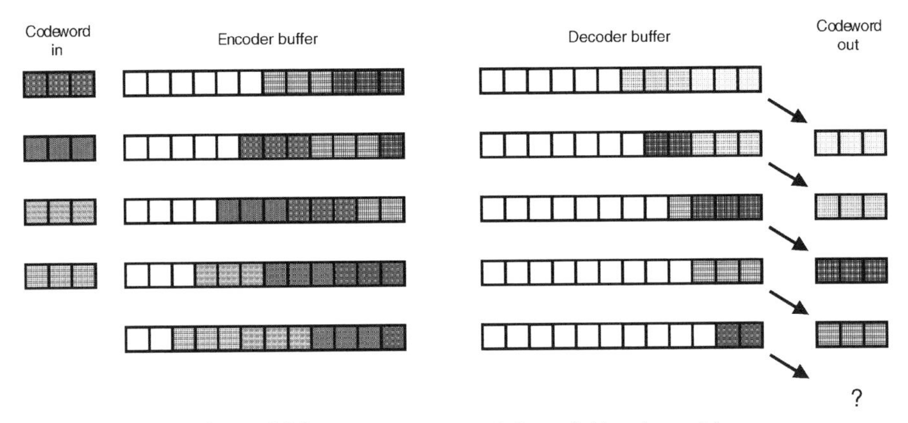

Figure 4.34 Rate control buffer fullness (3-bit code words).

As each code word arrives at the encoder buffer, three new bits are added into the buffer, while two bits are transmitted to the decoder buffer. The amount of data stored in the encoder buffer increases by one bit per code word. At the decoder buffer, two bits are received from the encoder while at the same time a 3-bit code word is read from the buffer. The amount of data stored in the decoder buffer decreases by one bit per code word. Eventually, the amount of data stored in the buffer is insufficient to allow a valid code word to be decoded.

This example is indicative of what happens when the amount of data being generated by the encoder is greater than the transmission rate for a period of time. The encoder buffer fills and the decoder buffer empties. Buffer overflow (at the encoder) and buffer underflow (at the decoder) can occur in this circumstance.

The sum of the bits in the two buffers is indicative of the delay introduced by buffering since

$$\text{Average delay} = \frac{(\text{Encoder buffer fullness} + \text{Decoder buffer fullness})}{\text{Average codeword length}}$$

In this case, the sum of the encoder buffer fullness and decoder buffer fullness is always constant (12 bits).

In the case of 1 bit code words, the situation is as shown in Figure 4.35

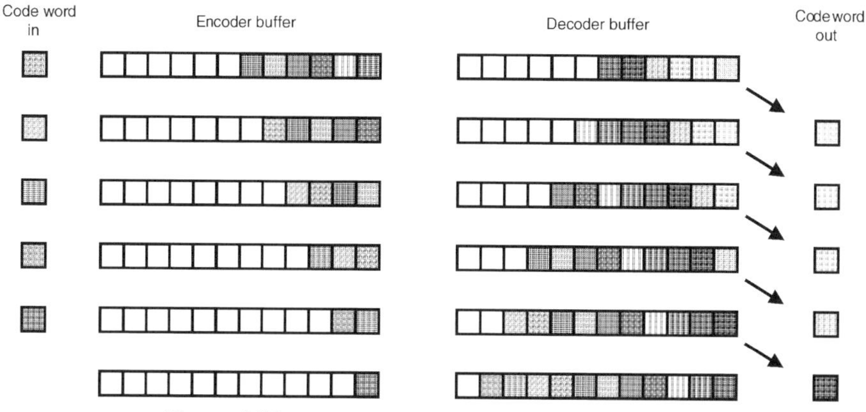

Figure 4.35 Rate control buffer fullness (1-bit code words).

In this case, the decoder buffer fills and the encoder buffer empties. Eventually, a stage is reached where there are insufficient bits in the encoder buffer for the next transmission. Buffer underflow has occurred. At the same time, the decoder buffer is approaching overflow.

Once again, the total number of bits in the two buffers at any time remains constant. ■

In general, the fullness of the encoder buffer at time, $(t+1)$, $F_e(t+1)$, can be calculated from the buffer fullness at time, t, $F_e(t)$, according to

$$F_e(t+1) = F_e(t) + R_{\text{coeff}} - R$$

where R_{coeff} is the number of coefficient bits received from the encoder and R is the number of bits transmitted to the decoder. A similar expression can be developed for the decoder buffer fullness F_d namely

$$F_d(t+1) = F_d(t) - T_{\text{coeff}} + R$$

where T_{coeff} is the number of coefficient bits passed to the decoder and R is the number of bits received from the encoder. It follows that the total buffer fullness at any time, F_{total}, can be calculated according to

$$F_{\text{total}}(t+1) = F_{\text{total}}(t) + R_{\text{coeff}} - T_{\text{coeff}}$$

This shows that the total buffer fullness can vary slightly with time. However, because coefficients placed into the encoder buffer are eventually read from the decoder buffer, it follows that over the long term

$$E(R_{\text{coeff}} - T_{\text{coeff}}) = 0$$

The average total buffer fullness is then just the buffer fullness initially. The likelihood of buffer overflow and underflow are minimized if each buffer starts half full. This implies an inherent delay associated with the two buffering processes. The larger the size of the buffers, the larger is this delay. For two-way video services (i.e., videoconferencing) this delay is crucial, as too much delay leads to substantial difficulties in communicating. For one-way services like digital television, delay is less important. However, the larger the buffer, the more the memory required in the decoder and hence higher the cost. For this reason, some limits are placed on buffer size. This topic is discussed further when we consider the profiles and levels of the MPEG-2 video standard in Chapter 6.

Because buffer overflow and buffer underflow are so undesirable, it is usual to take steps to ensure that they do not occur. One simplistic approach would be to use buffer fullness in determining the quantizer step size to use at the encoder. As the buffer empties, the quantizer step size would be reduced to increase the amount of data that was produced by the encoder. Similarly, as the buffer became full, the quantizer step size would be increased to reduce the amount of data produced by the encoder. In an extreme situation, entire pictures could be dropped completely to ensure that buffer overflow did not occur or dummy data could be transmitted to prevent buffer underflow. Video coding standards support both of these functions, although it is rare that they would need to be used in a high-quality application like digital television. They are commonly used in lower rate, lower quality videoconferencing applications. An encoder incorporating this simple rate control strategy is illustrated in Figure 4.36.

Such an approach, though likely to be successful in avoiding buffer overflow and underflow, is likely to produce a poor quality video service because no account is taken of the characteristics of the human visual system. A good rate controller would consider the likely impact of quantization on particular blocks of the picture from the point of view of a human observer.

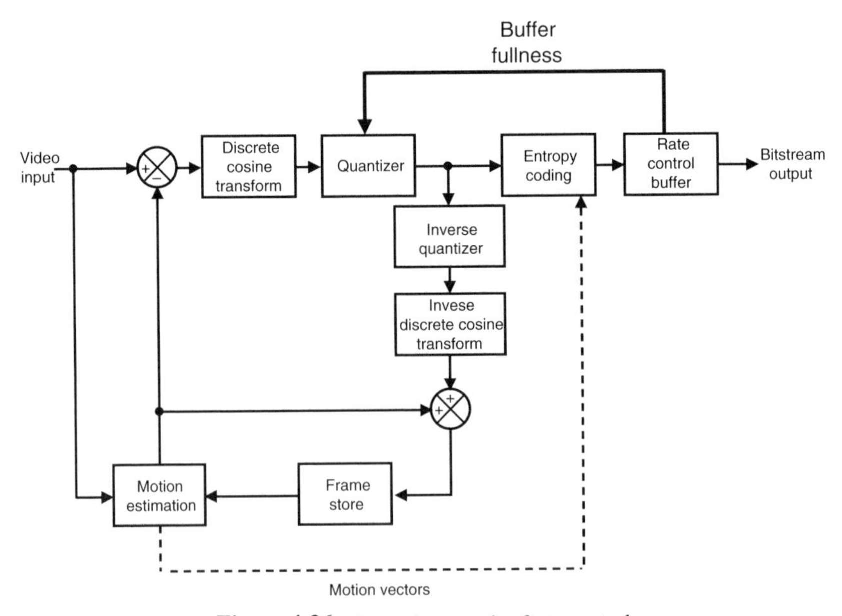

Figure 4.36 A simple example of rate control.

Characteristics of the human visual system that can be taken into account by an effective rate controller include

- *Frequency sensitivity.* As discussed in the section on transform coding, the human visual system is more sensitive to low to medium frequencies than to high frequencies. If an encoder is generating too much data, it makes sense to increase the quantizer step size for higher frequency DCT coefficients first.
- *Visibility threshold.* Coding artifacts are more visible in medium to dark regions of a picture than in brighter regions. Quantizer step size in brighter regions can therefore be increased faster than in darker regions.
- *Spatial masking.* Coding artifacts are far less visible near sharp luminance changes (edges) in pictures than in flat regions. Many rate controllers attempt to measure the "busyness" of a block (i.e., the amount of edge information) in determining the quantizer step size that should be employed.
- *Temporal masking—fast moving objects.* Although the human visual system is good at noticing coding artifacts in stationary objects or objects moving sufficiently slowly that the eye can track them, coding artifacts in fast moving objects are far less perceptible. Blocks containing such objects can be quantized more harshly.
- *Temporal masking—scene changes.* The ability of the human visual system to notice coding artifacts is significantly reduced after a scene change. The first picture of the new scene can therefore be quantized more harshly without affecting subjective quality. This is helpful because the amount of data required to represent the first picture of a new scene can be high due to the ineffectiveness of motion-compensated prediction in this situation.
- *Luminance masking of chrominance.* The human visual system is less able to resolve chrominance information than the luminance information. This is the reason that the two chrominance components are invariably subsampled prior to coding. The human visual system's ability to resolve chrominance is reduced still further at sharp changes in luminance (i.e., edges).

The effective exploitation of the characteristics of the human visual system is essential if a high-quality digital television service is to be achieved at an acceptable data rate. The topic of how best to do this would form the topic for a book in its own right. Although video compression standards provide tools that can be used to exploit human visual system characteristics, the best means of doing so is not standardized. This allows encoder manufacturers to differentiate their products in the marketplace.Standards bodies only standardize how decoders are to operate. This means that any two compliant decoders presented with the identical compliant bit stream produces an identical decoded video service. However, two compliant encoders when provided with the same video material to encode at the same data rate can produce quite different bit streams and quite different decoded service qualities depending largely on the effectiveness of their quantization and rate control strategies.

Over the years, much research has been performed in an attempt to define an objective algorithm that is able to measure the subjective quality of a video sequence after coding. The existence of such a measure would greatly ease the task of optimizing the various video coding tools employed by video encoders. Given the complexity of the human visual system, it is perhaps not surprising that this has proved to be an extremely complex task. Peak signal-to-noise ratio (PSNR), as defined in the previous chapter, is known to be inadequate in performing this task. However, its computational simplicity together with its ability to provide reasonable performance comparisons when comparing minor variations of a particular coding algorithm has meant that it is often used for this purpose.

4.7. CONCLUSION

This chapter has introduced the coding tools that form the basis of the MPEG-2 video compression algorithm that is used in current digital television broadcasting standards. Having coded the data, it needs to be transmitted to the decoder in a form that can be understood. This requires the definition of a syntax (basically a set of communication rules—almost a language) for the bit stream. Video syntax the topic of Chapter 5.

PROBLEMS

4.1 The pixel vector [35 17 18 44] is to be transformed using the basis functions shown in Figure 4.1. Calculate the resulting transform coefficients. Also perform the inverse transform and hence show that the transform is lossless.

4.2 Consider the set of four basis function shown in Figure 4.1.

(a) Use these basis functions first horizontally and then vertically to obtain the two-dimensional transform of the block of data given below.

60	121	222	247
105	227	240	172
74	186	246	206
184	254	191	212

Comment on the effectiveness of the transform.

(b) Calculate the pixel values that make up the reconstructed picture when only the top left 2×2 transform coefficients are retained. Also calculate the peak signal-to-noise ratio in decibels.

4.3 The horizontal covariance matrix of a picture is shown in Figure 4.37. Calculate the basis functions of the Karhunen–Loeve transform for this picture. Also determine the percentage of energy in each transform coefficient.

$$\begin{bmatrix} 3598 & 3445 & 3170 & 2932 \\ 3445 & 3601 & 3448 & 3172 \\ 3170 & 3448 & 3603 & 3449 \\ 2932 & 3172 & 3449 & 3604 \end{bmatrix}$$

Figure 4.37 Horizontal covariance matrix for Problem 4.3.

4.4 The vertical covariance matrix of the same picture used in Problem 4.3 is shown in Figure 4.38. Calculate the basis functions of the Karhunen–Loeve transform for this picture. Also determine the percentage of energy in each transform coefficient.

$$\begin{bmatrix} 3598 & 3121 & 2932 & 2566 \\ 3121 & 3599 & 3122 & 2933 \\ 2932 & 3122 & 3600 & 3123 \\ 2566 & 2933 & 3123 & 3600 \end{bmatrix}$$

Figure 4.38 Vertical covariance matrix for Problem 4.4.

4.5 The horizontal covariance matrix of a picture is shown in Figure 4.39.

$$\begin{bmatrix} 1 & a & a^2 & a^3 & a^4 & a^5 & a^6 & a^7 \\ a & 1 & a & a^2 & a^3 & a^4 & a^5 & a^6 \\ a^2 & a & 1 & a & a^2 & a^3 & a^4 & a^5 \\ a^3 & a^2 & a & 1 & a & a^2 & a^3 & a^4 \\ a^4 & a^3 & a^2 & a & 1 & a & a^2 & a^3 \\ a^5 & a^4 & a^3 & a^2 & a & 1 & a & a^2 \\ a^6 & a^5 & a^4 & a^3 & a^2 & a & 1 & a \\ a^7 & a^6 & a^5 & a^4 & a^3 & a^2 & a & 1 \end{bmatrix}$$

Figure 4.39 Horizontal covariance matrix for Problem 4.5.

Where a has the value 0.95. Calculate the basis functions of the Karhunen–Loeve transform for this picture. Compare the result with the basis function of the discrete cosine transform and comment on the result.

4.6 Explain why the Karhunen–Loeve transform, despite its optimum performance, is not used in practical image compression systems. Are there any circumstances where the Karhunen–Loeve transform might represent a practical solution?

4.7 The Karhunen–Loeve transform is known to concentrate energy into the smallest number of transform coefficients. Does this mean that the coefficients generated will require a minimum number of bits to represent them?

4.8 Calculate the basis functions for a 4-point DCT.

4.9 Calculate the basis functions for a 16-point DCT.

4.10 Calculate the two-dimensional 8-point DCT for the picture data shown in Figure 4.40. Comment on the effectiveness of the DCT in compressing the picture energy into a relatively small number of transform coefficients.

135	145	144	150	152	145	147	149
99	118	135	146	151	142	134	132
142	161	158	142	134	134	153	158
103	113	113	129	147	144	136	138
63	92	124	140	135	131	123	104
59	72	82	103	136	157	163	148
147	150	160	171	172	166	149	128
236	231	211	187	164	149	146	151

Figure 4.40 Picture data for Problem 4.10.

4.11 Consider the picture data given in Figure 4.40

(a) Perform the two-dimensional DCT and then plot the number of nonzero DCT transform coefficients as a function of quantizer step size.

(b) Use the weighting matrix given in Figure 4.20 as part of the quantization process and again plot the number of nonzero DCT coefficients as a function of quantizer step size.

(c) Comment on your results.

4.12 A block of picture data is processed using a two-dimensional DCT followed by quantization. The resulting quantized DCT coefficients are shown in Figure 4.41. Determine the (run,level) pairs that would need to be transmitted to the receiver in order to correctly represent this block.

28	–1	0	0	0	0	0	0
8	0	0	0	0	0	0	0
2	1	0	0	0	0	0	0
0	1	0	0	0	0	0	0
1	0	0	0	0	0	0	0
2	–1	0	0	0	0	0	0
–1	–1	0	0	0	0	0	0
–3	0	0	0	0	0	0	0

Figure 4.41 Quantized DCT data for Problem 4.12.

4.13 If the quantizer step size used to generate the data in Figure 4.41 was 16 and if quantization did not include weighting for the human visual system, calculate the value of the block after reconstruction at the decoder.

4.14 Repeat Problem 4.12 for the quantized DCT data shown in Figure 4.42

212	0	−10	2	10	−1	0	−1
0	0	0	3	−2	0	0	0
0	0	0	0	0	0	0	0
0	0	0	0	0	0	0	0
0	0	0	0	0	0	0	0
0	0	0	1	0	0	0	0
0	0	0	0	0	0	0	0
0	0	0	3	−1	−1	0	0

Figure 4.42 Quantized DCT data for Problem 4.14.

4.15 If the quantizer step size used to generate the data in Figure 4.42 was 6 and if quantization did include weighting for the human visual system as defined in Figure 4.20, calculate the value of the block after reconstruction at the decoder.

4.16 A decoder receives the (run,level) pair data shown in Figure 4.43 for a particular block of data that has been quantized with a two-dimensional DCT and then quantized using a quantizer step size of 10 with a quantizer that did include weighting for the human visual system as defined in Figure 4.20. Calculate the value of the reconstructed picture data.

(0,119)(0,–5)(0,–1)(0,3)(0,–2)(0,–1)(2,3)(0,1)(1,1)(0,1)(1,1)(33,–1)EOB

Figure 4.43 Received (run,level) pairs for Problem 4.16.

4.17 When motion-compensated prediction is combined with the two-dimensional DCT, it is quite common for all quantized DCT coefficients to take the value zero.

(a) How would such a block be transmitted to the decoder?

(b) In modern video coders, it is quite common for blocks in which all of the quantized transform coefficients are zero to not to be transmitted at all. They are called skipped blocks. How could such a scheme be made to work given the need for the decoder to be able to unambiguously understand all data transmitted by the encoder?

4.18 In what circumstances will a motion-compensated DCT encoder work inefficiently? Suggest methods that might be employed to improve performance. What impact will your improvements have on the way that the encoder and the decoder operate?

4.19 A sufficiently large rate control buffer can turn a variable bit rate stream into a constant bit rate stream irrespective of the quantizer used. Explain why such an arrangement is not suitable for real-world applications.

4.20 A constant bit rate video encoder is used to transmit a test pattern (i.e., a fixed picture that remains constant for a long period of time). How will the encoder quantizer and rate control buffer react to this situation? What techniques are available to overcome any difficulties?

4.21 Most practical video encoder systems spend some time (one or more picture times) analyzing a picture prior to passing it through the motion-compensated DCT process. Explain what is happening during this time. Why is this process important to the provision of high-quality video services?

MATLAB EXERCISE 4.1: EIGENVECTORS OF A PICTURE

In this exercise, we will generate the eigenvectors of a picture and compare the energy compaction ability with that of the discrete cosine transform. Each section should be repeated for a typical picture from the video sequences available to you.

Section 1 Average eigenvectors of a picture

- Calculate the 8×8 horizontal and vertical covariance matrices for the picture. Note that the matrix should be symmetrical about its leading diagonal. These matrices represent the average covariance of the picture and not the exact covariance of each 8×8 pixel block within the picture.
- Calculate the eigenvectors and corresponding eigenvalues that correspond to these covariance matrices. Make sure that you know the appropriate order of the eigenvectors so that maximum energy is compacted into a few coefficients.
- Divide the picture into 8×8 pixel blocks and apply the horizontal eigenvectors to each row within a block, a horizontal Karhunen–Loeve transform. Now apply the vertical eigenvectors to the result of the horizontal KLT. The result is a set of two-dimensional KLT coefficients.
- Plot the percentage of the total picture energy in the top left 1×1, 2×2, 3×3, ... , 8×8 two-dimensional KLT coefficients.
- Repeat the above four steps using the discrete cosine transform instead of the eigenvectors. Make sure that each of the DCT transform vectors is orthonormal. Compare the energy compaction ability of the DCT with that of the eigenvector approach.

Section 2 Individual eigenvectors for every block in a picture

- Repeat Section 1, but this time calculate new covariance matrices (and hence a new set of horizontal and vertical eigenvectors) for each 8×8 pixel block. This represents the optimum transform for the picture. Remember that this would imply the transmission of the eight horizontal and the eight vertical eigenvectors with every block in the picture—a very significant overhead.

In order to get sensible results, the mean of the entire picture (as opposed to the mean of each block) should be subtracted as part of the calculation of the covariance matrix. If the block mean is used, the average value of each block will be exactly zero, which implies that a basis function for the DC value might not be generated (or if it is generated have a relatively low position due to the small amount of energy it represents and thus the small eigenvalue generated). When a transform is applied to the original picture, in which each block has a nonzero mean, this leads to problems because most of the energy is deleted when the less significant transform coefficients are deleted.

Section 3 Eigenvectors of a difference picture
Repeat Section 1 for an interpicture difference picture produced by subtracting a prediction picture (picture N) from a current picture (picture $N+1$). Comment on the suitability of the DCT for the coding of prediction difference pictures. If time permits, repeat with a motion-compensated prediction difference picture of the type generated in MATLAB Exercises 3.3 and 3.4.

MATLAB EXERCISE 4.2: DISCRETE COSINE TRANSFORM

Despite the fact that it is suboptimal when compared to the Karhunen–Loeve transform, the discrete cosine transform is used as the basis of transform encoding in modern video encoders. The reasons relate to the fact that the basis functions are fixed and also that performance is close to that of the KLT. In this exercise, we will attempt to quantify the performance of the DCT. In performing this exercise, full use should be made of pictures from the video sequences available to you.

Section 1 The effect of block size on performance
Calculate the two-dimensional DCT of a picture for a range of block size from 2×2 pixels to 64×64 pixels. In each case, set all but the top left quarter of the coefficients to zero (i.e., for an 8×8 pixel block, set all but the top left 4×4 coefficients to zero). Reconstruct the picture and plot the peak signal-to-noise ratio (in decibels) versus block size. Also calculate the number of floating point operations required by MATLAB to perform the transform operation for the various block sizes using the "flops" command. Comment on the effect of block size on the performance of the discrete cosine transform both in terms of operations required and in terms of the reconstructed picture quality.

Section 2 Quantization of DCT coefficients

- Calculate the two-dimensional DCT of a picture using a block size of 8×8 pixels. Quantize the DCT coefficients with a linear quantizer with central dead zone of the type shown in Figure 4.18 for a range of quantizer step sizes and then reconstruct. Calculate the PSNR in decibels for each picture and also examine the quality of the reconstructed picture. Comment on the type of artifacts that are introduced as the quantizer step size increases. Also calculate the entropy of the DCT coefficients (i.e., without the run length coding described in the section on transform encoding) in each case and hence plot a rate distortion curve for the transform encoder.

- Repeat the previous part, but in this case always represent the DC DCT coefficient as a fixed length 8 bit number. This is usually done when a picture is being transform coded without the use of temporal prediction. Comment on the advantages of this approach.

MATLAB EXERCISE 4.3: DISCRETE COSINE TRANSFORM WITH MOTION COMPENSATION

In this MATLAB Exercise we will study the performance of the discrete cosine transform when combined with motion-compensated prediction. In performing this exercise, full use should be made of pictures from the video sequences available to you.

Section 1 The effect of block size on performance
Calculate the motion-compensated prediction of a picture from a picture that occurs temporally earlier in the sequence. Hence calculate the motion-compensated prediction difference.

Calculate the two-dimensional DCT of the motion-compensated prediction difference for a range of block size from 2×2 to 64×64 pixels. In each case, set all but the top left quarter of the coefficients to zero (i.e., for an 8×8 pixel block, set all but the top left 4×4 coefficients to zero). Reconstruct the picture (including reading the motion-compensated prediction) and plot the peak signal-to-noise ratio (in decibels) versus block size. Also calculate the number of floating point operations required by MATLAB to perform the transform operation for the various block sizes using the "flops" command. Comment on the effect of block size on the performance of the discrete cosine transform both in terms of operations required and in terms of the reconstructed picture quality.

Section 2 Quantization of DCT coefficients

- Calculate the two-dimensional DCT of the motion-compensated prediction difference using a block size of 8×8 pixels. Quantize the DCT coefficients with a linear quantizer with central dead zone of the type shown in Figure 4.18 for a range of quantizer step sizes and then reconstruct. Calculate the PSNR in decibels for each picture and also examine the quality of the reconstructed picture. Comment on the type of artifacts that are introduced as the quantizer step size increases. Also calculate the entropy of the DCT coefficients (i.e., without the run length coding described in the section on transform encoding) in each case and hence plot a rate-distortion curve for the transform encoder.

Chapter 5

Video Coder Syntax

5.1. INTRODUCTION

The previous two chapters have dealt with the signal processing tools that are used to compress the amount of information needed to represent video material. The result of the signal processing is a series of primarily Huffman code words that represent the information that needs to be transmitted to the decoder in order to allow the reconstruction of the video service. Examples of the information that needs to be transmitted include quantized DCT coefficients represented as (run,level) pairs, motion vectors and quantizer step sizes. In addition, there is other information that needs to be transmitted to the decoder to allow successful decoding. This includes picture resolution (i.e., pixels per line and lines per picture), interlace structure, picture rate, and data rate.

This information needs to be delivered to the decoder in a form that the decoder can understand. This implies careful formatting and ordering to ensure that the information can be correctly interpreted when received by the decoder. In this chapter, we will look at the structure of the information that needs to be transmitted. This is followed by a consideration of the aspects that go to make a suitable syntax. Finally, we design a simple video encoder syntax that incorporates a number of the features found in the full syntax for MPEG-2 video. A knowledge of the fundamentals of video coder syntax greatly simplifies the process of understanding the complete MPEG-2 video standard, which is discussed in the following chapter.

5.2. REPRESENTATION OF CHROMINANCE INFORMATION

At its most general, each pixel in an image would be represented by a value for each of the primary colors that are red (R), green (G), and blue (B). However, as explained in Chapter 1, it is more usual to represent these three signals by a luminace (Y) signal that represents the gray scale level of the image and two chrominance signals (U and V). These are calculated according to

$$Y = 0.30R + 0.59G + 0.11B$$

Digital Television, by John Arnold, Michael Frater and Mark Pickering.

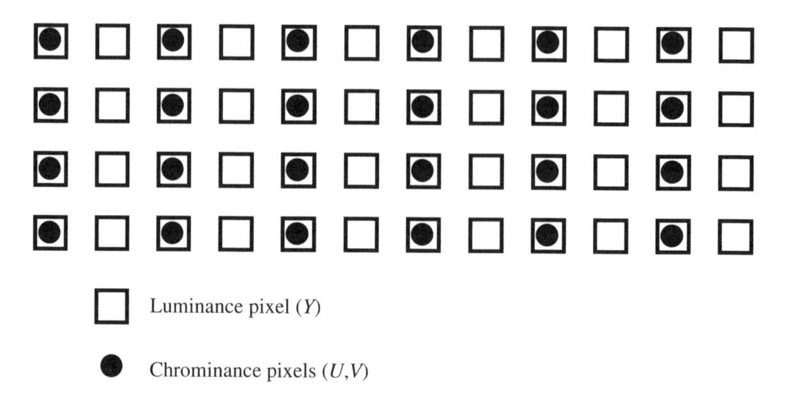

Figure 5.1 Luminance and chrominance positions in a 4:2:2 video picture.

$$U = \frac{B - Y}{2.03}$$

$$V = \frac{R - Y}{1.14}$$

If the R, G, and B components are sampled at CCIR Recommendation 601 resolution (i.e., 720 × 576 pixels at 25 Hz or 720 × 480 pixels at 30 Hz) then it is possible to have Y, U, and V components also represented at this resolution. However, the information content of the chrominance information is considerably less than that of the luminance information and thus some subsampling is possible. In CCIR Recommendation 601, the chrominance components are sampled horizontally at half the rate of the luminance information. The vertical chrominance sampling rate remains unchanged. This is known as the 4:2:2 video format and is shown in Figure 5.1.

The high chrominance accuracy of the 4:2:2 video format is important for specialized applications such as chromakeying.[1] However, for distribution television applications, further chrominance subsampling in the vertical direction is possible. The video format used in digital video broadcasting applications is usually the 4:2:0 format. The location of the luminance and chrominance pixels in the 4:2:0 video format are shown in Figure 5.2.

There is now exactly one pair of chrominance pixels for every four luminance pixels. Further, the chrominance pixels are located half way between each pair of horizontal lines. This can be achieved using a simple filtering procedure. This is explored further in a MATLAB exercise at the end of this chapter.

Figure 5.3 shows the position of the luminance and chrominance pixels in each field of an interlaced picture in 4:2:0 format.

[1]An example of chromakeying is when a reporter stands in front of a blue screen that is replaced during postprocessing by some active video material. This is a common practice in the television industry.

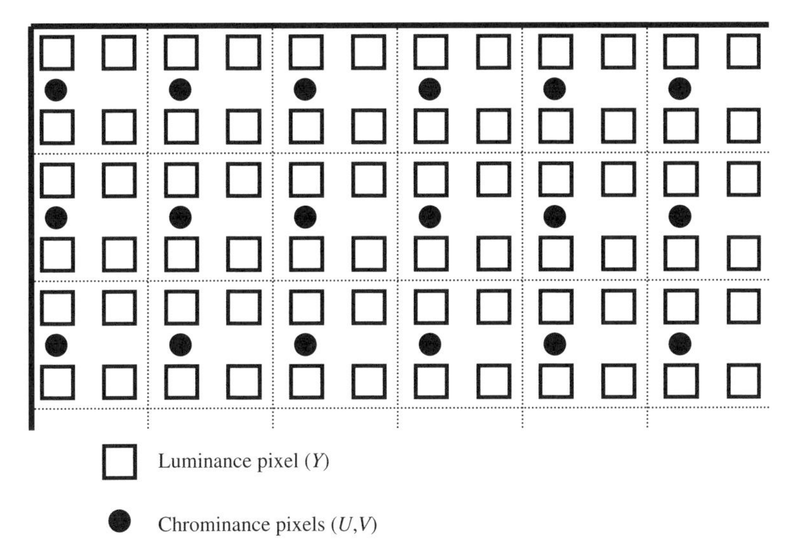

Figure 5.2 Luminance and chrominance positions in a 4:2:0 video picture.

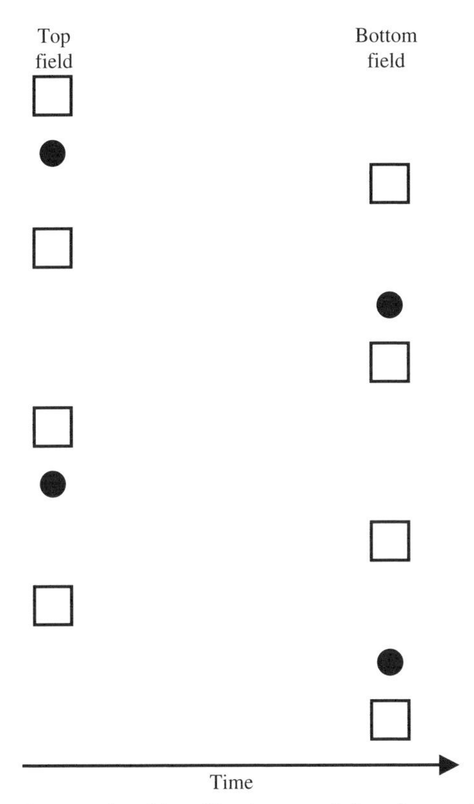

Figure 5.3 Vertical and temporal position of luminance and chrominance pixels in an interlaced 4:2:0 picture.

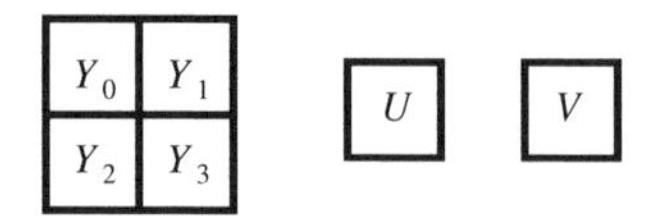

Figure 5.4 Blocks in a macroblock.

Note that in this case, the chrominance pixels do not lie vertically half way between the luminance pixels in a field. This approach means that the location of the chrominance pixels does not change depending upon whether the picture is coded as a whole (i.e., both fields at once) or one field at a time.

We saw in previous chapters that motion-compensated prediction is usually performed on blocks of size 16×16 pixels while the DCT is performed on blocks of 8×8 pixels. Using the 4:2:0 video format implies that for each 16×16 block of luminance pixels, there will be one 8×8 block of each of the two chrominance pixels. There will therefore be a total of six 8×8 pixel blocks, four luminance (Y_0, Y_1, Y_2, and Y_3), and two chrominance (U and V), to be coded using the DCT as shown in Figure 5.4. Each 8×8 group of pixels (whether luminance or chrominance) is usually called a *block*.

5.3. STRUCTURE OF A VIDEO BIT STREAM

Video encoder syntax is made up of a hierarchical tree of layers starting with the sequence layer at the top and finishing at the block layer at the bottom. This is shown in Figure 5.5.

So far, we have considered how blocks of quantized DCT coefficients are zig-zag scanned and then transmitted in a bit stream. We now study how we move from the coding of blocks of picture data to the coding of an entire sequence. For simplicity, we start at the level at which we are familiar, namely the block level, and then move up to the high levels in the hierarchy.

5.3.1. The Block Layer

A video encoder output bit stream is generated in a bottom up manner starting from the simplest element namely a Huffman code word that represents a single (run,coefficient) pair or, as they are more usually referred to in the MPEG standards, a (run,level) pair corresponding to a single nonzero DCT coefficient. A number of these (run,level) pairs, together with an end of block (EOB) code word, combine to represent an 8×8 block of quantized DCT coefficients. This structure is usually referred to as a block.

5.3.1.1. Quantization of DCT Coefficients

With the exception of the DC coefficient in intracoded macroblocks (which is discussed a little later), we assume that quantization is carried out according to the

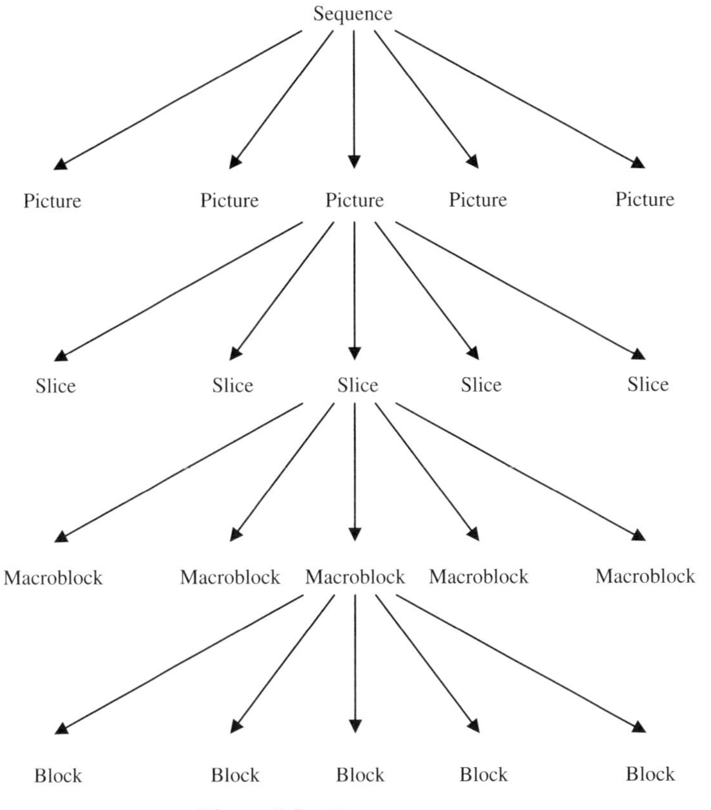

Figure 5.5 Bit-stream structure.

formula given below.

$$\hat{C}(i,j) = \frac{C(i,j)}{8 \times 2 \times \text{quantizer\_scale}}$$

where $C(i, j)$ is the original DCT coefficient value, $\hat{C}(i, j)$ is the quantized DCT coefficient value rounded to the nearest integer, and quantizer_scale is a value extracted from the bit stream that indicates the current quantizer step size.

5.3.1.2. Coding of (run,level) Pairs of DCT Coefficients

Each of the (run,level) pairs associated with each nonzero DCT coefficient is encoded using a variable length code word and this leads to a problem. Runs of zeros are in the range 0–63 (at total of 64 possibilities) while the level can be in the range ± 2047 (4094 possibilities[2]) for the video standard used for digital television

[2] A level of 0 cannot occur as all zero coefficients would be incorporated in to the run part of the (run,level) pair.

broadcasting. This leads to a total of 262,016 (64 × 4094) different variable length code words. Hardware capable of decoding such a huge number of variable length code words would be excessively complex. On the contrary, using fixed length code words would require at least 18 bits per code word (6 bits for run and 12 bits for level information in 2's complement form), and this is also highly inefficient. A compromise approach is therefore employed. Variable length code words are used for more common (run,level) pairs, with fixed length code words used for the remaining majority (in terms of number but not in terms of likelihood) of (run,level) pairs. Of course, the decoder needs to be able to determine whether a particular code word is from the variable or fixed length code word set. In our syntax, each fixed length code word is therefore preceded by a 6-bit ESCAPE code that is unique among the variable length code words. Upon receipt of an escape code, the decoder knows that the next 18 bits carry a fixed length code specifying the (run,level) pair.

A typical set of variable length codes for the coding of (run,level) pairs is given in Table 5.1, whereas the fixed length code words that apply only for those (run,level) pairs not included in Table 5.1 are shown in Figure 5.6. The code words in Table 5.1 deal with both positive and negative level values. In the case of a positive value of level, the last bit of the code word (indicated by an "x") is a 0 whereas for a negative value of level it is a 1. Thus a (0,+1) (run,level) pair is indicated by the code word 110 whereas a (0,−1) (run,level) pair is indicated by the code word 111.

For the variable length code words, as might be intuitively expected, the length of the code word increases as either the length of the run of zeros or the absolute level increases as each of these makes that particular (run,level) pair less likely. No valid variable length code word is ever a prefix for any other valid code word as expected with Huffman codes. The longest code word in the variable length code words is of length 17 bits compared to the 24 bits required when an ESCAPE code word is needed.

5.3.1.3. *Block with All Zero DCT Coefficients*

Successful motion-compensated prediction combined with quantization of the resulting DCT coefficients can often lead to a situation where all of the DCT coefficients in a block are quantized to zero. Using the variable length codes shown in Table 5.1, such a block would be coded with an end of block code word that requires two bits. A more efficient approach is to simply not transmit any information for this block—such a block is called a *skipped block*. However, this immediately leads to a problem, because the decoder needs to know which blocks have been coded and which blocks have been skipped. This is achieved by placing information in the header of each macroblock that indicates which blocks within the macroblock are actually coded (i.e., contain at least one nonzero quantized DCT coefficient).

5.3.2. The Macroblock Layer

A macroblock comprises a 16 × 16 pixel block of luminance pixels together with the associated chrominance information. For the 4:2:0 video format, there would

Table 5.1 Table of variable length codes for the coding of (run,level) pairs of DCT coefficients.

Run	Level	Code word	Run	Level	Code word
0	1	11x	0	4	0000 110x
1	1	011x	1	4	0000 0011 00x
2	1	0101 x	2	4	0000 0001 0100 x
3	1	0011 1x	3	4	0000 0000 1001 1x
4	1	0011 0x	0	5	0010 0110 x
5	1	0001 11x	1	5	0000 0001 1011 x
6	1	0001 01x	2	5	0000 0000 1010 0x
7	1	0001 00x	0	6	0010 0001 x
8	1	0000 111x	1	6	0000 0000 1011 0x
9	1	0000 101x	0	7	0000 0010 10x
10	1	0010 0111 x	1	7	0000 0000 1010 1x
11	1	0010 0011 x	0	8	0000 0001 1101 x
12	1	0010 0010 x	1	8	0000 0000 0011 111x
13	1	0010 0000 x	0	9	0000 0001 1000 x
14	1	0000 0011 10x	1	9	0000 0000 0011 110x
15	1	0000 0011 01x	0	10	0000 0001 0011 x
16	1	0000 0010 00x	1	10	0000 0000 0011 101x
17	1	0000 0001 1111 x	0	11	0000 0001 0000 x
18	1	0000 0001 1010 x	1	11	0000 0000 0011 100x
19	1	0000 0001 1001 x	0	12	0000 0000 1101 0x
20	1	0000 0001 0111 x	1	12	0000 0000 0011 011x
21	1	0000 0001 0110 x	0	13	0000 0000 1100 1x
22	1	0000 0000 1111 1x	1	13	0000 0000 0011 010x
23	1	0000 0000 1111 0x	0	14	0000 0000 1100 0x
24	1	0000 0000 1110 1x	1	14	0000 0000 0011 001x
25	1	0000 0000 1110 0x	0	15	0000 0000 1011 1x
26	1	0000 0000 1101 1x	1	15	0000 0000 0001 0011 x
27	1	0000 0000 0001 1111 x	0	16	0000 0000 0111 11x
28	1	0000 0000 0001 1110 x	1	16	0000 0000 0001 0010x
29	1	0000 0000 0001 1101 x	0	17	0000 0000 0111 10x
30	1	0000 0000 0001 1100 x	1	17	0000 0000 0001 0001 x
31	1	0000 0000 0001 1011 x	0	18	0000 0000 0111 01x
0	2	0100 x	1	18	0000 0000 0001 0000 x
1	2	0001 10x	0	19	0000 0000 0111 00x
2	2	0000 100x	0	20	0000 0000 0110 11x
3	2	0010 0100 x	0	21	0000 0000 0110 10x
4	2	0000 0011 11x	0	22	0000 0000 0110 01x
5	2	0000 0010 01x	0	23	0000 0000 0110 00x
6	2	0000 0001 1110 x	0	24	0000 0000 0101 11x
7	2	0000 0001 0101 x	0	25	0000 0000 0101 10x
8	2	0000 0001 0001 x	0	26	0000 0000 0101 01x

(continued)

Table 5.1 (*Continued*)

Run	Level	Code word	Run	Level	Code word
9	2	0000 0000 1000 1x	0	27	0000 0000 0101 00x
10	2	0000 0000 1000 0x	0	28	0000 0000 0100 11x
11	2	0000 0000 0001 1010 x	0	29	0000 0000 0100 10x
12	2	0000 0000 0001 1001 x	0	30	0000 0000 0100 01x
13	2	0000 0000 0001 1000 x	0	31	0000 0000 0100 00x
14	2	0000 0000 0001 0111 x	0	32	0000 0000 0011 000x
15	2	0000 0000 0001 0110 x	0	33	0000 0000 0010 111x
16	2	0000 0000 0001 0101 x	0	34	0000 0000 0010 110x
0	3	0010 1x	0	35	0000 0000 0010 101x
1	3	0010 0101 x	0	36	0000 0000 0010 100x
2	3	0000 0010 11x	0	37	0000 0000 0010 011x
3	3	0000 0001 1100 x	0	38	0000 0000 0010 010x
4	3	0000 0001 0010 x	0	39	0000 0000 0010 001x
5	3	0000 0000 1001 0x	0	40	0000 0000 0010 000x
6	3	0000 0000 0001 0100 x			
End of block		10	ESCAPE		0000 01

be four luminance blocks and two chrominance blocks (one of U pixels and one of V pixels) in each macroblock. This was shown in Figure 5.4. Because motion-compensated prediction is performed on 16×16 pixel blocks, each macroblock also contains a motion vector if motion-compensated prediction is employed.

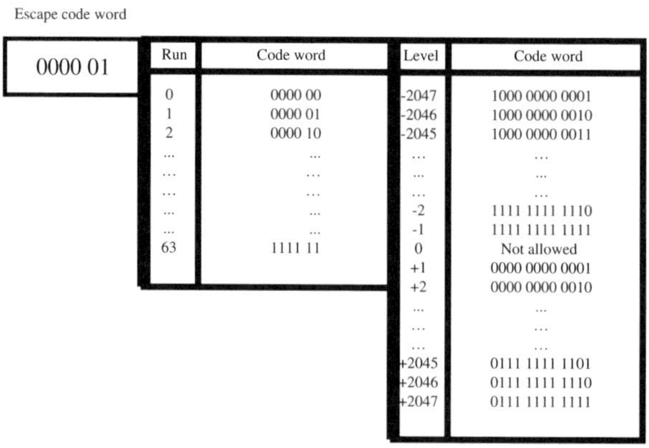

Escape code word

Escape code word
0000 01

Run	Code word
0	0000 00
1	0000 01
2	0000 10
...	...
...	...
...	...
...	...
...	...
63	1111 11

Level	Code word
-2047	1000 0000 0001
-2046	1000 0000 0010
-2045	1000 0000 0011
...	...
...	...
...	...
-2	1111 1111 1110
-1	1111 1111 1111
0	Not allowed
+1	0000 0000 0001
+2	0000 0000 0010
...	...
...	...
...	...
+2045	0111 1111 1101
+2046	0111 1111 1110
+2047	0111 1111 1111

Figure 5.6 Table of fixed length codes for the coding of (run,level) pairs of DCT coefficients.

5.3.2.1. Macroblock Type

Each macroblock can be coded in one of the following three modes: intramode, intermode, or motion-compensated prediction mode. The mode is used to inform the decoder which type of macroblock is being transmitted. In intramode, the macroblock is coded without temporal prediction from an earlier picture. For intermode, the macroblock is coded using temporal prediction by the macroblock in exactly the same position in an earlier (prediction) picture. In the motion-compensated prediction mode, the macroblock is coded using temporal prediction with a macroblock-sized block of data in an earlier (prediction) picture indicated by a motion vector.

Intermode is simply motion-compensated mode with a motion vector of (0,0). However, indicating intermode means that there is no requirement to transmit the motion vector. This results in an overall saving in the total number of bits required to represent the macroblock.

It is also possible to change the value of the quantizer step size at the macroblock level. Again, the decoder needs to know that a new quantizer is about to be specified. All of this information is contained in the macroblock header in a field called the *macroblock type*. An example of a typical macroblock type field is shown in Table 5.2.

The macroblock type information is variable length coded to minimize the overhead associated with its transmission. Because every block in an intracoded block requires at least a DC DCT coefficient, every block in an intra macroblock requires coding. In an intercoded macroblock, if all of the quantized DCT coefficients in each block are zero then the macroblock does not need to be transmitted—such a macroblock is called a *skipped macroblock*. Any skipped macroblocks are decoded as the reconstructed macroblock in the same spatial location in the reference picture. As with skipped blocks, there is a need to signal to the decoder the location of the current macroblock in case one or more of the macroblocks that immediately precedes it have been skipped. How this is achieved is described in detail a little later.

5.3.2.2. Macroblock Address

As for skipped blocks, it is also possible for a video encoder to skip an entire macroblock. In this case, the macroblock is assumed to have been encoded using interpicture prediction (i.e., the macroblock is predicted by the macroblock of information in the

Table 5.2 Macroblock type field.

Macroblock mode	Same quantizer	New quantizer	Not coded
Intra	0001 1	0000 01	Not allowed
Inter	01	0000 1	Skipped
Motion compensated	1	001	0010 0

same position in the reference picture) and all quantized DCT coefficients from this prediction difference are zero. The result is that the reconstructed macroblock in the current picture is simply the reconstructed macroblock in the same position in the reference picture. As with skipped blocks, the decoder needs to be able to detect when a macroblock has been skipped. One way that this could be achieved would be to number macroblocks from left to right across a picture. For a CCIR Recommendation 601 picture with 704 pixels/lines, there would be 44 macroblocks in a row of macroblocks thus requiring a 6-bit address field in the macroblock header.

In fact rather than transmitting the macroblock address, it is more usual to transmit the difference between the current macroblock and the last macroblock that was transmitted. This is called the *macroblock address increment.* Thus if the current macroblock is macroblock 3 and the last transmitted macroblock was macroblock 2 then the macroblock address increment would have value 1. Not all macroblock address increments are equally likely, and so variable length coding is once again possible. Table 5.3 shows typical variable length code words for the macroblock address increment. It is clear that small macroblock address increments are the most probable.

The **macroblock_escape** code word indicates that the macroblock address increment is greater than 33. It causes the value 33 to be added each time it occurs. Thus a macroblock address increment of 67 would be represented by the code word shown in Figure 5.7.

Table 5.3 Typical Huffman codes for macroblock address increment.

Macroblock address increment	Code word	Macroblock address increment	Code word
1	1	18	0000 0101 01
2	011	19	0000 0101 00
3	010	20	0000 0100 11
4	0011	21	0000 0100 10
5	0010	22	0000 0100 011
6	0001 1	23	0000 0100 010
7	0001 0	24	0000 0100 001
8	0000 111	25	0000 0100 000
9	0000 110	26	0000 0011 111
10	0000 1011	27	0000 0011 110
11	0000 1010	28	0000 0011 101
12	0000 1001	29	0000 0011 100
13	0000 1000	30	0000 0011 011
14	0000 0111	31	0000 0011 010
15	0000 0110	32	0000 0011 001
16	0000 0101 11	33	0000 0011 000
17	0000 0101 10	macroblock_escape	0000 0001 000

0000 0001 000	0000 0001 000	1
macroblock_escape	**macroblock_escape**	macroblock address increment = 1

Figure 5.7 Code word for a macroblock address increment of 67.

This is another example of a modified Huffman code.

5.3.2.3. Coding of Motion Vectors

Each macroblock that is coded in motion-compensated mode requires a motion vector to be included in its header information. This motion vector indicates the position in the reference picture of the macroblock that is to be used to predict the current macroblock. A positive value of the horizontal component of the motion vector means that the prediction is made from pixels in the reference picture that lie to the right of the pixels that are being predicted. Similarly, a positive value of the vertical component of the motion vector means that the prediction is made from pixels in the reference picture that lie below the pixels that are being predicted.

For a motion vector range of -16 to $+15$ both horizontally and vertically, five bits would be needed to represent each motion vector—a total of 10 bits per macroblock. Fortunately, motion vector fields are often highly correlated. Figure 5.8

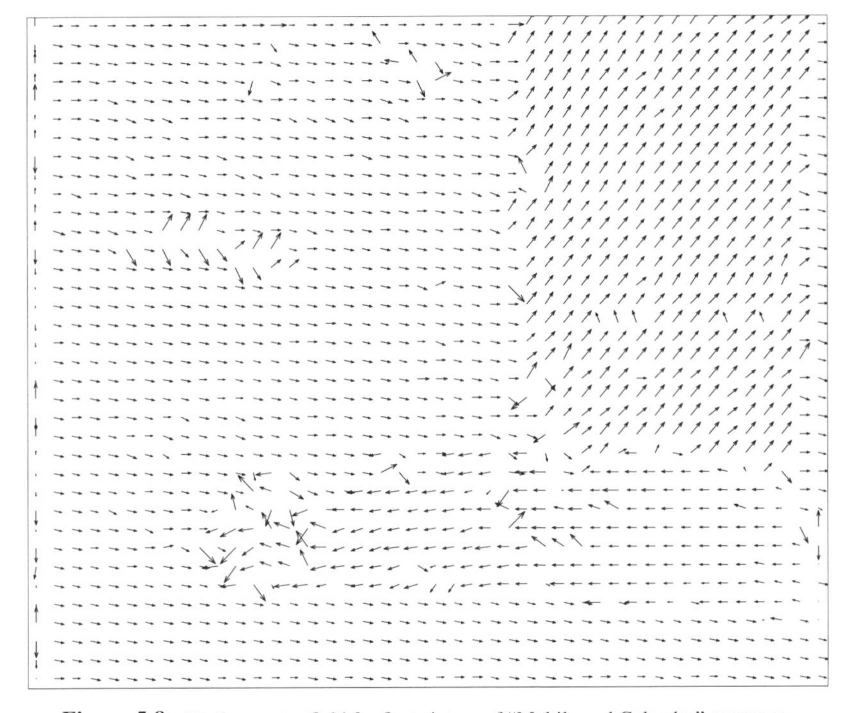

Figure 5.8 Motion vector field for first picture of "Mobile and Calendar" sequence.

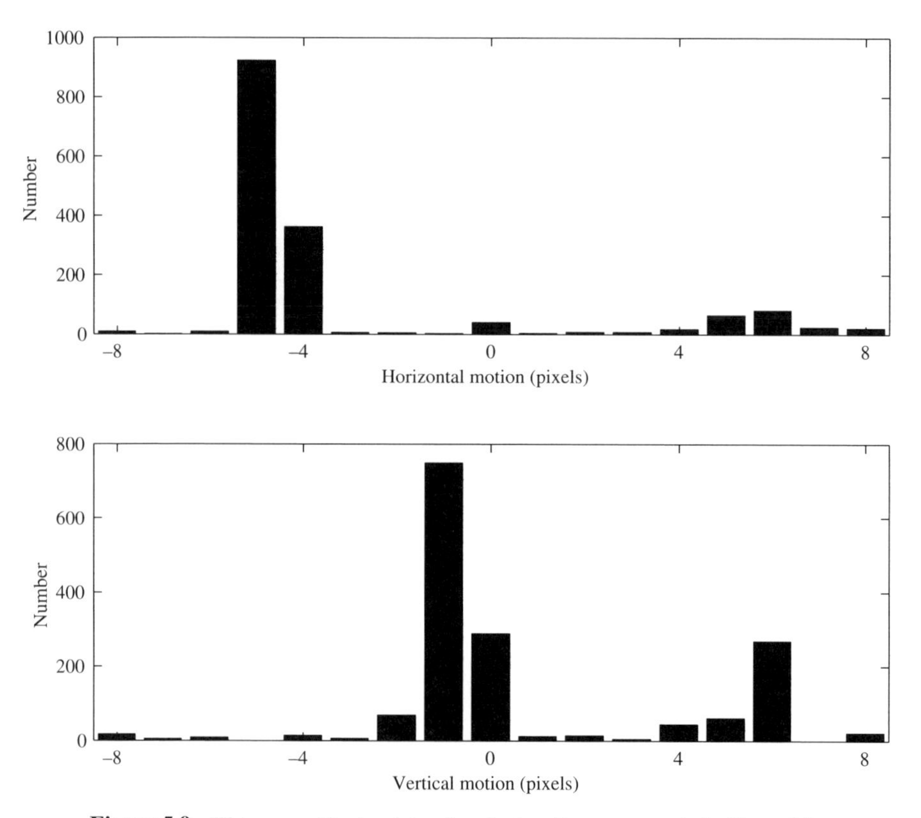

Figure 5.9 Histogram of horizontal and vertical motion components for Figure 5.8.

shows the motion vectors for a picture of the "Mobile and Calendar" sequence. The similarity in nearby motion vectors is clearly apparent. This is especially true in the calendar and also in the wallpaper background.

Figure 5.9 shows the histogram for the horizontal and vertical component of each motion vector. There are two dominant motions in the horizontal direction (-4 pixels and -5 pixels) corresponding to the motion in the background wallpaper. In the vertical direction, there is a pair of dominant motions close to zero (-1 pixels and 0 pixels) again corresponding to motion in the background and another dominant motion (at $+6$ pixels) corresponding to the motion of the calendar.

It is clear that there is strong correlation between nearby motion vectors. This correlation is exploited using lossless one-dimensional horizontal prediction of each component of the motion vector. Thus each component of the motion vector is predicted by the identical component of the motion vector immediately to its left and only the prediction difference is sent to the decoder. The first pixel is transmitted without prediction. If the first motion vector in a row was $(+3, -2)$ and the second motion vector was $(+2, -3)$, then the prediction difference that would be sent is $(+2 - +3, -3 - -2) = (-1, -1)$. Histograms of the prediction difference for the

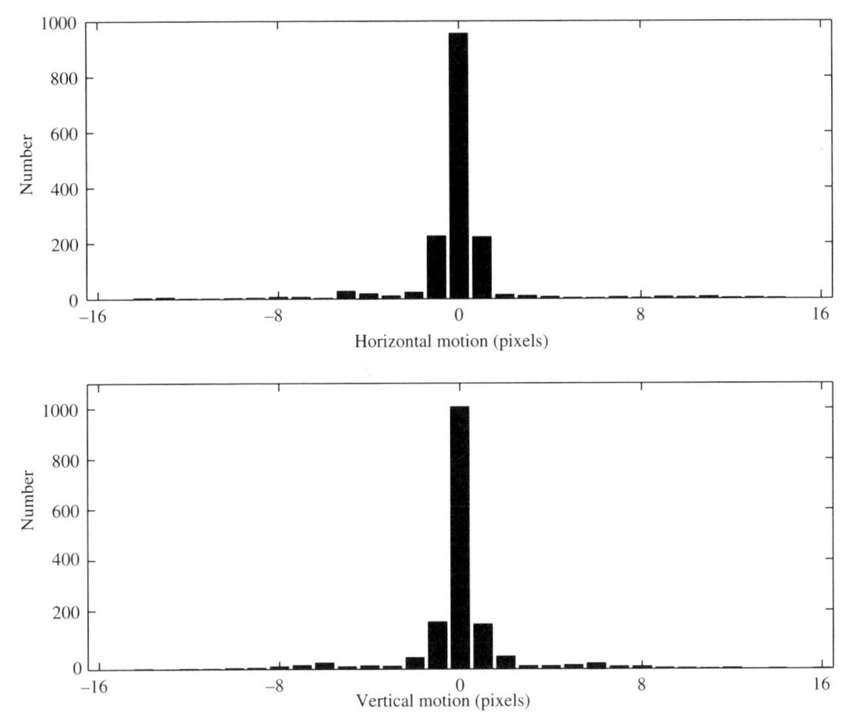

Figure 5.10 Histogram motion vector prediction differences for Figure 5.8.

motion vector field shown in Figure 5.8 are shown in Figure 5.10. Note that for motion vector components in the range of ±8, prediction differences will be in the range of ±16.

It is clear from Figure 5.10 that the prediction differences are closely packed around zero. A worthwhile saving therefore results if the prediction differences are coded using a variable length Huffman code. For the example just completed, the number of bits per motion vector component is reduced from 4 bits to approximately 2 bits—a saving of 50%.

A typical Huffman code for the motion vector prediction differences in the range of ±16 is shown in Table 5.4. Again, the last bit of the variable length code word indicates the sign of the motion vector prediction difference. In the case of a positive difference, the last bit of the code word (indicated by an "x") is 0 whereas for a negative value of the difference is 1. As expected, the code word becomes longer as the absolute size of the prediction difference increases.

5.3.2.4. *Coding of DC Coefficients in Intramacroblocks*

In intracoded macroblocks, the DC DCT coefficient always needs to be coded. In addition, the DC value of an intracoded block is particularly important as it represents the average luminance or chrominance value of the block. Significant quantization

Table 5.4 Typical motion vector prediction difference Huffman codes.

Motion vector prediction difference	Code word
0	1
1	01x
2	001x
3	0001 x
4	0000 11x
5	0000101x
6	0000 100x
7	0000 011x
8	0000 0101 1x
9	0000 0101 0x
10	0000 0100 1x
11	0000 0100 01x
12	0000 0100 00x
13	0000 0011 11x
14	0000 0011 10x
15	0000 0011 01x
16	0000 0011 00x

of this coefficient would be readily apparent in the decoded picture because it would lead to sharp intensity changes between adjacent blocks. These would show up as blocking artifacts (i.e., block boundaries would be clearly visible in the decoded picture). For this reason, the DC coefficient in intramacroblocks is not coded using the quantizer that is applied to the other coefficients in the block. Instead, a quantizer step size of 8 is always applied. Thus a DC coefficient of 601 would be quantized to a value of 75 irrespective of the quantizer used for the remainder of the macroblock.

In fact, the intra DC coefficients are coded separately from the other DCT coefficients that are coded as (run,level) pairs as shown in Table 5.1. We have seen already that there is strong correlation between nearby pixels in images. For this reason, there is also strong correlation between the DC DCT coefficients of nearby blocks within a picture because the DC value represents the average luminance or chrominance value of that block in the picture. Figure 5.11 shows the average luminance value of each 8×8 pixel block within the first picture of the "Mobile and Calendar" sequence.

An approach similar to that employed to code motion vectors can be used to code the DC DCT coefficients in intramacroblocks. For simplicity, prediction is made based on blocks in the current macroblock or from the previous macroblock as shown in Table 5.5. The block nomenclature shown in Figure 5.4 is used in Table 5.5.

When this lossless prediction is applied to the luminance data of the first picture of the "Mobile and Calendar" sequence, the prediction differences produce the

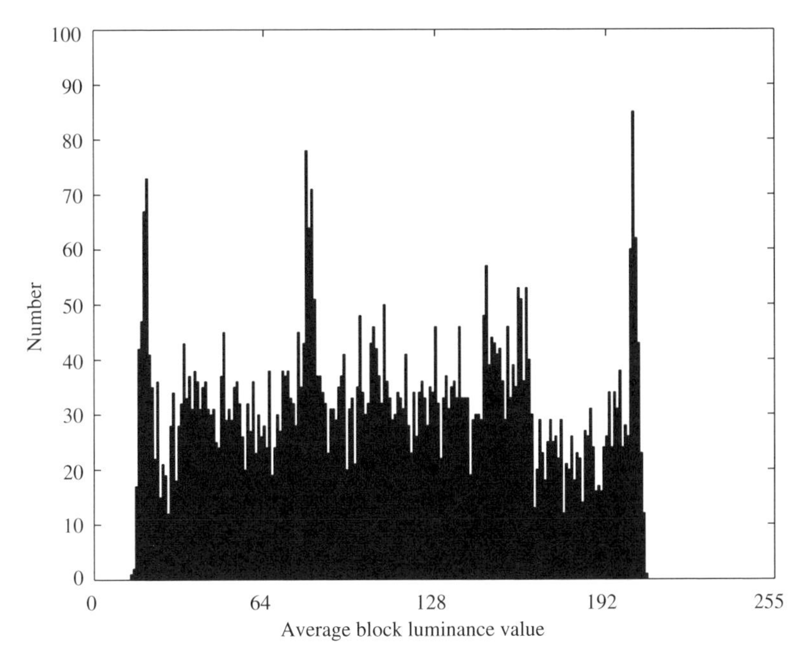

Figure 5.11 Histogram of average luminance values.

histogram shown in Figure 5.12. It is clear that some saving is possible using an appropriate variable length Huffman code. Note also that blocks used for prediction are all within the current row of macroblocks. This means that a row of macroblocks can be correctly decoded even if the previous row of macroblocks could not be decoded due to transmission errors, for example. The predictor is reset to 1024 (128×8) at the beginning of each row of macroblocks.

In fact the difference information is coded using two code words. The first code word specifies the length in bits of the difference (as shown in Table 5.6) whereas the second code word specifies the actual value (as shown in Table 5.7). Note that the code words used for length in bits are different for luminance and chrominance.

Table 5.5 Prediction of DC DCT coefficient in intramacroblocks.

Block	Prediction block
Y_0	Y_3 in previous macroblock
Y_1	Y_0 in current macroblock
Y_2	Y_1 in current macroblock
Y_3	Y_2 in current macroblock
U	U in previous macroblock
V	V in previous macroblock

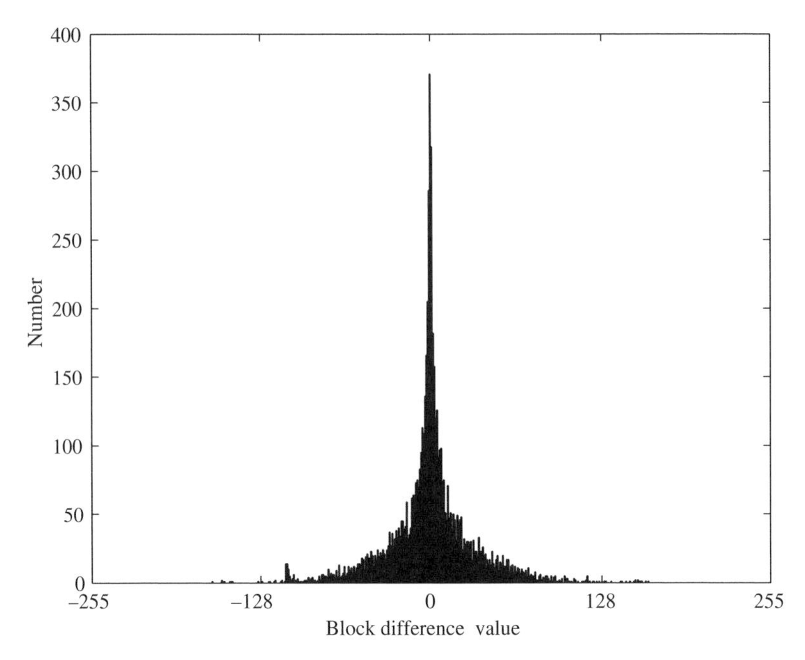

Figure 5.12 Histogram of DC block differences.

Thus a prediction difference of 29 in the luminance DC DCT value would be represented by the code words given in Figure 5.13.

1110	11101
(Absolute value in range 16–31)	(Actual value +29)

Figure 5.13 Calculation of luminance DC DCT code-word.

Table 5.6 Variable length codes for the differential DC length.

Absolute value of differential DC	Length (bits)	VLC code (luminance)	VLC code (chrominance)
0	0	100	00
1	1	00	01
2–3	2	01	10
4–7	3	101	110
8–15	4	110	1110
16–31	5	1110	1111 0
32–63	6	1111 0	1111 10
64–127	7	1111 10	1111 110
128–255	8	1111 110	1111 1110

Table 5.7 Value of differential DC.

Differential DC	Length (bits)	Code word
−255 to −128	8	00000000 to 01111111
−127 to −64	7	0000000 to 0111111
−63 to −32	6	000000 to 011111
−31 to −16	5	00000 to 01111
−15 to −8	4	0000 to 0111
−7 to −4	3	000 to 011
−3 to −2	2	00 to 01
−1	1	0
0	0	
1	1	1
2–3	2	10 to 11
4–7	3	100 to 111
8–15	4	1000 to 1111
16–31	5	10000 to 11111
32–63	6	100000 to 111111
64–127	7	1000000 to 1111111
128–255	8	10000000 to 11111111

5.3.2.5. *Coded Block Pattern*

As indicated when we were considering the block level, it is not necessary to code every block of DCT coefficients within a macroblock because some blocks, after quantization, will have all zero coefficients. In this case, the block is skipped. The decoder needs to know which blocks within a macroblock have been skipped. This is signaled in the macroblock header by a field called the coded block pattern (CBP). The value of the CBP is determined as follows with reference to Figure 5.4: If a block requires coding, that block is given the value 1 in the CBP. If the block is skipped, that block is given the value 0. Then

$$\text{CBP} = 32 \times Y_1 + 16 \times Y_2 + 8 \times Y_3 + 4 \times Y_4 + 2 \times U + 1 \times V$$

Thus, if all blocks needed to be coded then CBP would take the value 63 whereas CBP would take the value 60 if only the luminance blocks are needed to be coded. Some values of CBP are more likely than others, and so a saving can be achieved using Huffman coding. A typical Huffman code table for the CBP is shown in Table 5.8. Note that the CBP takes values in the range 1–63. A CBP of zero would imply either a skipped macroblock (i.e., no data is transmitted about the macroblock) or a motion-compensated not coded macroblock type (see Table 5.2) in which case a CBP is not transmitted.

These Huffman code words have been developed based on the statistics of real coded video sequences. The shortest code word and thus the highest probability symbol (CBP = 60) corresponds to the case of all four luminance blocks coded but

Table 5.8 Huffman code words for coded block pattern.

Coded block pattern	Pattern[i]						Code word
	Y_0 (i=0)	Y_1 (i=1)	Y_2 (i=2)	Y_3 (i=3)	U (i=4)	V (i=5)	
1						1	0101 1
2					1		0100 1
3					1	1	0011 01
4				1			1101
5				1		1	0010 111
6				1	1		0010 011
7				1	1	1	0001 1111
8			1				1100
9			1			1	0010 110
10			1		1		0010 010
11			1		1	1	0001 1110
12			1	1			1001 1
13			1	1		1	0001 1011
14			1	1	1		0001 0111
15			1	1	1	1	0001 0011
16		1					1011
17		1				1	0010 101
18		1			1		0010 001
19		1			1	1	0001 1101
20		1		1			1000 1
21		1		1		1	0001 1001
22		1		1	1		0001 0101
23		1		1	1	1	0001 0001
24		1	1				0011 11
25		1	1			1	0000 1111
26		1	1		1		0000 1101
27		1	1		1	1	0000 0001 1
28		1	1	1			0111 1
29		1	1	1		1	0000 1011
30		1	1	1	1		0000 0111
31		1	1	1	1	1	0000 0011 1
32	1						1010
33	1					1	0010 100
34	1				1		0010 000
35	1				1	1	0001 1100
36	1			1			0011 10
37	1			1		1	0000 1110
38	1			1	1		0000 1100
39	1			1	1	1	0000 0001 0
40	1		1				1000 0
41	1		1			1	0001 1000

(continued)

Table 5.8 (*Continued*)

Coded block pattern	Pattern[i]						Code word
	Y_0 (i=0)	Y_1 (i=1)	Y_2 (i=2	Y_3 (i=3)	U (i=4)	V (i=5)	
42	1		1		1		0001 0100
43	1		1		1	1	0001 0000
44	1		1	1			0111 0
45	1		1	1		1	0000 1010
46	1		1	1	1		0000 0110
47	1		1	1	1	1	0000 0011 0
48	1	1					1001 0
49	1	1				1	0001 1010
50	1	1			1		0001 0110
51	1	1			1	1	0001 0010
52	1	1		1			0110 1
53	1	1		1		1	0000 1001
54	1	1		1	1		0000 0101
55	1	1		1	1	1	0000 0010 1
56	1	1	1				0110 0
57	1	1	1			1	0000 1000
58	1	1	1		1		0000 0100
59	1	1	1		1	1	0000 0010 0
60	1	1	1	1			111
61	1	1	1	1		1	0101 0
62	1	1	1	1	1		0100 0
63	1	1	1	1	1	1	0011 00

with both chrominance blocks skipped. Next most likely are macroblocks containing just a single coded luminance block (CBP = 32, 16, 8, or 4) followed by macroblocks containing just two coded luminance blocks. It is clear that skipped blocks occur regularly in coded video sequences.

5.3.2.6. Summary

The macroblock header contains a large amount of information required to allow the decoder to correctly interpret the information contained in the macroblock. The information contained in the macroblock header is a function of the type of macroblock that is being coded. As shown in Table 5.2, there are seven valid macroblock types.[3] The format of each of these macroblock types is shown in Figure 5.14. For

[3] As we shall see in the next chapter, there are a considerably larger number of macroblock types available in the compression standard used for digital video broadcasting.

Macroblock address increment	MB type 0001 1	Block data for all of Y_1, Y_2, Y_3, Y_4, U, and V

Intramacroblock with same quantizer

Macroblock address increment	MB type 0001 01	New quantizer (5 bits)	Block data for all of Y_1, Y_2, Y_3, Y_4, U, and V

Intramacroblock with new quantizer

Macroblock address increment	MB type 01	Coded block pattern	Block data for selection of Y_1, Y_2, Y_3, Y_4, U, and V as indicated by CBP

Intermacroblock with same quantizer

Macroblock address increment	MB type 0000 1	New quantizer (5 bits)	Coded block pattern	Block data for selection of Y_1, Y_2, Y_3, Y_4, U, and V as indicated by CBP

Intermacroblock with new quantizer

Macroblock address increment	MB type 1	MV diff (*H*)	MV diff (*V*)	Coded block pattern	Block data for selection of Y_1, Y_2, Y_3, Y_4, U, and V as indicated by CBP

Motion-compensated macroblock with same quantizer—coded

Macroblock address increment	MB type 001	New quantizer (5 bits)	MV diff (*H*)	MV diff (*V*)	Coded block pattern	Block data for selection of Y_1, Y_2, Y_3, Y_4, U, and V as indicated by CBP

Motion-compensated macroblock with new quantizer—coded

Macroblock address increment	MB type 0010 0	MV diff (*H*)	MV diff (*V*)

Motion-compensated macroblock—not coded

Figure 5.14 Format of various macroblock types.

intramacroblocks, each block is always coded as it will contain at least a DC DCT coefficient. In motion-compensated macroblocks, MV Diff (*H*) and MV Diff (*V*) are the horizontal and vertical difference motion vectors respectively.

5.3.3. The Slice Layer

Macroblocks merge, commonly, to form a slice with a slice containing a complete row of macroblocks as shown in Figure 5.15. In this case, *n* macroblocks in the same row merge to form a slice.

Each slice needs a header. One important task of the slice header is to allow for decoder resynchronization in case the decoder reaches a stage where it is unable to

Figure 5.15 A row of macroblocks merges to form a slice.

decode the bit stream. This is usually caused by transmission errors affecting the bit stream received by the decoder. The digital bit stream is particularly vulnerable to transmission errors as a result of the large number of variable length Huffman code words used to reduce the amount of data that needs to be transmitted. This is illustrated in Figure 5.16 where a series of symbol is encoded using a simple Huffman code and then decoded after a single transmission error. The resulting decoded symbol stream is not the same as the transmitted symbol stream. Moreover, having some symbols decoded incorrectly, the total number of decoded symbols is also different.

When this occurs, the decoder often encounters some impossible situation such as attempting to decode more than 64 DCT coefficients in a block or trying to use a motion vector that points outside the reference picture. When this occurs, the decoder has no other alternative than to attempt to resynchronize to the incoming bit stream. This is achieved by looking for a resynchronization code word that can only occur in a nonerrored bit stream at the start of a slice (or at the start of a higher element in the syntax hierarchy as discussed a little later). Typically this is the 24-bit hexadecimal number 0x000001 and is called a start code. For this approach to work, care needs to be exercised in the design of the encoder syntax to ensure that the only time that a start code can occur is at the designated resynchronization points (i.e., 23 or more successive zero bits can never occur at other than the resynchronization point).

Because entire slices can be skipped in the same way as blocks and macroblocks, the slice header needs to contain information that allows the decoder to

Symbol	Huffman codeword
a	0
b	10
c	110
d	111

Encoded symbols	b	a	a	c	a	d
Transmitted Bit stream	10	0	0	110	0	111
Received Bit stream	10	**1**	0	110	0	111
Decoded symbols	b		b	c	a	d

Figure 5.16 The effect of a transmission error on Huffman decoding.

determine which slice in the current picture is about to be decoded. It would be possible to code this as an increment from the previous slice with a value of 1 indicating that the current slice is one on from the last slice. However, this would mean that when resynchronization was needed following a transmission error, the decoder could not be sure which slice was being decoded. For this reason (and noting that the number of slice headers is small), the actual slice number is incorporated into the slice header. Slice 1 is the first 16 lines of the picture, slice two is lines 17–32 and so on.

For resynchronization to be possible, it is essential that the decoder does not need to know any information that was transmitted before the current slice header (as this may have been lost). For this reason

- The current quantizer step size is included in the slice header.
- All elements that are usually coded with respect to a previous macroblock in the macroblock header are coded without prediction in the first macroblock after a slice header. This includes the macroblock address increment and the motion vectors.

A typical slice header is shown in Figure 5.17.

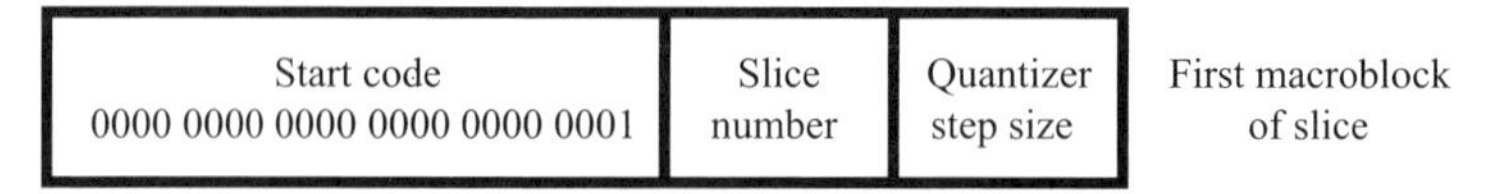

Figure 5.17 Typical slice header.

5.3.4. The Picture Layer

As we continue to move up the syntax hierarchy, the next level is the picture level. As indicated in Figure 5.15, the various slices merge to make up a complete picture, which is again preceded by a picture header. Resynchronization also needs to be available at the picture level, and so the picture header commences with a start code followed by a number that cannot be mistaken for a slice number (e.g., zero). This allows a picture header to be differentiated from a slice header by the decoder. As we shall see in the next chapter, the picture header can also be used to signal picture specific information. In order to minimize our syntax complexity, such information will be ignored here.

5.3.5. The Sequence Layer

The header at the sequence layer is used to signal information that applies to the entire video sequence that is to be decoded. Like the slice and picture layers, the sequence header begins with a start code followed by a unique number that differentiates it from slice and picture layer headers. Sequence specific information might include, inter alia, the following:

- The horizontal size of the pictures in the sequence (in pixels).
- The vertical size of the pictures in the sequence (in pixels).
- The picture rate (in pictures per second).
- The transmission bit rate (in bits per second).

5.4. BIT-STREAM SYNTAX

The task of specifying a bit-stream syntax is an important one because without a valid syntax, interoperability been encoders and decoders will be impossible to achieve. The syntax must be defined carefully—in the same way that a computer program needs to be specified carefully. This similarity has led to the use of a syntax description methodology that is quite similar to the C programming language. This section will provide an introduction to this method by using it to describe the simple video bit-stream syntax already developed in this chapter. These ideas will be extended in the next chapter when we consider the MPEG-2 video compression standard used for digital television broadcasting.

All video compression standards deal with the description of a decoder because it is mandated that all decoders that comply with the standard must be capable of decoding any compliant bit stream generated by an encoder. Although two different encoders presented with the same video material would most probably produce different encoded bit streams (both of which would comply with the standard), two different decoders presented with the same compliant bit stream should produce identical decoded outputs. The syntax is therefore written from the point of view of the decoding process.

5.4.1. Abbreviations

There are a number of abbreviations used to specify the type of data that is expected at the decoder at a particular point of the syntax. The most common of these are listed below.

bslbf	*b*it *s*tring with the *l*eftmost *b*it transmitted *f*irst
uimsbf	*u*nsigned *i*nteger with *m*ost *s*ignificant *b*it transmitted *f*irst
vlclbf	*v*ariable *l*ength *c*ode with *l*eftmost *b*it transmitted *f*irst

5.4.2. Start Codes

As indicated earlier, all start codes begin with the 24-bit hexadecimal code 0x000001. The syntax is written in such a way that this code will only occur at a start code in a nonerrored bit-stream. The next 8 bits (i.e, two hexadecimal numbers) specify the type of start code. As indicated earlier in this chapter, this is crucial information for the decoder particularly during resynchronization after a transmission error. The value corresponding to the various types of start codes described in this chapter are shown in Table 5.9. As we shall see later, the reserved start code types are used for other functions within the MPEG suite of standards. Slice start code field numbering is from 01 to AF hexadecimal (a total of 175 different values) with the number giving the vertical position of the slice within the picture. This means that there can be up to 2800 luminance lines per picture (175×16 rows per macroblock).

5.4.3. Describing the Bit-Stream Syntax

A number of constructs, similar to those of the C language, are used to describe the conditions under which the various syntactical units are present.

Table 5.9 Definition of start code types.

Type of start code	Start code type field (hexadecimal)
picture_start_code	00
slice_start_code	01 to AF
Reserved	B0 to B2
sequence_header_code	B3
Reserved	B4 to B6
sequence_end_code	B7
Reserved	B8
Reserved	B9 to FF

5.4.3.1. *The While Construct*

```
while (conditional_expression) {
  bitstream_element
     ⋮
}
```

If the conditional_expression is true, the bit-stream elements occur next in the bit stream. The bit-stream elements are repeated until the expression is not true.

5.4.3.2. *The Do–While Construct*

```
do {
  bitstream_element
     ⋮
} while (conditional_expression)
```

The bitstream_elements occur next in the bit stream and continue to be repeated until the conditional_expression is not true. Note that unlike the while construct, in the do-while construct the bitstream_elements must occur at least once.

5.4.3.3. *The If–Else Construct*

```
if (conditional_expression) {
    bitstream_element_a
⋮
}
else {
  bitstream_element_b
⋮
}
```

If the conditional_expression is true then the bit-stream elements beginning with bitstream_element_a occur next in the bit stream. If the conditional_expression is not true then the bit-stream elements beginning with bitstream_element_b occur next in the bit stream.

5.4.3.4. *The For Construct*

```
for(expression_1; conditional_expression;expression_2) {
  bitstream_element
⋮
}
```

The for construct is simply a shorthand version of the while construct that is useful in many circumstances. The equivalent while construct would be

```
expression_1
while(conditional_expression) {
  bitstream_element
    ⋮
  expression_2
}
```

In the for construct, expression_1 is used to initialize the loop and is often the initial state of a counter. Before each iteration of the loop the conditional_expression is tested and the loop is terminated if the conditional expression is false. At the end of each iteration of the loop, expression_2 is performed and is often used to increment a counter. A typical example of the use of the for construct is given below.

```
for (i=0; i<n; i++) {
  bitstream_element
    ⋮
}
```

In this case, the bit-stream elements occur n times. The expression i++ means that i is incremented by one.

5.4.4. Special Functions within the Syntax

A number of functions are used to define operations that need to be performed to correctly decode a bit stream.

5.4.4.1. The Bytealigned Function

The function bytealigned() returns the value 1 if the current position in the bit stream is on a byte (8-bit) boundary, that is, the next bit is the first bit of a byte. It is needed because start codes are typically bytealigned.

5.4.4.2. The Nextbits Function

The nextbits() function allows the next several bits in the bit stream to be compared to a particular bit string. This allows the syntax to move through the bit stream looking for a particular bit string. A typical example of this is when a start code is being sought.

5.4.4.3. The Next_Start_Code Function

The next_start_code() function shown in Table 5.10 uses both the bytealigned() and nextbits() functions described above to locate the next start code in the bit stream. Any zero bit stuffing (to reach the next byte boundary because start codes start on a byte boundary) and zero byte stuffing (which may have been added to increase the amount of transmitted data to meet rate control requirements) are ignored. This ensures that start codes can be preceded by any number of zero stuffing bits subject only to the start code being bytealigned.

Table 5.10 The next_start_code() function.

Syntax	Number of bits	Mnemonic
next_start_code() {		
while(!bytealigned()) {		
zero_bit	1	"0"
}		
while(nextbits != '0000 0000 0000 0000 0000 0001') {		
zero_byte	8	"0000 0000"
}		
}		

5.5. A SIMPLE BIT-STREAM SYNTAX

We will now define a simplified bit-stream syntax using the methods described in this chapter. This syntax greatly simplifies the understanding of the full MPEG-2 video syntax that is described in the next chapter. We will begin at the top layer (video sequence layer) and work our way down to the bottom layer (block layer).

5.5.1. The Video Sequence Layer

As explained earlier, this layer defines parameters that will apply to the entire video sequence. The syntax is given in Table 5.11.

Table 5.11 Video sequence syntax.

	Syntax	Bits	Mnemonic
1	video_sequence() {		
2	next_start_code()		
3	do {		
4	sequence_header()		
5	do {		
6	picture()		
7	}while (nextbits() == **picture_start_code**)		
8	}while (nextbits() == **sequence_header_code**)		
9	**sequence_end_code**	32	bslbf
10	}		

The function next_start_code() (line 2) is used to byte align the incoming bit stream. Next follows the sequence_header() (as defined below) followed immediately by a picture. Pictures then continue until nextbits() reveals that **picture_start_code** (line 7) is not present. The function nextbits() is then used to check if a new **sequence_header_code** is present (line 8). A new **sequence_header_code** followed by a new sequence_header() is possible before any picture of the sequence. This allows sequence parameters to be changed if necessary. If nextbits() reveals that there is neither a **picture_start_code** nor a **sequence_header_code**, then the sequence concludes with a **sequence_end_code**.

The function sequence_header() is defined in Table 5.12.

The sequence header() consists of

(a) The 32-bit start code **sequence_header_code** (000001B3 hexadecimal).

(b) The 12-bit horizontal size.

(c) The 12-bit vertical size.

(d) The 4-bit **aspect_ratio_information** defines the shape of each pixel in a picture. The horizontal and vertical size of the picture only define the aspect ratio of the picture if pixels can be assumed to be square. This is usually not the case. However, if we know the Display aspect ratio (DAR) and the horizontal and vertical size of the picture then the Sample aspect ratio (SAR) can be calculated according to the following equation.

$$\text{SAR} = \text{DAR} \times \frac{\text{horizontal\_size}}{\text{vertical\_size}}$$

The **aspect_ratio_information** defines either the SAR (for square pixels) or the DAR as shown in Table 5.13. The value 0000 is not used to avoid start code emulation.

Table 5.12 Sequence header syntax.

	Syntax	Bits	Mnemonic
1	sequence_header() {		
2	**sequence_header_code**	32	bslbf
3	**horizontal_size**	12	uimsbf
4	**vertical_size**	12	uimsbf
5	**aspect_ratio_information**	4	uimsbf
6	**picture_rate**	4	uimsbf
7	**bit_rate**	18	uimsbf
8	next_start_code()		
9	}		

Table 5.13 Definition of aspect_ratio_information.

aspect_ratio_information	Sample aspect ratio	Display aspect ratio
0000	Forbidden	Forbidden
0001	1.0 (Square)	—
0010	—	3÷4
0011	—	9÷16
0100	—	1÷2.21
0101 to 1111	—	Reserved

(e) The 4-bit **picture_rate** defines the picture rate of the video sequence. Allowed values are given in Table 5.14. The value 0000 is not used to avoid start code emulation.

(f) The 18 bit **bit_rate** field gives the bit rate of the channel in multiples of 400 bits/s and is assumed constant. Any bit rate that is not a multiple of 400 bits/s is rounded up to the next multiple of 400 bits/s. Thus the bit rate 999,800 bits/s would be rounded up to 1,000,000 bits/s and represented by a bit_rate value of 2500 (1,000,000 ÷ 400). The value 0x3FFFF is used to indicate variable bit-rate operation.

5.5.2. The Picture Layer

The picture layer provides information relating to the current picture only. An example of a simple syntax for this layer is given in Table 5.15.

Table 5.14 Allowable picture rates.

picture_rate	Pictures per second
0000	Forbidden
0001	24,000÷1001 = 23.976
0010	24
0011	25
0100	30,000÷1001 = 29.97
0101	30
0110	50
0111	60,000÷1001 = 59.94
1000	60
1001 to 1111	Reserved

Table 5.15 Picture syntax.

	Syntax	Bits	Mnemonic
1	picture() {		
2	**picture_start_code**	32	bslbf
3	**temporal_reference**	10	uimsbf
4	next_start_code()		
5	do {		
6	slice()		
7	}while (nextbits() == **slice_start_code**)		
8	}		

The header information at the picture layer consists of the following fields:

(a) The 32-bit **picture_start_code** (00000100 hexadecimal).

(b) A 10-bit **temporal_reference**, which is an unsigned integer that is incremented by 1 (modulo 1024) for each input picture.

The next_start_code() function is then called to look for a bytealigned start code that should be a slice header. Slices are then processed until the next start code is not a slice header.

5.5.3. The Slice Layer

The slice layer provides information about the current slice. This includes the quantizer step size to be used initially for the decoding of the slice. A simple slice syntax is given in Table 5.16.

Table 5.16 Slice syntax.

	Syntax	Bits	Mnemonic
1	slice() {		
2	**slice_start_code**	32	bslbf
3	**quantizer_scale**	5	uimsbf
4	do {		
5	macroblock()		
6	} while (nextbits() != 000 0000 0000 0000 0000 0000)		
7	}		

The slice begins with a **slice_start_code**. The first 24-bits are the hexadecimal code 000001 whereas the last 8 bits give the vertical position of the slice and take values in the range 01 to AF hexadecimal.

This is followed by the **quantizer_scale** that defines the quantizer step size to be used for the quantization of DCT coefficients. This is a 5-bit number in the range 1–31 that corresponds to quantizer step sizes in the range 2–62, respectively. This quantizer value is used for the entire slice unless changed in a subsequent macroblock header.

Macroblocks are then processed (lines 4–6) until a start code (23 successive zeros) is encountered indicating a new slice, picture or sequence level start code.

5.5.4. The Macroblock Layer

The macroblock layer provides detailed information about each 16 × 16 block of luminance (and associated chrominance) pixels. This includes the way that the macroblock is to be coded (intra, inter, or motion compensated), a new quantizer step size (if needed) and motion vector (if required). A simplified syntax for the macroblock layer is shown in Table 5.17.

The first piece of information in the macroblock (lines 2–4) is the **macroblock_address_increment**. This indicates the difference between the current macroblock address and the address of the previously coded macroblock. The code word **macroblock_address_increment** is limited to the range 1–33. Because the

Table 5.17 Macroblock syntax.

	Syntax	Bits	Mnemonic
1	macroblock() {		
2	while(nextbits() == '0000 0001 000') {		
3	**macroblock_escape**}	11	vlclbf
4	**macroblock_address_increment**	1–11	vlclbf
5	**macroblock_type**	1–6	vlclbf
6	if(macroblock_quant) {		
7	**quantizer_scale** }	5	uimsbf
8	if(macroblock_motion) {		
9	**motion_horizontal_code**	1–11	vlclbf
10	**motion_vertical_code**	1–11	vlclbf
11	}		
12	if(macroblock_pattern) {		
13	**coded_block_pattern**}	3–9	vlclbf
14	for(i=0; i<6; i++) {		
15	block(i) }		
16	}		

Table 5.18 Definition of Internal Syntax Variable Based on **macroblock_type**.

macroblock_type	1	01	001	0001 1	0001 0	0000 1	0000 01
Macroblock type	MC	Inter	MC	Intra	MC	Inter	Intra
Coded	Yes	Yes	No	Yes	Yes	Yes	Yes
New quantizer	No	No	No	No	Yes	Yes	Yes
macroblock_quant	0	0	0	0	1	1	1
macroblock_motion	1	0	1	0	1	0	0
macroblock_pattern	1	1	0	0	1	1	0
macroblock_intra	0	0	0	1	0	0	1

previously coded macroblock may be more than 33 macroblocks back, the **macroblock_escape** code word can be used to allow for larger increments. Each time the **macroblock_escape** code word is included 33 is added to the total increment. Thus an increment of 70 macroblocks would comprise the **macroblock_escape_code word** twice followed by a **macroblock_address_increment** of four macroblocks.

Next (line 5) comes the **macroblock_type** code word. As discussed earlier in this chapter, this defines which other elements are contained in the macroblock. Based on the value of the **macroblock_type** code word, various internal syntax variables are set as shown in Table 5.18.

The meaning of these internal variables is listed below.

- **macroblock_quant** when set indicates that the macroblock contains a new value for **quantizer_scale**.
- **macroblock_motion** when set indicates that the macroblock contains a motion vector.
- **macroblock_pattern** when set indicates that the macroblock contains a coded block pattern.
- **macroblock_intra** when set indicates that the macroblock is coded in intramode.

It is now a simple task to interpret the remainder of the macroblock syntax. If **macroblock_quant** is set (line 6) then a new 5 bit fixed length value of **quantizer_scale** appears next in the bit stream (line 7).

If **macroblock_motion** is set (line 8) then the two components of the motion vector appear next in the bit stream (lines 9–10). Note that the motion vector difference between the motion vector in the current macroblock and the motion vector in the previous macroblock is transmitted. The prediction motion vector is reset to zero if:

- the macroblock immediately follows a slice header.
- the previous macroblock was coded in intermode because a zero motion vector is assumed in this case. This applies also if the previous macroblock was a skipped macroblock.

- the previous macroblock was coded in intramode because no motion vector information is available from an intracoded macroblock.

The motion vector differences are variable length coded as described earlier in this chapter and are of length 1–11 bits each.

If **macroblock_pattern** is set (line 13) then the coded block pattern appears next in the bit stream (line 14).

We now have all the information needed to decode the blocks of DCT coefficients associated with the macroblock, and this is now performed (lines 14–15) for each of the six possible blocks in a macroblock.

5.5.5. The Block Layer

A simple syntax for the block layer is given in Table 5.19.

The coded block pattern read at the macroblock level produces a 6-bit pattern code indicating which of the blocks 0 (Y_0) to 5 (V) are to be coded with a value of 1, indicating that the block is to be coded (see Table 5.8). If the block is not coded, as indicated by **pattern[i]** (line 2) then the block syntax is skipped as required for a skipped block. Next a check is made to see if the block is coded in intramode

Table 5.19 Block syntax.

	Syntax	Bits	Mnemonic
1	block(i) {		
2	if(pattern_code[i]) {		
3	if (macroblock_intra) {		
4	if (i < 4) {		
5	**dct_dc_size_luminance**	2–7	vlclbf
6	if (dc_size_luminance !=0) {		
7	**dct_dc_differential}**	1–8	uimsbf
8	}		
9	else {		
10	**dct_dc_size_chrominance**	2–8	vlclbf
11	if (dc_size_chrominance != 0) {		
12	**dct_dc_differential }**	1–8	uimsbf
13	}		
14	}		
15	while (nextbits() != ‘10’) {		
16	**dct_coeff_next }**	3–28	vlclbf
17	**end_of_block**	2	vlclbf
18	}		
19	}		

(line 3). If this is the case then the DC DCT coefficient is treated separately from the other DCT coefficients. As described earlier, each DC coefficient is predicted from an earlier DC coefficient. For the first four blocks where the block count, *i*, is in the range of 0–3 (line 4), the variable length codes for **dct_dc_size_luminance** and **dct_dc_differential** appear next in the bit stream (lines 5–7) whereas for a block count in the range 4–5 the **dct_dc_size_chrominance** and **dct_dc_differential** appear next in the bit stream. Then follows the remaining (run,level) pairs of DCT coefficients (in the case of intra macroblocks) or all DCT coefficients represented as (run,level) pairs (in the case of motion-compensated or inter macroblocks) until the end of block code word ("10") is encountered.

5.6. CONCLUSION

This chapter has provided an introduction to the syntax used by video encoders to transmit compressed video information to the decoder in a form that it can understand. We have also introduced a number of tools for the efficient representation of this information.

In the next chapter, detailed consideration will be given to the MPEG-2 video compression standard that forms the basis of all current digital television broadcasting systems worldwide.

PROBLEMS

5.1 Briefly explain the reasons that a bit-stream syntax is required for video encoders. What are the major advantages?

5.2 Most of the information contained in a video bit stream is encoded using variable word length codes. Why is this the case? What disadvantages does the use of these variable word length codes introduce? How does the use of a video encoder syntax allow these problems to be minimized?

5.3 A camera produces signals representing the red, green, and blue components of the scene that it is imaging. This information is invariably transformed into *YUV* form. Explain briefly why this is the case.

5.4 Table 5.1 and Figure 5.6 show that the (run,level) pairs representing quantized DCT coefficients are sometimes represented by variable length code words and sometimes by fixed length code words. Given that even in the (run,level) pairs represented by the fixed length code words some will be more common than others, it would seem desirable to use variable length codes here as well. Explain briefly why this is not done.

5.5 Why is it that the macroblock header for an intracoded macroblock does not require a field for the coded block pattern?

5.6 Explain briefly why motion vectors are coded differentially. What are the advantages and disadvantages of doing this?

5.7 A quantized block of DCT coefficients from a macroblock coded in intramode is given in Figure 5.18. Calculate the code words that would need to be transmitted to represent this block of data. Assume that this is the first block in the macroblock and that the macroblock immediately follows a slice header.

+ 20	0	0	0	0	0	0	0
0	0	− 1	0	+ 1	0	− 1	0
0	0	0	0	0	0	0	0
0	0	+ 1	0	0	− 1	0	0
0	0	0	− 1	0	0	0	0
0	0	0	− 1	− 1	0	+ 1	0
0	0	0	+ 1	0	0	− 1	0
0	0	0	0	0	0	0	0

Figure 5.18 Quantized DCT coefficients for Problem 5.7.

5.8 A quantized block of DCT coefficients from a macroblock coded using motion compensated prediction is given in Figure 5.19. Calculate the code words that would need to be transmitted to represent this block of data.

− 2	0	0	0	0	0	0	0
0	− 5	0	− 1	0	0	0	0
+ 13	− 7	+ 2	0	0	0	0	0
0	0	0	0	0	0	0	0
0	0	− 1	0	0	0	0	0
0	0	0	0	0	0	0	0
− 1	+ 2	0	0	0	0	+ 1	0
− 3	0	0	0	0	0	0	0

Figure 5.19 Quantized DCT coefficients for Problem 5.8.

5.9 The seventh macroblock of a particular slice is to be coded. The macroblock is coded using motion compensated prediction and the motion vector employed is (+2, −1). After motion-compensated prediction, application of the DCT, and quantization with the same quantizer as used in the previous macroblock, all blocks in the macroblock contain all zero coefficients with the exception of block Y_2 that is shown in Figure 5.20.

+ 2	0	0	0	0	0	0	0
0	− 1	0	0	0	0	0	0
0	0	0	0	0	0	0	0
0	0	0	0	0	0	0	0
0	0	0	0	0	0	0	0
0	0	0	0	0	0	0	0
0	0	0	0	0	0	0	0
0	0	0	0	0	0	0	0

Figure 5.20 Quantized DCT data for Problem 5.9.

The previous macroblock coded was the fourth macroblock in the slice and was also coded with motion-compensated prediction using a motion vector of (+1, 0). Using the syntax described in this chapter, determine the bit stream required to represent this macroblock.

5.10 A macroblock is coded in intramode. The first quantized block (Y_1) in the macroblock contains only a quantized DC DCT coefficient with value 120. The second block (Y_2) is shown in Figure 5.21.

+ 95	+ 1	+ 1	0	0	0	0	0
– 1	– 1	– 1	0	0	0	0	0
+ 1	+ 1	0	0	0	0	0	0
0	0	0	0	0	0	0	0
0	0	0	0	0	0	0	0
0	0	0	0	0	0	0	0
0	0	0	0	0	0	0	0
0	0	0	0	0	0	0	0

Figure 5.21 Quantized DCT data for Problem 5.10.

Using the syntax described in this chapter, determine the bit stream required to represent this block.

5.11 The two-dimensional DCT of a luminance block in an intracoded macroblock produces the data shown in Figure 5.22.

+ 755.6	+ 141.3	+ 78.6	+ 29.1	+ 2.1	– 7.6	– 2.7	0.9
– 119.2	– 177.1	– 107.2	+ 39.6	+ 21.2	– 18.0	– 8.1	0.0
+ 107.5	+ 136.7	+ 54.9	– 42.1	– 32.0	+ 6.4	– 1.5	0.0
– 45.0	– 41.8	+ 30.8	+ 67.6	+ 21.3	+ 2.6	+ 5.0	+ 3.7
– 1.1	– 13.7	– 32.9	– 20.5	+ 4.4	+ 7.6	+ 3.2	+ 0.6
+ 15.1	+ 15.4	+ 8.8	– 15.9	– 23.0	– 8.4	– 5.0	0.0
– 13.8	– 22.0	+ 1.3	+ 13.7	– 5.8	– 11.7	– 8.9	+ 0.9
+ 14.2	+ 14.5	+ 8.2	+ 1.2	– 10.4	0.0	+ 6.5	+ 4.0

Figure 5.22 DCT data for Problem 5.11.

(a) If the **quantizer_scale** value is 10 and the DC DCT coefficient of the previous luminance block was 480, calculate the values of the quantized DCT coefficients.

(b) Determine the bit stream required to represent these quantized DCT coefficients using the syntax developed in this chapter.

5.12 The bit stream shown in Figure 5.23 represents the first block of an intracoded macroblock which immediately follows a slice header.

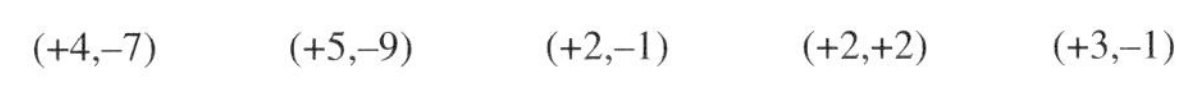

Figure 5.23 Bit stream values for Problem 5.12.

(a) Determine the decoded quantized DCT values.

(b) If the quantizer step size used was 8, determine the DCT values after inverse quantization.

(c) Determine the decoded pixel values for this block.

5.13 The first five motion vectors in a slice are shown in Figure 5.24.

(+4,–7) (+5,–9) (+2,–1) (+2,+2) (+3,–1)

Figure 5.24 Motion vectors for Problem 5.13.

(a) Calculate the differential motion vectors that need to be transmitted to the receiver.

(b) Calculate the code words used to represent these motion vectors.

5.14 Repeat Problem 5.13 for the case where the motion vectors given in Figure 5.24 apply to macroblocks 1, 2, 4, 5, and 6, respectively. Macroblock 3 is a skipped macroblock.

5.15 The DC DCT coefficients for two successive macroblocks coded in intramode are shown in Figure 5.25. Determine the variable length code words that would be transmitted to represent the DC coefficients in the second macroblock.

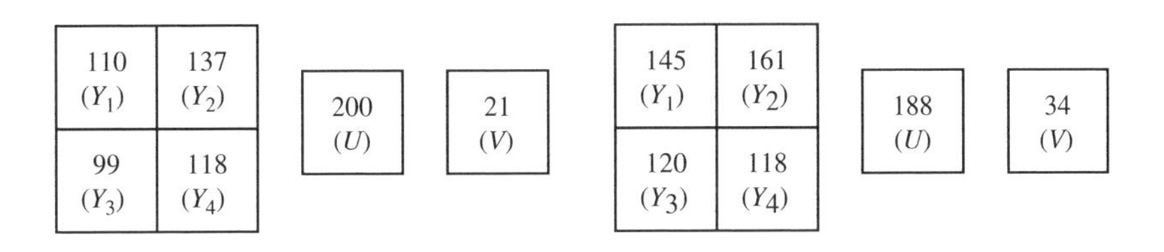

Figure 5.25 DC DCT coefficient values for Problem 5.15.

5.16 Repeat Problem 5.15, but this time assume that the first macroblock in Figure 5.25 is preceded by a slice header. Calculate the variable wordlength code words that would be used to represent each of the DC coefficients in both macroblocks.

5.17 Explain the meaning of the following piece of syntax shown in Table 5.20. What functionality does this syntax allow?

Table 5.20 Syntax for Problem 5.17.

	Syntax	Bits	Mnemonic
1	while (nextbits() == '1'){		
2	**extra_bit_picture** /* with value 1 */	1	uimsbf
3	**extra_information_picture**	8	uimsbf
4	}		
	extra_bit_picture /* with value 0 */	1	uimsbf

5.18 Suppose that it was decided to always transmit the DC DCT coefficients in all blocks (as opposed to just intrablocks).

(a) Rewrite the block syntax shown in Table 5.19 to reflect this change.

(b) Would this change have any impact on any of the higher level syntax elements?

5.19 If we are to be able to decode a bit stream stored in a file using MATLAB, we need to be able to read bits from the file. The MATLAB command given below reads one byte from the file defined by fid.

```
fread(fid,1,'uchar');
```

Develop a MATLAB function bit_in(length) that will read a code word of "length" bits from the file. For example, suppose that the file contained the first four bytes shown in Figure 5.26.

00011100 10010100 01110011 10010010

Figure 5.26 Data for Problem 5.19.

The succession of function calls listed below would then return the binary code words shown in Figure 5.27.

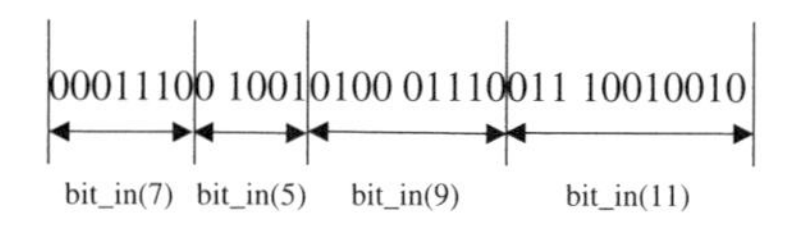

Figure 5.27 Reading variable length code words from a file.

bit_in(7) would return 0001110
bin_in(5) would return 01001
bit_in(9) would return 010001110
bit_in(11) would return 01110010010
This is illustrated in Figure 5.27.

5.20 An important part of the video syntax is the nextbits() function that allows the next several bits of the bit stream to be checked for a particular value. Write and test a MATLAB function that will carry out this function. Note that your routine needs to be able to check the next several bits in the bit stream while still leaving the bits in the bit stream to be subsequently read by another part of the syntax if required. Line 15 of the block syntax shows an example of where this might be necessary. Note that the MATLAB functions fseek and ftell might prove useful in carrying out this task.

MATLAB EXERCISE 5.1: EFFICIENT CODING OF MOTION VECTOR INFORMATION

In this chapter we saw an example of how the coding of motion vectors using horizontal prediction. In this exercise, you will verify that these results are correct and study the efficiency of this approach for other video sequences.

Section 1 Coding of motion vectors for a video sequence

(a) Load the first and fifth picture from a video sequence. Form the motion-compensated prediction of the fifth picture using the first picture as the reference picture using a search range of ± 8 both horizontally and vertically. Save the motion vectors generated.

(b) Display the histogram of each component of the motion vector information.

(c) Predict each component of each motion vector by the same component of the motion vector immediately to its left. The first motion vector of each row of macroblocks should be predicted using a zero motion vector. Draw the histogram of the prediction differences for each motion vector component.

(d) Compare the entropy of the original motion vector information and the entropy of the motion vector prediction differences. Comment on the result.

Section 2
Repeat Section 1 for pairs of pictures taken from other video sequences. Comment on the results achieved. What types of sequences achieve the biggest saving when motion-compensated prediction is employed?

Section 3
Extend the motion-compensated prediction process developed in Section 1 to include motion-compensated prediction to half pixel accuracy. Use your improved motion-compensated prediction scheme for various pairs of pictures from the sequences provided. Comment on the performance compared to motion-compensated prediction to single pixel accuracy.

MATLAB EXERCISE 5.2: A SIMPLE VIDEO ENCODER

In this exercise you are required to write an encoder that can be used to code pictures in a video sequence. For simplicity, we will code every macroblock in every picture in intramode. Although not terribly effective from a coding efficiency point of view, this exercise will give you experience in working with a reasonably detailed syntax. In running the encoder, you can assume the following.

(a) The shape of each pixel in the picture is square (because this is almost certainly the case for your computer monitor).

(b) For constant bit-rate operation, it is necessary to include rate control into the encoder. This is too complex in this case so constant quantizer, variable bit-rate operation should be employed.

(c) The quantizer step size is to remain the same for each picture in the sequence.
(d) If only luminance information is available, the chrominance blocks should be assumed to be always zero.

Use your encoder to encode a variety of short video sequences. Comment on the quality and bit rate requirement for each sequence over a range of quantizer step sizes.

MATLAB EXERCISE 5.3: A SIMPLE VIDEO DECODER

In this exercise you will decode a video sequence that has been encoded using the syntax described in this chapter. Given that the full syntax presented is quite complex, we will simplify the task by assuming that all macroblocks in all pictures are coded in intramode. This means that there will be only one set of variable length code words in the bit stream for the (run,level) pairs of quantized DCT coefficients. The variable word length codes for motion vector differences, macroblock increment, and coded block pattern are not needed in intra macroblocks. The bit stream to be decoded is one of the bitstreams that you generated in MATLAB Exercise 5.2.

Display the decoded sequence on your computer monitor using the MATLAB movie function.

MATLAB EXERCISE 5.4: A VIDEO ENCODER

In this exercise you are required to write an encoder that can be used to code pictures in a video sequence. In running the encoder, you can assume the following.

(a) The first picture in the sequence is coded in intramode with all the remaining pictures coded using motion-compensated prediction from the previous picture.
(b) All macroblocks in a picture that uses motion-compensated prediction are coded using motion-compensated prediction (i.e., intramode and intermode are not used for these pictures).
(c) The shape of each pixel in the picture is square (because this is almost certainly the case for your computer monitor).
(d) For constant bit-rate operation, it is necessary to include rate control into the encoder. This is too complex in this case so constant quantizer, variable bit-rate operation should be employed.
(e) The quantizer step size is to remain the same for each picture in the sequence.
(f) The motion vector search range should be ± 8 pixels both horizontally and vertically.
(g) If only luminance information is available, the chrominance blocks should be assumed to be always zero.

Use your encoder to encode a variety of short video sequences. Comment on the quality and bit rate requirement for each sequence over a range of quantizer step sizes.

This exercise is very challenging and requires significant effort.

MATLAB EXERCISE 5.5: A VIDEO DECODER

In this exercise you will decode a video sequence that has been encoded using the syntax described in this chapter. The bit stream to be decoded is one of the bit streams generated by your encoder in MATLAB Exercise 5.4. As might be expected, all macroblocks in the first picture of the sequence are encoded in intramode. Display the decoded sequence on your computer monitor using the MATLAB movie function.

Note that this exercise is quite challenging and requires significant effort.

MATLAB EXERCISE 5.6: INTRA/INTER/MOTION-COMPENSATED CODING OF MACROBLOCKS

In this chapter we saw that macroblocks can be coded in either intramode, intermode, or motion-compensated mode. It is, of course, necessary for the encoder to determine the appropriate manner of encoding for each macroblock. Like many encoder issues, this is not the subject of standardization because decoders only need to know the result of the choice and this is specified in the **macroblock_type** field. This gives encoder manufacturers the ability to differentiate their products in the marketplace because the more efficiently this choice is made, the better will be the performance of the encoder. In this exercise you will look at the success of a simple technique for deciding which of the three modes should be chosen. For this exercise, you will need first to have completed MATLAB Exercise 4.3.

The choice is made in two steps. First a decision is made whether to code in intermode or using motion compensation. Following this, a decision is made as to whether to use the inter/MC decision made in the first step or to use intracoding. The approach described was used in the development of the MPEG video standard.

Stage 1 Inter – motion-compensated decision

Calculate the average absolute difference for the luminance macroblock using the chosen motion vector (M) and using the zero motion vector (Z). The decision as to whether to use intermode or motion compensated mode is made according to Figure 5.28. Points on the line dividing the two regions are defined as being inter.

Stage 2 Intra – nonintra decision

The mode chosen in stage 1 (inter or MC) needs to be compared with intracoding. This is done by calculating the variance of the luminance information in the macroblock to be encoded (var_current) and the variance of the luminance difference macroblock produced after the appropriate prediction macroblock is subtracted (var_difference). Note that the macroblock mean luminance value needs to be

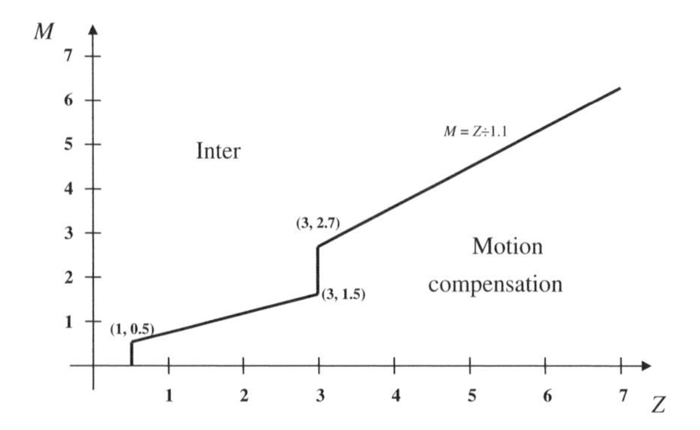

Figure 5.28 Inter motion-compensated decision.

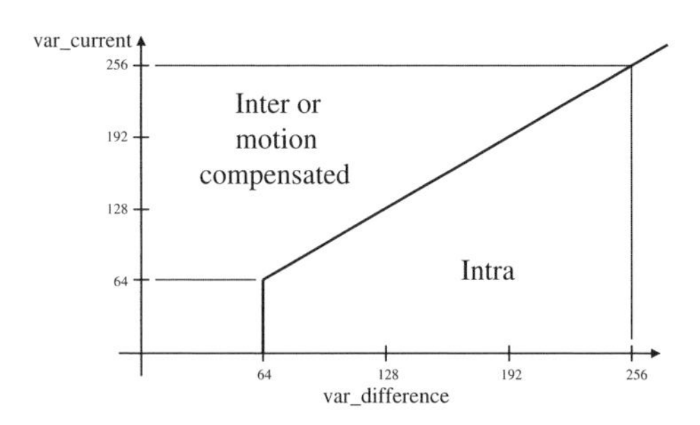

Figure 5.29 Intra – Inter/MC Decision.

subtracted from the current macroblock before var_current can be calculated. The difference macroblock is always assumed to have a zero mean for the calculation of var_difference. The decision between intra or inter/MC coding is then made according to Figure 5.29.

Modify the encoder developed in MATLAB Exercise 5.4 to incorporate this method of decision between inter – motion compensated – intra macroblocks. Apply the encoder to a variety of short video sequences. Comment on the change in performance compared to the encoder of MATLAB Exercise 5.4.

Chapter 6

The MPEG-2 Video Compression Standard

6.1. INTRODUCTION

In Chapter 5 we examined the syntax of a simple MPEG-like encoder and decoder. This coder is, in fact, quite close to an MPEG-1 video encoder. In this chapter we consider the additional signal processing and syntactical elements that are used in the full MPEG-2 video standard that forms a central part of all modern digital television systems. Although the aim of the previous chapter was to provide sufficient information to allow the reader to implement an encoder and a decoder, the complexity of the full MPEG-2 video standard is such that this aim is beyond the reasonable expectations of this chapter. The chapter aims instead to give the reader a full understanding of the capabilities, power, and flexibility of the MPEG-2 video standard. Armed with a full understanding of the information provided in this chapter as well as in Chapter 5 on syntax, it should be possible for a reader to consult the relevant international standard[1] and carry out a full implementation of a compliant encoder or decoder.

After a brief introduction to the MPEG-2 video standard, we introduce the picture types used in the MPEG-2 standard. Then follows a detailed discussion of the syntax of MPEG-2. The discussion is limited to those elements of the standard that are used in current digital television applications. We then look at the video buffer verifier (VBV), which is important in ensuring the compliance of any MPEG-2 video encoder. Finally, we discuss the currently defined profiles and levels for MPEG-2.

The development of MPEG-2 followed and built on the successful earlier development of the MPEG-1 suite of standards. MPEG-1 was designed to code moving pictures and the associated audio for digital storage media applications at about 1.5 Mbit/s.[2] The primary limitation of MPEG-1 was that it was not designed to handle interlaced material. This does not mean that it could not code interlaced material—just that a specification that included specific coding tools to deal with

[1] ISO/IEC 13818-2.

[2] Although optimised for operation at around 1.5 Mbit/s, MPEG-1 can operate at rates up to 100 Mbit/s and so meet the requirements for many applications.

Digital Television, by John Arnold, Michael Frater and Mark Pickering.

interlace would provide a better quality of service at a given bit rate than MPEG-1. The most important difference between MPEG-1 and MPEG-2 is the introduction of these coding tools designed to handle interlace. MPEG-2 was also designed with higher quality applications in mind than MPEG-1,[3] and this led to the inclusion of a number of improved coding techniques as well as extensions to the approach used in MPEG-1. Finally, the likelihood of data errors when information is read from digital storage media is low. However, transmission errors are more common in terrestrial broadcast digital television applications. Noting this, MPEG-2 gave special consideration to error resilience in the development of the standard.

The MPEG-2 standard is designed to be extremely flexible and thus be able to handle a wide variety of possible applications. For example, the standard allows for pictures made up of more than $16{,}000 \times 16{,}000$ pixels and transmitted at bit rates in excess of 100 Gbit/s. For an encoder or a decoder to be able to claim full compliance with the standard, it would need to be able to handle these picture sizes and bit rates. For digital television applications, it is known that the picture sizes and bit rates are far less than the maximum allowed by the standard. The maximum picture size defines the size of the internal frame stores[4] that the encoder requires whereas the bit rate defines the speed at which the decoder front end (i.e., the variable-length code (VLC) decoder) needs to operate. Requiring full compliance with the standard would therefore greatly increase the cost of encoders and, more importantly, of decoders that are required by every user.

Recognizing this problem, the committee that designed the MPEG-2 video standard has provided a number of less complex compliance points that are suitable for a wide range of applications. Using these compliance points, which are defined by profiles and levels, it is possible to design encoders and decoders that comply with the standard but require only the hardware complexity necessary to service a particular application. Further details on the currently defined profiles and levels of MPEG-2 are provided later in this chapter after the definition of the MPEG-2 syntax.

MPEG-2 video is designed to code both interlaced and progressive material. For progressive material, all information is coded as frame pictures. For interlaced material, information can be coded as a frame picture (i.e., two fields merged to form a single frame) or as two separate field pictures. The two field pictures are referred to as the top field picture and the bottom field picture. Each line of the top field picture is spatially located immediately above the corresponding line in the bottom field picture. For compatibility with interlaced displays, it is required that the bottom field picture always follows the top field picture and the top field picture always follows the bottom field picture after decoding.

The hierarchy of layers in the MPEG-2 syntax is shown in Figure 6.1. The only change in the structure used in Chapter 5 is the introduction of the group of pictures (GOP) layer. This is a compulsory layer in MPEG-1 although optional in MPEG-2. Its functions are described later in the chapter.

[3] It was envisaged that a standard-definition television would require a bit rate around 4 Mbit/s and a high-definition television around 20 Mbit/s.

[4] Although it may seem sensible to refer to these stores as picture stores, they are more commonly referred to as frame stores and so this terminology is used.

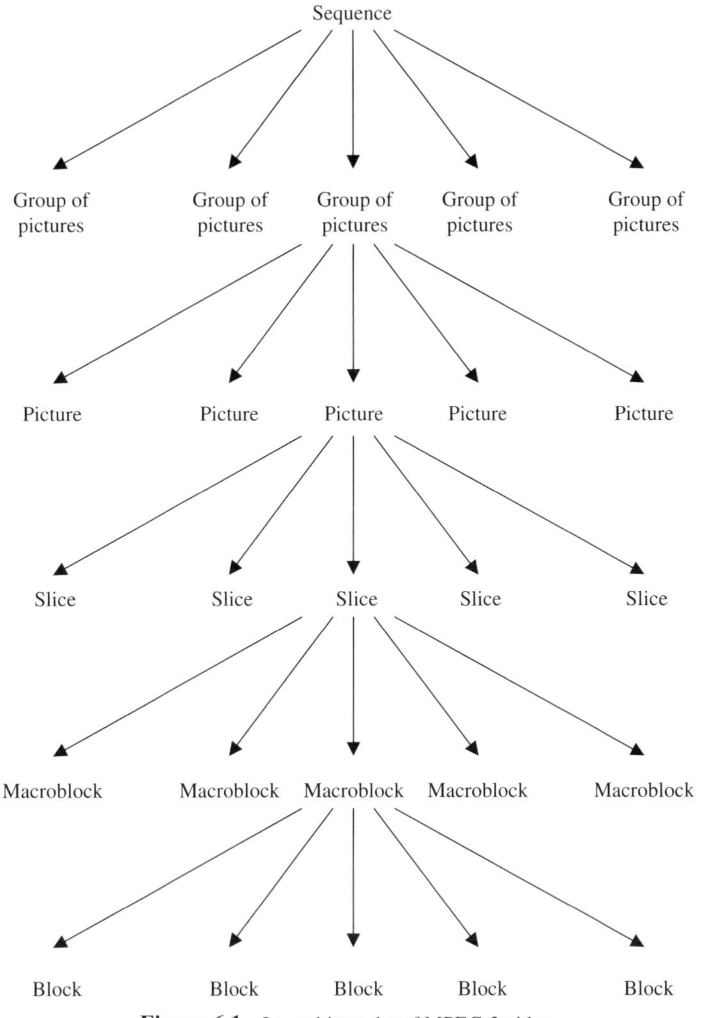

Figure 6.1 Layer hierarchy of MPEG-2 video.

6.2. PICTURE TYPES IN MPEG-2

Although a number of new coding tools are introduced in this chapter at the layer in which they appear in the syntax, there is one new coding tool that is so important for an overall understanding of the standard and has an impact at so many different levels within the syntax that it needs to be introduced immediately. This is the picture type.

In the MPEG-2 video standard, pictures are classified as one of the three types. The first of these is an *intrapicture* (I picture). In an I picture, all macroblocks are coded as *intra macroblocks* (i.e., without any prediction from any other picture).

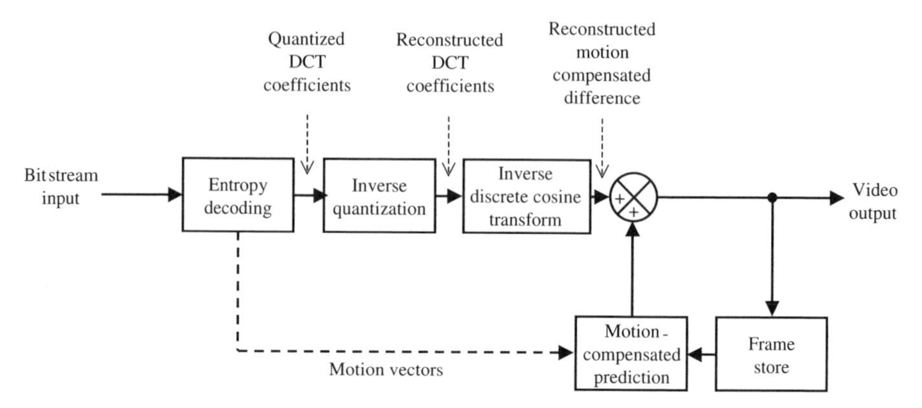

Figure 6.2 Motion-compensated DCT decoder.

This means that an I picture usually requires a large amount of data to represent it. However, in digital television applications, the regular inclusion of I pictures is essential. The reason for this becomes clear when we consider the architecture of the video decoder shown in Figure 6.2. Consider the situation when a user changes from one channel to another. For convenience, assume that the change of channel is forced to occur on a picture boundary. For the first picture after the switch, the frame store in Figure 6.2 contains a reconstruction of the last picture received from the previous channel. However, the bit stream provides coded difference information from the new channel. The video output is therefore of very poor quality, and it takes a long time before a reasonable quality video service is restored. This problem can be overcome if the first picture received from the new channel is an I picture, as I pictures can be decoded without reference to the current contents of the frame store. The decoded I picture then replaces the current contents of the frame store. Subsequent pictures can then be successfully predicted in the normal way using motion-compensated prediction.

Decoder set-top boxes are therefore invariably constrained to change to a new channel only when an I picture is being received from the new channel. This implies that I pictures need to occur regularly within the bit stream as a large I picture spacing would mean that there would be a significant delay before a change from one channel to another could occur. This would be unacceptable to users. Typically,[5] the spacing between I pictures is around half a second. This means that a channel change can occur on average after a quarter of second and the upper threshold on the delay is half a second. Such a delay is unlikely to be noticed by users. I pictures are also useful for error resilience as they remove the effect of any transmission errors from the reconstructed picture in the frame store. As mentioned earlier, I pictures generate a large amount of data and so the inclusion of I pictures decreases the video quality of a given service bit rate.

[5] This is not a matter of standardization but is left up to the service provider.

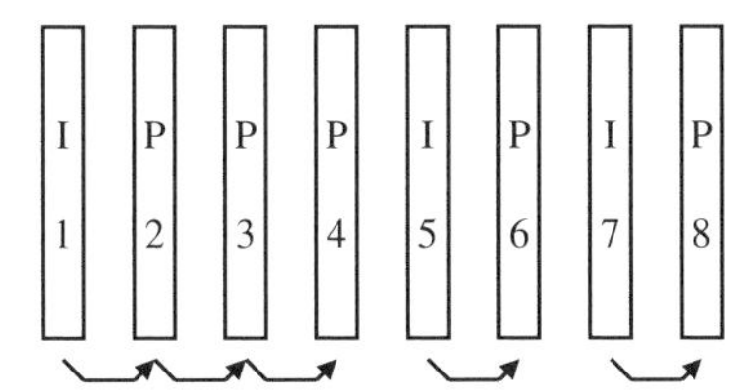

Figure 6.3 Video sequence comprising I and P pictures.

The second type of picture is the *predicted picture* (P picture). P pictures are coded using motion-compensated prediction from an immediately past I picture or an immediately past P picture. This is referred to as forward motion-compensated prediction. Although every macroblock in a P picture can be predicted using forward motion-compensated prediction, there is no requirement for every macroblock to be predicted using forward motion-compensated prediction. Macroblocks can be coded using interprediction (i.e., forming the difference between the current macroblock and the macroblock in the same position in the reference picture, which is useful when there has been little change between pictures) or intraprediction (i.e., coding without using any form of temporal prediction, which is useful for macroblocks where no suitable prediction is possible from the previous picture). A video sequence made up of I and P pictures is shown in Figure 6.3. The arrows illustrate the forward prediction. Each P picture is predicted from the I or P picture that immediately precedes it, and I pictures do not use prediction. P pictures can usually be represented by a significantly smaller amount of data than I pictures.

The final type of picture is the *bidirectionally predicted picture* (B picture). B pictures are coded using motion-compensated prediction from an immediately past I or P picture (forward prediction), an immediately future I or P picture (backward prediction), or an average of the prediction from both immediately future I or P picture and immediately past I or P picture (interpolated prediction). Thus, if three successive pictures in a video sequence are coded as an I picture followed by a B picture followed by a P picture, then the P picture is predicted from the reconstruction of the I picture and then the B picture is predicted from the reconstructions of both the I picture and the P picture. This situation is illustrated in Figure 6.4.

In Figure 6.4, the B picture (2) is predicted from the preceding I picture (1) and the following P picture (3). As motion-compensated prediction is based on reconstructed pictures, the B picture cannot be coded until after the P picture is coded. This means that the display order of the pictures (1, 2, 3) is different from the

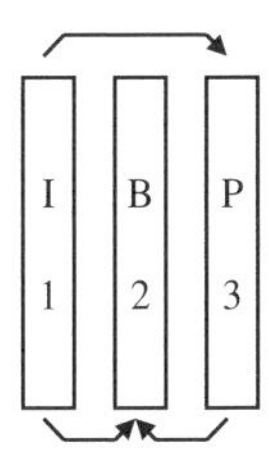

Figure 6.4 Prediction for an I picture followed by a B picture followed by a P picture.

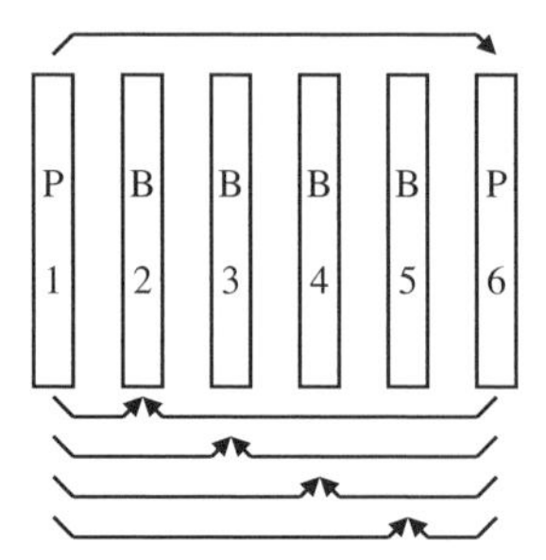

Figure 6.5 A video sequence containing several successive B pictures.

order of encoding and decoding (1, 3, 2) and that the encoder must store the original B picture until after the following P picture is coded.

The situation becomes even more complex when several B pictures lie between the I or P pictures used to form the prediction. This situation is shown in Figure 6.5, where four B pictures are predicted from a pair of P pictures.

In this case, the four pictures to be coded as B pictures would need to be stored in the encoder until the second P picture has arrived and been encoded. In digital television applications, there is only one encoder transmitting to a very large number of decoders and so the added memory requirement at the encoder is not a serious problem.

Initially, it might seem sensible to rearrange the coded pictures back into the correct display order before transmission from the encoder. For this to be done, it would be necessary for the encoder to internally store the bit-stream representation of the future reference picture (I or P) for transmission after the bit-stream representation of the B pictures. This would add slightly to encoder complexity.

At the decoder, if the coded pictures are transmitted in their original (i.e., display) order, then the picture (I or P) preceding the B pictures is received and can be immediately decoded as its reference picture (a previous I or P picture) occurred earlier in the bit stream and so has already been received by the decoder. The data representing the four B pictures need to be held at the decoder until after the arrival of the second reference picture (again I or P). After that reference picture has been decoded, the four B pictures that preceded it can finally be decoded.

However, there is a simpler approach that can be employed, which eases the task for both the encoder and the decoder: Pictures are transmitted in the same order that they are encoded. The encoder has to wait for the arrival of the second reference picture (I or P) before it can encode any B pictures that precede it. When the second reference picture is encoded, then it is transmitted immediately followed by the encoded B pictures that occurred immediately before it. This means that the encoder no longer needs to store the coded bit-stream representation of the future reference picture.

When a B picture arrives at the decoder, both of the reference pictures needed for its prediction have already arrived. Thus, when a coded B picture arrives, it can immediately be decoded. There is never a need to store the data representing a B picture at the decoder while a later reference picture is awaited. This means that for video sequences employing B pictures, the picture transmission order is different from the picture display order. The picture order rearrangement done by the encoder

1	2	3	4	5	6	7	8	9	10	11	12	13	14	15	16
I	B	B	P	B	B	P	B	B	P	B	B	P	B	B	I

Figure 6.6 Pictures to be transmitted in display order with picture types shown.

is easily undone by the decoder, thereby ensuring that pictures are always displayed in the correct order.

Consider the case of 16 pictures that are to be coded with the picture types shown in Figure 6.6.

The transmission order after encoding is shown in Figure 6.7. The different sizes used for each picture are to indicate that I pictures usually require the largest amount of data to represent them, whereas P pictures require less data and B pictures the smallest amount of data.

The decoder requires just two frame stores (each able to store one decoded picture) irrespective of the number of consecutive B pictures that occur in the video sequence. The decoding procedure is shown in Table 6.1.

Every time an I or a P picture arrives at the decoder, it is decoded and overwrites the oldest I or P picture currently stored in the pair of frame stores. When a B picture arrives, the two reference pictures from which it was predicted are already available in the decoder's frame stores and so the B picture can be immediately decoded. The process of reordering the received pictures into the correct order for display is therefore straightforward.

The use of B pictures greatly increases the overall coding efficiency since motion-compensated prediction is almost always possible. There are a number of situations where forward-only motion compensation does not work effectively. We now consider how the use of bidirectional motion-compensated prediction effects these situations.

Forward-only motion-compensated prediction is ineffective at a scene cut because the reference picture bears little or no resemblance to the current picture to be coded. With bidirectional prediction, the reference picture used for forward prediction is still of little use. However, the picture used for backward prediction is highly likely to be suitable for motion-compensated prediction.

When a foreground object moves and uncovers a previously hidden background material, forward-only motion-compensated prediction is ineffective since a match to the uncovered background cannot be found in the past reference picture. With bi-directional motion-compensated prediction, performance is improved, as it is likely that the background is still uncovered in the future reference picture.

I	P	B	B	P	B	B	P	B	B	P	B	B	I	B	B
1	4	2	3	7	5	6	10	8	9	13	11	12	16	14	15

Figure 6.7 Transmission order of pictures shown in Figure 6.6.

Table 6.1 Decoding process for bit stream of Figure 6.6.

Picture decoded and picture type	Contents of frame store 1	Contents of frame store 2	Picture to be displayed
1 (I)	1	—	
4 (P)	1	4	1
2 (B)	1	4	2
3 (B)	1	4	3
7 (P)	7	4	4
5 (B)	7	4	5
6 (B)	7	4	6
10 (P)	7	10	7
8 (B)	7	10	8
9 (B)	7	10	9
13 (P)	13	10	10
11 (B)	13	10	11
12 (B)	13	10	12
16 (I)	13	16	13
14 (B)	13	16	14
15 (B)	13	16	15

When a new material enters a picture (e.g., a new object or new background due for example to camera movement), forward-only motion-compensated prediction is ineffective in predicting this new material since it is not available in the past reference picture. With bidirectional motion-compensated prediction, performance improves, as it is likely that the new material can still be seen in the future reference picture.

If an object moves between pictures by an amount greater than the search area, then forward-only motion-compensated prediction is not effective. Unless the object slows down, this problem also exists for bidirectional motion-compensated prediction and can only be addressed by increasing the size of the search area.

A further advantage of B pictures is that they are not used to predict other pictures in the video stream. This means that if a B picture is quantized coarsely, it will have no impact on the efficiency of coding the subsequent pictures in the sequence. Coding I pictures or P pictures coarsely would mean that the subsequent pictures would likely require additional bits as the quality of the motion-compensated prediction from the coarsely coded I picture or P picture would be reduced.

It is clear that bidirectional motion-compensated prediction can improve coding performance in a number of situations. It needs to be remembered that these improvements do come with some significant costs. The first of these costs is that all pictures to be encoded using motion-compensated prediction (i.e., P or B pictures) and the relevant reference picture or pictures are often separated by several picture times when B pictures are used. This means that the moving objects have moved by a greater amount than would have been the case if the current and reference pictures were adjacent. Larger search areas are therefore required with a consequential increase in computational requirements at the encoder. The wider spacing also leads to a decrease in coding efficiency. This is a particular problem for P pictures since the separation between a P picture and its reference picture is greater than the separation between an associated B picture and its reference pictures. This is shown in Figure 6.6.

In addition, for B pictures, motion-compensated prediction needs to be performed twice—once in the forward direction and once in the backward direction. A decision also needs to be made as to whether to use forward prediction, backward prediction, or an average of the two to code the current macroblock. This decision needs to be signaled to the decoder together with the appropriate motion vector(s). If the average of the two predictions is to be used, then two motion vectors need to be transmitted.

A further problem introduced by the use of B pictures is an increase in delay. This problem arises because a B picture cannot be transmitted until the future I or P picture from which it is to be predicted has arrived and been encoded. If there are several consecutive B pictures, there can be several picture times. Although not a problem for distribution television services, this can be a serious problem in two-way interpersonal communication services where round trip delay is very important. For these latter applications, B pictures are often not used.

6.3. THE SYNTAX OF MPEG-2

We now discuss the special coding tools and semantic definitions that are used in the full MPEG-2 video coding standard. These build directly on the tools and semantics described in Chapter 5. As in the previous chapter, our discussions are ordered by the structure of the video bit stream.

Like the syntax described in Chapter 5, MPEG-2 makes use of start codes to allow resynchronization within the bit stream. Each start code begins with the 24-bit value 0x000001 followed by the appropriate start-code type as shown in Table 6.2.

Table 6.2 MPEG-2 start codes.

Type of start code	Start-code values
picture_start_code	0x00
slice_start_code	0x01 to 0xAF
Reserved	0xB0 to 0xB1
user_data_start_code	0xB2
sequence_header_code	0xB3
sequence_error_code	0xB4
extension_start_code	0xB5
Reserved	0xB6
sequence_end_code	0xB7
group_start_code	0xB8
Reserved	0xB9 to 0xFF[a]

[a] These start code values are used in MPEG-2 system specification.

MPEG-2 also makes use of extension start codes to expand the number of headers available. Before moving to a full discussion of the syntax, we first introduce the definition and functions of extension start codes.

6.3.1. Extension Start Code and Extension Data

Extension start codes identify extensions of the video syntax beyond those elements that were provided in the MPEG-1 video syntax. Like all start codes, extension start codes begin with the 24-bit pattern 0x000001 followed by the 8-bit start-code identifier 0xB5. The type of extension is identified by a 4-bit **extension_start_code_identifier** field that immediately follows the extension start code and is defined in Table 6.3.

Extensions considered in this chapter include sequence extension, sequence display extension, quantizer matrix extension, copyright extension, picture display extension, and picture coding extension. Other currently defined extensions are provided for the various scalable[6] modes of MPEG-2. As these are not used in any current digital television applications, they are not considered in detail here.

We are now in a position to discuss the MPEG-2 syntax. We begin at the top layer called the sequence layer and then move through the group of pictures, picture,

Table 6.3 Codes for **extension_start_code_identifier**.

extension_start_code_identifier	Extension data type
0000	Reserved
0001	Sequence extension
0010	Sequence display extension
0011	Quantizer matrix extension
0100	Copyright extension
0101	Sequence scalable extension
0110	Reserved
0111	Picture display extension
1000	Picture coding extension
1001	Picture spatial scalable extension
1010	Picture temporal scalable extension
1011	Reserved
1100	Reserved
1101	Reserved
1110	Reserved
1111	Reserved

[6] Scalability in video compression is the ability to code a sequence into several bit streams. Decoding just the base layer bit stream gives a basic service whereas decoding the base layer bit stream plus one or more of the enhancement bit streams provides an improvement in the received service (e.g., higher quality, higher spatial resolution, higher frame rate).

slice, macroblock, and block levels. Notice that the MPEG-2 syntax is really just a superset of the syntax described in Chapter 5.

6.3.2. Sequence Layer

The complete syntax for a video sequence in MPEG-2 is shown in Table 6.4.

For a new video sequence, the first step is to search for a **start_code** (line 2) that should be a sequence_header() (line 3) as discussed in Section 6.3.2.1. If this is followed by an **extension_start_code** (line 4), then the bit stream is compliant with the MPEG-2 video standard. Otherwise, it is compliant with the MPEG-1 video standard (line 24).[7] For MPEG-2 video, the next item in the bit stream is the sequence_extension() (line 5) (Section 6.3.2.2). This is followed by extension_and_user_data(0)

Table 6.4 Video sequence syntax for MPEG-2 video.

	Syntax	Bits	Mnemonic
1	video_sequence() {		
2	next_start_code()		
3	sequence_header()		
4	if(nextbits() = = extension_start_code) {		
5	sequence_extension()		
6	do {		
7	extension_and_user_data(0)		
8	do {		
9	if(nextbits() = = group_start_code) {		
10	group_of_pictures_header()		
11	extension_and_user_data(1)		
12	}		
13	picture_header()		
14	picture_coding_extension()		
15	extension_and_user_data(2)		
16	picture_data()		
17	} while ((nextbits() = = picture_start_code)||(nextbits() == group_start_code))		
18	if (nextbits() != sequence_end_code) {		
19	sequence_header()		
20	sequence_extension()		
21	}		
22	} while (nextbits() != sequence_end_code)		
23	} else {		
24	/* ISO/IEC 11172-2 MPEG-1*/		
25	}		
26	**sequence_end_code**	32	bslbf
27	}		

[7] All MPEG-2 video decoders need to be able to decode MPEG-1 encoded video (ISO/IEC 11172-2).

(line 7), which can include user data as discussed in Section 6.3.2.3 or a sequence_display_extension() as discussed in Section 6.3.2.4. Next in the bit stream there can be a **group_start_code** indicating the presence of a group_of_pictures_header() followed by extension_and_user(1) (lines 10–11) as discussed in Sections 6.3.3.1 and 6.3.3.2. With or without the group of pictures information, next follows picture_header() (Section 6.3.4.1), picture_coding_extension() (Section 6.3.4.2), extension_and_user_data(2) (Sections 6.3.4.3–6.3.4.5) followed by the coded slice and macroblock information in picture data() (lines 13–16) (Sections 6.3.5–6.3.7).

Groups of pictures and pictures continue until a start code arrives, which is neither a **group_start_code** nor a **picture_start_code** (line 17). At this point, if the next start code is not a **sequence_end_code**, then a new sequence_header() and sequence_header_extension() are read (lines 18–20) and the loop starting at line 6 is repeated. Alternatively, a **sequence_end_code** occurs (line 26) indicating the end of the current sequence. As the chapter unfolds, we look in more detail at each of these major structures as well as the elements that make them up.

6.3.2.1. The Sequence Header

In MPEG-2, information about a video sequence is contained in the sequence_header(), the sequence_extension(), the sequence_display_extension(), and the extension_and_user_data(0).[8] The sequence_header() is designed to be similar to the sequence header used in MPEG-1. The additional information required for MPEG-2 sequences (e.g., greater number of pixels in a line or lines in a picture, higher bit rates, wider range of frame rates) is achieved by including extension data in the sequence_extension() header with the least significant bits (LSB) of the parameter contained in the sequence_header() and the most significant bits (MSB) in the sequence_extension(). For convenience, in this section we consider those parts of the sequence_extension() that augment values in the sequence_header().

The next section on the sequence_extension() will only cover the new elements contained in the sequence_extension().

The sequence_header() begins with a **sequence_header_code**, which is a 32-bit start code. Other parameters fully or partly defined in the sequence_header() are listed below.

Horizontal and Vertical Picture Size The picture size in pixels is defined by the variables **horizontal_size** and **vertical_size**. Both are 14-bit values of which the 12 LSB (**horizontal_size value** and **vertical_size_value**) are contained in the sequence_header() and the two MSB are contained in the sequence_extension(). The maximum allowed horizontal and vertical picture sizes for pictures in MPEG-2 are thus in excess of 16,000 pixels. In order to avoid start-code emulation, the values of **horizontal_size value** and **vertical_size_value** cannot be zero. This implies

[8] This structure has been used to maintain compatibility with MPEG-1 standard that has only a sequence header.

that values of **horizontal_size** and **vertical_size** that are multiples of 4096 are not allowed.

Picture Aspect Ratio Aspect ratio information is identical to that defined in Chapter 5.

Frame Rate The 4-bit **frame_rate_code** in the sequence header serves the same purpose as described in Chapter 5 in defining the **frame_rate_value**. In the full MPEG-2 standard, the range of possible frame rates can be expanded using the 2-bit **frame_rate_extension_n** field (range 0–3) and the 5-bit **frame_rate_extension_d** field (range 0–31) defined in the sequence_extension(). The equation below shows how these can be used to vary the frame rate.

$$\text{frame\_rate} = \text{frame\_rate\_value} \times \frac{\text{frame\_rate\_extension\_}n + 1}{\text{frame\_rate\_extension\_}d + 1}$$

Although a part of the full MPEG-2 specification, the existing profiles do not make use of **frame_rate_extension_n** and **frame_rate_extension_d**, both of which must take the value zero.

Sequence Bit Rate The bit rate of the sequence is defined by a 30-bit value **bit_rate** with the 18 LSB (**bit_rate_value**) in the sequence_header() and the 12 MSB (**bit_rate_extension**) in the sequence_extension(). The bit rate is defined in multiples of 400 bits/s. It is therefore possible to specify bit rates in excess of 400 Gbits/s for MPEG-2 video bit streams. The variable **bit_rate** may not take the value zero and so the allowed range is 400–400 $\times$ (2^{30} − 1) bits/s.

Video Buffer Verifier Buffer Size The **vbv_buffer_size** variable defines the size of the video buffer verifier rate control buffer. The role of this buffer is described further in Section 6.4. It is an 18-bit field with the 10 LSB (**vbv_buffer_size_value**) in the sequence_header() and the eight MSB (**vbv_buffer_size_extension**) in the sequence_extension(). It specifies the minimum size rate control buffer needed by a decoder to be able to successfully decode the video sequence. The buffer size in bits is derived from **vbv_buffer_size** according to the equation given below.

$$\text{VBV\_BufferSize} = 16 \times 1024 \times \text{vbv\_buffer\_size}$$

The larger the buffer, the better the smoothing and the higher the video quality for a given bit rate. However, both delay and cost increase as the buffer size increases. The maximum buffer size is in excess of 4 Gbits (range 0 to $16 \times 1024 \times (2^{18} - 1)$).

Constrained Parameter Flag The constrained parameter flag was used in the specification of MPEG-1 to perform a limited but similar function to that achieved by the definition of levels in MPEG-2. For MPEG-2 video bit streams, it must be set to zero.

8	16	19	22	26	27	29	34
16	16	22	24	27	29	34	37
19	22	26	27	29	34	34	38
22	22	26	27	29	34	37	40
22	26	27	29	32	35	40	48
26	27	29	32	35	40	48	58
26	27	29	34	38	46	56	69
27	29	35	38	46	56	69	83

Figure 6.8 Default quantizer matrix for intrablocks. © This Figure is based on AS/NZS 13818.2:2002. Permission to reprint has been granted by SAI Global Ltd. The standard can be purchased online at http://www.sai-global.com.

Quantizer Matrix Definition In the MPEG video compression standards, DCT coefficients can be quantized using different quantizers depending upon their relative frequency using the method described in Section 4.4.3. This allows the quantization to better match the characteristics of the human visual system. The quantizer matrix definition allows the user to specify how this quantization is performed.

The quantizer matrices given in Figures 6.8 and 6.9 are the defaults defined in the standard for intrablocks and inter (including motion-compensated) blocks, respectively. They are used if no other quantizer matrices are specified in the bit stream and apply equally to luminance and chrominance blocks. However, it is possible to download new quantizer matrices both at the sequence and picture layers. At the sequence layer, a 1-bit flag is used to indicate whether a new intrablock quantizer matrix or a new interblock quantizer matrix is to be transmitted as part of the sequence header. If a flag is set to one, it is immediately followed by the new quantizer matrix represented by sixty-four 8-bit numbers ordered in the default zig-zag scan order used for

16	16	16	16	16	16	16	16
16	16	16	16	16	16	16	16
16	16	16	16	16	16	16	16
16	16	16	16	16	16	16	16
16	16	16	16	16	16	16	16
16	16	16	16	16	16	16	16
16	16	16	16	16	16	16	16
16	16	16	16	16	16	16	16

Figure 6.9 Default quantizer matrix for interblocks. © This Figure is based on AS/NZS 13818.2:2002. Permission to reprint has been granted by SAI Global Ltd. The standard can be purchased online at http://www.sai-global.com.

Table 6.5 Profile indication.

Profile indication	Profile
111	Reserved
110	Reserved
101	Simple
100	Main
011	SNR scalable
010	Spatially scalable
001	High
000	Reserved

DCT coefficients. This means that the values in the quantizer matrix need to be in the range of 1–255. A zero quantizer value is, of course, not allowed. Any quantizer matrix downloaded at the sequence layer applies to both luminance and chrominance blocks.

6.3.2.2. The Sequence Extension

As already described in Section 6.3.2.1, a number of parameters in the sequence_header() are defined by entries in both the sequence_header() and the sequence_extension(). Several other parameters are defined only in the sequence_extension() and so are relevant to MPEG-2 video bit streams but not to MPEG-1 video bit streams.

Profile and Level Indication An 8-bit **profile_and_level_indication** field defines the profile and level of the current video sequence. For profiles and levels[9] defined in the original MPEG-2 standard, which includes all of the profiles and levels used in digital television applications, the MSB in this field (called an escape bit) is 0 and is followed by a 3-bit profile indication and finally a 4-bit level indication. The profile indication field is defined in Table 6.5, whereas the level indication is shown in Table 6.6. Other profiles have been defined subsequently and have been defined with the escape bit set to 1. We will not consider these further here as they do not relate to digital television services. Further discussion of currently defined profiles and levels is included later in this chapter.

Progressive Sequence The 1-bit **progressive_sequence** field defines that the sequence is progressive (i.e., each picture is fully imaged at the same time) when set to one. Otherwise, the sequence is interlaced (i.e., each picture is imaged twice to produce two fields separated in time by half the period between pictures). Most current digital television applications use interlaced sequences.

[9] Profiles define a subset of the complete syntax of the standard, whereas levels place constraints on the values that may be taken by parameters in the bit stream. See Section 6.5 for more details.

Table 6.6 Level indication.

Level indication	Level
1011 to 1111	Reserved
1010	Low
1001	Reserved
1000	Main
0111	Reserved
0110	High 1440
0101	Reserved
0100	High
0000 to 0011	Reserved

Chrominance Format MPEG-2 allows three chrominance formats: 4:2:0, 4:2:2, and 4:4:4. The format used in the current video sequence is signaled in the 2-bit **chroma_format** field in the sequence_extension(). All current distribution digital television applications use the 4:2:0 chrominance format and so this field takes the value 01.

Low Delay The 1-bit **low_delay** flag in the sequence extension indicates that the video sequence is a low-delay sequence and therefore does not contain B pictures that, as explained earlier, introduce delay due to the need for picture reordering after decoding.

6.3.2.3. User Data at the Sequence Layer

User data can be specified at the sequence layer and is preceded by a start code and a unique start-code value as shown in Table 6.1. Then follows the user data in 8-bit bytes until a new start code is encountered. Because start-code emulation must be avoided, user data must not contain a string of 23 or more consecutive zeros. As the name implies, user data is user defined for particular applications. As is shown later, user data can also be present at lower levels of the syntax.

6.3.2.4. The Sequence Display Extension

The final type of information that is available at the sequence layer is the sequence_display_extension(). It is important to realize that the MPEG-2 video specification does not define the display process. Thus, the information contained in the sequence_display_extension() may be helpful to some decoders while it may be completely ignored by others. All compliant decoders must be capable of decoding bit streams in which these values are defined. The elements contained in the sequence_display_extension() are described briefly below.

Video Format The 3-bit **video_format** field indicates the format of the analog video before coding. Possible formats include component, PAL, NTSC, SECAM, and MAC.

Color Description The 1-bit **color_description** flag when set to one indicates the presence of the following three elements next in the bit stream:

- **color_primaries**: an 8-bit number that defines the chromaticity coordinates of the source primaries.
- **transfer_characteristics**: an 8-bit number that specifies the optoelectronic characteristics of the source picture. This is commonly referred to as the gamma.
- **matrix_coefficients**: an 8-bit number that specifies the matrix coefficients used to derive the luminance and chrominance signals from the red, green, and blue primaries in the original source material.

Display Horizontal and Vertical Size The two 14-bit parameters **display_horizontal_size** and **display_vertical_size** define a rectangle that may be considered as the active region of the intended display. If the rectangle is smaller than the decoded picture size, the display would be expected to show only a part of the decoded picture. Alternatively, if the rectangle is larger than the decoded picture size, the display would be expected to show the decoded picture in a portion rather than all of the display.

6.3.3. The Group of Pictures Layer

A *group of pictures* is defined as one or more coded pictures with the first coded picture (in transmission order) as an intracoded picture. The situation is illustrated in Figure 6.10, which shows the position of the group_of_pictures_header() for a sequence in both transmission and display order.

A GOP can contain more than one I picture as well as any number of P pictures and B pictures. The last coded picture (in display order) of a GOP is either an I picture or a P picture. The GOP layer provides a convenient structure for random access into a video sequence. It was developed as part of the MPEG-1 video standard that was designed to handle video from digital storage media such as CDs. Clearly, random access to the video material is essential for such applications. In the MPEG-2 video standard, the use of the group of pictures layer is optional.

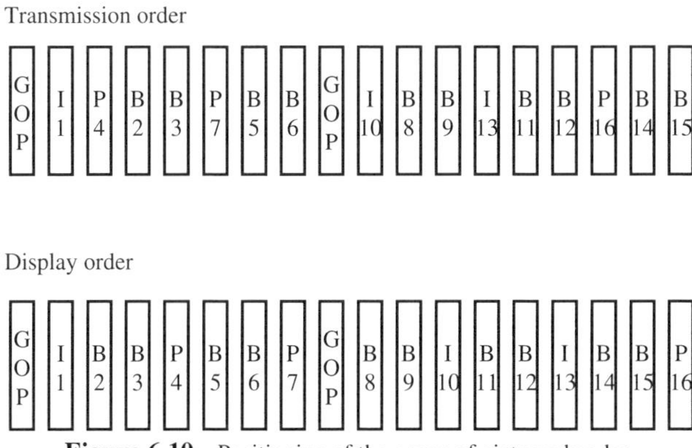

Figure 6.10 Positioning of the group of pictures header.

6.3.3.1. Group of Pictures Header

The group_of_pictures() header begins with a **group_start_code** as defined in Table 6.2. Other items defined in the group_of_pictures_header() are listed below.

Time Code The 25-bit **time_code** field is in the format used for the transmission of time and control codes for video tape recorders.[10] The information carried in this field plays no part in the video decoding process.

Closed GOP and Broken Link Flags As shown in Figure 6.10, in transmission order the first I picture in a GOP may be followed by a number of B pictures calculated using a reference picture in the previous GOP. In general, these B pictures need both the I picture in the current GOP and the last I picture or P picture from the previous GOP for correct decoding. In television applications, independent pieces of video material are often edited together. As a result of this editing, the previous GOP may no longer be present in the bit stream having been replaced by a GOP from a different sequence. The 1-bit **closed_gop** flag indicates that the B pictures that follow the first I picture in the GOP have been encoded in intramode or using motion-compensated prediction only from the first I picture of the GOP. They can therefore still be correctly decoded even if the previous GOP is no longer available in the bit stream. The 1-bit **broken_link** flag indicates that the B pictures following the first I picture in the GOP cannot be correctly decoded. It is set to one during the editing process if the previous GOP is removed unless the current GOP is a closed GOP.

6.3.3.2. User Data

User data can also be included in the GOP layer. It begins with the unique **user_data_start_code** and has a format identical to that used for user data at the sequence layer.

6.3.4. The Picture Layer

The picture layer includes the picture_header() and the picture_coding_extension(). It may also contain the optional extensions quant_matrix_extension(), picture_display_extension(), and copyright_extension().[11] The picture layer is organized in a similar manner to the sequence layer in that the picture_header() is almost identical to MPEG-1, whereas new additions at the picture layer required for MPEG-2 are contained in the picture_coding_extension(). The picture header begins with the 32-bit **picture_start_code**. Other items defined at the picture layer are listed below.

[10] See IEC Publication 461, Time and Control Code for Video Tape Recorders.

[11] Extensions for temporal scalability and spatial scalabilty can also occur at the picture layer. As these are not used for current distribution television applications, they are not considered here.

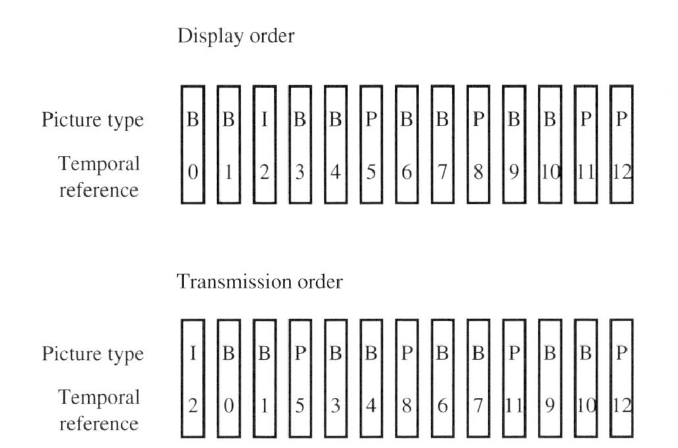

Figure 6.11 Use of the **temporal_reference** in display and transmission orders.

6.3.4.1. Picture Header

In this section we discuss parameters defined in the picture_header().

Temporal Reference The **temporal_reference** is a 10-bit integer that is associated with each coded picture and defines the display order of the pictures. This information can be useful to a decoder as pictures do not necessarily arrive at the decoder in display order when B pictures are used. The first picture in display order in each group of pictures has a **temporal_reference** value of zero, and the temporal reference is incremented as described below for each successive picture in the group of pictures.

For sequences that are not coded in low-delay mode (i.e., **low_delay** at the sequence layer is set to zero),[12] the **temporal_reference** for each coded picture increases by one modulo 1024 in display order. For a frame picture coded as two field pictures, as is allowed for interlaced sequences, each field is assigned the same temporal reference. Figure 6.11 shows the use of the temporal reference in a typical group of pictures where each picture is a frame picture. Both display and transmission orders are shown.

Picture Coding Type The 3-bit **picture_coding_type** field specifies the prediction method used to code a particular picture as defined by Table 6.7.

Although every macroblock in an I picture is coded in intramode, it is still possible to have intracoded macroblocks in both P pictures and B pictures. In addition and as described earlier, not every macroblock in a B picture needs to be coded using bidirectional prediction, as prediction from only one of the two reference pictures is also possible.

VBV Delay The *video buffer verifier* is a hypothetical decoder that is connected directly to the output of the encoder. Its role is to ensure that buffer overflow or

[12] Distribution digital television services do not use low-delay mode and so the case of low delay mode is not considered. Details can be found in the international standard.

Table 6.7 Definition of **picture_coding_type**.

picture_coding_type	Prediction method used for coding
000	Forbidden
001	Intrapicture (I)
010	Predictive picture (P)
011	Bidirectionally predicted picture (B)
100	Not to be used[a]
101–111	Reserved

[a] This type was used for D pictures (pictures that only contained the DC DCT coefficient after coding in intramode) in MPEG-1. It is not allowed in MPEG-2.

buffer underflow does not occur at the decoder. The VBV is described in detail later in this chapter. The **vbv_delay** field sets the number of periods of a 90-kHz clock (which is derived from the 27-MHz system clock[13]) that the VBV waits after the receipt of the final byte of **picture_start_code** before decoding the picture. The **vbv_delay** can be used by the decoder to correct for any variation between the timing (such as the clock speeds) at the encoder and the decoder.

Redundant Fields The picture header next contains four redundant fields, namely, **full_pel_forward_vector** (1 bit), **forward_f_code** (3 bits), **full_pel_backward_vector** (1 bit), and **backward_f_code** (3 bits). All were used in the picture header for the MPEG-1 video standard to specify information about the way that motion vectors had been coded. They are not used in MPEG-2 and are set to predefined values (the 1-bit fields to 0 and the 3-bit fields to 111). Instead, data on the coding of motion vectors are included in the picture_coding_extension() for MPEG-2 bit streams.

Extra Picture Information The 1-bit **extra_bit_picture** field when set to one is followed immediately by the 8-bit **extra_information_picture** field. This field can be used for future extensions of the standard. It is currently not used and must not be used in compliant bit streams. After this field, the **extra_bit_picture** field is again checked and if set to one, a further 8-bit **extra_information_picture** field is read. This continues until **extra_bit_picture** is not set to one. If the **extra_bit_picture** is initially zero, then no **extra_information_picture** data is read.

6.3.4.2. Picture Coding Extension

The picture_coding_extension() begins with a 32-bit **extension_start_code** and a 4-bit **extension_start_code_identifier** as defined in Table 6.3. Other parameters are defined below.

[13] See Chapter 11 on MPEG-2 systems for further details.

Motion Vector Range Motion vector ranges are defined by the values of the four fields **f_code[0][0], f_code[0][1], f_code[1][0],** and **f_code[1][1]** that refer to forward horizontal, forward vertical, backward horizontal, and backward vertical motion vector components, respectively. Each field is of length 4-bits, and not all fields are relevant for every picture type. Thus, no **f_code[ ][ ]** fields are needed in an I picture,[14] and so all 4-bit fields take the value 1111. Only forward **f_code[ ][ ]** fields are required for P pictures, and so the two 4-bit backward fields take the value 1111.

The four values are needed to allow a separate **f_code[ ][ ]** value for forward and backward motion vectors (both of which could be required for B pictures) and for the vertical and horizontal component of each motion vector. The **f_code[ ][ ]** value defines the motion vector range as defined in Table 6.8. The range for the vertical component of field motion vectors in frame pictures is half of that for all other components. This is because a one-pixel vertical displacement in a field picture is equivalent to a two-pixel horizontal displacement in a frame picture because a line from the alternate field lies between the position of the current pixel and the pixel displaced by one pixel vertically in the same field. A smaller motion vector range is therefore appropriate.

From the data in Table 6.8, it can be seen that very large motion vector ranges are allowed and that motion-compensated prediction is performed to half pixel accuracy.

Intra-DC Precision In an intracoded macroblock, the value of the DC DCT coefficient for luminance and chrominance defines the average color of the macroblock. Coding this information with insufficient accuracy means that only a limited range of colors are possible. This can show up as blocking artifacts in flat regions of a picture.

Table 6.8 Motion vector ranges for various values of **f_code**.

f_code[ ][ ] value	Range for vertical component of field motion vectors in frame pictures	Range for all other motion vectors
0	Not allowed	
1	−4 to +3.5	−8 to +7.5
2	−8 to +7.5	−16 to +15.1
3	−16 to +15.5	−32 to +31.5
4	−32 to +31.5	−64 to +63.5
5	−64 to +63.5	−128 to +127.5
6	−128 to +127.5	−256 to +255.5
7	−256 to +255.5	−512 to +511.5
8	−512 to +511.5	−1024 to +1023.5
9	−1024 to +1023.5	−2048 to +2047.5
10–14	Reserved	Reserved
15	Used when a particular **f_code**[ ][ ] is not needed	

[14]Unless the I picture makes use of concealment motion vectors for error concealment. These are described later in the picture_coding_extension().

Table 6.9 Definition of **intra_dc_precision**.

intra_dc_precision	Precision of DC DCT coefficient in I blocks (bits)
00	8
01	9
10	10
11	11

For this reason, particular care is taken in coding these DC DCT coefficients using 8, 9, 10, or 11-bit precision. This precision can be set at the picture layer using the 2-bit **intra_dc_precision** field defined in Table 6.9.

Picture Structure The field or frame structure of a picture is defined by the 2-bit **picture_structure** field as defined in Table 6.10.

Top Field First This **top_field_first** flag has a number of meanings, depending upon the values of the **progressive_sequence** flag (sequence layer) and the **picture_structure** and **repeat_first_field** flags (picture layer).

For interlaced sequences (**progressive_sequence** set to zero), this flag indicates which field is to be output first by the decoder. For field pictures, it always takes the value zero since whether the field is the top field or the bottom field is indicated by the **picture_structure** field. For frame pictures, if **top_field_first** is set to one then the first field output from the decoding process is the top field. Otherwise, the first field output is the bottom field.

For progressive sequences (**progressive_sequence** set to one), the **top_field_first** and **repeat_first_field** flags indicate how many times (one, two, or three) the reconstructed progressive picture is to be output by the decoding process as defined in Table 6.11.

Frame-Only Prediction and DCT Frame pictures in MPEG-2 can still be predicted using field prediction. In addition, the DCT can also be carried out in either field or frame mode. Both of these are discussed further at the macroblock level later

Table 6.10 Definition of **picture_structure**.

picture_structure	Type of picture
00	Reserved
01	Top field of a pair of field pictures
10	Bottom field of a pair of field pictures
11	Frame picture

Table 6.11 Use of **top_field_first** and **repeat_first_field** for progressive sequences.

top_field_first	**repeat_first_field**	Decoder output
0	0	One progressive picture
0	1	Two identical progressive pictures
1	0	Not allowed
1	1	Three identical progressive pictures

in this chapter. However, if the flag **frame_pred_frame_dct** is set to one in a frame picture from an interlaced video sequence, then only frame prediction and frame DCT are used in the coding of this picture. This allows some savings in overhead bits to be achieved at the macroblock layer. The flag is set to zero in field pictures since field prediction is always employed. It is always set to one in progressive sequences since only frame prediction and frame DCT are employed.

Concealment Motion Vectors The MPEG-2 syntax allows motion vectors to be transmitted in I macroblocks even though motion-compensated prediction is not used for the coding of these macroblocks. The availability of these motion vectors allows improved concealment if transmission errors occur. Transmitting the motion vector in the same macroblock which has the data for that macroblock is of little use because if the data is lost, then it is highly likely that the motion vector is also lost. For this reason, a concealment motion vector sent in a particular macroblock refers to the macroblock directly below the current macroblock. For the last row of macroblocks in a picture, concealment motion vectors serve no useful purpose and so should be set to zero to reduce unnecessary overhead.

Quantizer Scale Type The quantizer to be used for the coding and decoding of DCT coefficients is defined by a 5-bit **quantizer_scale_code** field, which is defined at the slice and macroblock layers. This can be decoded in one of the two ways depending on the value of the **q_scale_type** flag as defined in Table 6.12.

The first column of quantizer values is the set used in MPEG-1 with the quantizer directly proportional to the **quantizer_scale_code** value. The second column is an alternative set introduced for MPEG-2, which allows both an increased range of possible quantizers (1–112 as opposed to 2–62) as well as a large range of small quantizer values (i.e., 1–8 in steps of 1), which is useful for high-quality applications.

Intra-Variable-Length Code Format The variable-length code words used to represent run-level pairs of DCT coefficients are the same as those given in Chapter 5, with the exception that if the **intra_vlc_format** flag is set then an alternate set of variable-length code words are used for the coding of intramacroblocks. Since intra macroblocks can occur in any type of picture (I, P, or B), the variable-length code word table to be used needs to be specified for every picture. The exact specification of the alternate table can be found in the standard.

Table 6.12 Definition of quantizer step size for different values of **q_scale_type**.

	Quantizer step size	
quantizer_scale_code	**q_scale_type** = 0	**q_scale_type** = 1
0	Not allowed	
1	2	1
2	4	2
3	6	3
4	8	4
5	10	5
6	12	6
7	14	7
8	16	8
9	18	10
10	20	12
11	22	14
12	24	16
13	26	18
14	28	20
15	30	22
16	32	24
17	34	28
18	36	32
19	38	36
20	40	40
21	42	44
22	44	48
23	46	52
24	48	56
25	50	64
26	52	72
27	54	80
28	56	88
29	58	96
30	60	104
31	62	112

Zig-Zag Scanning of DCT Coefficients The zig-zag scan order for DCT coefficients used in MPEG-1 has already been defined in Chapter 4. An alternate zig-zag scan order has been introduced in MPEG-2 and is used when the **alternate_scan** flag is set to one. This alternate zig-zag scan pattern is shown in Figure 6.12.

The alternate scan order scans the coefficients toward the bottom of the first two columns of DCT coefficients considerably earlier in the scan than the other original

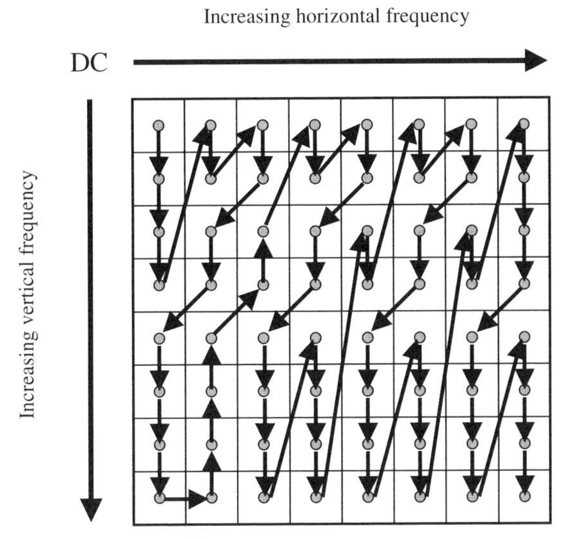

Figure 6.12 Alternate zig-zag scan order.

scan order. It would therefore be suitable for blocks with faster vertical than horizontal changes. This can be expected to occur in field pictures in interlaced sequences, as the 16 vertical pixels in a macroblock effectively span 32 pixels in the equivalent frame picture.

Repeat First Field The 1-bit **repeat_first_field** flag causes the first field of an interlaced frame picture to be repeated. The first field (either top or bottom) is followed by the alternate field, and then the first field is repeated. At the first glance, this might not seem a particularly useful functionality. Its usefulness becomes apparent when we consider the transmission of film material over digital television.

Film material is usually recorded at 24 frames/s. For transmission over a 50-field/s system, the film material is speeded up to 25 frames/s. This results in a barely noticeable change in the pitch of the accompanying audio material. However, for a 60-field/s systems, a process known as 3:2 pulldown is employed. Using this approach, one film frame produces three fields of television information whereas the next film frame produces just two fields. This is illustrated in Figure 6.13.

The first film frame is scanned into three fields, which are then merged to form a single frame for coding and transmission (Fields 1 and 3 can be identical or alternatively averaged to form the top field of Frame 1). For correct display, Frame 1 needs both **repeat_first_field** and **top_field_first** flags set to one. The second film scene is scanned to two fields that are merged to form a single frame for coding and transmission. To keep the required alternate order of fields, which is needed by interlaced displays, the **top_field_first** flag is set to zero in this case as it is the bottom field that needs to be displayed first. The third film frame is treated in identical manner to the first frame, except that the bottom field needs to be displayed first and is the field that is repeated. The fourth frame is treated like the second, except

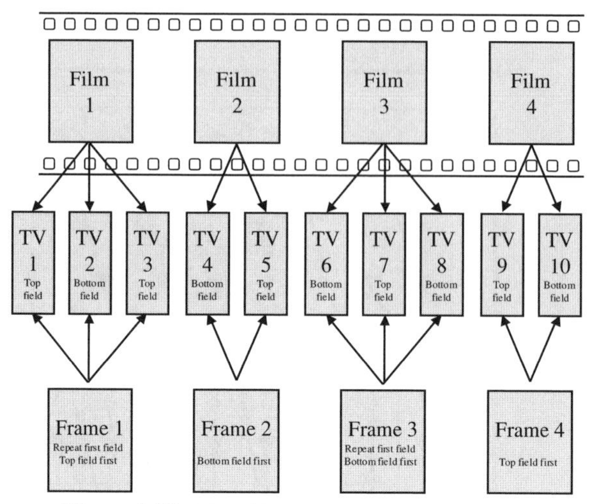

Figure 6.13 Use of 3:2 pulldown in 60-Hz systems.

that the **top_field_first** flag is set to one. This allows a 30-picture/s movie to be displayed while only transmitting 24 pictures/s. This greatly enhances the coding performance for movie material. For field pictures, the **repeat_first_field** flag is always set to zero.

As described in the section on "Top field first", for progressive sequences the **repeat_first_field** flag is used together with the **top_field_first** flag to indicate the number of times that a decoded progressive frame should be output by the decoder.

Composite Display Flag This flag indicates that the bit stream contains information about the analog composite video prior to encoding. This includes the field sequence, burst amplitude, and subcarrier phase. This information is not needed to correctly decode the digital bit stream but can be useful if the decoded video signal is to be returned to the analog form.

6.3.4.3. *Quantizer Matrix Extension*

We have already seen at the sequence layer that new intrablock and interblock quantizer matrices can be defined by the user and downloaded in the bit stream. It is possible to redefine these quantizer matrices using the quantizer matrix extension at the picture layer. In addition, the extra flexibility of having different quantizer matrices for luminance and chrominance is also provided at the picture layer for chrominance formats other than 4:2:0. A 1-bit flag indicates whether a particular quantizer matrix is present or not. Thus, up to four quantizer matrices can be loaded in the following order: intraquantizer matrix, interquantizer matrix, chrominance intraquantizer matrix, and chrominance interquantizer matrix.[15]

[15] For a 4:2:0 chrominance structure, separate luminance and chrominance quantizer matrices are not available.

When the intraquantizer matrix is loaded, its values are also used for the chrominance intraquantizer matrix, unless this is subsequently defined in the extension. Similarly, when the interquantizer matrix is loaded, its values are also used for the chrominance interquantizer matrix, unless this is subsequently defined in the extension. Similar to the sequence layer, the quantizer matrices are downloaded as sixty-four 8-bit values (range 1–255) in the default zig-zag scan order.

6.3.4.4. Picture Display Extension

The picture_display_extension() allows the position of the display rectangle defined in the sequence_display_extension() to be moved on a picture-by-picture basis. It begins with a start code followed by the appropriate **extension_start_code_identifier** value as defined in Table 6.3. The position of the display rectangle is defined relative to the center of the complete reconstructed picture as shown in Figure 6.14. The two offsets are 16-bit signed integers giving each offset in units of 1/16th of a pixel.[16] A positive value for the **frame_center_horizontal_offset** indicates that the center of the reconstructed picture lies to the right of the center of the display rectangle. A positive value for the **frame_center_vertical_offset** indicates that the center of the reconstructed picture lies below the center of the display rectangle. In Figure 6.14, both of these offsets take negative values.

The frame center offsets can be used to implement pan and scan so that the display rectangle can be moved around the entire reconstructed picture to ensure that the most "interesting" area is always displayed. As an example, suppose that we have a wide-screen (16 × 9) image of two people talking at either end of a long table. One person is therefore at the extreme left in the reconstructed picture while the other person is at the extreme right of the picture. For a viewer watching on a standard-definition (4 × 3) display, looking at the central portion of the reconstructed picture

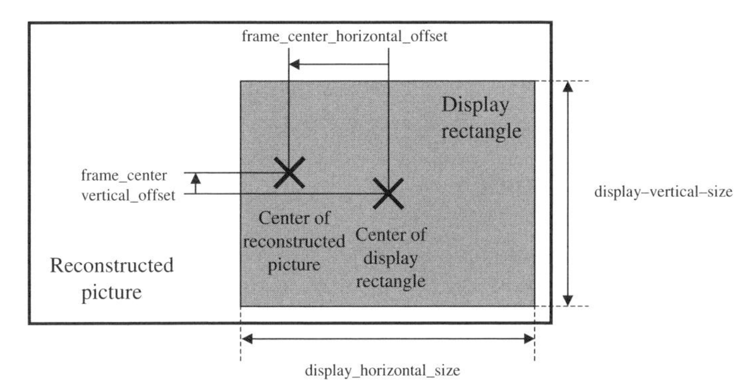

Figure 6.14 Offset of displayed picture relative to reconstructed picture.

[16] Specification to 1/16th of a pixel accuracy is required for scalable applications where higher layers may contain more pixels than lower layers and still require at least pixel accuracy. Scalability is not covered in detail in this text as it is not used in current digital television applications.

would show the table, but possibly neither of the speakers. The use of the frame center offsets allows the displayed area to move both left and right to show each person while they are speaking.

Earlier in this chapter we saw that for interlaced sequences, a field picture always contains exactly one field whereas a frame picture can contain two or three fields depending on the value of the **repeat_first_field** flag. If there is more than one field, then separate values of the **frame_center_horizontal_offset** and the **frame_center_vertical_offset** are required for each displayed field.

Similarly for progressive sequences, the picture can be displayed one, two, or three times depending upon the values of the **repeat_first_field** and **top_field_first** flags. Again, if a picture is displayed more than once then separate values for the two offsets are required for each displayed picture.

6.3.4.5. Copyright Extension

The copyright extension allows materials to be marked as copyright. It again consists of a start code followed by the appropriate **extension_start_code_identifier** value shown in Table 6.3. A 1-bit flag indicates whether the material is copyright or not. When set to one, this indicates that all the coded pictures following the copyright extension up to the next copyright extension or the end of sequence start code are copyright. This is followed by an 8-bit **copyright_identifier**, which indicates the registration authority for the copyright, an **original_or_copy** flag, which indicates that the material is original when set to one and a copy when set to zero, and a 64-bit **copyright_number**, which uniquely identifies the copyright material. To remove the possibility of start-code emulation (i.e., 23 or more successive zero bits), the copyright number is transmitted as a 20-bit field and two 22-bit fields with each field separated by a marker bit, which is set to one.

6.3.4.6. User Data

User data can also be included at the picture layer. It begins with the unique **user_data_start_code** and has a format identical to that used for user data at the sequence layer.

6.3.5. The Slice Layer

A *slice* consists of a number of consecutive macroblocks all in the same row of macroblocks. A slice can contain all of the macroblocks in a particular row of macroblocks in a picture or a series of macroblocks in the same row of macroblocks. This is shown in Figure 6.15. Each row of macroblocks starts with a new slice header. The first row of macroblocks in Figure 6.15 forms a single slice while the second row of macroblocks forms two slices. There is no limit to the number of macroblocks that can be included in a slice, except that each slice must contain at least one macroblock and all macroblocks in a slice must be in the same row of macroblocks. The first and last macroblocks in a slice cannot be skipped macroblocks.

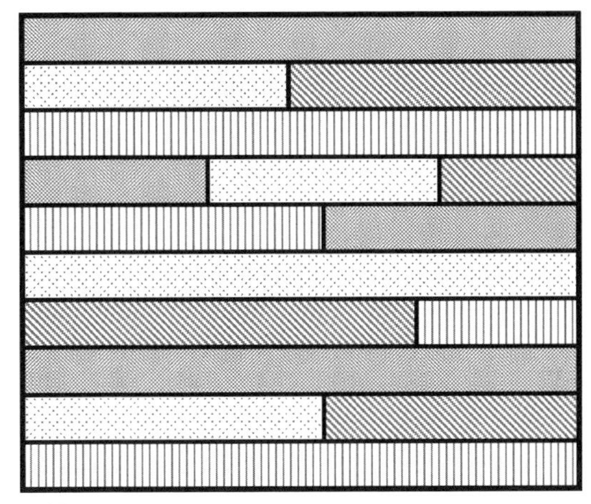

Figure 6.15 Arrangement of slices in a picture (restricted slice structure).

Requiring all macroblocks in a picture to be covered by a slice is called the *restricted slice structure.* This structure is used in all current digital television applications. The *unrestricted slice structure* allows gaps of unencoded macroblocks between slices as shown in Figure 6.16. In this case, it is clear that the first and last macroblocks in each slice need to be coded, so that the parts of the picture coded can be uniquely defined.

Each slice begins with a slice start code that consists of the start code 0x000001 followed by an 8-bit **slice_vertical_position**, which takes values in the range 0x01–0xAF (in decimal 1–175). The **slice_vertical_position** defines which row of macroblocks is contained in the slice with the first row of macroblocks in a picture numbered one, the second row of macroblocks numbered two, and so on. If a row of macroblocks is made up of more than one slice, then each of these slices has the same value for **slice_vertical_position**. The horizontal position of the first

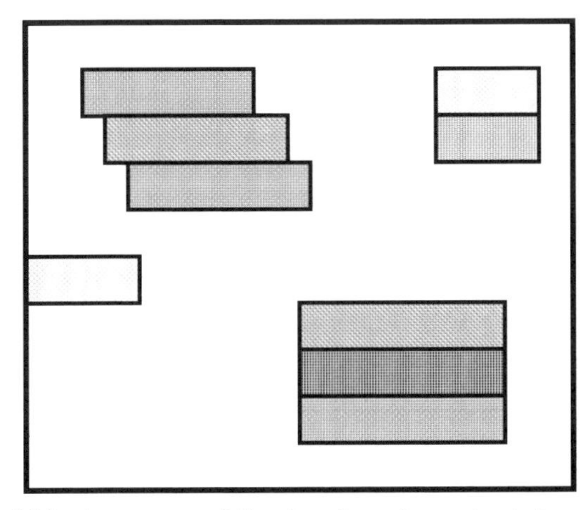

Figure 6.16 Arrangement of slices in a picture (unrestricted slice structure).

macroblock in each slice is defined in the macroblock header as is discussed in the section on 'Macroblock address'.

6.3.5.1. *Slice Number Extension for Pictures With More Than 2800 Lines*

A picture with more than 2800 lines of pixels (i.e., more than 175 rows of macroblocks) requires more slices than are allowed by the range of **slice_vertical_position**. For pictures of this size (as defined at the sequence layer), a further 3-bits immediately follows the **slice_start_code**. This field is called the **slice_vertical_position_extension** and extends the range of slice numbers from 1 to 1023 by acting as the three MSBs of the **slice_vertical_position**. This range is sufficient for all picture sizes permitted by the MPEG-2 specification.

6.3.5.2. *Quantizer Step Size*

Next follows the 5-bit **quantizer_scale_code** field that indicates the quantizer to be used for macroblocks in the slice. The mapping between **quantizer_scale_code** and quantizer value is given in Table 6.12 for each value of **q_scale_type** (defined at the picture layer). The quantizer value can be redefined in the macroblock header of any macroblock that is not skipped. In practice, it is not usually desirable to do this for every macroblock because of the large overhead involved.

6.3.5.3. *Intraslice*

The **intra_slice_flag** field when set to one indicates the presence in the bit stream of an **intra_slice** flag as well as further **reserved_bits** and extra slice information. The **intra_slice** flag, when set to one, indicates that all macroblocks in the slice are coded in intramode. This can be useful in allowing a decoder to perform fast forward and fast reverse operations. The **intra_slice** flag is followed by a 7-bit field (**reserved_bits**), which is reserved for future use and must currently take the value zero. The extra slice information consists of an **extra_bit_slice** flag, which if set to one, indicates that it is followed by an 8-bit **extra_information_slice** field. Then follows another **extra_bit_slice** flag, and the process continues until the **extra_bit_slice** flag takes the value zero. In the current systems, the **extra_bit_slice** flag should be set to zero and any information following the **extra_bit_slice** flag, if set to one, should be ignored by the decoder.

The slice header is followed by a series of coded macroblocks of data until a further start code is encountered.

6.3.6. The Macroblock Layer

The macroblock header is used to define the address of the macroblock, the macroblock type, the quantizer to be used, the motion vector(s), and the coded block pattern. We consider each of the elements one by one. Not all of these elements are

macroblock_escape

macroblock_escape

macroblock_address_increment = 22

Figure 6.17 Representation of a **macroblock_address_increment** of 88.

present in every macroblock header. However, the decoder can correctly determine which elements to expect from the macroblock-type information.

6.3.6.1. Macroblock Header

Macroblock Address The macroblock address defines the horizontal position of the macroblock within a row of macroblocks. The row of macroblocks in which the macroblock is located can be inferred from the slice header. As explained in Chapter 5, the data transmitted is actually the difference between the address of the current macroblock and the address of the last transmitted macroblock and is called the **macroblock_address_increment**. A variable-length code is used to represent these differences with a range of 1–33. As discussed when considering the sequence layer, MPEG-2 can deal with very large pictures. It is therefore possible that the **macroblock_address_increment** might need to be larger than 33. In this case, it is preceded by one or more fixed-length 11-bit code words called the **macroblock_escape**. Each time the **macroblock_escape** is included in the bit stream, the **macroblock_address_increment** is increased by 33. Thus, if the difference in address between the last-coded and the current macroblocks was 88 macroblocks, then this would be encoded as shown in Figure 6.17.

After the slice header, the address of the first macroblock is coded as its actual value rather than the difference from the last-encoded macroblock in the previous slice. This allows the decoder to correctly determine the position of this macroblock even if transmission errors have made it impossible for the previous slice to be correctly decoded. The slice header gives the vertical position of the macroblock while the macroblock header gives the horizontal position. All the coded macroblocks contain a **macroblock_address_increment**.

Macroblock Type The macroblock-type information depends on the type of picture (I, P, or B) that is currently being decoded. We consider each picture type separately.

I Pictures All macroblocks in I pictures are coded in intramode. As shown in Table 6.13, there are only two types of macroblocks in I pictures, namely macroblocks where the quantizer step size is redefined and macroblocks where the quantizer remains unchanged. Motion vectors are not needed for the decoding of I macroblocks.[17] Since every block in an I macroblock needs to be coded as it must contain

[17] Other than concealment motion vectors if specified at the picture layer. All intramacroblocks in a picture that uses concealment motion vectors contain a motion vector and so this does not need to be determined at the macroblock layer.

Table 6.13 Macroblock types for I pictures.

Macroblock-type VLC code word	New quantizer defined	forward motion vector	Backward motion vector	Coded block pattern	Intramacroblock	Description of macroblock
1	No	No	No	No	Yes	Intra, no new quantizer
01	Yes	No	No	No	Yes	Intra, new quantizer

at least a nonzero DC DCT coefficient, there is no need for a coded block pattern to indicate which blocks are coded.

P Pictures The types of macroblocks that can occur in a P picture are shown in Table 6.14.

Table 6.14 Macroblock types for P pictures.

Macroblock-type VLC code word	New quantizer defined	forward motion vector	Backward motion vector	Coded block pattern	Intra macroblock	Description of macroblock
1	No	Yes	No	Yes	No	MC, coded blocks, old quant
01	No	No	No	Yes	No	Inter, coded blocks, old quant
001	No	Yes	No	No	No	MC, no coded blocks, old quant
00011	No	No	No	No	Yes	Intra, old quant
00010	Yes	Yes	No	Yes	No	MC, coded blocks, new quant
00001	Yes	No	No	Yes	No	Inter, coded blocks, new quant
000001	Yes	No	No	No	Yes	Intra, new quant

The macroblock types can be divided into three groups: macroblocks that utilize forward motion-compensated prediction from the previous I or P reference picture (motion-compensated mode), macroblocks that use the macroblock in the same position in the previous I or P reference picture for prediction (intermode), and macroblocks that do not use any form of temporal prediction (intramode).

For macroblocks coded using motion-compensated prediction, the possibilities are as follows:

- One or more blocks contains coded data and the quantizer is unchanged.
- No block contains coded data and the quantizer is unchanged. The macroblock still needs to be transmitted to send the appropriate motion vector.
- One or more blocks contain coded data and a new quantizer value is used.

There is no need to define a macroblock type for the case where no block contains coded data and a new quantizer value is used, because if no blocks contain coded data then all reconstructed blocks are zero irrespective of the quantizer value used.

For macroblocks coded using interprediction, there are two possibilities

- One or more blocks contain coded data and the quantizer is unchanged.
- One or more blocks contain coded data and a new quantizer value is used.

An intercoded macroblock in which no block has a coded data is a skipped macroblock. The reconstructed value of a skipped intermacroblock (all zeros) is the same irrespective of the quantizer. As with motion-compensated macroblocks, there is therefore no need to define an intermacroblock that only defines a new quantizer value but for which all blocks of data are zero.

The case for intracoded macroblocks is the same as that for I pictures, that is, the macroblocks are coded using the current quantizer or have a new quantizer value defined in the macroblock header. Intramacroblocks are always coded.

B Pictures The types of macroblocks that can occur in a B picture are shown in Table 6.15.

In this case, there are four different classes of macroblocks. Interpolated macroblocks require two motion vectors: a forward motion vector that points to a previous I or P reference picture and a backward motion vector that points to a future I or P reference picture (interpolated prediction). Two motion vectors are transmitted in the macroblock header. The final prediction is the average of the two predictions generated. The result of motion-compensated prediction and coding can result in all blocks being zero. In this case, the quantizer is assumed to be unchanged because decoding produces the same result irrespective of the quantizer used. Alternatively, one or more blocks might contain the coded data in which case the quantizer can remain the same or be redefined in the macroblock header.

The same three macroblock types are used for macroblocks coded with forward motion-compensated prediction only (forward prediction) and for macroblocks coded with backward motion-compensated prediction only (backward prediction).

Table 6.15 Macroblock types for B pictures.

Macroblock-type VLC code word	New quantizer defined	forward motion vector	Backward motion vector	Coded block pattern	Intramacroblock	Description of macroblock
10	No	Yes	Yes	No	No	Interp, no coded blocks, old quant
11	No	Yes	Yes	Yes	No	Interp, coded blocks, old quant
010	No	No	Yes	No	No	Back, no coded blocks, old quant
011	No	No	Yes	Yes	No	Back, coded blocks, old quant
0010	No	Yes	No	No	No	Fwd, no coded blocks, old quant
0011	No	Yes	No	Yes	No	Fwd, coded blocks, old quant
00011	No	No	No	No	Yes	Intra, old quant
00010	Yes	Yes	Yes	Yes	No	Interp, coded blocks, new quant
000011	Yes	Yes	No	Yes	No	Fwd, coded blocks, new quant
000010	Yes	No	Yes	Yes	No	Back, coded blocks, new quant
000001	Yes	No	No	No	Yes	Intra, new quant

In each of these cases, only one motion vector is transmitted in the macroblock header.

Finally, there can still be intracoded macroblocks in B pictures. As for I and P pictures, an intracoded macroblock can use the current quantizer or have a new quantizer defined in the macroblock header. Intramacroblocks in B pictures are always coded.

In B pictures, a skipped macroblock is a macroblock that uses exactly the same form of prediction with exactly the same motion vectors as the macroblock that immediately precedes it. For this reason, the macroblock immediately after an intracoded macroblock cannot be a skipped macroblock in a B picture.

Types of Motion-Compensated Prediction For macroblocks that are coded using motion-compensated prediction, the next field in the bit stream specifies the prediction type, unless this has been already determined for a frame picture by the fact that the **frame_pred_frame_dct** flag has been set to one at the picture layer. The field specifying the prediction type is the **frame_motion_type** for frame pictures or the **field_motion_type** for field pictures. Each of these 2-bit fields defines the type of motion compensation used to code the macroblock.

Table 6.16 Definition of **frame_motion_type**.

frame_motion_ type	Prediction type	Number of motion vectors	Motion vector format	Differential motion vector
00	Reserved	—	—	—
01	Field	2	Field	No
10	Frame	1	Frame	No
11	Dual prime	1	Field	Yes

Motion-Compensated Prediction for Frame Pictures The type of motion-compensated prediction used in frame pictures is specified by the **frame_motion_type** field and defined in Table 6.16. There are three possible forms of motion-compensated prediction: field prediction, frame prediction, and dual-prime prediction.

The simplest form of prediction is the frame prediction where a macroblock in the current frame picture is predicted by the best matching macroblock in a previous I or P frame picture (P pictures) or from a previous and/or a future I or P frame picture (B pictures). The frame prediction process for a P picture is illustrated in Figure 6.18[18] for a pair of vertically adjacent macroblocks. For both current and reference pictures, the two fields are merged to form a single frame reference picture. This reference frame picture may originally have been transmitted as two field pictures. A macroblock in the current frame picture is then predicted from the reference picture. Thus, in the top macroblock shown in Figure 6.18, the top field pixels in the current picture are predicted from top field pixels in the reference picture whereas in the bottom macroblock top field pixels in the current frame picture are predicted from bottom field pixels in the reference picture. The vertical component of the motion vector relative to a pixel in the current picture (shown surrounded by a square) is also shown. Thus, in the top macroblock, the vertical component of the motion vector is +2 whereas in the bottom macroblock it is −5.

The situation is similar for frame prediction in B pictures and is illustrated in Figure 6.19. Once again the fields of the current picture and the two reference pictures are merged to form frame pictures. One or both of the reference frame pictures may originally have been transmitted as field pictures. In the top macroblock shown in Figure 6.19, interpolative prediction is employed. Thus, both forward and backward motion vectors are used. In the bottom macroblock, only backward prediction is employed, and so only a single motion vector is required. The values of the vertical component of the motion vectors are again shown relative to a pixel in the current picture (shown surrounded by a square).

As shown in Table 6.16, only a single motion vector is required for frame prediction of a P picture. Since prediction is required in up to two directions for interpolated prediction in a B picture, up to two motion vectors may be required.

[18] All of the diagrams in this section of the text show motion vectors with integer components. This is for clarity. In reality, motion-compensated prediction is performed to half-pixel accuracy in MPEG-2.

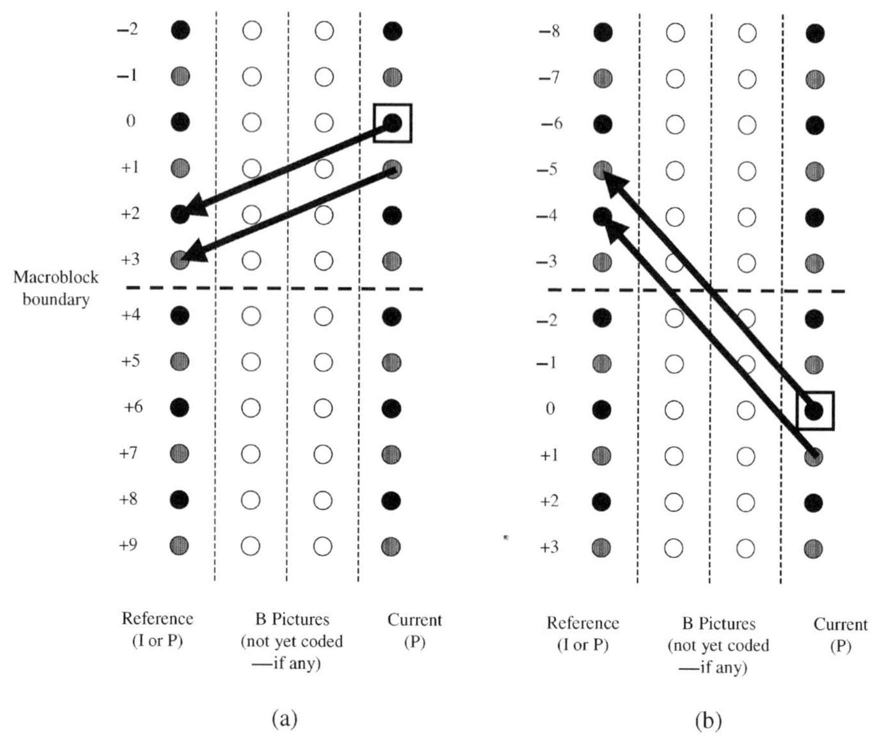

Figure 6.18 Frame prediction of a P frame picture: (a) top macroblock and (b) Bottom macroblock.

Now let us consider field prediction in frame pictures. In this case, the pixels in a macroblock are split according to whether they come from the top field or the bottom field. A separate motion vector is then used to encode each field. The reference fields used for prediction are the most recently decoded reference top field and the most recently decoded reference bottom field prior to the current frame picture. Each of the fields in the frame picture can be predicted from either of the reference fields. There is no requirement for the top field to be predicted from the top reference field, the bottom field from the bottom reference field, or even each field from a different reference field.

Figure 6.20 shows the situation for field prediction in a P frame picture. In the top macroblock, both fields are predicted from the bottom reference field. In the bottom macroblock, the top field is again predicted from the bottom reference field whereas the bottom field is predicted from the top reference field. The vertical motion vector displacements are again shown in the diagram and are identical for the two reference pixels in the two fields of the current frame picture (indicated by a square). Thus, a vertical displacement of zero when the top field is predicted from the top reference field refers to the pixel on the same line as the pixel in the current picture. A vertical displacement of zero when the top field is predicted from the bottom reference field is actually predicted from the line in the frame picture directly below (and therefore in the other field from) the line containing the current pixel.

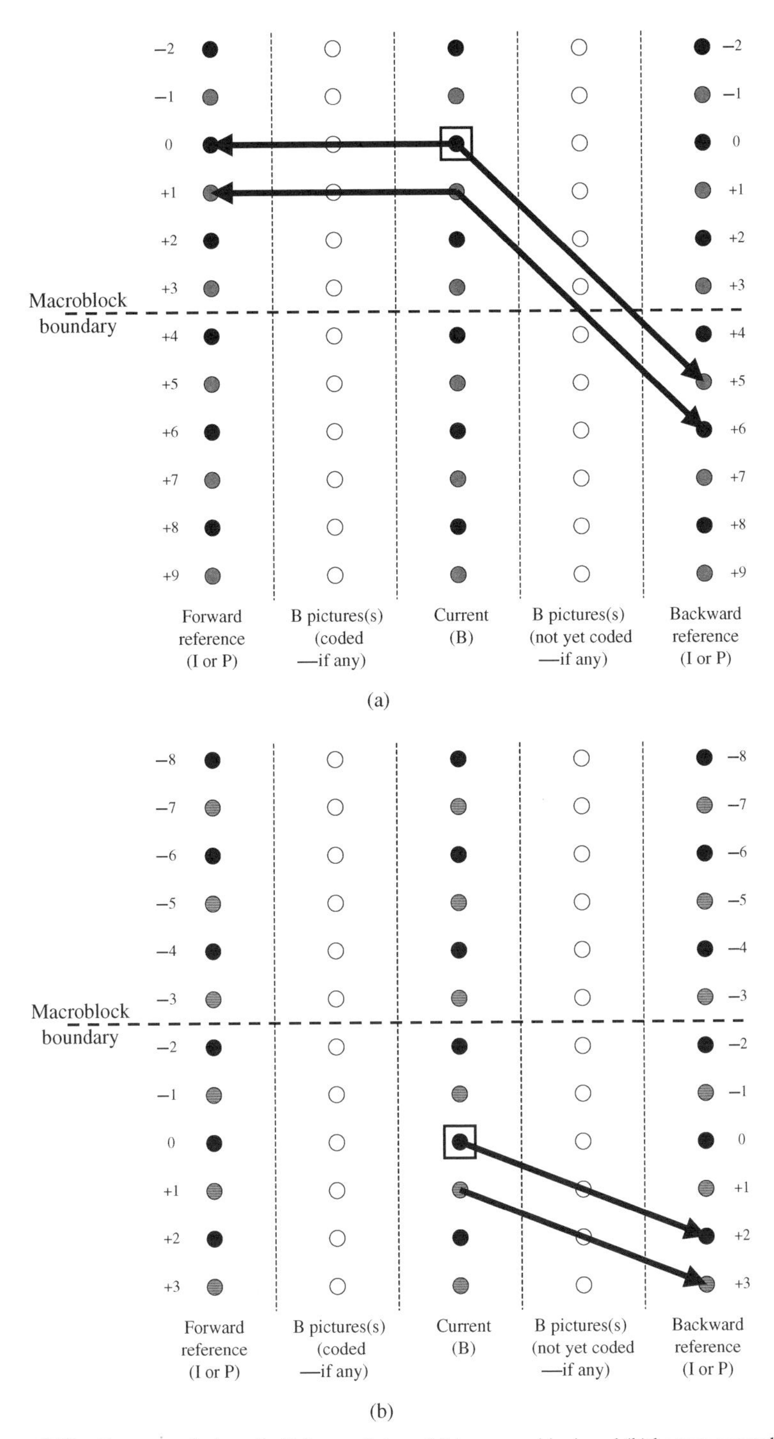

Figure 6.19 Frame prediction of a B frame picture: (a) top macroblock and (b) bottom macroblock.

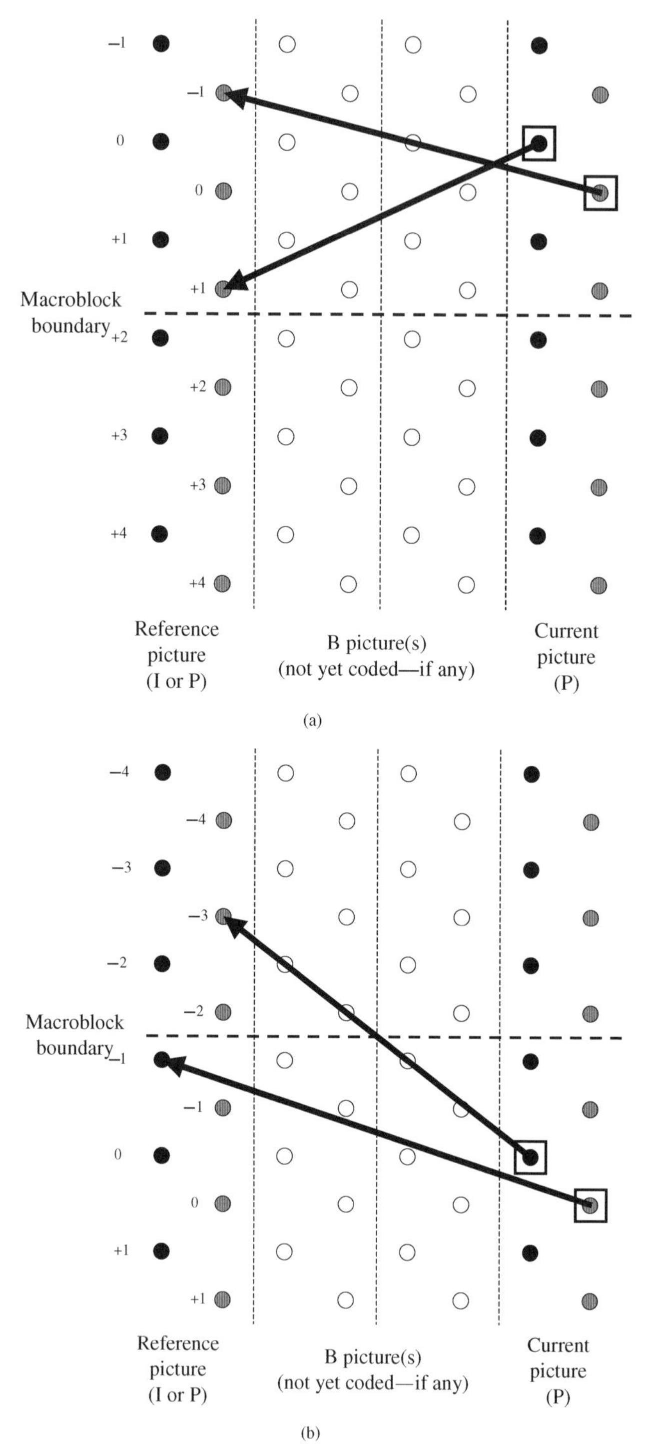

Figure 6.20 Field prediction in a P frame picture: (a) top macroblock and (b) bottom macroblock.

A similar situation applies when the bottom field is being predicted, except that in the case of prediction from the top field, a vertical displacement of zero refers to the pixel in the line directly above (and therefore in the other field from) the line containing the current pixel. In all cases, a zero vertical displacement refers to the same line number in the field as the line number in the field of the current picture.

Comparing Figure 6.20 with Figure 6.18, we can see that a vertical displacement of +1 using field prediction is the same as a vertical displacement of +2 using frame prediction. This is the reason, as shown in the section on "Motion vector range", that the vertical motion vector range used for field prediction is half that for frame prediction as well as half that for the horizontal motion vector range in both modes.

The situation for the use of field prediction in a B frame picture is shown in Figure 6.21. In this case, the top macroblock is coded using interpolative prediction. As shown, a total of four motion vectors are required. In the bottom macroblock, only backward prediction is employed, and so only two motion vectors are required in this case.

In summary, field prediction in frame pictures requires two motion vectors for macroblocks in P pictures and two motion vectors (backward prediction only or forward prediction only) or four motion vectors (interpolative prediction) for B pictures. In the latter case, an encoder needs to carefully balance the additional overhead associated with transmitting additional motion vectors with the saving in bits required to represent the prediction difference.

The final form of motion-compensated prediction used in frame pictures is dual-prime motion-compensated prediction. This is a special motion-compensated prediction technique developed during the MPEG-2 standardization process. Dual-prime motion-compensated prediction can only be used for P pictures and only when the current picture and the reference picture from which prediction is to occur are concurrent in display order (i.e., not separated by any intervening B pictures).

Dual prime operates by sending a single full motion vector and a small differential motion vector. Each component of the differential motion vector can only take one of the three values: −0.5, 0, or +0.5. The prediction mode for a frame picture is shown in Figure 6.22.

Each field within the macroblock to be predicted is treated separately. A single motion vector that is read from the bit stream is used to perform field prediction for each field with the top field being predicted from the top field of the reference picture and the bottom field being predicted from the bottom field of the reference picture. This motion is then interpolated or extrapolated to make a prediction from the other field in the reference frame picture. In the case of the top field shown in Figure 6.22, the motion vector used to form a prediction from the bottom field in the reference frame picture is half the motion vector used to form a prediction from the top reference field since the temporal distance is halved. Similarly, for the bottom field shown in Figure 6.22, the motion vector used to form a prediction from the top reference field is one and a half times the motion vector used to form a prediction from the bottom reference field since the temporal distance is one and a half times as great.

Where a component of the field motion vector is not an integer, the motion vector calculated by extrapolation or interpolation contains a motion vector to quarter-pixel

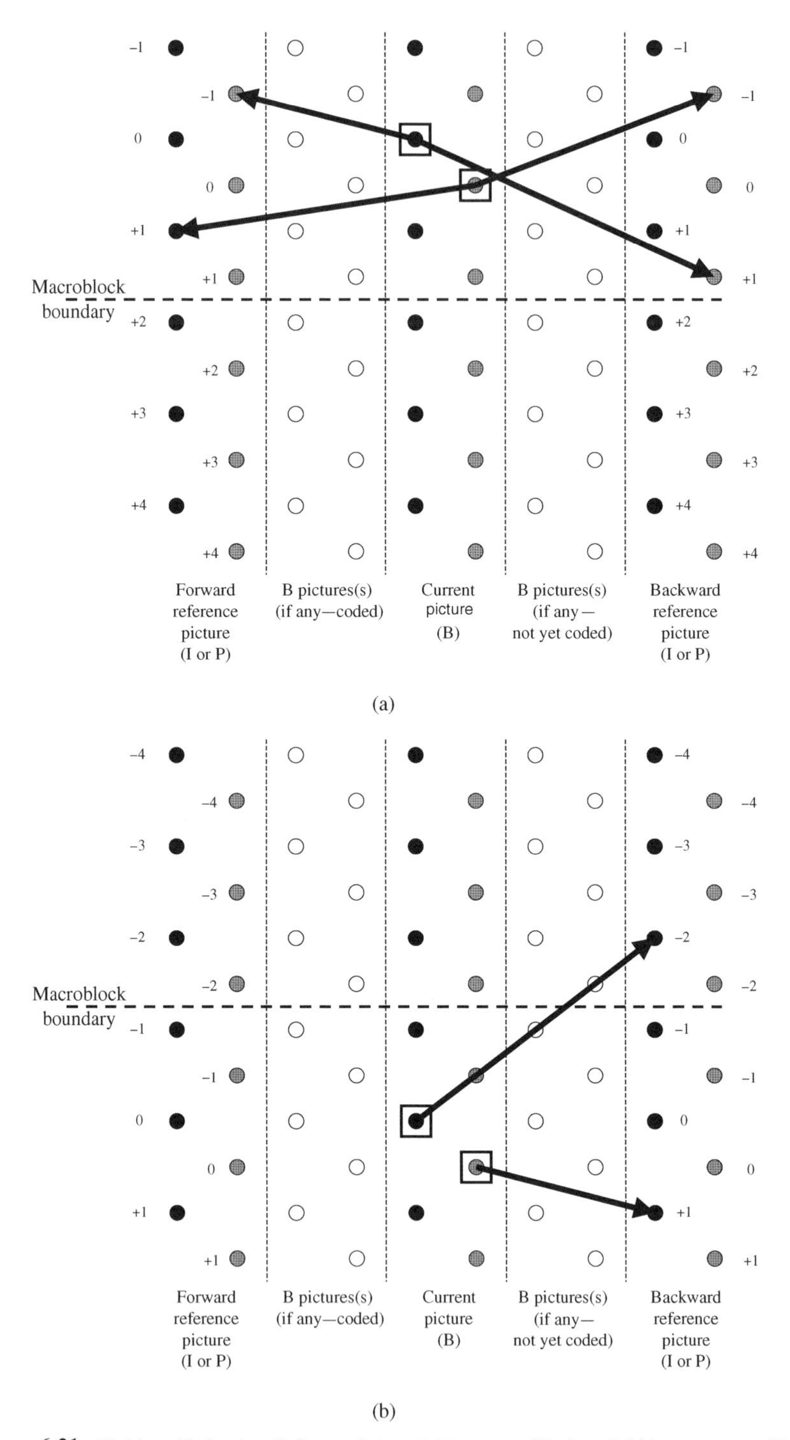

Figure 6.21 Field prediction in a B frame picture: (a) top macroblock and (b) bottom macroblock.

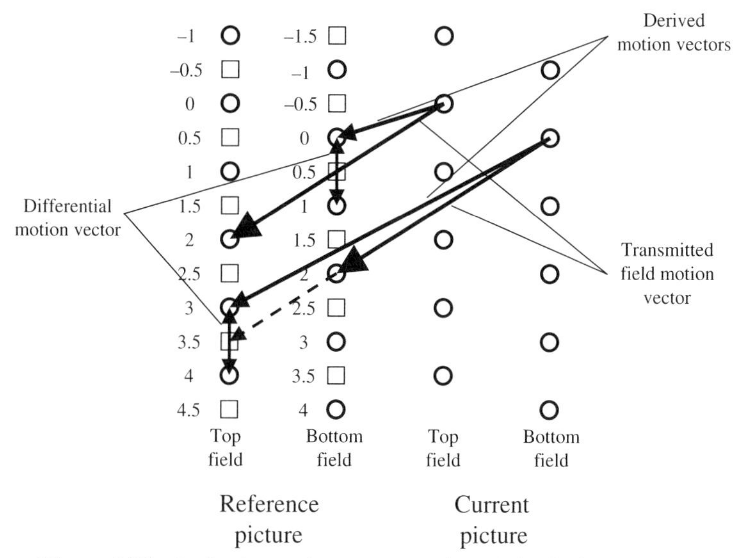

Figure 6.22 Dual-prime motion-compensated prediction for frame pictures.

accuracy. For example, a field motion vector of (+2.5, −3.5) would be interpolated to (+1.25, −1.75) and extrapolated to (+3.75, −5.25). Rounding is performed to the nearest half-pixel value away from zero. Thus, in the case just considered, the interpolated motion vector would be (+1.5, −2.0) and the extrapolated motion vector (+4.0, −5.5). As shown in Figure 6.22, for the vertical component of the motion vector, a 0.5-pixel offset needs to be included to account for the vertical displacement the between top and bottom fields.

The motion vectors calculated by extrapolation and interpolation are further modified by the differential motion vector, each component of which is limited to the three values: −0.5, 0, or +0.5. The two predictions for each field within the macroblock are then averaged to form the final prediction. The following example illustrates the operation of dual-prime motion-compensated prediction.

EXAMPLE 6.1

The field motion vector for dual-prime motion-compensated prediction in a frame picture is (+2, −2). The differential motion vector is (−0.5, +0.5). Calculate the displacements of the information to be used to predict each field in the macroblock. Assume that the top field is transmitted first for each picture.

We consider the prediction of the top and bottom fields within the macroblock separately.

Top Field

Predicted from top field of reference picture (+2, −2)

Predicted from bottom field of reference picture

Vertical component = Half vertical component of field motion vector + Field offset (− 0.5 in this case) + Vertical component differential motion vector
$= 0.5 \times (+2) - 0.5 - 0.5 = 0$

Horizontal component = Half horizontal component of field motion vector + Horizontal component differential motion vector
$= 0.5 \times (-2) + 0.5 = -0.5$

Bottom Field

Predicted from bottom field of reference picture (+2, −2)

Predicted from top field of reference picture

Vertical component = 1.5 times vertical component of field motion vector + Field offset (+0.5 in this case) + Vertical component differential motion vector
$= 1.5 \times (+2) + 0.5 - 0.5 = 3$

Horizontal component = 1.5 times horizontal component of field motion vector + Horizontal component differential motion vector
$= 1.5 \times (-2) + 0.5 = -2.5$ ■

Motion-Compensated Prediction for Field Pictures The type of motion-compensated prediction used in field pictures is specified by the **field_motion_type** field and defined in Table 6.17. There are three possible forms of motion-compensated prediction: field prediction, 16 × 8 prediction, and dual-prime prediction.

The simplest of these is field prediction, where the macroblock in the current field picture is predicted from a displaced macroblock in a previous I or P field picture (P pictures) or a previous and/or a future I or P field picture (B pictures). The reference pictures used are the most recently decoded I or P field pictures. These reference field pictures may themselves have been coded as frame pictures. The situation for the first P field picture (which is assumed to be the top field) is shown in Figure 6.23. In the top macroblock, prediction is made from the previous top reference field whereas in the bottom macroblock prediction is made from the bottom reference field.

The situation for the second P field picture is shown in Figure 6.24. Remember that prediction is performed from the two most recently received reference fields. This means that one of the reference fields is the P field picture just received and decoded together with the last received reference field of the type opposite (top or bottom) to the P field picture just decoded. In Figure 6.24, the top macroblock is

Table 6.17 Definition of **field_motion_type**.

field_motion_type	Prediction type	Number of motion vectors	Motion vector format	Differential motion vector
00	Reserved	—	—	—
01	Field	1	Field	No
10	16 × 8 MC	2	Field	No
11	Dual prime	1	Field	Yes

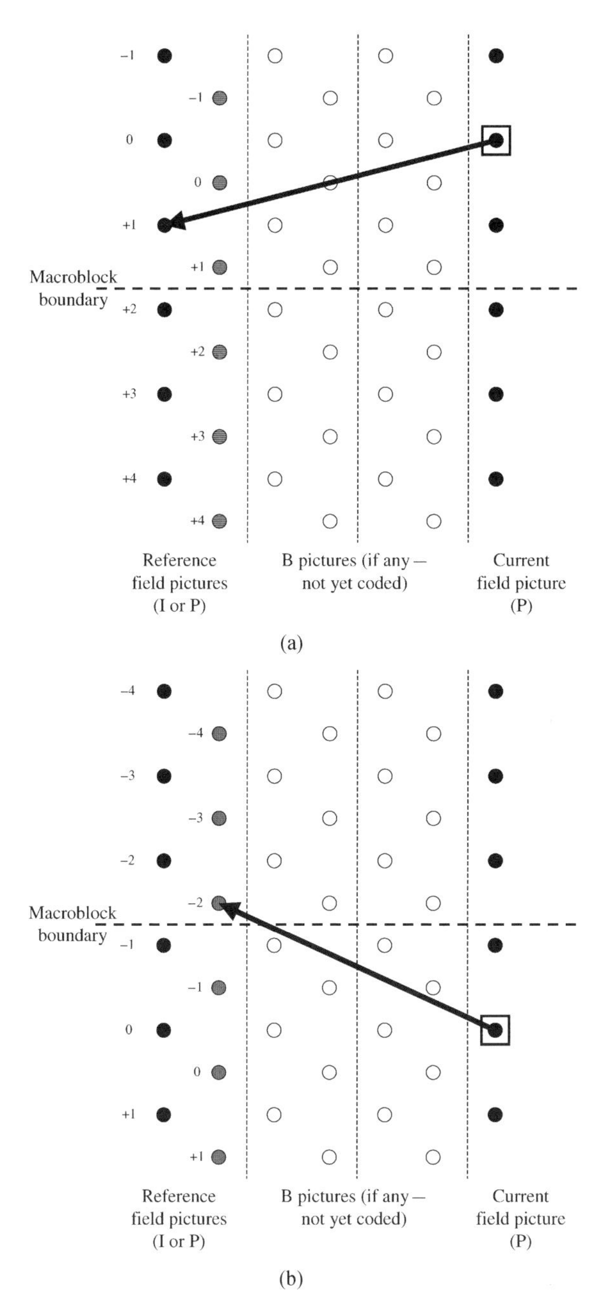

Figure 6.23 Field prediction of the first P field picture: (a) top macroblock and (b) bottom macroblock.

predicted from the P top field picture just received whereas the bottom macroblock is predicted from the bottom reference field received some time earlier.

The situation for the prediction of a B field picture is shown in Figure 6.25. Interpolative motion-compensated prediction is used for the top macroblock requiring

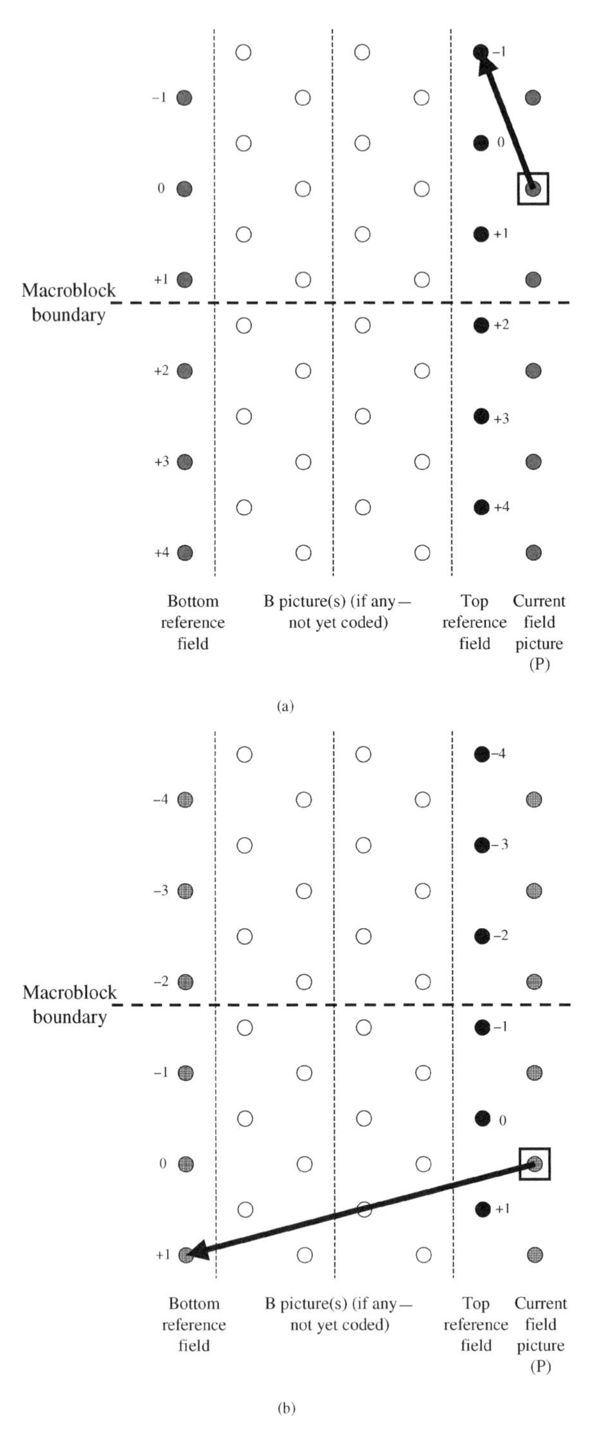

Figure 6.24 Field prediction of the second P field picture: (a) top macroblock and (b) bottom macroblock.

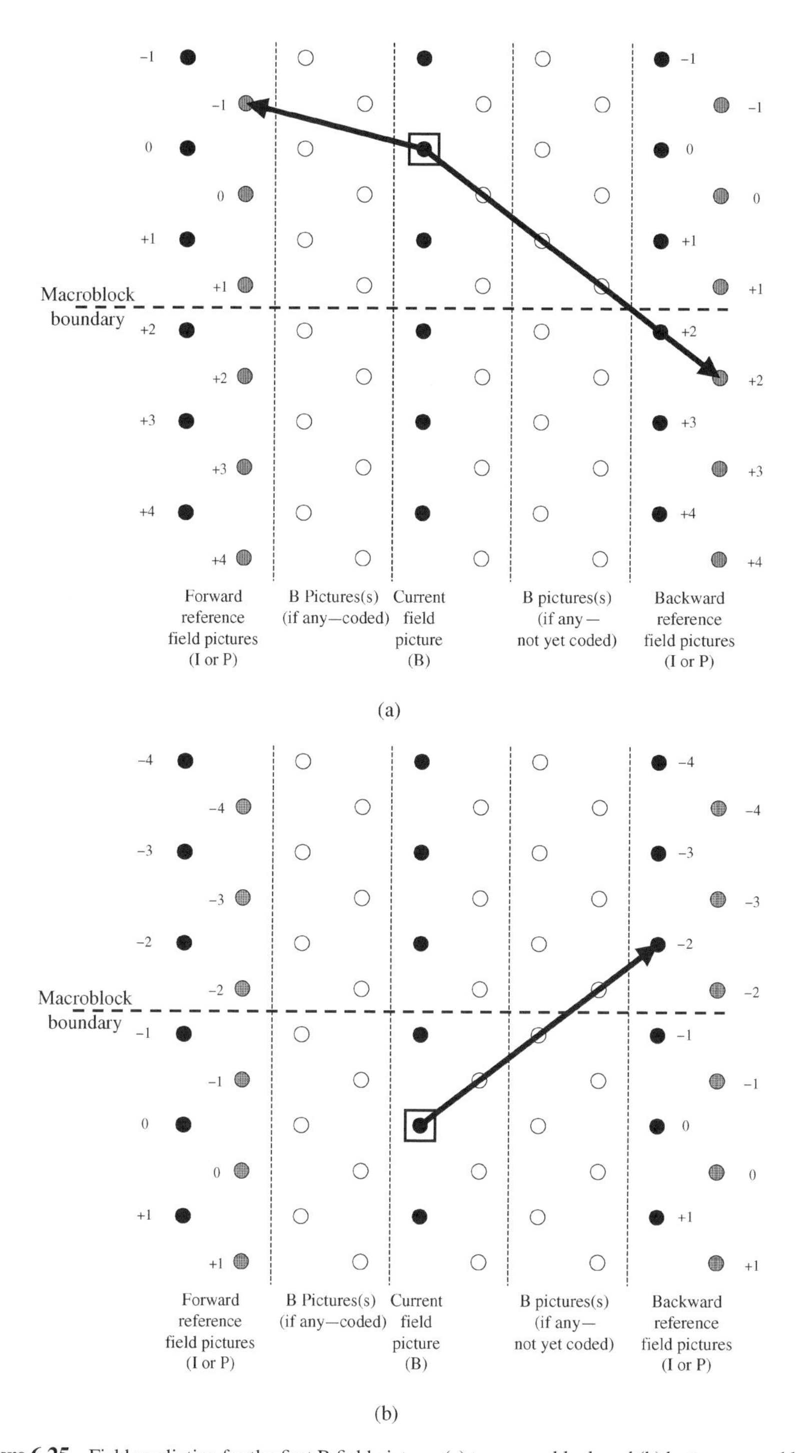

Figure 6.25 Field prediction for the first B field picture: (a) top macroblock and (b) bottom macroblock.

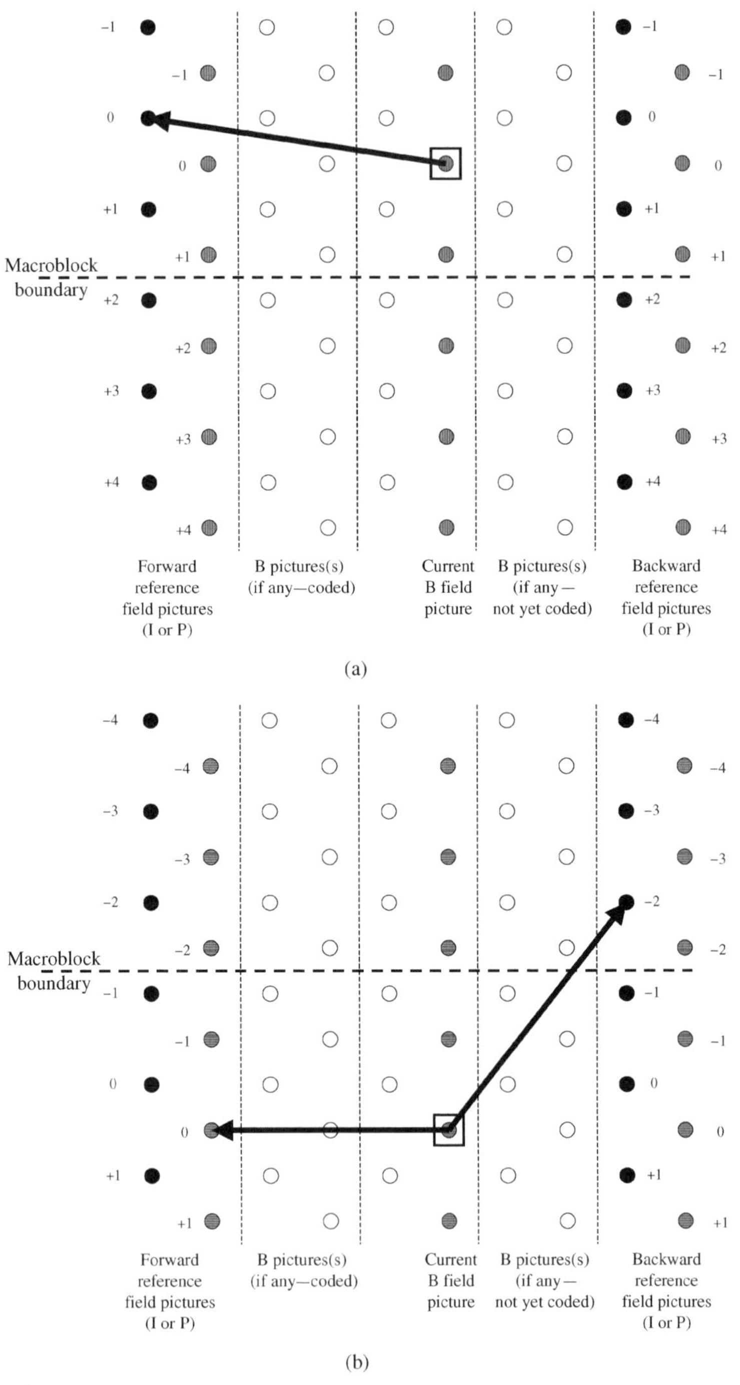

Figure 6.26 Field prediction for a second B field picture: (a) top macroblock and (b) bottom macroblock.

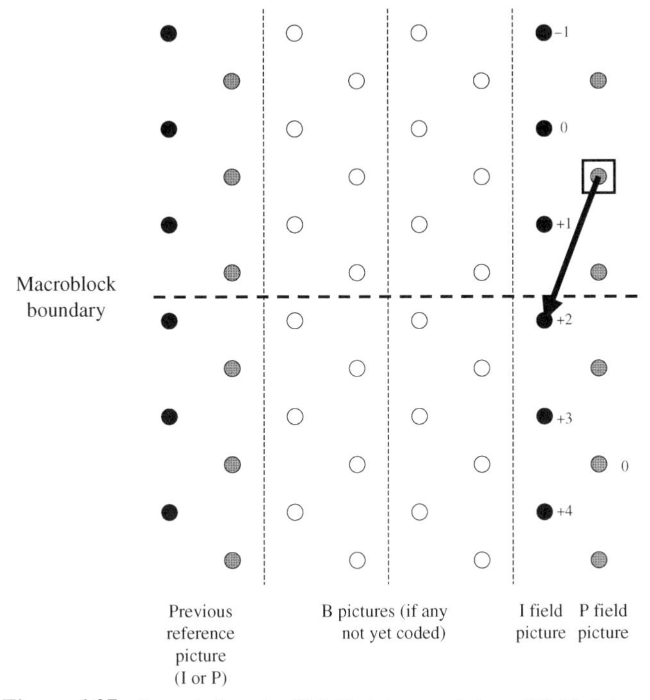

Figure 6.27 Second of a pair of I field pictures coded as a P field picture.

two motion vectors. Only backward prediction is used for the bottom macroblock, and so only a single motion vector is required. The case for the second B picture is shown in Figure 6.26. Unlike P pictures, the appropriate reference pictures for the second B picture are identical to those for the first, and so the prediction process is also identical.

There is another special type of prediction possible in field pictures. This occurs in the case of a pair of field pictures that make up an I frame picture. In this case, the second field picture can be coded as a P field picture. However, all predictions must be from the immediately preceding I field picture as shown in Figure 6.27.

As shown in Table 6.17, only a single motion vector is required for field prediction for macroblocks in P field pictures. For B field pictures, either one motion vector (forward prediction or backward prediction) or two motion vectors (interpolative prediction) are required.

Figure 6.28 shows the top field of the first picture of the "Mobile and Calendar" sequence. When compared to the full frame picture, it is clear that the picture changes twice as quickly vertically in a field picture as in the equivalent frame picture (i.e., the entire image is contained in half as many lines). This means that the chance of more than one type of motion within a macroblock (which is still 16×16 luminance pixels in a field picture) is increased. For this reason, a new motion compensated prediction mode has been introduced for field pictures and is called 16×8 motion compensation. In this mode, two motion vectors (four in the case of a macroblock in a B picture

Figure 6.28 Top field of the first picture of the mobile and calendar sequence.

where both forward and backward predictions are employed) are included in the macroblock. The first motion vector (pair of motion vectors) is used for the prediction of the top 16×8 pixels in the macroblock whereas the second motion vector (pair of motion vectors) is used for the prediction of the bottom 16×8 pixels in the macroblock.

Finally, dual-prime prediction is also available for field pictures. As shown in Figure 6.29, the field motion vector is used to form a prediction from the reference field of the same parity (top or bottom) as the current field picture. A further prediction is formed from the reference field of opposite parity in exactly the same manner as that for dual-prime prediction for frame pictures.

Motion Vector Prediction As was the case in the syntax described in Chapter 5, the values of the transmitted motion vectors in MPEG-2 are predicted using the values of the previously transmitted motion vectors. This prediction is set to zero after a slice header. This allows correct decoding of a new slice even if transmission errors

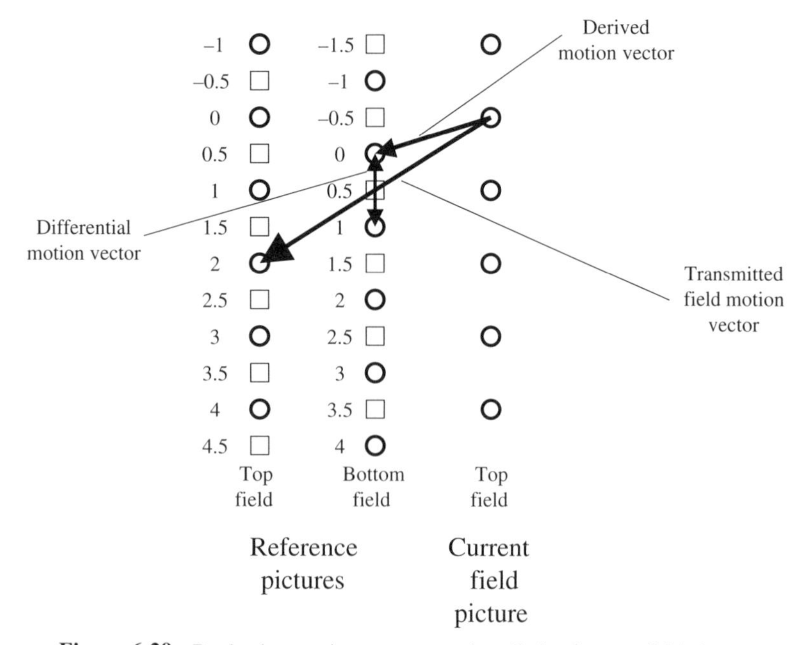

Figure 6.29 Dual-prime motion-compensated prediction for a top field picture.

prevented the correct decoding of the previous slice. For P pictures, the motion vector in the current macroblock is predicted from the motion vector in the previous macroblock, if it made use of motion-compensated prediction (i.e., was not coded as either an intramacroblock or an intermacroblock). Otherwise a prediction of zero is used. This implies that the motion vector predictor is set to zero after a skipped macroblock in a P picture.

For B pictures, there needs to be separate predictors for forward and backward motion vectors. In each case, the predictor is the most recently transmitted motion vector of that type in the current slice, if any. A skipped macroblock in a B picture does not set to zero the predictor because a skipped macroblock in a B picture is assumed to use the same form of motion-compensated prediction as the macroblock that came immediately before it.

This has given a brief introduction of the way that motion vectors are coded. There is considerably more detailed information contained in the standards documentation.

Type of DCT The **dct_type** flag indicates whether a macroblock in a frame picture uses frame DCT coding or field DCT coding. The difference between these two modes is shown in Figure 6.30.

In frame-based DCT, the 16×16 pixel macroblock is simply split into four 8×8 pixel blocks prior to transformation by the DCT. As shown, each block still contains pixels from both fields. In field-based DCT, the lines in the macroblock are rearranged so that all the lines from one field are contained in the top eight lines of the macroblock with the eight lines from the other field contained in the bottom eight lines. The rearranged macroblock is then split into 8×8 pixel blocks and then transformed by the DCT. As shown in Figure 6.30, this rearrangement ensures that all pixels within a block are from the same field.

Field-based DCT is useful for blocks where movement has occurred between fields. Figure 6.31 shows a simple macroblock containing an 8×8 pixel black block

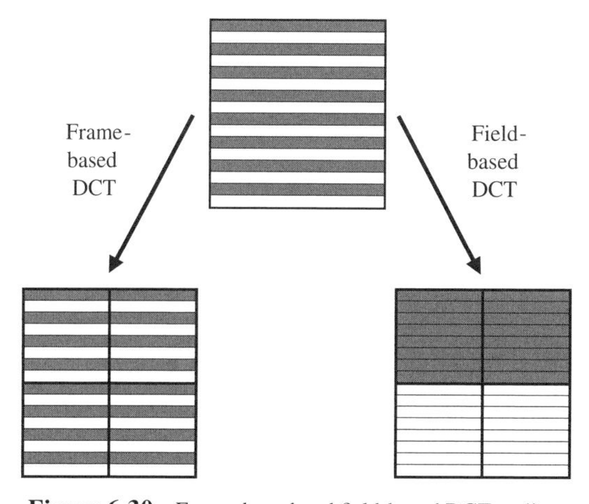

Figure 6.30 Frame-based and field-based DCT coding.

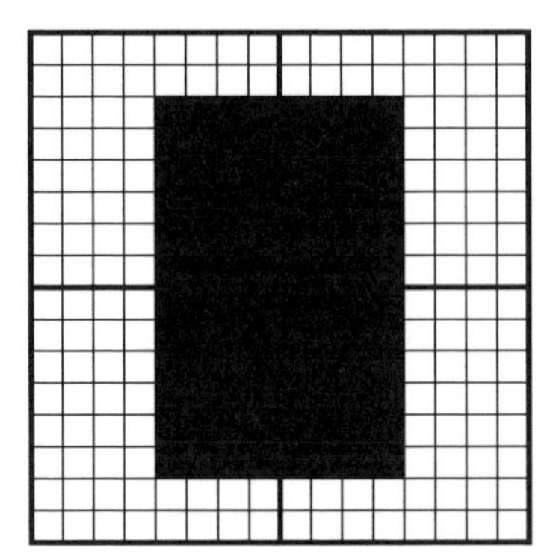

Figure 6.31 Example macroblock.

on a white background. If we assume that the block starts to move to the right at the rate of 2 pixels/field, then after the first field of motion, the macroblock would look as shown in Figure 6.32.

Looking at the pixels in columns 5, 6, 13, and 14, we observe high-frequency vertical changes of intensity. After the application of the DCT, considerable energy appears near the bottom left of the block of DCT coefficients. This leads to inefficient coding as a large number of bits are required to represent DCT coefficients in this area.

The macroblock after rearrangement for field-based DCT coding is shown in Figure 6.33. The high-frequency vertical changes are no longer present, and so improved coding efficiency after the DCT can be expected.

Coded Block Pattern The macroblock header can also contain a coded block pattern to indicate which blocks within the macroblock are coded where this needs to be specified. For digital television, which uses the 4:2:0 color sampling structure, this is identical to the use of the coded block pattern as described in Chapter 5.

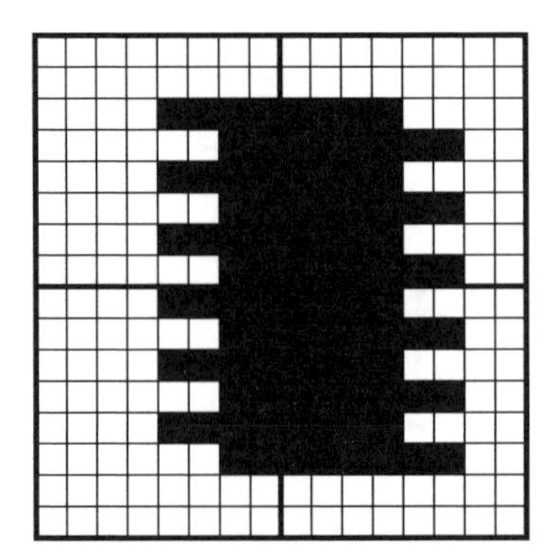

Figure 6.32 Object shown in Figure 6.31 with motion of 2 pixels/field.

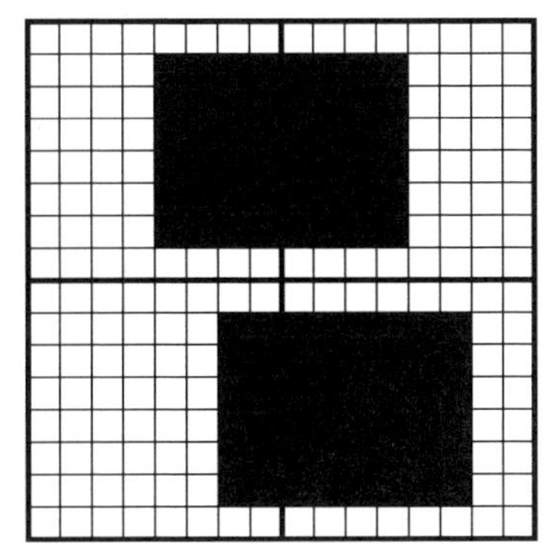

Figure 6.33 Macroblock shown in Figure 6.32 after rearrangement for field DCT coding.

6.3.7. The Block Layer

Coding at the block layer is very similar to what was discussed in Chapter 5. Again, rather than coding the actual value of DC DCT coefficients in intracoded blocks, the coefficient is predicted by the value of the previous intracoded block of the same type (i.e., *Y*, *U*, or *V*). The prediction difference is then transmitted. The predictor is reset at the start of a slice, when a non-intra-macroblock is decoded or when a macroblock is skipped. The last two of these only occur in P and B pictures. The value to which the predictor is reset is defined in Table 6.18. The reset value is the midpoint of the relevant DC range.

Figure 6.34 shows the prediction of DC DCT coefficients for *Y*, *U*, and *V* blocks. In the case of *Y* blocks, the top left block is predicted from the bottom right block of the previous macroblock, the top right block is predicted from the top left block, the bottom left block is predicted from the top right block, and the bottom right block is predicted from the bottom left block. For *U* and *V* blocks, the DC DCT coefficient is predicted from the *U* or *V* block in the previous macroblock.

As was the case in Chapter 5, the value of the prediction difference is coded in two parts. The first part gives the number of bits used for the prediction difference that immediately follows and is called **dct_dc_size_luminance** for luminance

Table 6.18 DC coefficient reset value for various DC coefficient precisions.

intra_dc_precision	Precision (bits)	Reset value
0	8	128
1	9	256
2	10	512
3	11	1024

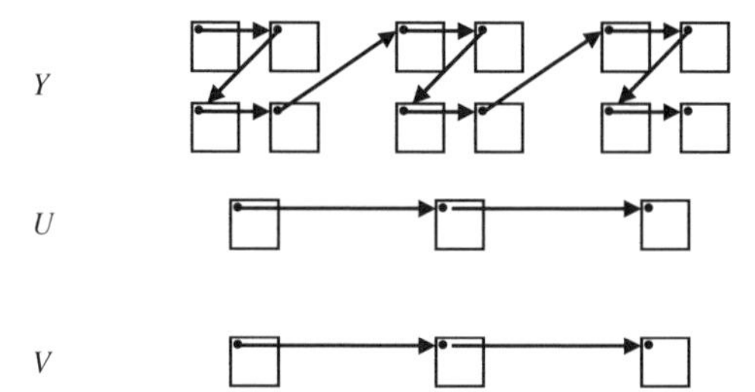

Figure 6.34 Prediction of DC DCT coefficients in intramacroblocks.

blocks and **dct_dc_size_chrominance** for chrominance blocks. The variable-length codes are given in Table 6.19.

So, a value of zero indicates that the difference has zero bits, that is, it has the value zero and so the DC DCT coefficient is the same as the relevant predictor. A value of one indicates that the prediction difference (**dct_dc_differential**) is represented by a single bit and so can take values −1 or +1. A value of two indicates that the prediction difference is represented by 2 bits and so can take the values −3, −2, +2, or +3. The full definition of **dct_dc_differential** is given in Table 6.20. The large range of possible values comes from the fact that DC coefficients can be represented by up to 11-bit accuracy (i.e., values from 0 to 2047) in MPEG-2 video. This means that the prediction difference can take values in the range of ±2047.

All other DCT coefficients in intrablocks as well as all coefficients in inter- or motion-compensated blocks are coded in the manner described in Chapter 5. However, there are two possible variable-length codes that can be used to encode the

Table 6.19 VLCs for **dct_dc_size_chrominance** and **dct_dc_size_luminance**.

Luminance		Chrominance	
dct_dc_size_ luminance	VLC	**dct_dc_size_chrominance**	VLC
0	100	0	00
1	00	1	01
2	01	2	10
3	101	3	110
4	110	4	1110
5	1110	5	11110
6	11110	6	111110
7	111110	7	1111110
8	1111110	8	11111110
9	11111110	9	111111110
10	111111110	10	1111111110
11	111111111	11	1111111111

Table 6.20 Definition of **dct_dc_differential**.

DC prediction difference	**dct_dc_differential**	Code word
−2047 to −1024	11	00000000000 to 01111111111
−1023 to −512	10	0000000000 to 0111111111
−511 to −256	9	000000000 to 011111111
−255 to −128	8	00000000 to 01111111
−127 to −64	7	0000000 to 0111111
−63 to −32	6	000000 to 011111
−31 to −16	5	00000 to 01111
−15 to -8	4	0000 to 0111
−7 to −4	3	000 to 011
−3 to −2	2	00 to 01
−1	1	0
0	0	None
1	1	1
2 to 3	2	10 to 11
4 to 7	3	100 to 111
8 to 15	4	1000 to 1111
16 to 31	5	10000 to 11111
32 to 63	6	100000 to 111111
64 to 127	7	1000000 to 1111111
128 to 255	8	10000000 to 11111111
256 to 511	9	100000000 to 111111111
512 to 1023	10	1000000000 to 1111111111
1024 to 2047	11	10000000000 to 11111111111

(run, level) pairs defining nonzero DCT coefficients. Blocks not coded in intramode are always coded using the VLC codes defined in Chapter 5. Intracoded blocks can also be coded in this way if the **intra_vlc_format** flag at the picture layer is zero. An alternate VLC code word set is used for intrablocks when the **intra_vlc_format** flag is set to one. Full details of the alternate VLC code words can be found in the standard documentation.

6.4. VIDEO BUFFER VERIFIER

The video buffer verifier is a hypothetical decoder that is connected directly to the output of a compliant encoder. The input buffer to the decoder, referred to as the VBV buffer, receives coded data from the encoder. Data is removed from the buffer at regular intervals. On each occasion, the data removed is that corresponding to an entire coded picture. In order to conform to the MPEG specification, a bit stream is not permitted to cause the VBV buffer either to overflow or to underflow. As described earlier, the VBV buffer size and the bit rate of the encoder are specified at the sequence layer.

The **vbv_delay** field given in the picture header specifies the number of periods of a 90-kHz clock (which is derived from the 27-MHz system clock) that the VBV waits after receiving the final byte of the picture start code before decoding that picture.

For constant bit-rate operation,[19] the value to be placed in the **vbv_delay** field can be calculated according to the following equation.

$$\text{vbv\_delay}_n = \frac{90{,}000 \times B_n}{R}$$

where B_n is the VBV buffer occupancy immediately before removing picture n from the buffer but after removing any headers, user data, and stuffing that immediately precede the data elements of picture n and R is the actual bit rate as opposed to the value given in the sequence header that is rounded up to the next multiple of 400 bits/s.

Picture data of the nth coded picture enters the VBV buffer at a rate $R(n)$ as defined below.

$$R(n) = \frac{d_n^*}{\left(\tau_n - \tau_{n+1} + t_{n+1} - t_n\right)}$$

where d^*_n is the number of bits from the final bit of the nth picture start code to the final bit of the $(n+1)$th picture start code, τ_n the decoding delay coded in the **vbv_delay** field for the nth coded picture in seconds, and t_n is the time when the nth coded picture is removed from the VBV buffer.

For constant bit-rate operation, the value of $R(n)$ remains constant throughout the sequence to the accuracy allowed by the quantization of **vbv_delay**.

Prior to the receipt of the first picture start code of a video sequence and following the final picture start code of a video sequence, $R(n)$ is assumed to take the value specified in the **bit_rate** field.

At the start of a video sequence, the VBV buffer is filled with all of the header data up to and including the first picture start code. The buffer is then further filled with the incoming bit stream for the time specified in the **vbv_delay** field of this first picture header at which time decoding (i.e., the removal of the first picture from the VBV buffer) begins. This is illustrated in Figure 6.35.

From the time when decoding begins, the VBV buffer is regularly examined.[20] At the time when the VBV buffer is examined and prior to the removal of any picture data, the total number of bits in the buffer lies between zero and the maximum VBV buffer size. All of the data for the picture that has been in the buffer for the maximum time at the time of examination is then instantaneously removed. This must not result in buffer underflow (i.e., the number of bits in the buffer cannot be negative).

The time at which the VBV buffer is examined depends upon whether the sequence is interlaced or progressive, whether the picture structure is field or frame,

[19] This is the usual situation in digital television applications.

[20] The exact times are defined a little later.

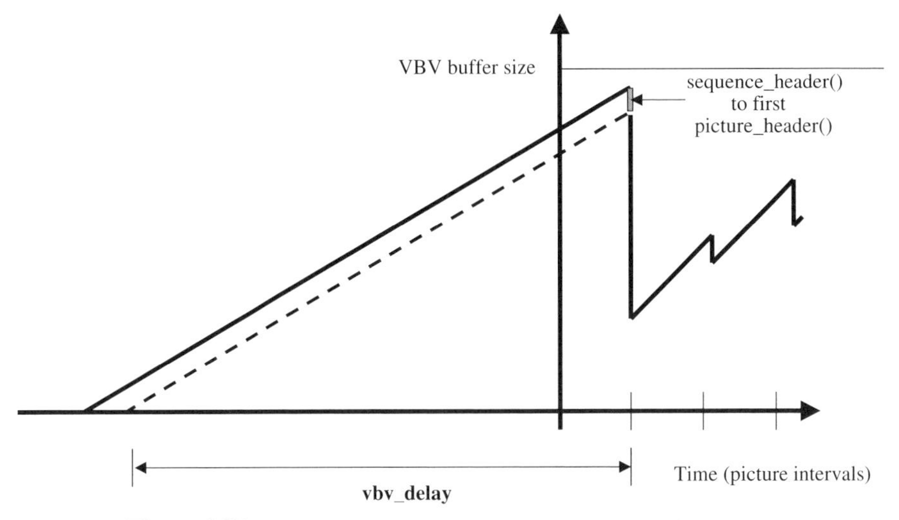

Figure 6.35 VBV buffer occupancy up to the removal of the first picture.

whether the value of the **repeat_first_field** field in the picture coding extension is set to one or not, and on the type of picture (field or frame) that is being decoded, the frame/field structure, and the state of the **repeat_first_field** flag in a previous picture.

The time for which the VBV buffer is examined is determined by the pictures in display order as opposed to the order in which they are removed from the buffer. Although B pictures are decoded and displayed immediately upon their removal from the buffer, I and P pictures are decoded immediately but may not be displayed for some time as they are transmitted before B pictures, which precede them in display order.

We use some examples to make the situation clear.[21] Let us first consider the case of a video sequence that does not contain B pictures. In this case, the picture transmission order is the same as the picture display order. The time for which the VBV is examined is then determined by the display time of each picture. Thus, if the current picture is a field picture, the VBV is examined after one field time. Alternatively, if the current picture is a frame picture then the VBV is examined after two field times if **repeat_first_field** is not set to one and after three field times if **repeat_first_field** is set to one. In each case, the time between VBV examinations is just the display time of the current picture. This is illustrated in Figure 6.36 where it is assumed that all pictures are frame pictures and that pictures I_0, P_2, and P_4 have the **repeat_first_field** field set to one and so are displayed for three field times each.

The situation is slightly more complex when B pictures are included. However, the time between examinations is still determined by the display time of the picture currently being displayed. Since B pictures are displayed immediately,

[21]Only interlaced video sequences are considered here. For progressive sequences, there are no field pictures and any field repetition implies that the entire progressive picture is repeated.

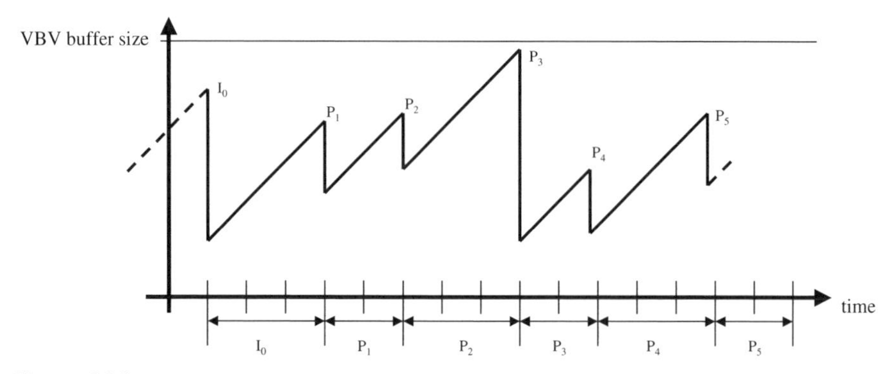

Figure 6.36 Typical VBV for interlaced video sequence without B pictures.

when a B picture is being decoded, the time between VBV examinations is one field time in the case of a field-structured B picture, two field times in the case of a frame-structured B picture when **repeat_first_field** is not set to one, and three field times in the case of a frame-structured B picture when **repeat_first_field** is set to one.

When B pictures are included, any subsequent I or P picture that is needed to predict the B pictures is transmitted before these B pictures. Thus, when an I or P picture is received, it implies that all the B pictures prior to the previously received I or P pictures have been received and decoded, and therefore, it is time for this previously received I or P picture to be displayed. The time between VBV examinations when I or P pictures are being decoded is therefore the time required to decode this previously received I or P picture. The time between VBV examinations is two field times if the previous I or P picture is frame structured and **repeat_first_field** is not set to one, three field times if the previous I or P picture is frame structured and

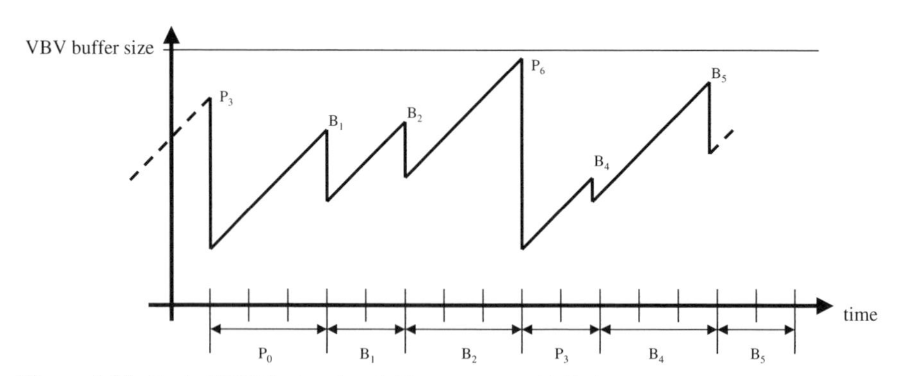

Figure 6.37 Typical VBV for interlaced video sequence with B pictures.

repeat_first_field is set to one, one field time if the current I or P picture is the first field of a field-structured picture, one field time if the current I or P picture is the second field of a field-structured picture and **repeat_first_field** is not set to one, and two field times if the current I or P picture is the second field of a field-structured picture and **repeat_first_field** is set to one. This is illustrated in Figure 6.37 for a simple case with only frame pictures. In Figure 6.37, P_0, B_2, and B_4 have display durations of three fields.

6.5. PROFILES AND LEVELS

The complete MPEG-2 video standard is designed to cater to a large number of different applications ranging from videoconferencing to high-definition television and beyond. For an encoder or decoder to fully comply with the standard, it would need to be able to handle all of these applications. Such encoders and decoders would be far more complex than necessary for their intended applications and would involve significant unnecessary expense. For this reason, MPEG has defined a number of profiles and levels for the video standard. These allow compliance with the standard at a point that is more suitable for a particular application.

A profile specifies a defined subset of the complete syntax of the standard. This means that profiles do not use all of the coding tools provided by the standard. For example, it would be possible to define a profile that did not allow temporal prediction (i.e., either interprediction or motion-compensated prediction) between pictures. Such a profile would lead to inexpensive encoders and decoders because interpicture predictions (and in particular, motion-compensated prediction) are computationally complex operations. In fact, no such profile currently exists.

A level places constraints on the values that may be taken by parameters in the bit stream. Parameters include values such as picture size, picture rate, and bit rate. Placing a limit on the maximum values of these parameters also leads to simpler encoders or decoders. For example, limiting the maximum picture size reduces the memory requirement for internal frame stores whereas reducing the bit rate reduces the speed of operation of the variable-length code encoders and decoders.

At the time of writing, MPEG has defined seven profiles and four levels. We now briefly discuss each of these profiles and levels.

6.5.1. Profiles

6.5.1.1. Simple Profile

The simple profile is intended for real-time applications. As such, round-trip delays are important, and so additional delays in encoding and decoding need to be avoided. For this reason, B pictures are not allowed in the simple profile. In addition, the simple profile uses the restricted slice structure and only the 4:2:0 color sampling structure. The simple profile is defined only for main level.

6.5.1.2. Main Profile

Main profile is the profile for digital television services. The only difference between simple profile and main profile is that main profile allows B pictures. Main profile is defined for all four levels (low, main, high 1440, and high).

6.5.1.3. Signal-to-Noise Ratio (SNR) Scalable Profile

The scalable profiles in MPEG allow a service to be made up of more than one transmitted bit stream. Receiving just the base-level bit stream provides a service of a particular quality. Receiving the base and one (or more) enhancement bit streams provides a service of enhanced quality. In the case of SNR scalability, the enhancement is in the picture quality of the service. Using this profile, a maximum of two bit streams (base plus a single enhancement) are allowed. The SNR scalable profile is defined for low and main levels.

6.5.1.4. Spatial Scalable Profile

In this case, the enhancement of quality produced by the enhancement bit stream is an increase in the spatial resolution of the displayed pictures. This might mean that the base-level service is at standard-definition television resolution whereas the enhanced service could be at high-definition television resolution. A single SNR scalable bit stream is also permitted, and this may be applied at either resolution level. The spatial scalable profile is defined only for the high-1440 level.

6.5.1.5. High Profile

The high profile is identical to the spatial scalable profile except that both 4:2:0 and 4:2:2 color sampling structures are allowed. The high profile is defined for main, high 1440, and high levels.

6.5.1.6. Professional Profile

The professional profile has been defined for use within television stations where higher quality levels are required so that subsequent postprocessing of the video material is possible. It needs to allow higher bit rates and is defined primarily for main-level applications. However, as it can encode all the active lines of video (as opposed to those defined by CCIR Recommendation 601), it does not strictly conform to any of the currently defined levels.

6.5.1.7. Multiview Profile

The multiview profile allows for the efficient coding of a service comprising more than one view such as a stereo television. It is defined for the four existing levels (low, main, high 1440, and high).

Table 6.21 Parameter values for various levels of MPEG-2.

Parameter	Low level	Main level	High-1440 level	High level
Horizontal size (pixels)	352	720	1,440	1,920
Vertical size (pixels)	288	576	1,152	1,152
Picture rate (Hz)	30	30	60	60
Pixel rate (s^{-1})	3,041,280	10,368,000	47,001,600	62,668,800
Bit rate (Mbits/s)	4	15	60	80
Bits/picture (kbits)	167	626	2,503	3,337
VBV buffer size (bits)	475,136	1,835,008	7,340,032	9,781,248
Motion vector range	−64 to +63.5	−128 to + 127.5	−128 to + 127.5	−128 to + 127.5
Intra-DC precision (bits)	8–10	8–10	8–11	8–11

6.5.2. Levels

The parameter restrictions for the various defined levels of MPEG-2 are shown in Table 6.21. For profiles that include scalability, there is a need for separate definition of some of the parameters for each bit stream (base or enhancement). As scalability is not used in current digital television services, this additional information has been omitted for clarity.

6.6. SUMMARY

This chapter has provided an overview of the MPEG-2 video compression standard that forms the basis of current digital television services worldwide. The aim of the chapter has been to introduce the reader to the main features and functionalities of the MPEG-2 standard. It has not attempted to provide sufficient depth to allow the reader to completely implement the standard. However, armed with the information contained in this chapter and a copy of the standard itself, a competent student should be capable of implementing a compliant encoder or decoder.

PROBLEMS

6.1 Calculate the compression ratio required to represent a 720 × 480 pixel, 30 frames/s interlaced service with 4:2:0 chrominance structure at a total rate of 4 Mbit/s. Repeat your calculation for a 720 × 576 pixel, 25 frames/s interlaced service again with 4:2:0 chrominance structure.

6.2 List the advantages and disadvantages of each of the picture types (I, P, and B) used in the MPEG-2 video standard.

6.3 Why are profiles and levels important for MPEG-2 encoder and decoder manufacturers?

6.4 A series of pictures in a video sequence are received by an encoder and coded as the picture types shown in Figure 6.38. Determine the order in which the pictures will be transmitted to the decoder.

1	2	3	4	5	6	7	8	9	10	11	12	13	14	15	16	17	18	19	20
I	B	P	B	B	P	P	I	B	B	B	B	P	I	I	B	P	B	B	I

Figure 6.38 Picture types for Problem 6.4.

6.5 For the series of pictures and picture types given in Figure 6.38, indicate which picture(s) are used as reference picture(s) for the motion-compensated prediction of each picture.

6.6 A series of coded pictures is received by an MPEG-2 decoder with the picture types shown in Figure 6.39.

1	2	3	4	5	6	7	8	9	10	11	12	13	14	15	16	17	18	19	20
I	P	B	B	B	I	P	P	B	B	I	I	P	P	B	B	I	B	B	I

Figure 6.39 Picture types for Problem 6.6.

(a) In what order should these pictures be displayed?

(b) What pictures are held in the decoder frame stores as each picture is received?

(c) What pictures are used as reference pictures for each received picture?

6.7 Although some MPEG-2 encoders use a regular sequence of picture types for simplicity, there is no requirement for this. Thus, an intelligent encoder is able to choose the picture type according to the characteristics of each picture to be encoded.

(a) What characteristics would the pictures have that would make them particularly suitable for encoding as a particular picture type?

(b) Suggest some simple algorithms by which an encoder might go about making this decision.

6.8 In MPEG-2 video, great care is taken to ensure that not more than 23 consecutive zeros can occur. Explain briefly why this is important.

6.9 In the production of services for digital television, bit streams from different sources often need to be edited together.

(a) Why is it desirable to perform editing at the bit stream level rather than decoding and then editing the decoded pictures prior to a further encoding stage?

(b) Explain how the use of the **closed_gop** and **broken_link** flags in the GOP header can be used to assist the bit stream editing process.

6.10 The MPEG-2 video standard allows a number of motion vector ranges to be defined. Describe the advantages and disadvantages of using a large motion vector range.

6.11 Concealment motion vectors can be transmitted in intramacroblocks. Describe how these concealment motion vectors might be used if transmission errors occur.

6.12 The first four macroblocks after a slice header are all coded as P macroblocks with the prediction modes and motion vectors given in Table 6.22. Calculate the differential motion vector values transmitted for each macroblock.

Table 6.22 Prediction modes and motion vectors for Problem 6.12.

Macroblock	1	2	3	4
Prediction mode	Forward	Skipped	Forward	Forward
Motion vector	(−1.0, −1.0)	—	(−1.5, +1.0)	(−2.0, +2.5)

6.13 The first four macroblocks after a slice header are all coded as B macroblocks with the prediction modes and motion vectors given in Table 6.23. Calculate the differential motion vector values transmitted for each macroblock.

Table 6.23 Prediction modes and motion vectors for Problem 6.13.

Macroblock	1	2	3	4
Prediction mode	Backward	Interpolated	Forward	Interpolated
Forward motion vector	—	(−2.0, −4.0)	(−3.5, −4.0)	(−3.0, −3.5)
Backward motion vector	(+3.5, −2.5)	(+2.0, −2.0)	—	(+3.0, +4.5)

6.14 Under what circumstances is 3:2 pulldown useful in MPEG-2 video systems? How does it work and what are the advantages?

6.15 What is the use of pan and scan in MPEG-2 video? How is it implemented and in what circumstances is it useful?

6.16 Two coded macroblocks in the same slice are separated by 35 skipped macroblocks. What is the variable-length coded value of **macroblock_address_increment** that is transmitted to the decoder?

6.17 The field motion vector for dual-prime prediction in a P frame picture is (−8.5, + 6.5). The differential motion vector is (+0.5, +0.5). Calculate the displacement of the pixels to be used to predict each field in the macroblock. You may assume that the top field is transmitted first for each picture.

6.18 The first eight coded macroblocks in a particular slice taken from a P picture have the **macroblock_address_increment**, macroblock type (I or P), and raw motion vectors as shown in Table 6.24. What are the unencoded values of the motion vectors transmitted to the decoder after the appropriate motion vector prediction?

Table 6.24 Macroblock types and motion vectors for Problem 6.18.

Macroblock address increment	Macroblock type	Raw motion vector
3	P	(+7.5, −9.0)
2	P	(+8.0, −9.5)
1	I	—
1	P	(+6.0, −5.0)
2	P	(+6.0, −5.0)
1	I	—
1	P	(+6.0, −5.0)
1	P	(+6.0, −5.0)

6.19 The first eight coded macroblocks in a particular slice taken from a B picture have the **macroblock_address_increment**, macroblock type (I, B forward prediction, B backward prediction, and B interpolated prediction), and raw motion vectors as shown in Table 6.25. What are the unencoded values of the motion vectors transmitted to the decoder after the appropriate motion vector prediction?

Table 6.25 Macroblock types and motion vectors for Problem 6.19.

Macroblock address increment	Macroblock type	Raw forward motion vector	Raw backward motion vector
3	B interpolated	(+4.5, −5.0)	(−3.5, +4.0)
2	B forward	(+4.0, −6.5)	—
1	B backward	—	(−4.5, +3.5)
1	B backward	—	(−5.0, +2.0)
2	B interpolated	(+5.0, −5.0)	(−1.0, +3.5)
1	I	—	—
1	B backward	—	(0.0, +1.0)
1	B forward	(+6.0, −5.0)	—

6.20 The DC coefficients of the first two macroblocks of an I picture are shown below. Calculate the variable-length code words transmitted to the decoder to represent each of these coefficients. Assume that DC coefficients are represented to 10-bit accuracy.

Macroblock 1	Macroblock 2
$Y_1 = 643$	$Y_1 = 685$
$Y_2 = 659$	$Y_2 = 732$
$Y_3 = 801$	$Y_3 = 847$
$Y_4 = 894$	$Y_4 = 922$
$U = 140$	$U = 184$
$V = 278$	$V = 278$

MATLAB EXERCISE 6.1: BIDIRECTIONAL MOTION-COMPENSATED PREDICTION

In this exercise we study the effectiveness of bidirectional motion-compensated prediction.

Section 1 Forward and backward prediction
Write a MATLAB program that will perform forward motion estimation and backward motion estimation from a current picture to a past and a future picture, respectively. Use minimum absolute error as the matching criteria and generate a motion-compensated prediction picture from both the past and future pictures.

It is now necessary to decide whether forward or backward prediction is to be employed. Do this by choosing as the predictor the forward or backward prediction macroblock that has the minimum-squared error to the macroblock in the current picture.

Summarize your results under the following circumstances:

- The past, current, and future pictures are all from the same video sequence.
- The past picture is from one video sequence and the current and future pictures are from a different video sequence. This simulates a cut prior to the current picture.
- The past and current pictures are from one video sequence whereas the future picture is from a different video sequence. This simulates a cut after the current picture.

Consider the circumstances where an inappropriate prediction (i.e., forward instead of backward) is made. Suggest and apply tests that might improve the accuracy of this decision.

Section 2 Interpolative prediction
Repeat Section 1 of this exercise, but this time allow for interpolated prediction as well as forward and backward predictions. The interpolated prediction is formed by averaging the forward and backward motion-compensated prediction pictures.

In what circumstances is interpolative prediction useful?

MATLAB EXERCISE 6.2: DUAL-PRIME MOTION-COMPENSATED PREDICTION

In this exercise you will implement dual-prime motion-compensated prediction and compare its performance to other forms of motion-compensated prediction.

Section 1 Implementation of dual-prime motion-compensated prediction
Using the description contained in the chapter, implement dual-prime motion-compensated prediction for frame pictures. Use a full search motion estimation approach with a search range from -8.0 to $+7.5$ pixels and for each field motion vector. For each possible field motion vector you should consider each of the nine possible differential motion vectors.

Test your dual-prime program on the first two pictures of a video sequence. Compare the result obtained with that obtained using frame-based motion-compensated prediction with the same search range.

If time permits, also experiment with pictures from other video sequences.

Section 2 Speeding up dual-prime motion-compensated prediction
Rather than considering the differential motion vector in association with each possible field motion vector, simply calculate the best field motion vector and then calculate the appropriate differential vector just for this "best" field motion vector. Comment on the change in performance.

MATLAB EXERCISE 6.3: FIELD AND FRAME MOTION-COMPENSATED PREDICTION

In this exercise we will consider the advantages of using adaptive field/frame motion-compensated prediction in frame pictures.

Section 1 Comparison of field- and frame-based motion-compensated prediction
Use MATLAB to implement frame-based motion-compensated prediction and field-based motion-compensated prediction on a pair of frame pictures. Choose the prediction type according to which of the two produces the lower value of squared error for each macroblock in the current picture.

Experiment with pictures from the available sequences, and comment on the characteristics of macroblocks that require field as opposed to frame coding.

Section 2 Realistic field/frame decision
The use of field motion-compensated prediction carries a penalty in that two motion vectors are required as opposed to a single motion vector for frame motion-compensated prediction. For this reason, the decision whether to use field or frame-based prediction should be biased slightly toward frame prediction.

One way to do this is to introduce a threshold so that if

$$\mathrm{MSE}_{\mathrm{frame}} < \mathrm{MSE}_{\mathrm{field}} + \mathrm{threshold}$$

then frame prediction is employed. Incorporate this threshold into the simulation of Section 1 of this exercise and comment on the results achieved.

MATLAB EXERCISE 6.4: FIELD AND FRAME DCT CODING

In this exercise, we examine the implementation of adaptive field/frame DCT coding of pictures both with and without motion-compensated prediction.

Section 1 Field and frame DCT
Develop MATLAB code to implement both field-based DCT and frame-based DCT.

It is now necessary to find a criterion to determine whether to use field- or frame-based DCT. One simple approach would be to look at the amount of energy packed in low-frequency DCT coefficients in each case. Use the energy in the low-frequency 4×4 DCT coefficients of each block in a macroblock to determine whether the macroblock should be coded in field or frame mode.

Section 2 Field/frame decision prior to DCT
In the development of the MPEG-2 video standard, a decision was made in the pixel domain as to whether frame- or field-based DCT should be employed. The method employed is described below.

1. Determine the sum-squared difference between lines 1 and 3, 2 and 4, 3 and 5, . . . 14 and 16 and call this var1,
2. Determine the sum-squared difference between lines 1 and 2, 2 and 3, 3 and 4, . . . 14 and 15 and call this var2.
3. If var1 is less than var2, then use field DCT coding. Otherwise use frame DCT.

This is a simple method for estimating whether correlation is greatest between every second line (as would be expected in macroblocks suited to field DCT) or between adjacent lines (as would be the case in macroblocks suited for frame DCT).

Implement this approach and comment on its ability to correctly choose between frame and field DCT.

Section 3 Incorporation of motion-compensated prediction
The previous two sections have concentrated on intra macroblocks only. Extend your simulation to include field and frame motion-compensated prediction as well. MATLAB Exercise 6.3 provides details on how to do this.

There are now four possible types of macroblocks:

- Frame motion-compensated prediction with frame DCT coding.
- Frame motion-compensated prediction with field DCT coding.
- Field motion-compensated prediction with frame DCT coding.
- Field motion-compensated prediction with field DCT coding.

What characteristics do macroblocks that fall under each of these categories possess?

Chapter 7

Perceptual Audio Coding

The term *audio coding* usually refers to the source coding of digital audio signals, including high fidelity music, with bandwidths of up to 24 kHz. The first digital audio coding algorithms were developed in the early 1970s and used techniques such as adaptive differential pulse code modulation (ADPCM) and logarithmic quantization. These algorithms achieved typical compression ratios of 4:1 while maintaining high-quality output.

Since the early 1980s audio compression techniques have employed models of the human auditory system to achieve higher compression ratios while still maintaining good perceptual audio quality. This technique is known as *perceptual audio coding* (PAC), and a typical block diagram of an audio coder employing this technique is shown in Figure 7.1.

The *analysis* stage involves converting the time-domain signals of the sampled audio signal into frequency-domain coefficients. Because there is usually a high degree of correlation between successive audio samples, audio signals can be generally represented more efficiently in the frequency domain.

The *psychoacoustic model* stage is used to identify frequency components of the audio signal that do not contribute to the perceived quality of the reconstructed signal. These perceptually redundant components can then be encoded using fewer bits than other components that are necessary to reproduce an audio signal with a high perceptual quality.

The *dynamic quantizer* stage then uses the information produced by the psychoacoustic model to calculate the required quantizer step size for each frequency band in the audio signal. Each frequency coefficient produced by the analysis stage is then quantized using the required step size for that band.

The *bit-stream formatting* stage takes the quantized coefficients and transforms them into a binary representation. This transformation into binary data will also usually include entropy coding techniques such as Huffman coding. Any formatting overhead bits are also added at this stage such as frame headers and error correction bits.

Digital Television, by John Arnold, Michael Frater and Mark Pickering.

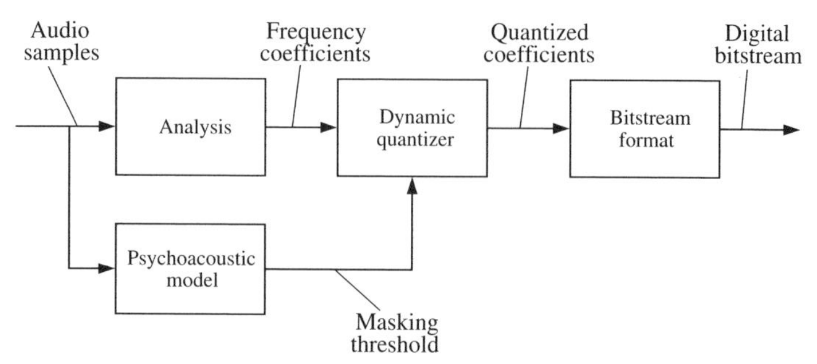

Figure 7.1 Typical block diagram of a perceptual audio coder.

The psychoacoustic model forms an integral part of any perceptual audio coder, and it is this component that is responsible for providing high-quality audio output at compression ratios of up to 12:1. In order to learn about this stage however, it is first necessary to gain a basic understanding of the human auditory system. In the remainder of this chapter, the basic anatomy and psychoacoustic properties of the human auditory system are explained.

7.1. THE HUMAN AUDITORY SYSTEM

The human auditory system can be divided into three main areas: the *outer ear*, the *middle ear*, and the *inner ear* as shown in Figure 7.2. The *pinna*, *ear canal* (*meatus*), and *eardrum* (*tympanic membrane*) form the outer ear. Sound waves in the air travel down the ear canal and cause the eardrum to vibrate. The middle ear is an air-filled cavity that contains a group of three small bones called the *ossicles*. The ossicles

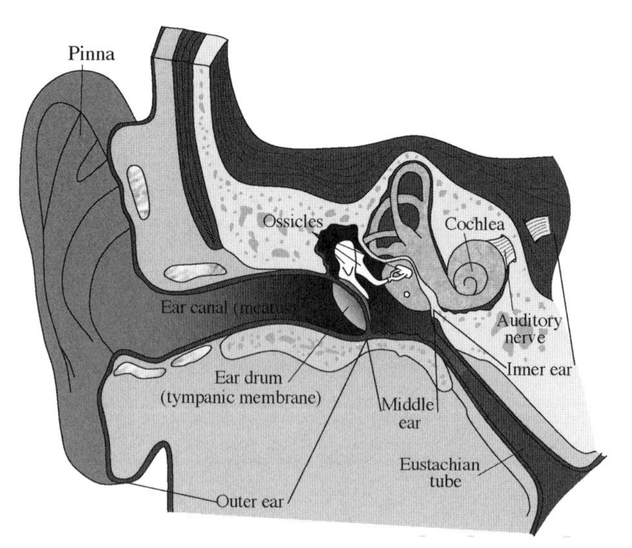

Figure 7.2 Parts of the ear.

convert the vibration of the eardrum to pressure waves in the fluid contained in the cochlea. The major components of the inner ear are the cochlea and the auditory nerve. The cochlea is responsible for transforming the mechanical vibrations of the ossicles into nerve impulses in the auditory nerve.

7.1.1. Outer Ear

The most outer part of the ear is called the pinna as shown in Figure 7.2. The shape of the pinna significantly modifies the sound waves arriving at the entrance to the ear canal. The changes in the sound waves made by the pinna depend on the direction at which they arrive. It is the shape of the pinna that allows us to determine the location of sound sources. The ear canal is a tube about 2.7 cm in length and 0.7 cm in diameter that is terminated at one end by the eardrum or tympanic membrane. The eardrum is a thin membrane containing a layer of radial fibers that form a stiff cone with an angle at its apex of about 135°.

7.1.2. Middle Ear

The individual names for the three bones that form the ossicles are the *malleus*, *incus*, and *stapes* as shown in Figure 7.3. They are more commonly called the *hammer*, *anvil*, and *stirrup* and are famous for being the smallest bones in the body. The hammer is connected directly to the eardrum at one end and makes contact with the anvil at the other. When the eardrum vibrates, this mechanical motion is passed on to the stirrup via the hammer and anvil. The head of the stirrup is connected to the anvil in a ball and socket joint and the *footplate* of the stirrup is attached to a membrane covered opening in the wall of the cochlea called the *oval window*. As

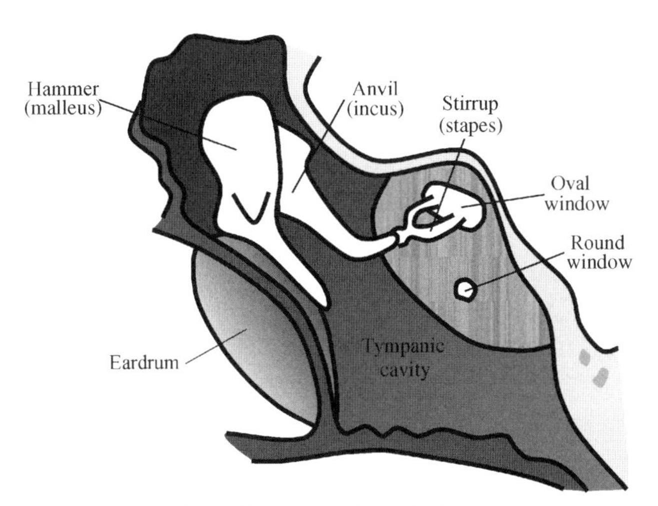

Figure 7.3 Parts of the middle ear.

the stirrup vibrates, the oval window is pushed inward and pulled outward by the footplate causing a pressure wave to be induced in the fluid of the cochlea.

The mechanical impedance of this fluid is much greater than for the eardrum, and if sound waves were to operate directly on the inner ear, they would have little effect. Because the effective area of the oval window is much smaller than that of the eardrum, the middle ear acts as a mechanical impedance matching device and allows vibrations of the eardrum to be transferred efficiently to the inner ear.

The ossicles of the middle ear also protect the inner ear from damage caused by loud sounds. The bones of the ossicles are connected to the walls of the tympanic cavity by a series of ligaments that limit their range of motion in response to vibrations in the eardrum. Once the intensity of the sound reaches a certain level, these ligaments limit the movement of the anvil, and the intensity of the pressure wave in the cochlea is also limited.

7.1.3. Inner Ear

The *cochlea* and *auditory nerve* perform the major functions of the inner ear. The cochlea forms a cone-shaped spiral of approximately two and a half turns, and the auditory nerve is intertwined throughout the center of this spiral. The main coiled structure of the cochlea is approximately 35 mm in length and has a cross-sectional area of approximately 4 mm^2 at the base and 1 mm^2 at the apex. This structure consists of two fluid-filled chambers, the *scala vestibuli* and the *scala tympani*, that are separated by the *cochlea partition*. Figure 7.4 shows a simplified illustration of the cochlea as it would look if the chambers were uncoiled. The two chambers are filled with a colorless liquid called *perilymph*. Note that the cochlea partition does not completely separate the two halves of the cochlea. There is a small opening at the apex of the cochlea partition called the *helicotrema* that allows the perilymph to flow between the two chambers.

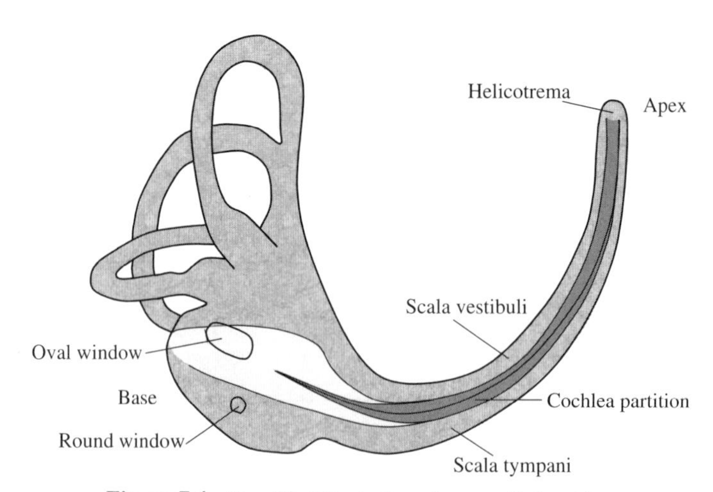

Figure 7.4 Simplified illustration of an uncoiled cochlea.

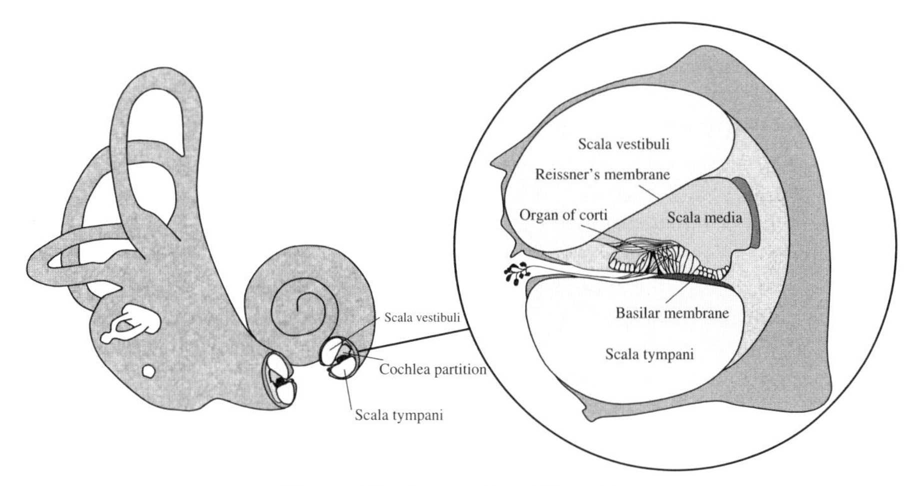

Figure 7.5 Cross section of the cochlea.

Figure 7.5 shows an expanded view of the cross section of the cochlea. The cochlea partition mainly consists of a self-contained chamber called the *scala media*. The scala media is filled with a fluid called *endolymph* and is separated from the scala vestibuli by *Reissner's membrane* and from the scala tympani by the *basilar membrane*.

The basilar membrane is attached to a complex system of hair cells and supporting structures called the *organ of Corti*. The organ of Corti contains approximately 30,000 sensory cells and is responsible for translating the mechanical movements of the basilar membrane into neural impulses. Figure 7.6 shows a more detailed view of the organ of Corti.

When the stirrup in the middle ear vibrates in response to the eardrum, the footplate induces a pressure wave in the fluid of the scala vestibuli. As this pressure wave travels down the length of the chamber, a fixed point on the cochlea partition experiences alternating levels of high and low pressure in the fluid at that location. An increase in pressure in the fluid of the scala vestibuli causes the cochlea partition to bend toward the scala tympani and hence transfer the pressure increase to the fluid in the scala tympani. This pressure increase is absorbed by the expansion of the membrane covering the *round window*. Conversely, a decrease in pressure causes the cochlea partition to bend away from the scala tympani.

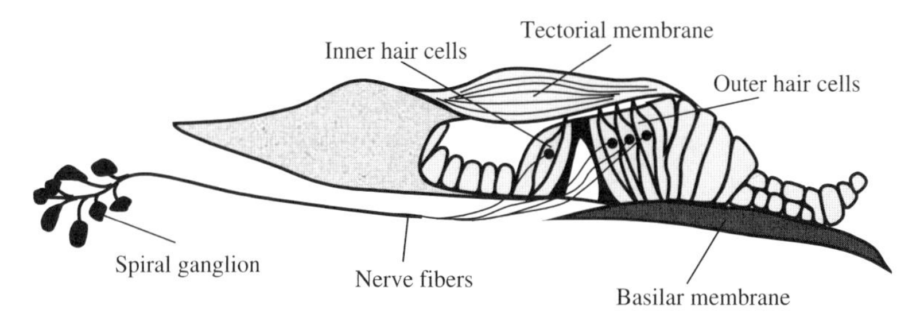

Figure 7.6 Parts of the organ of Corti.

As the cochlea partition bends in response to increasing and decreasing pressure in the scala vestibuli, the basilar membrane is displaced from its original stationary position. The displacement of the basilar membrane causes the tips of the hair cells in the organ of Corti to brush against the underside of the *tectorial membrane*. This brushing action bends the hair cells and causes nerve impulses to be sent along the auditory nerve. In this way, the inner ear converts the mechanical vibrations of the stirrup into nerve impulses that the brain interprets as sounds.

The basilar membrane is approximately 32 mm long and varies in width from approximately 0.5 mm at the apex of the cochlea to 0.1 mm at the base. As its width narrows, the basilar membrane also becomes much thicker and stiffer so that it is approximately 100 times stiffer at the base than at the apex. This variation in width and stiffness means that the basilar membrane has varying resonant properties along its length. Consequently, each point on the basilar membrane has a particular *resonant frequency*. This resonant frequency behavior can be visualized by imagining the basilar membrane as a series of very closely spaced "guitar strings" stretched across the width of the cochlear and joined together by an elastic membrane. The guitar strings at the base of the cochlear are quite short and stiff, and if one of these strings were "plucked," it would vibrate at a high frequency. On the contrary, the guitar strings at the apex are longer and more elastic and would vibrate at a much lower resonant frequency when plucked.

For a sinusoidal pressure wave, all parts of the basilar membrane oscillate at the same frequency as the input sound. However, due to the varying resonant properties of the basilar membrane, the amplitude of the oscillations at each point along the membrane depends on the frequency. There is a point on the basilar membrane where the frequency of the input sound matches the resonant frequency of the membrane and the oscillations have maximum amplitude. The location of this point of maximum displacement is very close to the stirrup for higher frequencies and gradually moves toward the apex as the frequency decreases. Figure 7.7 shows the location of the maximum displacement for various frequencies ranging from 20 kHz to 200 Hz.

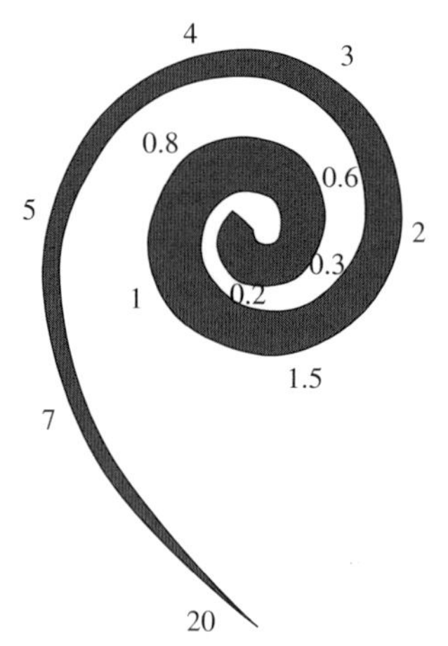

Figure 7.7 Location of the maximum displacement of the basilar membrane for various frequencies (kHz).

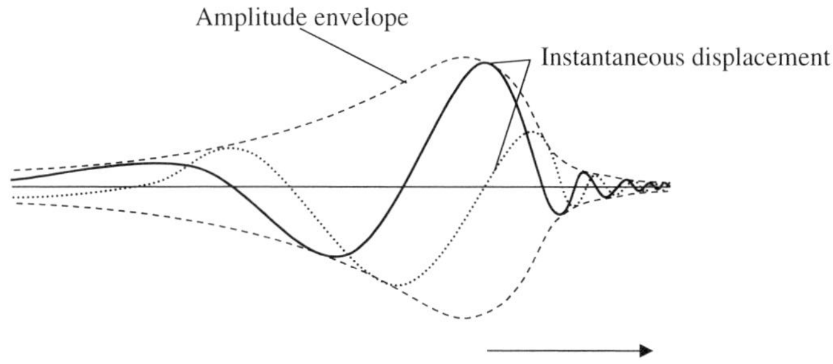

Figure 7.8 General shape of the displacement of the basilar membrane for a sinusoidal input.

As the pressure wave travels along the basilar membrane, the amplitude of the oscillations gradually increases until a maximum is reached and then rapidly decreases at points further along the cochlea. This point of maximum displacement corresponds to the point on the basilar membrane where its resonant frequency matches the frequency of the pressure wave. At this point most of the energy of the wave is absorbed by the basilar membrane, and consequently there is a rapid decrease in the amplitude of the oscillations as the wave travels further on toward the apex of the cochlea. Figure 7.8 shows the general shape of the instantaneous displacement of the basilar membrane for two successive moments in time as well as the general shape of the amplitude *envelope*, which is the line joining the amplitude peaks.

Figure 7.9 shows the instantaneous displacement of the basilar membrane for two specific cases. The double-headed arrow in each case indicates the location of maximum displacement. Figure 7.9(a) shows the displacement of the basilar membrane at one instant in time for a 5-kHz sinusoidal input, and Figure 7.9(b) shows the displacement of the basilar membrane at one instant in time for a 1-kHz sinusoidal input. Note that the point of maximum displacement is much closer to the base of the basilar membrane for the 5-kHz input than for the 1-kHz input.

The variable width and stiffness of the basilar membrane allows the cochlea to discriminate between different frequency sound waves. Sound waves of different frequencies cause different parts of the basilar membrane to oscillate with maximum amplitude, and hence the auditory nerve receives nerve impulses from different parts of the cochlea. As the frequency of an input sound varies, the brain receives nerve impulses from different parts of the auditory nerve and perceives these changes in location as variations in the *pitch* of the sound.

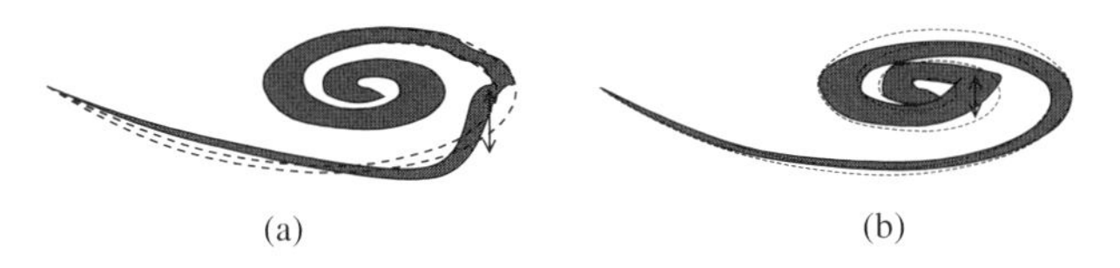

Figure 7.9 Instantaneous displacement of the basilar membrane for (a) a 5-kHz sinusoidal input and (b) a 1-kHz sinusoidal input.

7.2. PSYCHOACOUSTICS

The field of psychoacoustics deals with how we perceive the physical stimuli that affect the human auditory system. The air pressure waves that reach our ears are the physical stimulus, and the sound that we hear is the perceived response. The relationship between the physical properties of the stimulus and the perceptual qualities of the sound that we hear is the subject of the following sections.

7.2.1. Sound Pressure Level

A vibrating object produces a sound wave by causing air molecules to be compressed and then rarefied. As this sound wave travels past a point in space, the air pressure at this point increases and decreases at a frequency equal to the original vibrations of the sound source. Recall that pressure is a measure of force per unit area and is measured in N/m^2. For a sinusoidal sound wave, the air pressure oscillates between a peak value above and a peak value below the ambient atmospheric pressure. The difference between these two peak values is the *peak-to-peak* amplitude of the sound wave. However, the *sound pressure* of a wave is defined as the *root-mean-square* (rms) amplitude of the sound wave, which is equal to $1/2\sqrt{2}$ times the peak-to-peak amplitude.

The human auditory system detects these pressure variations in the air. As a sound wave reaches the eardrum, the variations in air pressure cause the eardrum to vibrate, and the rest of the auditory system converts this vibration into nerve impulses that our brain interprets as sound. The human auditory system can detect a huge range of sound pressure values. The smallest sound pressure that results in an audible sound is usually given as $2 \times 10^{-5}\,N/m^2$, and the loudest sound that can be tolerated has a sound pressure of $200\,N/m^2$.

Because of the large range of audible sound pressure values, it is more convenient to express sound pressure values using a logarithmic scale. For this reason, sound pressure is usually converted to *sound pressure level* (SPL) and expressed in decibels using the formula in Equation (7.1).

$$\mathrm{SPL} = 20\log_{10}\left(\frac{P}{P_0}\right)\mathrm{dB} \tag{7.1}$$

In Equation (7.1), P is the sound pressure and P_0 is the reference pressure value of $2 \times 10^{-5}\,N/m^2$. Some examples of typical sounds and their corresponding sound pressure levels are shown in Table 7.1.

7.2.2. Auditory Thresholds

The threshold for the quietest sound that can be heard when no other sound is present is known as the *threshold in quiet* or the *absolute threshold*. Figure 7.10 shows a plot of the threshold in quiet for a typical young adult at frequencies ranging from 100 Hz to 20 kHz. This graph shows that the threshold in quiet varies appreciably with the frequency of the sound. Note that the human auditory system is most sensitive to

Table 7.1 Typical examples of sounds and their SPL values.

Example sound	SPL (dB)
Gunshot at close range	140
Jackhammer	120
Shouting at close range	100
Traffic noise	80
Normal conversation	70
Quiet conversation	50
Soft whisper	30
Watch ticking	20

sound pressure at frequencies between about 1 and 5 kHz. This sensitive range is not surprising because these are the frequencies that occur most often in human speech.

The threshold for the loudest sound that can be tolerated is known as the *terminal threshold*. This threshold does not vary significantly with frequency. A sound pressure level of 120 dB causes slight discomfort, and a sound pressure level of 140 dB causes a tickling sensation and pain at any frequency.

As the frequency of an input sound varies, the brain receives nerve impulses from different parts of the auditory nerve and perceives these changes in location as variations in the *pitch* of the sound. The band of audible frequencies extends from approximately 20 Hz to 20 kHz. Below 20 Hz the variations in air pressure are too slow for the human auditory system to detect a tonal quality of the sound. Sounds above 20 kHz can be heard if the sound pressure is increased to uncomfortably high levels. However, at about 23 kHz, the threshold in quiet coincides with the pain threshold, so this is effectively the upper frequency limit for hearing.

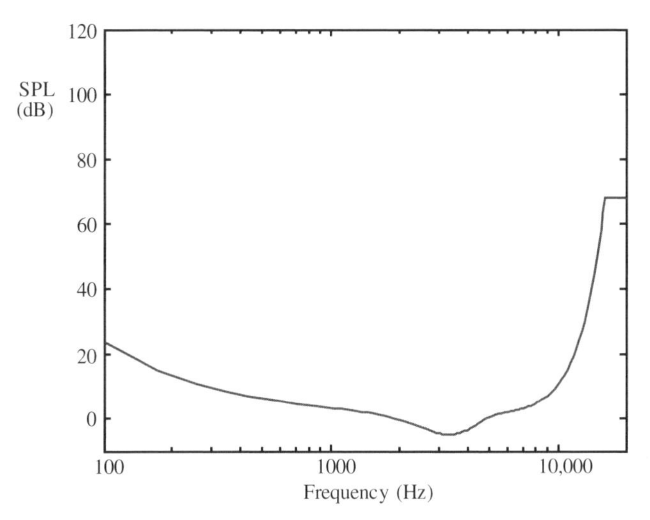

Figure 7.10 Threshold in quiet for a typical young adult.

7.2.3. The Critical Bandwidth and Auditory Filters

Early researchers observed that there was a *critical bandwidth* associated with the human auditory system at which our perception of sounds changed abruptly. Harvey Fletcher first reported this critical bandwidth in 1940 [1]. Fletcher observed the critical bandwidth by listening to a tone and band-limited noise simultaneously. To find the critical bandwidth, the frequency of a tone was set to the center frequency of the noise, and the bandwidth of the noise was then gradually widened. As the bandwidth of the noise increased, the intensity of the tone was adjusted so that it was just audible above the noise. Fletcher observed that when the bandwidth of the noise was less than the critical bandwidth, the just audible level of the tone increased as the bandwidth of the noise was increased. However, as the bandwidth of the noise was increased above the critical bandwidth, the just audible level of the tone remained the same, regardless of the bandwidth of the noise.

Another example of the critical bandwidth phenomena can be observed by listening to two tones of different frequencies simultaneously. If the frequency of the two tones differs by more than the critical bandwidth, then two distinct tones are perceived. However, if the frequency differs by less than a critical bandwidth, the sound perceived has a harsh quality, and the two distinct tones can no longer be heard.

It was also observed that the critical bandwidth is not constant for all frequencies but in fact increases with increasing frequency. Figure 7.11 shows a plot of the critical bandwidth for frequencies within the audible range.

The critical bandwidth phenomena led researchers to believe that the human auditory system acts like a bank of band-pass *auditory filters*, with continuously overlapping center frequencies. It is generally thought that the basilar membrane provides the physical basis for these auditory filters because each location on the basilar membrane has a maximum response to a particular frequency sound wave. It is also easy to see that, as the frequency is shifted above and below this resonant frequency, the amplitude of the oscillations of the basilar membrane at the original

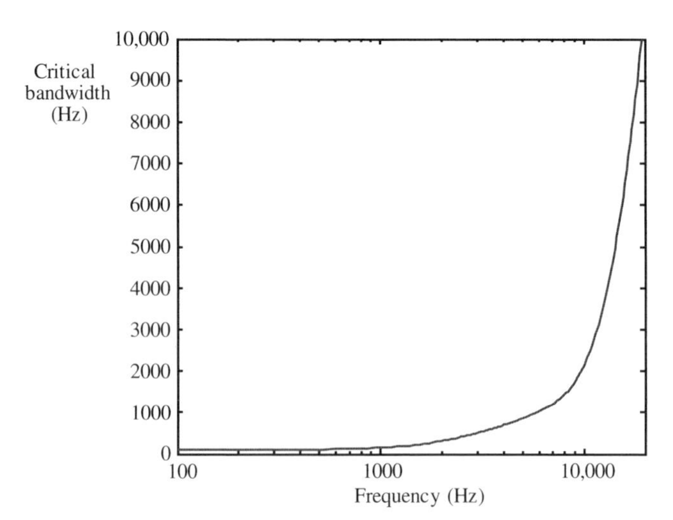

Figure 7.11 Critical bandwidth versus center frequency.

location is reduced. In this way, each point on the basilar membrane effectively acts like a band-pass filter with a different center frequency.

For all frequencies below about 16 kHz, the range of frequencies that produce significant neural activity at a fixed location on the basilar membrane corresponds to a *constant physical distance* around this fixed location. This constant physical distance determines the bandwidth of the auditory filter at each location. Remember though that, as we move along the basilar membrane from apex to base, a constant physical distance corresponds to an increasing range of frequencies. The basilar membrane therefore effectively acts like a band-pass filter bank with a continuous range of center frequencies and with the bandwidth of each filter increasing as the center frequency increases.

Empirical measurements of the critical bandwidth at various frequencies by Bertram Scharf in 1970 [2] resulted in the definition of 25 *critical bands*. The frequencies that define the center and edges of these critical bands are shown in Table 7.2. It is important to note that these critical band boundaries do not represent

Table 7.2 Center and edge frequencies for the critical bands of the human auditory system.

Band number	Lower edge (Hz)	Center (Hz)	Upper edge (Hz)
1	0	50	100
2	100	150	200
3	200	250	300
4	300	350	400
5	400	450	510
6	510	570	630
7	630	700	770
8	770	840	920
9	920	1,000	1,080
10	1,080	1,170	1,270
11	1,270	1,370	1,480
12	1,480	1,600	1,720
13	1,720	1,850	2,000
14	2,000	2,150	2,320
15	2,320	2,500	2,700
16	2,700	2,900	3,150
17	3,150	3,400	3,700
18	3,700	4,000	4,400
19	4,400	4,800	5,300
20	5,300	5,800	6,400
21	6,400	7,000	7,700
22	7,700	8,500	9,500
23	9,500	10,500	12,000
24	12,000	13,500	15,500
25	15,500	19,500	24,000

discrete nonoverlapping filters but instead represent the critical bandwidth at a set of sample frequencies.

The boundaries of the critical bands shown in Table 7.2 were then used to generate a new scale of measurement called the *critical band rate*. The unit for the critical band rate is the *Bark*,[1] and the critical band rate is sometimes called the *Bark scale*. The formula given in Equation (7.2) can be used to perform an approximate conversion from frequency to critical band rate.

$$z = \frac{28f}{f + 2200} - 0.5 \text{ Bark} \tag{7.2}$$

In Equation (7.2), z is the critical band rate and f is the frequency in Hertz. Figure 7.12 shows the relationship between the critical band rate and frequency.

The critical band rate of a sound is effectively a measure of the location on the basilar membrane that will be affected by the sound. A difference in critical band rate of less than 1 Bark indicates that two sounds are within a critical bandwidth of each other regardless of their absolute frequency. The critical band rate is also very useful in determining if one sound will affect the perception of another sound. This psychoacoustic effect is the subject of the following section.

7.2.4. Auditory Masking

Auditory masking is the general term used to describe a situation where the perception of one sound is affected by the presence of another sound. We experience forms of auditory masking regularly in everyday life. For example, when the noise from our car engine cannot be heard above the sounds from the car radio or when the

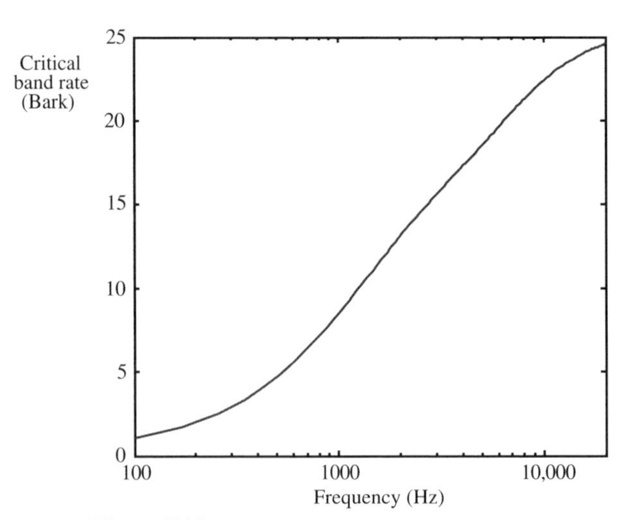

Figure 7.12 Critical band rate versus frequency.

[1]The unit of Bark is named after Heinrich Georg Barkhausen, a German Physicist who introduced the phon as a unit of measurement for loudness.

music from our home stereo is inaudible because of the sound of construction work from the house next door.

In specific terms auditory masking occurs when one sound, the *masker*, raises the threshold of audibility for another sound, the *maskee*. If sound A is the masker and sound B is the maskee, then the raised threshold of sound B is called the *masked threshold* or *masking threshold*. We say that sound B has been *masked* by sound A, and the amount of *masking* is equal to the difference between the threshold in quiet for sound B alone and the masked threshold for sound B in the presence of sound A.

There are two main types of auditory masking: *frequency* masking and *temporal* masking.

7.2.4.1. Frequency Masking

Frequency masking occurs when two or more sounds are presented simultaneously, and the amount of masking depends on the difference in frequency between the masker and the maskee and the sound pressure level of the two sounds. Figure 7.13 shows a typical example of frequency masking with a 3-kHz masker and a 4-kHz maskee. The masking threshold of the 3-kHz tone indicates the sound pressure level required by a second tone before it can be heard in the presence of the masker. Because the sound pressure level of the 4-kHz tone is below the masking threshold, it is inaudible, and a human listener only hears the 3-kHz masker. For sounds that have frequencies that are quite different to the masker, the masking threshold corresponds to the threshold in quiet, and the masker has no effect on the audibility of these tones.

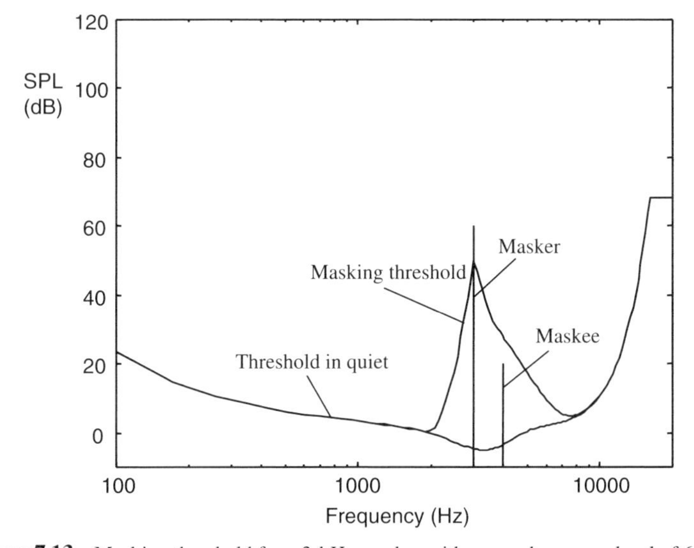

Figure 7.13 Masking threshold for a 3-kHz masker with a sound pressure level of 60 dB.

It is commonly thought that the underlying physical process involved in masking is the swamping of maskee neural activity by the masker. If the masker produces a significant amount of activity at the location on the basilar membrane that would normally be activated by the maskee, then the small amount of extra activity caused by the maskee may be undetectable.

Note that the slope of the masking threshold is less steep for frequencies above the masker frequency. This phenomenon is known as the *upward spread of masking* and is generally thought to occur because of the shape of the oscillations that are induced in the basilar membrane. Recall from Section 7.1.3 that oscillations of the basilar membrane decay away much more rapidly for parts of the basilar membrane that have resonant frequencies below the frequency of the current sound stimulus. This means that a masker tone has the best effect on parts of the basilar membrane that have a resonant frequency above the masker frequency.

In the previous section it was shown that it is the distance between the stimulated locations on the basilar membrane that will determine how one sound will affect the perception of the other. For this reason, the masking threshold for a masker is usually estimated using the critical band rate rather than absolute frequency.

7.2.4.2. Temporal Masking

Temporal masking is the term used to describe the masking effect that occurs when the masker and the maskee are not presented to the ear simultaneously. Instead, there is a small time delay between the presentation of the masker and the maskee. If the masker is presented before the maskee, this is known as *postmasking* or *forward masking*, and if the masker is presented after the maskee, it is known as *premasking* or *backward masking*. The amount of masking that occurs will depend on the time delay between the masker and the maskee, the order in which they are presented, and

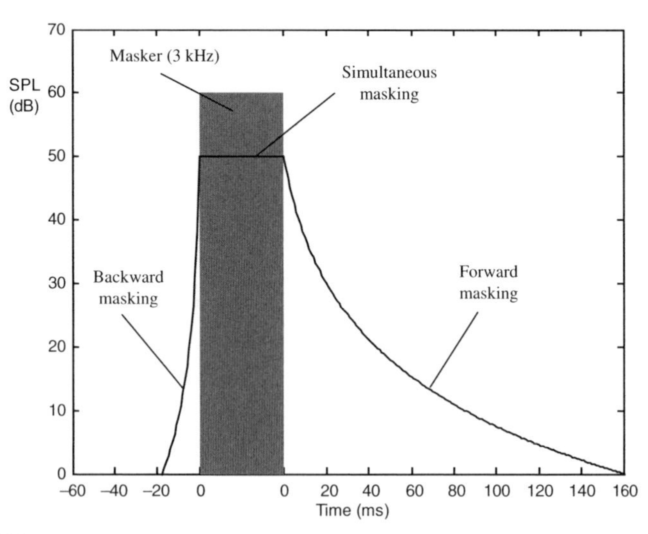

Figure 7.14 The temporal masking threshold for a 3-kHz, 60-dB masker and a 3-kHz maskee.

the difference in frequency between the two sounds. Figure 7.14 shows the temporal masking threshold for both forward and backward masking by a 60-dB, 3-kHz masker of a 3-kHz maskee.

Forward masking can occur with much longer time delays between the masker and the maskee than backward masking. Typically, backward masking cannot be observed once the maskee occurs more than 20 ms before the masker. However, forward masking can be observed with a time delay of up to 200 ms between the masker and maskee. The amount of temporal masking reduces logarithmically with the time delay between the maskee and masker. It is also interesting to observe that the maximum time delay at which temporal masking can be detected does not vary significantly with sound pressure level of the masker. As the masker sound pressure level increases, the masking threshold reduces more quickly, so the masking effect persists for the same amount of time regardless of the initial masker level. From the previous section it is also easy to see that the maximum temporal masking effect will occur when both the masker and the maskee have the same frequency.

The physical basis for temporal masking is not well understood. Two possible causes have been suggested. One is that the oscillations in the basilar membrane from the masker may not have completely decayed away before the onset of the maskee. The other suggests that recently stimulated hair cells may be reduced in sensitivity for a short time and hence emit less neural activity than would normally be the case for the maskee.

7.3. SUMMARY

In this chapter it was shown how the physical properties of the human auditory system can alter the way we perceive sounds. In particular, it was demonstrated how some sounds can mask the perception of other sounds that have a similar frequency or follow closely in time. In the following audio chapters, how these psychoacoustic limitations of the human auditory system have been exploited in the design of perceptual audio coding algorithms is explained.

PROBLEMS

1. Briefly explain the function of the following parts of the human auditory system as it converts a sound wave into neural impulses:
   - **(a)** Eardrum
   - **(b)** Ossicles
   - **(c)** Basilar membrane
   - **(d)** Hair cells
2. Explain how the variable width and stiffness of the basilar membrane allows variations in the frequency of a sound wave to be perceived as changes in the pitch of the sound.
3. If the peak-to-peak variation in air pressure for a sound wave is held constant at $4 \times 10^{-5}\,N/m^2$ and the frequency is varied between 100 Hz and 20 kHz,

(a) Determine the frequency range for which this sound will be audible to a typical young adult.

(b) Repeat part (a) for a sound wave with a peak-to-peak variation in air pressure of $1 \times 10^{-4}\,\text{N/m}^2$.

REFERENCES

1. H. Fletcher, Auditory patterns, *Rev. Modern Phy.*, **12**, 1940, 47–66.
2. B. Scharf, Critical bands, in *Foundations of Modern Auditory Theory* (J. Tobias, ed.), New York and London: Academic Press, 1970, pp. 159–202.

Chapter 8

Frequency Analysis and Synthesis

A conversion from time-domain samples to frequency-domain coefficients is the first stage in all of the current standard digital audio compression algorithms. In the MPEG suite of algorithms, this conversion is done using an efficient implementation of a subband filterbank approach, whereas in the Dolby AC-3 algorithm, a cosine transform is used that incorporates time-domain aliasing cancellation. In this chapter, the basic theory behind each of these techniques is explained and the actual implementations used in the standard algorithms are discussed.

8.1. THE SAMPLING THEOREM

The first stage in digital audio coding is usually the conversion of the analog waveform into a series of digital samples. Probably the most important concept to consider during this conversion is the introduction of *aliasing* errors. In this section, the mathematical explanation for aliasing, which is more commonly known as *the sampling theorem*, is introduced.

Consider the signal $x(t)$ that has a spectrum $X(f)$ and a bandwidth of B Hz as shown in Figure 8.1(a) and (b).

If this signal is sampled at a rate of one sample every T_S seconds, this is equivalent to multiplying the signal by a unit impulse train with period T_S. The sampled signal $x(nT_S)$ is given by the following equation:

$$x(nT_S) = x(t) \sum_{n=-\infty}^{\infty} \delta(t - nT_S) \tag{8.1}$$

By taking the exponential Fourier series of an impulse train, this equation can be rewritten to give

$$x(nT_S) = \frac{1}{T_S} \sum_{n=-\infty}^{\infty} x(t) e^{jn\omega_S t} \tag{8.2}$$

Digital Television, by John Arnold, Michael Frater and Mark Pickering.

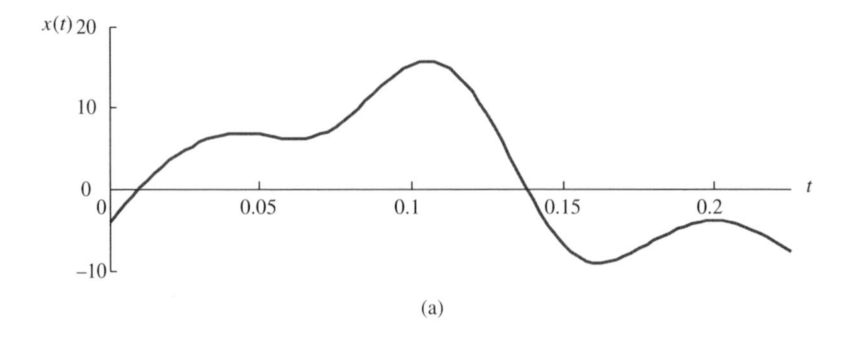

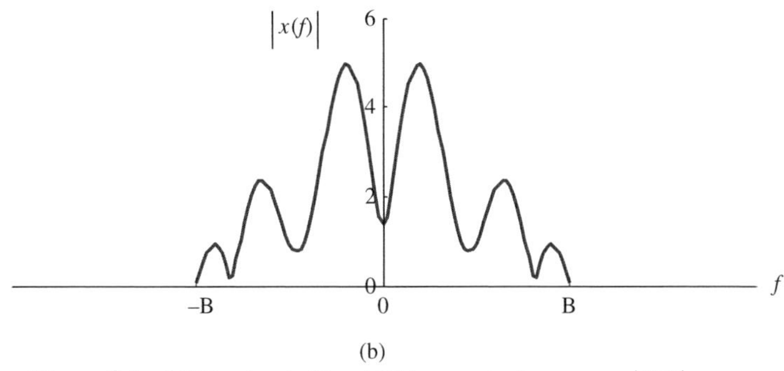

Figure 8.1 (a) The signal $x(t)$ and (b) its magnitude spectrum $|X(f)|$.

and taking the Fourier transform of $x(nT_S)$ gives

$$X_S(\omega) = \frac{1}{T_S} \sum_{n=-\infty}^{\infty} X(\omega - n\omega_S) \tag{8.3}$$

The spectrum of the sampled signal therefore consists of the spectrum of the original signal repeating itself indefinitely at intervals of $\omega_S = 2\pi f_S$, where $f_S = 1/T_S$ is the *sampling frequency*. Figure 8.2(a) shows the sampled signal (in future, for ease of notation, the sampling period term T_S will be dropped and this signal will simply be referred to as $x(n)$), and Figure 8.2(b) shows the magnitude spectrum of the sampled signal.

Figure 8.2(b) shows that there is no overlap between successive cycles of $X_S(\omega)$ provided that the sampling frequency ω_S is greater than twice the maximum frequency of the signal. If there is no overlap between successive cycles, the original signal can be recovered from the sampled signal by passing it through a low-pass filter. The sampling theorem states that a signal with a bandwidth of B Hz can be reconstructed perfectly from samples taken at a rate of not less than $2B$ samples per second. This sampling rate of $2B$ samples per second is known as the *Nyquist sampling rate*.

If a signal is sampled at a rate less than the Nyquist rate, the spectrum of the sampled signal consists of overlapping repetitions of $X(\omega)$. If the sampled signal

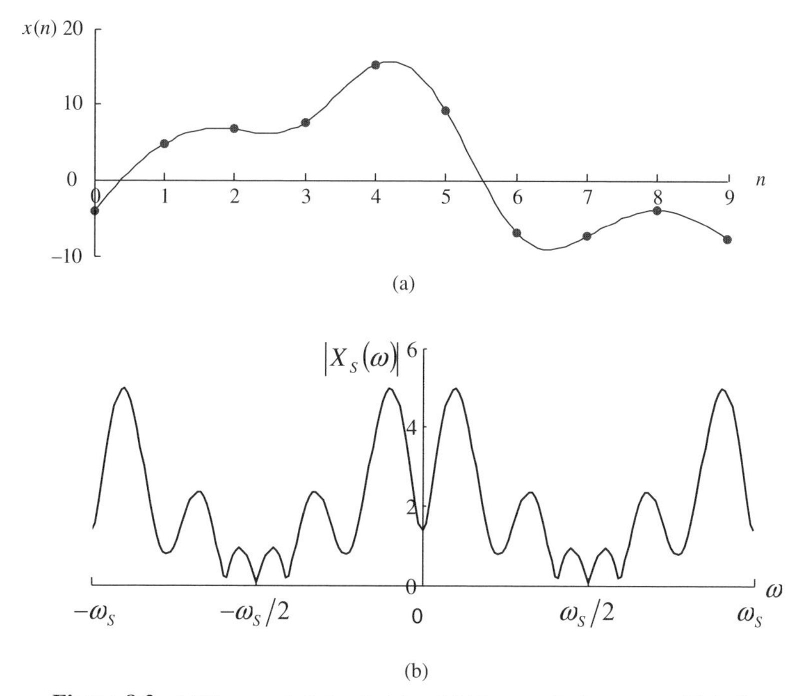

Figure 8.2 (a) The sampled signal $x(n)$ and (b) its magnitude spectrum $|X_S(\omega)|$.

is then passed through a low-pass filter, the result is a distorted version of the original signal. This distortion is known as *aliasing error* and is caused by the loss of frequency components above the frequency of $\omega_S/2$ and frequency components from the next repetition of $X(\omega)$ adding to the original signal below the frequency of $\omega_S/2$.

8.2. DIGITAL FILTERS

For the MPEG suite of audio coding algorithms, the conversion from time-domain samples to frequency coefficients is performed using a bank of *digital filters*. A digital filter is specified by its discrete impulse response $h(k)$, $k = 0, \ldots, N$. A sampled signal $x(n)$ is *filtered* by performing a discrete convolution of the filter impulse response with the sampled signal. Hence the output of a digital filter with impulse response $h(k)$ is given by the equation

$$s(n) = \sum_{k=0}^{N} h(k)x(n-k) \tag{8.4}$$

The frequency response of a digital filter can be found by taking the Fourier transform of the filter impulse response.

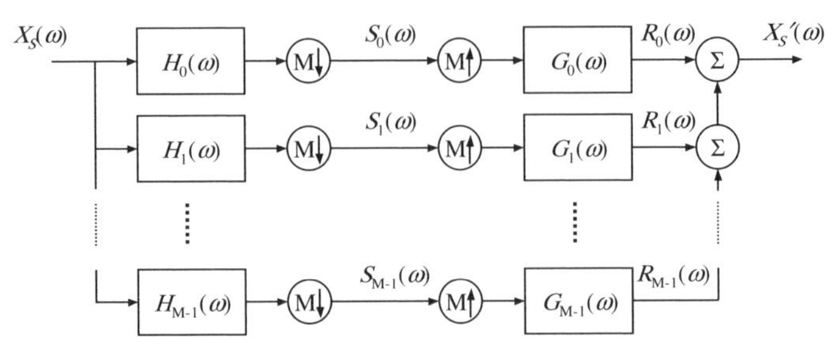

Figure 8.3 *M*-channel subband analysis and synthesis filters.

8.3. SUBBAND FILTERING

Subband filtering is the process of dividing the original signal into a series of frequency bands. This results in a set of frequency coefficients or *subband samples* that represent the amplitude of the component with the frequency of the corresponding subband. These subband samples are then transmitted and added together to reconstruct the original signal. The subband filtering process requires two steps: First, the analysis filter bank is used to convert time-domain samples of the original signal to frequency-domain subband samples. The synthesis filter bank is then used to reconstruct the time-domain samples of the signal from the transmitted subband samples. The block diagram of an M-channel analysis–synthesis filter bank is shown in Figure 8.3.

8.3.1. The Analysis Filter Bank

The analysis stage uses a set of bandpass digital filters to divide the frequency spectrum of the sampled signal into M subbands. The output of each of these filters is then subsampled by a factor of M so that the number of subband samples produced is the same as the original number of time-domain samples. From the sampling theorem, because the output of each bandpass filter has a bandwidth of B/M, the sampling rate can also be reduced to $2B/M$ while still guaranteeing perfect reconstruction.

8.3.1.1. The Effect of Subsampling

Subsampling by a factor of M is effectively dividing the sampling frequency by M. This means that the sampling frequency is now not high enough to eliminate aliasing errors and the spectrum becomes distorted. This distorted spectrum is a combination of the M repetitions of the original spectrum and can be represented by the equation

$$X_{SS}(\omega) = \frac{1}{M}\sum_{j=0}^{M-1} X_S\left(\frac{\omega + j\pi}{M}\right) \tag{8.5}$$

Equation (8.5) shows that, after filtering and subsampling, the spectrum of the outputs of the analysis filter bank are given by the equation

$$S_i(\omega) = \frac{1}{M}\sum_{j=0}^{M-1} X_S\left(\frac{\omega + j\pi}{M}\right) H_i\left(\frac{\omega + j\pi}{M}\right) \quad \text{for } i = 0,1,2,\ldots,M-1 \tag{8.6}$$

EXAMPLE 8.1

Consider the sampled signal $x(n)$ with the magnitude spectrum $|X_S(\omega)|$ as shown in Figure 8.2. Now if this signal is subsampled by a factor of 2, the result is the signal $x_{SS}(n)$ as shown in Figure 8.4.

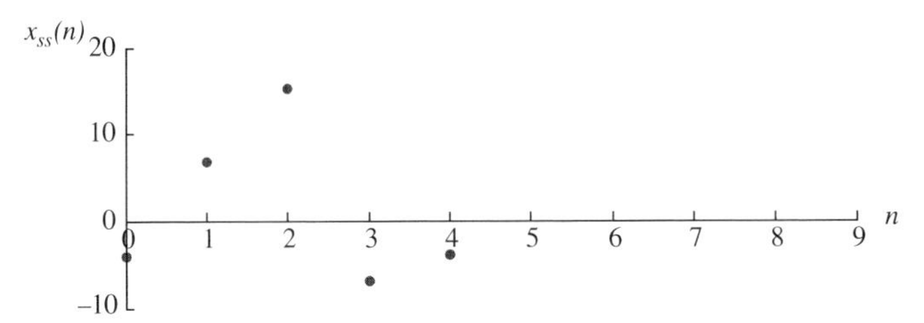

Figure 8.4 The subsampled signal $x_{SS}(n)$.

However, because the sampling frequency has been halved, the aliased spectrum is given by the following equation.

$$X_{SS}(\omega) = \frac{1}{2}\left[X_S\left(\frac{\omega}{2}\right) + X_S\left(\frac{\omega+\pi}{2}\right)\right] \tag{8.7}$$

The magnitude spectrum of this subsampled signal is shown as the full line in Figure 8.5; the dashed line is $|X(\omega/2)|$ and the dotted line is $|X((\omega + \pi)/2)|$. Remember that the magnitude spectra of the two terms in Equation (8.7) cannot simply be added together. Figure 8.5 shows the magnitude spectrum of the subsampled signal after the two terms have been added as complex signals with both magnitude and phase information.

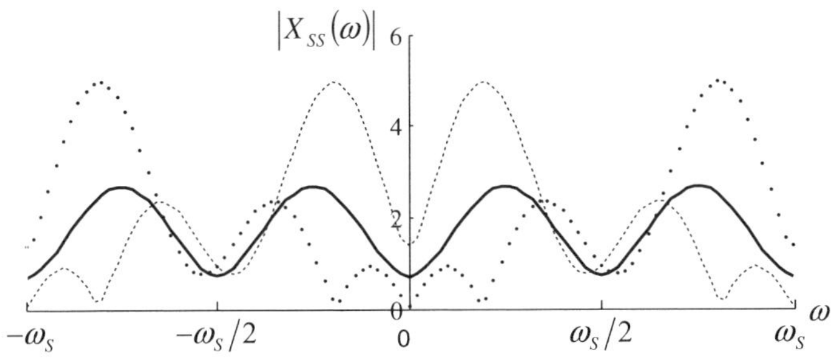

Figure 8.5 The magnitude spectrum of the subsampled signal $|X_{SS}(\omega)|$ (full line) with the magnitude spectrum of $|X_S(\omega/2)|$ (dashed line) and the magnitude spectrum of $|X_S((\omega + \pi)/2)|$ (dotted line). ■

8.3.2. The Synthesis Filter Bank

The synthesis stage also uses a set of bandpass filters to reconstruct the samples of the time-domain signal. The input to each of these synthesis filters is upsampled by M so that the output of each filter has the same number of samples as the original time-domain signal. This upsampling process is implemented by inserting $M - 1$ zero-valued samples between each sample of the subsampled signal. The output samples from each synthesis filter are then added together to form the reconstructed signal.

8.3.2.1. The Effect of Upsampling

Upsampling by a factor of M is effectively dividing the frequency of each component by M. The spectrum of the upsampled signal is therefore given by the following equation:

$$X_{US}(\omega) = X_S(M\omega) \tag{8.8}$$

Equation (8.8) shows that, after upsampling and filtering, the spectrum of the outputs of the synthesis filter bank in Figure 8.3 is given by the equation

$$R_i(\omega) = \frac{1}{M} G_i(\omega) \sum_{j=0}^{M-1} X_S(\omega + j\pi) H_i(\omega + j\pi) \quad \text{for } i = 0,1,2,\ldots,M-1 \tag{8.9}$$

EXAMPLE 8.2

Consider the sampled signal $x(n)$ with the magnitude spectrum $|X_S(\omega)|$ as shown in Figure 8.2. If this signal is upsampled by a factor of 2, the result is the signal $x_{US}(n)$ as shown in Figure 8.6, and the magnitude spectrum of this upsampled signal is shown in Figure 8.7.

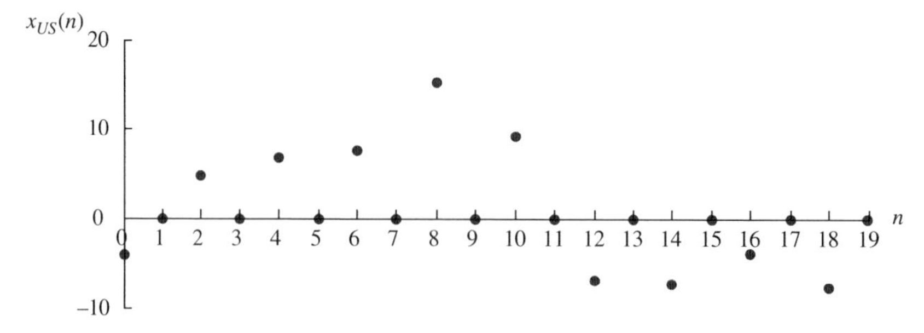

Figure 8.6 The upsampled signal $x_{US}(n)$.

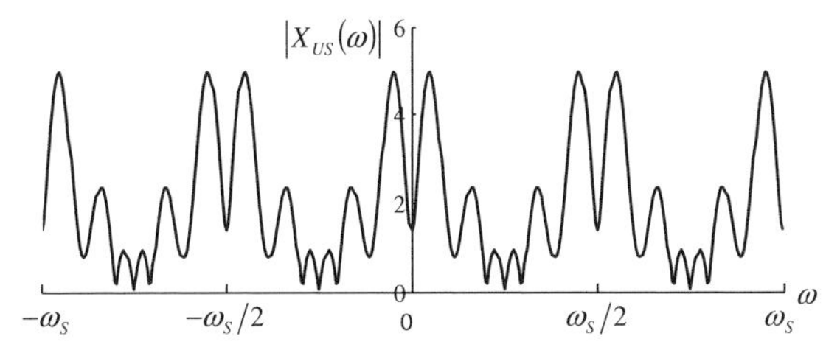

Figure 8.7 The magnitude spectrum of the upsampled signal $|X_{US}(\omega)|$. ■

8.3.3. Filters for Perfect Reconstruction

The reconstructed signal from the synthesis filterbank in Figure 8.3 is found by adding the outputs of the synthesis filters and is given by

$$X_S'(\omega) = \frac{1}{M}\sum_{i=0}^{M-1} G_i(\omega)\sum_{j=0}^{M-1} X_S(\omega + j\pi)H_i(\omega + j\pi) \tag{8.10}$$

and rearranging this equation to separate the terms where $j = 0$ gives the following equation:

$$X_S'(\omega) = \frac{1}{M}X_S(\omega)\sum_{i=0}^{M-1} G_i(\omega)H_i(\omega) + \frac{1}{M}\sum_{j=1}^{M-1} X_S(\omega + j\pi)\sum_{i=0}^{M-1} G_i(\omega)H_i(\omega + j\pi) \tag{8.11}$$

Equation (8.11) has one term that contains the original sampled signal $X_S(\omega)$, and the remaining terms contain frequency-shifted versions of the original signal $X_S(\omega + j\pi)$. To perfectly reconstruct the original signal, a set of filters that satisfy the following equations must be found:

$$\left|\sum_{i=0}^{M-1} G_i(\omega)H_i(\omega)\right| = M \tag{8.12}$$

$$\sum_{j=1}^{M-1}\sum_{i=0}^{M-1} G_i(\omega)H_i(\omega + j\pi) = 0 \tag{8.13}$$

Equation (8.12) tells us that if the frequency response for the analysis and synthesis filters for each subband are multiplied together and these combined frequency responses are added for all subbands, then the overall frequency response should be constant over the bandwidth of the original sampled signal. Equation (8.13) tells us that if the analysis filter for one subband is multiplied with the synthesis filters for all other subbands and these combined frequency responses are added for all subbands, then this overall frequency response should be zero. For a set of filters that satisfy these equations, the aliasing terms from Equation (8.11) are eliminated, and the overall frequency response of the filterbank is constant over the bandwidth of the signal.

8.4. COSINE-MODULATED FILTERS

One technique that can be used to determine a set of filters that satisfy the perfect reconstruction equations is to use a set of *cosine-modulated filters*. In this technique a prototype filter is determined, and each of the analysis and synthesis filters is derived from this prototype filter using cosine modulation.

For an M-channel subband filter bank, the impulse responses for the analysis and synthesis filters are given by the following equations:

$$h_i(k) = h(k)\cos\left(\frac{(2i+1)\left(k-\frac{M}{2}\right)\pi}{2M}\right) \tag{8.14}$$

$$g_i(k) = h(k)\cos\left(\frac{(2i+1)\left(k+\frac{M}{2}\right)\pi}{2M}\right) \tag{8.15}$$

where $h(k)$ is the impulse response of the prototype filter.

EXAMPLE 8.3

The impulse response $h(k)$ of a prototype filter for a two-channel filterbank is shown in Figure 8.8(a), and the magnitude of the frequency response of this filter $|H(\omega)|$ is shown in Figure 8.8(b).

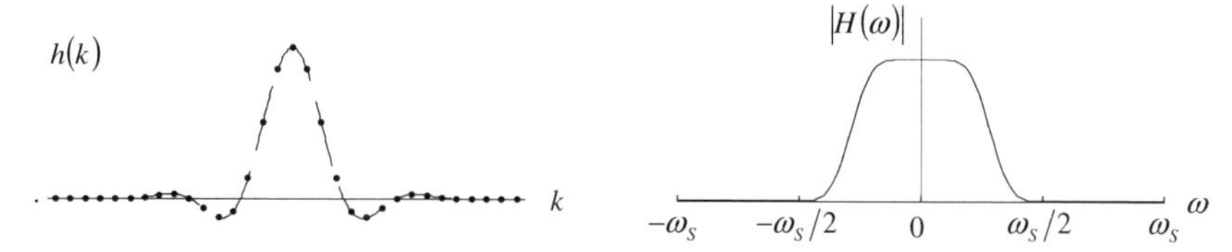

Figure 8.8 (a) The impulse response $h(k)$ and (b) the magnitude of the frequency response $|H(\omega)|$ of a prototype filter for a two-channel perfect reconstruction filterbank.

The two analysis filters are determined using the following equations:

$$h_0(k) = h(k)\cos\left(\frac{(k-1)\pi}{4}\right) \tag{8.16}$$

$$h_1(k) = h(k)\cos\left(\frac{3(k-1)\pi}{4}\right) \tag{8.17}$$

The filter coefficients of the analysis filters are found by multiplying the coefficients of the prototype filter by a sampled cosine wave as shown in Figure 8.9(a). In the frequency domain this is equivalent to the convolution of the spectrum of the prototype filter with the spectrum of the cosine wave as shown in Figure 8.9(b).

Figure 8.10 shows the same process for $h_1(k)$, and Figures 8.9 and 8.10 show that the two analysis filters divide the spectrum of the sampled input signal into two equal width subbands. ■

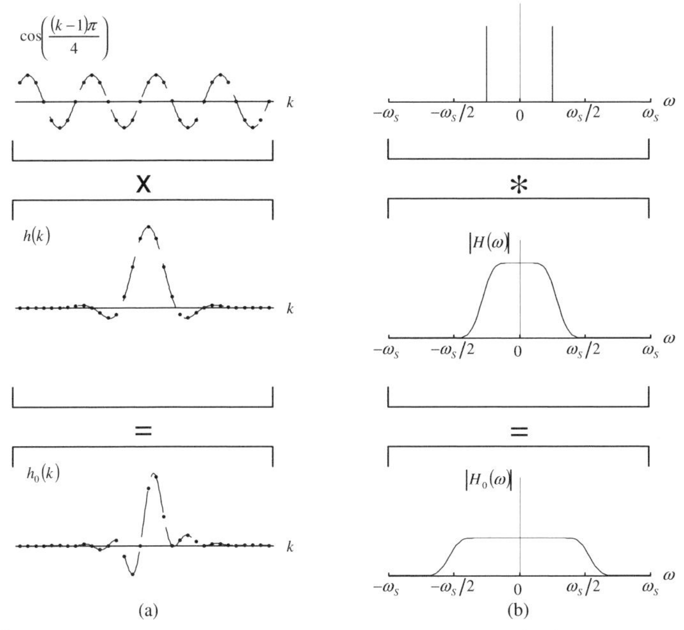

Figure 8.9 (a) The coefficients of the low-pass filter are found by multiplying the prototype filter coefficients by a sampled cosine wave. (b) In the frequency domain, this is equivalent to the convolution of the prototype frequency response with an impulse function at the frequency of the cosine.

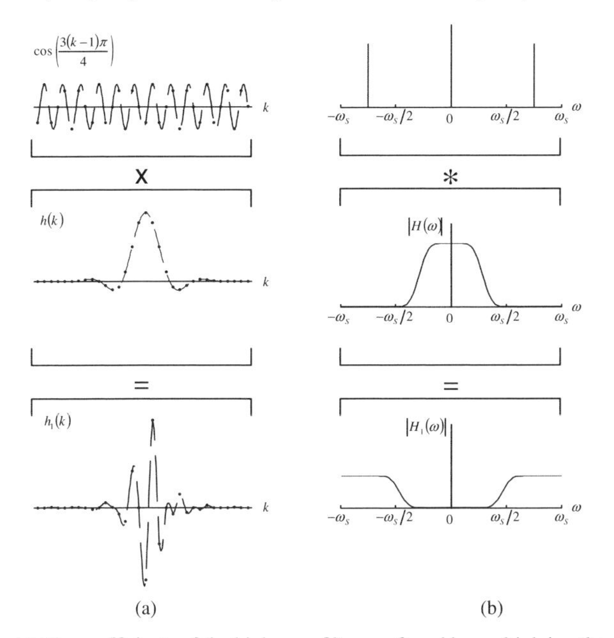

Figure 8.10 (a) The coefficients of the high-pass filter are found by multiplying the prototype filter coefficients by a sampled cosine wave. (b) In the frequency domain this is equivalent to the convolution of the prototype frequency response with an impulse function at the frequency of the cosine.

The two synthesis filters are determined using the following equations:

$$g_0(k) = h(k)\cos\left(\frac{(k+1)\pi}{4}\right) \tag{8.18}$$

$$g_1(k) = h(k)\cos\left(\frac{3(k+1)\pi}{4}\right) \tag{8.19}$$

The sampled cosine waves and the resulting filter coefficients for the synthesis filters are shown in Figure 8.11. Note from Figure 8.11 that the synthesis filters are simply the analysis filters reversed in time.

The output of each of the analysis filters is the digital convolution of the impulse response of the filters with the input signal that is given by the following equations:

$$s_0(n) = \sum_{k=0}^{N} h_0(k)x(n-k) \tag{8.20}$$

$$s_1(n) = \sum_{k=0}^{N} h_1(k)x(n-k) \tag{8.21}$$

A simple implementation of an M-channel analysis filterbank with filters of N coefficients requires MN multiplications and $M(N - 1)$ additions for each sample of the input signal.

Similarly, the output of each of the synthesis filters is the digital convolution of the impulse response of the filters with the upsampled subband samples. The outputs of these filters are then added to produce the reconstructed output samples. A simple implementation of an M-channel synthesis filterbank with filters of N coefficients requires MN multiplications and $MN - 1$ additions for each sample of the reconstructed signal.

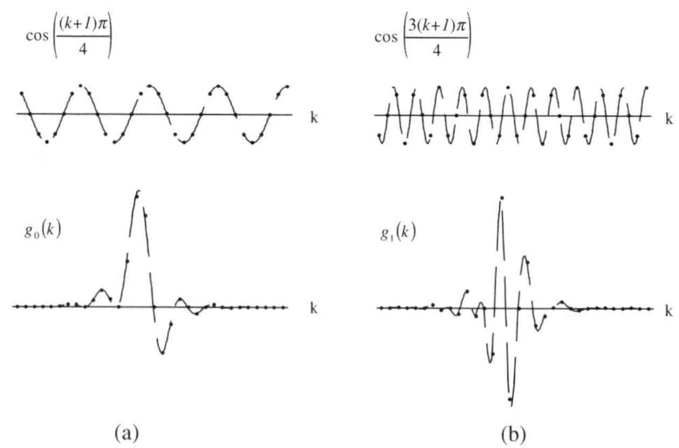

Figure 8.11 The sampled cosine waves and the resulting filter coefficients for (a) the low-pass synthesis filter and (b) the high-pass synthesis filter.

EXAMPLE 8.4

Figure 8.12(a) shows a graphical representation of the calculations required for an output subband sample of the low-pass filter in the analysis filter bank. The filter has a length of 32, so each set of 32 input samples is first reversed in time and then multiplied by the corresponding value of the impulse response. The sum of all these product terms is then calculated to obtain the single output subband sample $s_0(n)$ (shown as a full circle). For a two-channel filterbank using the analysis filters from the previous example, each output subband sample requires 32 multiplications and 31 additions. These same calculations are also required for the high-pass filter. For each input sample, a total of 64 multiplications and 62 additions are required to obtain each pair of low-pass and high-pass subband samples.

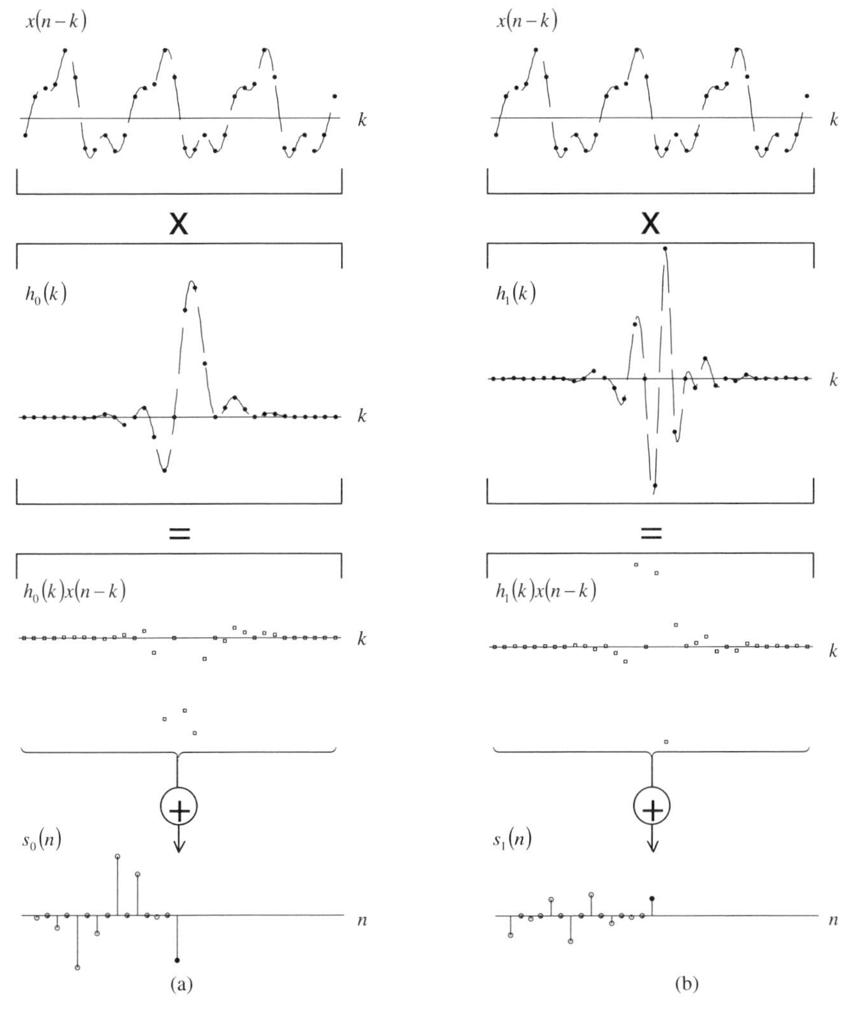

Figure 8.12 Multiplications and additions required for one output sample of the analysis filter (shown as a filled circle). (a) Low-pass filter and (b) high-pass filter (zero samples have been added to the output sample sequence to maintain the same timescale as for the input samples).

Figure 8.13(a) shows a graphical representation of the calculations required for an output sample of the low-pass filter in the synthesis filter bank (shown as a full circle). The input to the filter, $s_0\,(n - k)$, is the set of 32 time-reversed subband samples from the low-pass analysis filter that have been subsampled by a factor of 2 and then upsampled by inserting a zero-valued sample between each sample. Each set of 32 samples of $s_0\,(n - k)$ is multiplied by the corresponding value of the impulse response of the low-pass synthesis filter $g_0\,(k)$. Then the sum of all these product terms is calculated to obtain the single reconstructed low-pass output sample $r_0\,(n)$. For a two-channel filter bank using the synthesis filters from the previous example, each reconstructed output sample requires 32 multiplications and 31 additions. These same calculations are also required for the high-pass filter, and a further addition is required to add the low-pass and high-pass samples together. Therefore, for each output sample, a total of 64 multiplications and 63 additions are required.

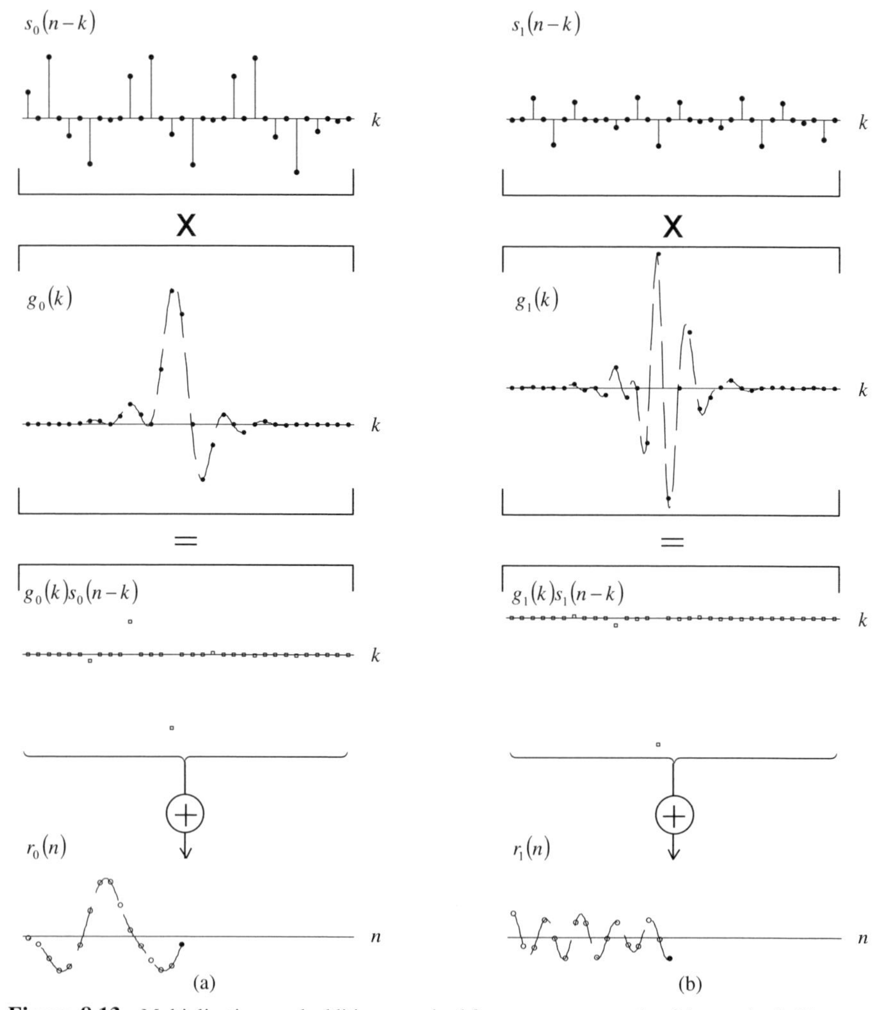

Figure 8.13 Multiplications and additions required for one output sample of the synthesis filter (shown as a filled circle). (a) Low-pass filter and (b) high-pass filter. ■

8.5. EFFICIENT IMPLEMENTATION OF A COSINE-MODULATED FILTERBANK

The standard implementation of a subband filterbank requires many more calculations than is necessary. A more efficient implementation of a set of cosine-modulated filters was proposed by Joseph Rothweiler in 1983 [1]. This efficient implementation takes advantage of two things: (a) the prototype filter coefficients are common to all filters with the only difference being the frequency of the sampled cosine waves and (b) the periodic nature of the sampled cosine waves.

8.5.1. Analysis Filter

In order to explain the basic principles behind Rothweiler's proposal, first consider the following example that shows how the two-channel analysis filterbank from Section 8.4 can be implemented more efficiently.

EXAMPLE 8.5

Figure 8.14(a) shows the conventional method for calculating the low-pass filter impulse response by multiplying the prototype filter impulse response with a sampled cosine wave. Figure 8.14(b) shows how the same impulse response can be calculated by inverting every second set of four samples in both the cosine wave and the prototype filter response. By inverting every second set of four samples, the sampled cosine wave becomes a repeating set

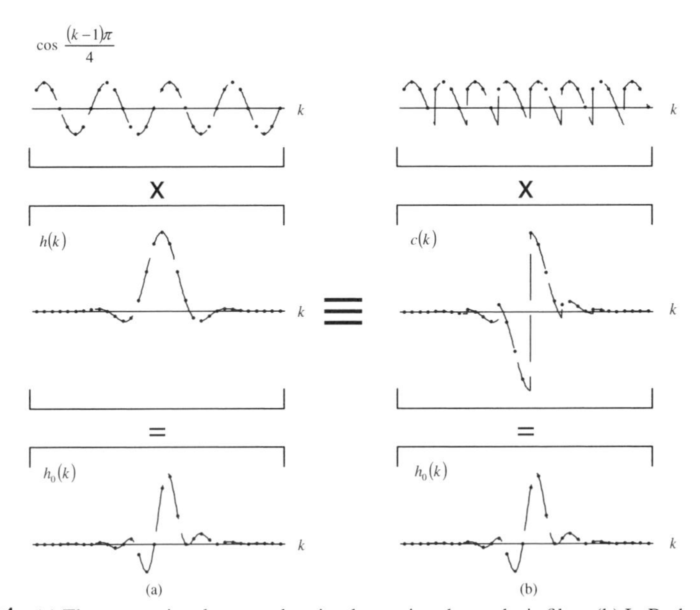

Figure 8.14 (a) The conventional approach to implementing the analysis filter. (b) In Rothweiler's approach the cosine-modulated filters can be constructed by multiplying $c(k)$ by a repeating set of samples from the original sampled cosine wave.

of the same four samples of the original wave and the prototype filter impulse response, $h(k)$, becomes the *modified impulse response*, $c(k)$.

In the discrete convolution process that would take place in a conventional implementation of the analysis filterbank, 32 time-reversed samples of the audio input are multiplied with the 32 samples of the low-pass filter impulse response as shown in Figure 8.15(a). This is equivalent to multiplying by the modified impulse response, $c(k)$, and then by the repeating set of the same four samples of the original cosine wave as shown in Figure 8.15(b).

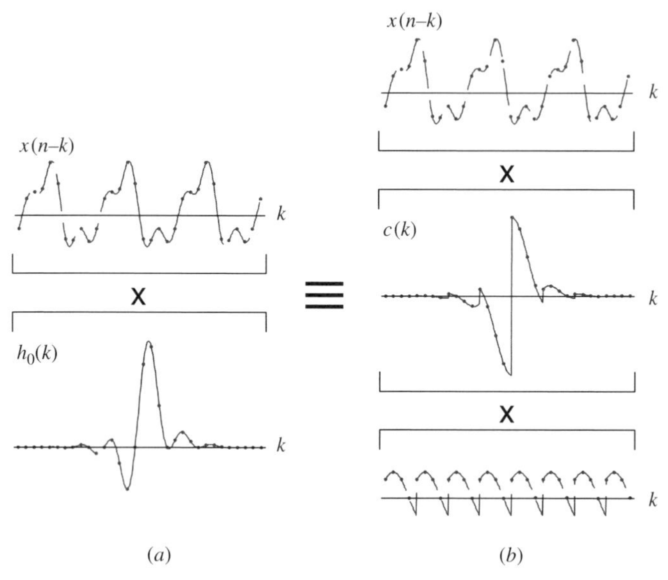

Figure 8.15 (a) The conventional approach to filtering the input signal involves multiplying by the low-pass filter impulse response. (b) In Rothweiler's approach, multiplying by the low-pass filter impulse response is equivalent to multiplying by $c(k)$ and then by a repeating set of samples from the original sampled cosine wave.

If $c(k)$ is substituted for $h(k)$, the next step in the convolution process is simplified. The corresponding samples of $c(k)x(n - k)$ are simply multiplied by the repeating set of four cosine samples, and then the sum of these products is calculated to determine the current output sample value as shown in Figure 8.16(a). However, in the calculations performed in Figure 8.16(a), all samples of $c(k)x(n - k)$ that are four samples apart are multiplied by the same cosine value. This set of four cosine samples can be factored out as shown in Figure 8.16(b). To achieve this factorization, the 32 samples of $c(k)x(n - k)$ are divided into eight groups of four samples, and the corresponding samples in each of these groups are added to determine the intermediate sum, $y(j)$. Each sample of this intermediate sum is then multiplied by its corresponding cosine value, and the sum of these products is calculated to determine the current output sample value.

In this approach, instead of performing 32 multiplications and 31 additions, 32 multiplications and 28 additions are performed to calculate the intermediate sum, and then further four multiplications and three additions are performed to calculate the output value.

For the low-pass filter on its own, this approach does not produce any significant saving in calculations, but there are more savings to be made for the entire analysis filterbank. Because the only difference between the values of the low-pass and high-pass impulse response is the

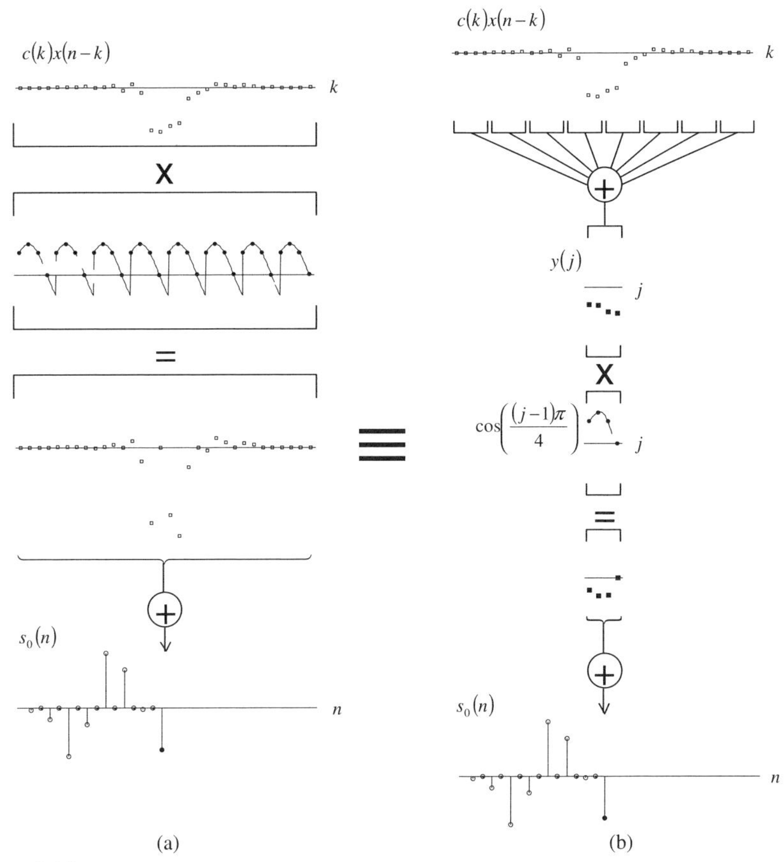

Figure 8.16 (a) The corresponding samples of $c(k)x(n - k)$ are simply multiplied by the repeating set of four cosine samples, and then the sum of these products is calculated to determine the current output sample value (shown as a full circle). (b) The process shown in (a) is simplified by dividing the 32 samples of $c(k)x(n - k)$ into eight groups of four samples, and the corresponding samples in each of these groups are added to determine the intermediate sum, $y(j)$. Each sample of this intermediate sum is then multiplied by its corresponding cosine value, and the sum of these products is calculated to determine the current output sample value (shown as a full circle).

frequency of the sampled cosine wave, the intermediate sum is reused in the calculations for the output sample of the high-pass filter. The four samples of the low-pass cosine wave in Figure 8.16(b) are replaced with the first four samples of the high-pass cosine wave to determine the current output sample for the high-pass filter.

Figure 8.17 shows the efficient implementation of the two-channel analysis filter bank from the previous examples using Rothweiler's method. Figure 8.17(a) shows the calculation of the intermediate sum, and Figure 8.17(b) shows how this intermediate sum is multiplied by four samples from both the low-pass and high-pass cosine waves to obtain the output samples of the low-pass and high-pass filters, respectively.

Instead of the 64 multiplications and 62 additions required in a conventional implementation of this filterbank, 32 multiplications and 28 additions are required for the intermediate sum, and then four multiplications and three additions are required for each of the low-pass

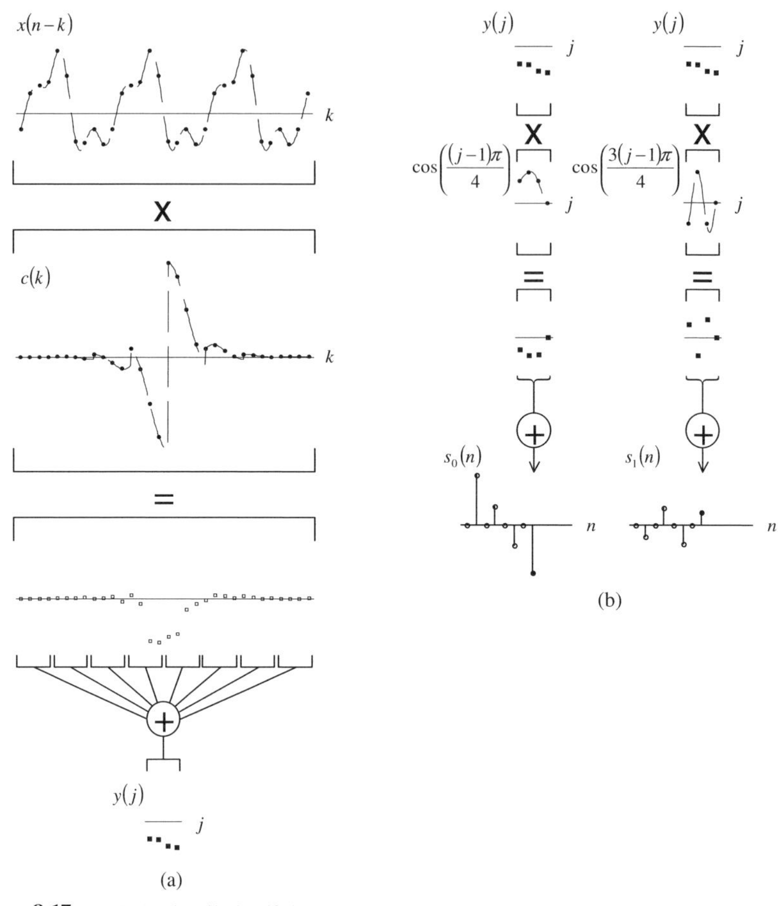

Figure 8.17 (a) In Rothweiler's efficient implementation of a two-channel cosine-modulated analysis filterbank, the product of the time-reversed input samples, $x(n - k)$, and the modified impulse response, $c(k)$, is divided into eight groups of four samples, and the samples in each of these groups are added to determine the intermediate sum, $y(j)$. (b) Each sample of this intermediate sum is then multiplied by its corresponding high-pass and low-pass cosine values, and the sum of these products is calculated to determine the current high-pass and low-pass output sample values (shown as a full circle).

and high-pass outputs giving a total of 40 multiplications and 34 additions. It is clear from this example that Rothweiler's approach becomes much more efficient than a conventional implementation as the number of bandpass filters in the filterbank increases. ■

For the general case of an M-channel filterbank using a prototype filter with length N, the following MATLAB code defines the operations performed in the analysis filter bank. The array x acts as a FIFO buffer to hold the audio samples in time-reversed order; the modified impulse response is contained in the array c, and the set of M PCM samples to be converted to subband samples is contained in the array next_pcm_input. The array subband_samples contains the set of M subband samples corresponding to the current set of M PCM samples.

```
% shift in the next M audio samples in time-reversed order
for k = N:−1:M+1
  x(k) = x(k−M);
end
for k = M:−1:1
  x(k) = next_pcm_input(M−k+1);
end

% multiply the audio samples by the modified impulse response c(k)
for k = 1:N
  z(k) = c(k)*x(k);
end

% calculate the intermediate sum y(j)
for j = 1:2*M
  y(j) = 0;
  for q = 0:N/(2*M)−1;
    y(j) = y(j)+z(j+2*M*q);
  end
end

% multiply the intermediate sum by each sampled cosine wave
% to produce the next M subband samples
for i = 1:M;
  subband_samples(i) = 0;
  for j = 1:2*M
    basis(j,i) = cos(((2*i−1)* (j−(M/2)−1)* pi)/(2*M));
    subband_samples(i) = subband_samples(i)+y (j)*basis(j,i);
  end
end
```

This implementation requires N multiplications and $N - 2M$ additions to calculate y and then $2M^2$ multiplications and $2M^2 - M$ additions to calculate the output samples of the M analysis filters.

In the analysis filterbank used in the MPEG audio coding algorithms, the number of bandpass filters in the filterbank is 32, and the prototype impulse response has a length of 512, that is, $M = 32$ and $N = 512$. A conventional implementation would therefore require $512 \times 32 = 16{,}384$ multiplications and $511 \times 32 = 16{,}352$ additions for each set of 32 output samples. An implementation using Rothweiler's method requires 512 multiplications and 448 additions to calculate the intermediate sum and then 64 multiplications and 63 additions for each of the 32 output samples, giving a total of 2560 multiplications and 2464 additions.

In addition, the equation for calculating the M subband samples is effectively describing a modified discrete cosine transform (DCT) with $2M$ inputs and M outputs, so a further reduction in computational complexity is possible by using a fast DCT implementation of this equation.

8.5.2. Synthesis Filter

As for the analysis filter, to explain Rothweiler's proposal it is best to first consider the following example that shows how the two-channel synthesis filterbank from Section 8.4 is implemented more efficiently.

EXAMPLE 8.6

In the discrete convolution process that takes place in a conventional implementation of the synthesis filterbank, 32 time-reversed subband samples are multiplied with the 32 samples of the low-pass filter impulse response as shown in Figure 8.18(a). This is equivalent to multiplying by a repeating set of the same four samples of the original cosine wave and then by the modified impulse response, $c(k)$, as shown in Figure 8.18(b).

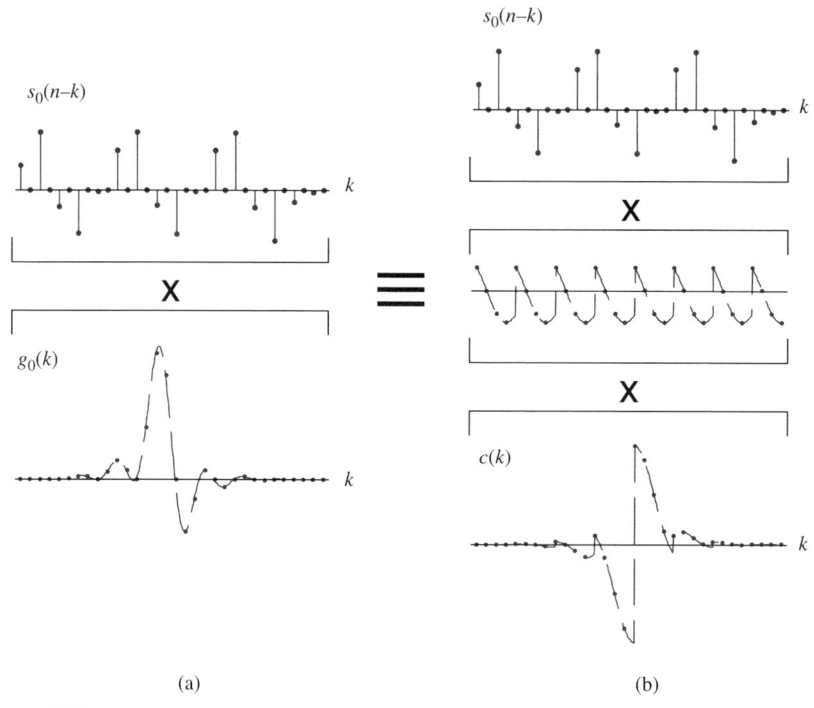

Figure 8.18 (a) The conventional approach to reconstructing the time-domain signal involves multiplying by the low-pass synthesis filter impulse response. (b) In Rothweiler's approach, multiplying by the low-pass filter impulse response is equivalent to multiplying by a repeating set of samples from the original sampled cosine wave and then by $c(k)$.

During the course of the process shown in Figure 8.18(b), each subband sample is only multiplied by four different cosine sample values. To explain this further, consider the diagram in Figure 8.19(a) that shows the relative position of the subband samples and the first four samples of the cosine wave for four consecutive output sample times. This diagram shows that, because every second subband sample is zero, the same set of input subband samples is used in the convolution process for the output samples at times n and $n + 1$. The only difference between times n and $n + 1$ is the relative position of the subband samples and the cosine samples. Then for output samples at times $n + 2$ and $n + 3$, the next subband sample is included in the convolution process, and this new set of subband samples is used for both output sample times. Hence, during the entire convolution process, one individual subband sample is multiplied by the first two cosine samples for times n and $n + 1$, the last two cosine samples for times $n + 2$ and $n + 3$, the first two cosine samples for times $n + 4$ and $n + 5$, the last two cosine samples for times $n + 6$ and $n + 7$, and so on.

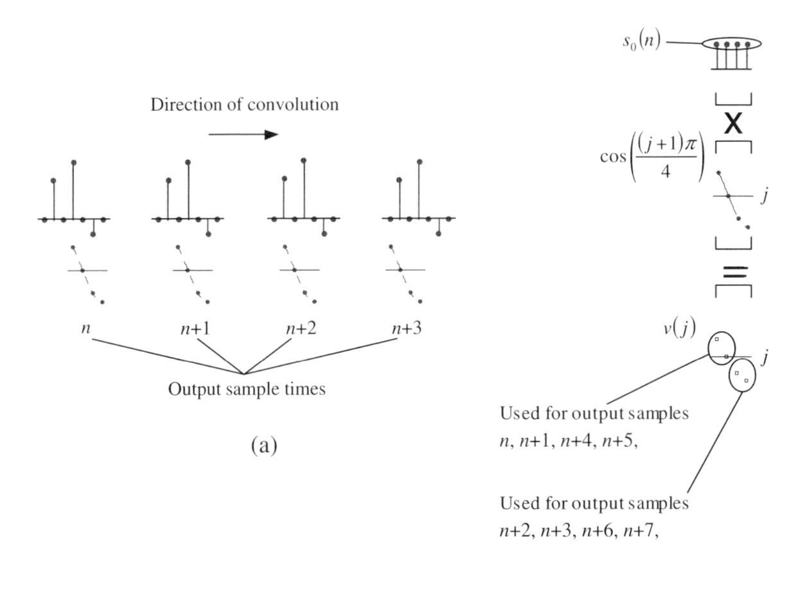

Figure 8.19 (a) One individual subband sample during the convolution process is multiplied by the first two repeating cosine samples for times n and $n + 1$ and the last two cosine samples for times $n + 2$ and $n + 3$. (b) The product of one individual subband sample with each of the four cosine sample values is denoted by $v(j)$. The first two values of $v(j)$ are included in the convolution process for output samples at times $n, n + 1, n + 4, n + 5, \ldots$, and the last two values are included in the convolution process for output samples at times $n + 2, n + 3, n + 6, n + 7, \ldots$.

Now consider the product of one individual subband sample with each of the four cosine sample values, $v(j)$, as shown in Figure 8.19(b). The first two values of $v(j)$ are included in the convolution process for output samples at times n and $n + 1$, and the last two values are included in the convolution process for output samples at times $n + 2$ and $n + 3$. Then for output samples at times $n + 4$ and $n + 5$, the first two values of $v(j)$ are reused, and for output samples at times $n + 6$ and $n + 7$, the last two values of $v(j)$ are reused, and so on.

For the case of the two-channel synthesis filter from the previous example, the number of calculations required is reduced by first calculating $v_0(j)$ using the first four low-pass cosine samples and $v_1(j)$ using the first four high-pass cosine samples as shown in Figure 8.20(a). The corresponding values of $v_0(j)$ and $v_1(j)$ are then added to produce four intermediate values, $v(j)$. For each two input subband samples, four new $v(j)$ values are produced. The current set of output samples only requires the first two values, but the last two $v(j)$ values are needed for the next two output samples. To save recalculating values, a FIFO buffer is constructed, and each time four new values for $v(j)$ are calculated, the current contents of the buffer is shifted by four positions and the current set of $v(j)$ values is placed at the start of the buffer. Note that there are 64 values stored in the FIFO buffer for any one output sample time, so the complete set of values in the buffer is not shown in Figure 8.20(a).

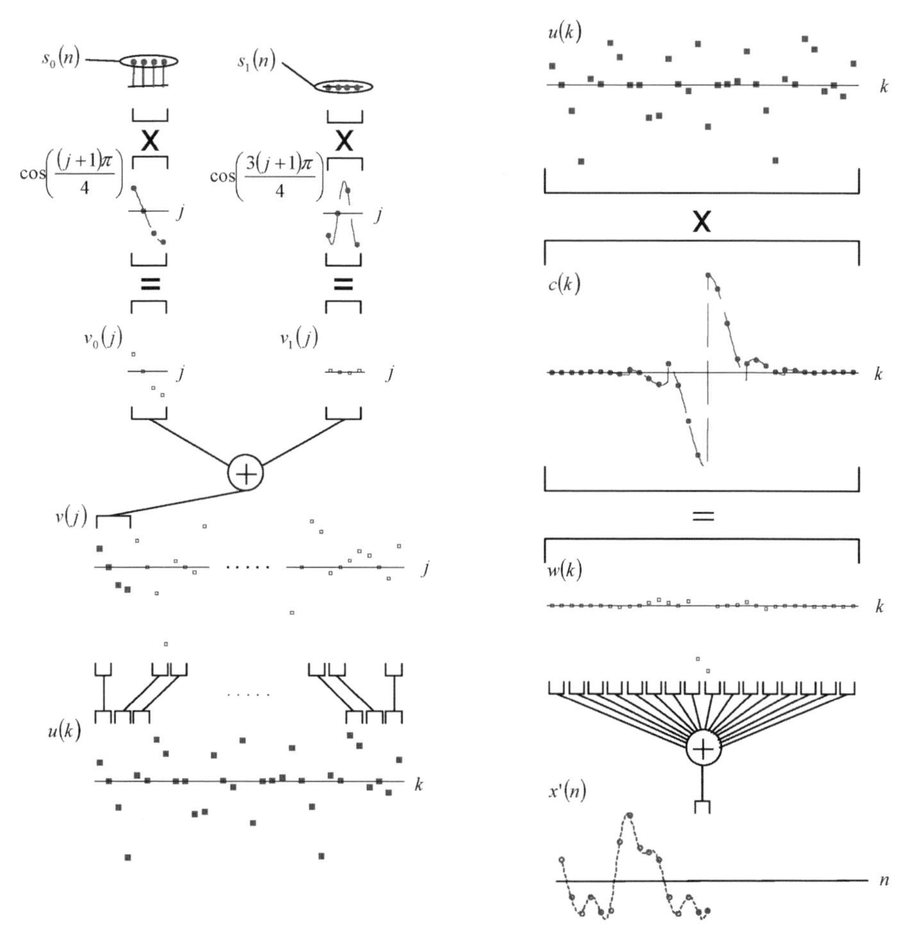

Figure 8.20 (a) In Rothweiler's efficient implementation of a two-channel cosine-modulated synthesis filterbank, $v_0(j)$ is calculated using the first four low-pass cosine samples and $v_1(j)$ using the first four high-pass cosine samples. The corresponding values of $v_0(j)$ and $v_1(j)$ are then added to produce four intermediate values, $v(j)$. A FIFO buffer is constructed and, each time four new values for $v(j)$ are calculated, the current contents of the buffer is shifted by four positions, and the current set of $v(j)$ values is placed at the start of the buffer. The set of values $u(k)$ is constructed by taking the first two values from the current set of four $v(j)$ values, the second two samples from the previous set of four $v(j)$ values, and so on. (b) The values of $u(k)$ are multiplied by the corresponding samples of the modified prototype filter $c(k)$ to give $w(k)$. The even values in $w(k)$ are added together to produce the output sample at time n, and the odd values are added together to produce the output sample at time $n + 1$ (shown as full circles).

The set of values $u(k)$ is then constructed as shown in Figure 8.20(a) by taking the first two values from the current set of four $v(j)$ values, the second two samples from the previous set of four $v(j)$ values, and so on.

The values of $u(k)$ are then multiplied by the corresponding samples of the modified prototype filter $c(k)$ to give $w(k)$ as shown in Figure 8.20(b). The even values in $w(k)$ are then added together to produce the output sample at time n, and the odd values are added together to produce the output sample at time $n + 1$.

Instead of the 64 multiplications and 63 additions required in a conventional implementation of this filterbank, four multiplications and three additions are required for each of the low-pass and high-pass outputs, four additions are required for the intermediate values of $v(j)$, and 32 multiplications and 30 additions are required for the two output sample values giving a total of 40 multiplications and 40 additions. Again, it is clear from this example that Rothweiler's approach becomes much more efficient than a conventional implementation as the number of bandpass filters in the synthesis filterbank increases. ■

For the more general case of an M-channel filterbank using a prototype filter with length N, the following MATLAB code defines the operations performed in the synthesis filter bank. The array v acts as a FIFO buffer to hold intermediate values $v(j)$; the modified impulse response is contained in the array c, and the set of M subband samples to be converted to PCM samples is contained in the array subband_samples. The array next_pcm_output contains the set of M output PCM samples corresponding to the current set of M subband samples.

```
% shift the values in the FIFO buffer v(j) by 2M elements
for j = 2*N:-1:2*M+1
  v(j) = v(j-2*M);
end

% calculate the next 2M values and place them at the start of the FIFO buffer v(j)
for j = 1:2*M
  v(j) = 0;
  for i = 1:M
    basis(i,j) = cos(((2*i-1)*(j+M/2-1)*pi)/(2*M));
    v(j) = v(j)+subband_samples(i)*basis(i,j);
  end
end

% calculate the intermediate sum u(k)
for q = 0:N/(2*M)-1
  for i = 1:M
    u(2*M*q+i) = v(4*M*q+i);
    u(2*M*q+M+i) = v(4*M*q+3*M+i);
  end
end

% multiply the intermediate sum by the modified impulse response c(k)
for k = 1:N
  w(k) = u(k)*c(k);
end
% calculate the next M output PCM samples
```

```
for n = 1:M
  next_pcm_output(n) = 0;
  for p = 0:N/M-1
    next_pcm_output(n) = next_pcm_output(n)+w (n+M*p)*M;
  end
end
```

It is important to note that for each M output samples of the synthesis filterbank, it is only necessary to calculate the first $2M$ values of $v(j)$ because the remaining values have already been calculated for previous output samples of the filterbank.

This implementation requires $2M^2$ multiplications and $2M^2 + M$ additions to calculate the first $2M$ samples of v and then N multiplications and $N - M$ additions to calculate the M samples of the reconstructed signal.

In the synthesis filterbank used in the MPEG audio coding algorithms, the number of bandpass filters in the filterbank is 32, and the prototype impulse response has a length of 512, that is, $M = 32$ and $N = 512$. Therefore, a conventional implementation would require $512 \times 32 = 16{,}384$ multiplications and $512 \times 32 - 1 = 16{,}383$ additions for each set of 32 output samples. But an implementation using Rothweiler's method requires 2048 multiplications and 2080 additions to calculate the first 64 samples of $v(j)$ and then 512 multiplications and 480 additions to calculate the 32 samples of the reconstructed signal, giving a total of 2560 multiplications and 2560 additions.

In addition, the equation for calculating each $2M$ values of $v(j)$ is effectively describing a modified inverse DCT with M inputs and $2M$ outputs, so a further reduction in computational complexity is possible by using a fast IDCT implementation of this equation.

8.6. TIME-DOMAIN ALIASING CANCELLATION

An alternative method to conducting subband filtering with aliasing cancellation in the frequency domain is to construct a set of filters that overlap in both time and frequency and provide aliasing cancellation in the time domain. This system was first proposed by John Princen and Alan Bradley in 1986 [2] and is commonly referred to as time-domain aliasing cancellation (TDAC).

This system is effectively a transform-based approach to conversion between the time and frequency domains. However, with TDAC transforms a critically sampled system is designed using overlapping transform windows in the time domain. In all filterbank designs there is some trade-off between window size in the time domain and resolution in the frequency domain. The main disadvantage of orthonormal transform-based techniques is that the number of samples in the time-domain corresponds directly to the number of frequency-domain coefficients. This means that relatively small time-domain windows are used, and the resulting coefficients represent strongly overlapping subbands in the frequency domain. One solution to this problem is to use a large strongly overlapping window in the time domain such as that used in the cosine-modulated filter bank of the previous section. The outputs of this filterbank are then subsampled, and aliasing is introduced in the frequency domain. If the frequency responses of these filters are designed correctly, a perfect reconstruction filter bank is produced and the frequency-domain aliasing can be removed.

Time-domain aliasing cancellation takes a different approach. In this system an overlapped window in the time domain is used, but this window typically only overlaps by half the size of the window. By extending the size of the window, the amount of overlap in the corresponding frequency-domain subbands is reduced compared to a critically sampled transform. However, there are now more frequency-domain coefficients than time-domain samples in the analysis stage. To overcome this problem, in a TDAC system, only half the frequency-domain coefficients are transmitted, and distortion is introduced into the reconstructed time-domain samples. If the transform is designed correctly, the distortion introduced by discarding coefficients is directly canceled out by overlapping and adding successive windows in the time domain.

Such a TDAC system uses certain properties of the DCT to guarantee perfect reconstruction. The general equation for an N-point DCT of the sequence $x(n)$ is given by

$$X(k) = \sum_{n=0}^{N-1} x(n) \cos\left(\frac{2\pi k}{N}(n + n_0)\right), \quad \text{for } k = 0 \ldots N-1 \tag{8.22}$$

and the inverse of this transform is given by

$$x'(r) = \frac{1}{N}\sum_{k=0}^{N-1} X(k) \cos\left(\frac{2\pi k}{N}(r + n_0)\right), \quad \text{for } r = 0 \ldots N-1 \tag{8.23}$$

This set of equations produces a distorted version of the original sequence. Note that the variable r has been substituted for the variable n to indicate the sample time of the outputs of the inverse transform. This substitution is necessary for the remainder of the derivation as the equations for the forward and inverse transforms are combined.

An example of the type of distortion produced is shown by substituting Equation (8.22) into Equation (8.23) to give

$$x'(r) = \frac{1}{N}\sum_{k=0}^{N-1}\sum_{n=0}^{N-1} x(n) \cos\left(\frac{2\pi k}{N}(n + n_0)\right) \cos\left(\frac{2\pi k}{N}(r + n_0)\right) \tag{8.24}$$

This equation can be rearranged using the simple trigonometric identity, $\cos(A)\cos(B) = 1/2\,[\cos(A - B) + \cos(A + B)]$, to give

$$x'(r) = \frac{1}{2N}\sum_{n=0}^{N-1} x(n)\left[\sum_{k=0}^{N-1} \cos\left(\frac{2\pi k}{N}(n - r)\right) + \sum_{k=0}^{N-1} \cos\left(\frac{2\pi k}{N}(n + r + 2n_0)\right)\right] \tag{8.25}$$

The following equation holds for the general form of the terms in the square brackets

$$\sum_{k=0}^{N-1}\cos\left(\frac{2\pi k}{N}n\right)=\begin{cases}N, & \text{if } n=0,N,2N,\ldots\\ 0, & \text{otherwise}\end{cases} \tag{8.26}$$

Because the values of n in Equation (8.25) are restricted to the range $0\cdots N-1$, the first term in the square brackets in Equation (8.25) reduces to a value of N when $n=r$ (i.e., when $n-r=0$), and the second term reduces to a value of N when $n=N-r-2n_0$ (i.e., when $n+r+2n_0=N$). Hence Equation (8.25) can be rewritten as

$$x'(r)=\frac{1}{2}\left[x(r)+x(N-r-2n_0)\right] \tag{8.27}$$

From Equation (8.27), it is evident that the distorted output signal is the sum of the original sequence and a time-reversed and shifted replica of the original sequence. This time-reversed replica of the original sequence is effectively a time-domain aliased version of the original signal.

EXAMPLE 8.7

Determine the output signal obtained from the general DCT described in Equations (8.22) and (8.23) when the input signal is $x(n)$ = [0 1 2 3 4 5 6 7] and the phase term is $n_0 = 2$. Verify that this output signal is equal to $x'(r) = 1/2\,[x(r) + x(N - r - 2n_0)]$.

SOLUTION If Equations (8.22) and (8.23) are implemented directly, the output signal is given by $x'(r)$ = [2 2 2 2 2 6 6 6]. Then if $N = 8$, the second term of Equation (8.27) is given by $x(8 - r - 4)$ = [4 3 2 1 0 7 6 5], and substituting these values into Equation (8.27) gives the desired result.

Figure 8.21 shows a graphical representation of the transform process with the original signal shown as the dotted line, the time-domain aliasing error as the dashed line, and the output signal of the transform process as the full line with dots showing the actual sample values.

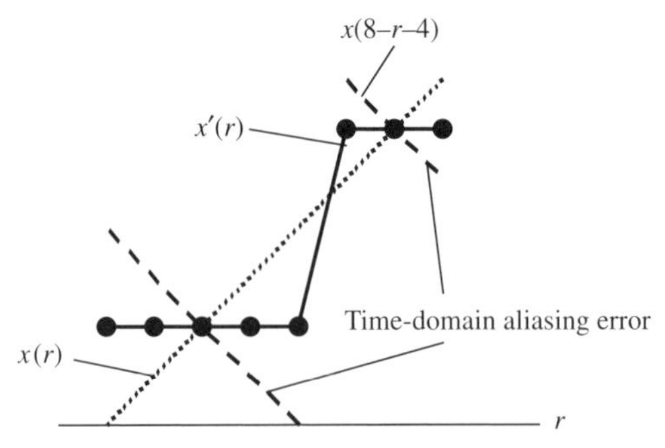

Figure 8.21 The output signal of the transform process is the sum of the original signal and the time-domain aliasing error. ■

Now, using this same basic idea, the general equation of the DCT is modified to give the following transform equations:

$$X(k)=\frac{2}{N}\sum_{n=0}^{N-1}x(n)\cos\left(\frac{2\pi}{4N}(2k+1)\left(2n+\tfrac{N}{2}+1\right)\right),\quad \text{for } k=0\cdots N/2-1 \tag{8.28}$$

and

$$x'(r)=\sum_{k=0}^{N/2-1}X(k)\cos\left(\frac{2\pi}{4N}(2k+1)\left(2r+\tfrac{N}{2}+1\right)\right),\quad \text{for } r=0\cdots N-1 \tag{8.29}$$

Note that this modified transform only requires half as many transform coefficients as there are samples of the input signal. In this way, the transform is implemented on an extended set of input samples, but a critically sampled system is still maintained. Then substitute Equation (8.28) into Equation (8.29) and rearrange using the same simple trigonometric identity to give the following equation:

$$\begin{aligned}x'(r)=\frac{2}{N}\sum_{n=0}^{N-1}x(n)\Bigg[&\sum_{k=0}^{N/2-1}\cos\left(\frac{2\pi}{4N}(2k+1)(2n-2r)\right)\\&+\sum_{k=0}^{N/2-1}\cos\left(\frac{2\pi}{4N}(2k+1)(2n+2r+N+2)\right)\Bigg]\end{aligned} \tag{8.30}$$

The following equation holds for the general form of the terms in the square brackets:

$$\sum_{k=0}^{N/2-1}\cos\left(\frac{2\pi}{4N}(2k+1)2n\right)=\begin{cases}N/2, & \text{if } n=0,2N,4N,\ldots\\ -N/2, & \text{if } n=N,3N,5N,\ldots\\ 0, & \text{otherwise}\end{cases} \tag{8.31}$$

Because the values of n in Equation (8.30) are restricted to the range $0 \cdots N-1$, the first term in the square brackets in Equation (8.30) reduces to a value of $N/2$ when $n = r$. However, the second term in the square brackets has two different values depending on the value of r. When $r = 0 \cdots N/2 - 1$, the second term has the value $-N/2$ because this range of values of r corresponds to $n = N/2 - 1 \cdots 0$ when the equation $n + r + N/2 + 1 = N$ is satisfied. Similarly, when $r = N/2 \cdots N - 1$, the second term has the value $N/2$ because this range of values of r corresponds to $n = N - 1 \cdots N/2$ when the equation $n + r + N/2 + 1 = 2N$ is satisfied. Hence Equation (8.30) can be rewritten as

$$x'(r)=\begin{cases}\frac{1}{2}\left[x(r)-x(N/2-r-1)\right], & \text{for } r=0\cdots N/2-1\\[2ex] \frac{1}{2}\left[x(r)+x(3N/2-r-1)\right], & \text{for } r=N/2\cdots N-1\end{cases} \tag{8.32}$$

EXAMPLE 8.8

Determine the output signal obtained from the DCT described in Equations (8.28) and (8.29) when the input signal is $x(n) = [0\ 1\ 2\ 3\ 4\ 5\ 6\ 7]$. Verify that this output signal is equal to the signal obtained using Equation (8.32).

SOLUTION If Equations (8.28) and (8.29) are implemented directly, the output signal is given by $x'(r) = [-1.5\ -0.5\ 0.5\ 1.5\ 5.5\ 5.5\ 5.5\ 5.5]$. Then if $N = 8$ the second term of Equation (8.32) is given by $-x(4 - r - 1) = [-3\ -2\ -1\ 0]$ for $r = 0 \cdots 3$ and $x(12 - r - 1) = [7\ 6\ 5\ 4]$ for $r = 4 \cdots 7$. Substituting these values into Equation (8.32) gives the desired result. Figure 8.22 shows a graphical representation of the transform process with the original signal shown as the dotted line, the time-domain aliasing error as the dashed line, and the output signal of the transform process as the full line with dots showing the actual sample values.

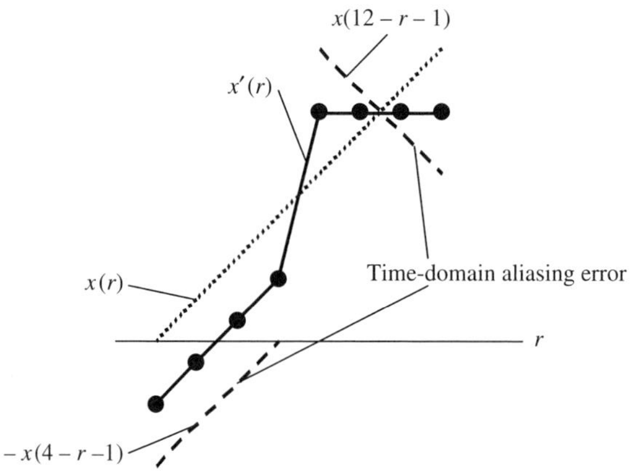

Figure 8.22 The output signal of the transform process is the sum of the original signal and the time-domain aliasing error. ■

Now using the transform process defined by Equations (8.28) and (8.29) a critically sampled overlapped transform with perfect reconstruction can be implemented. However, the properties of this transform process need to be combined with a suitable window function. This window function should have similar properties to the frequency response of perfect reconstruction bandpass filters, that is, when the weighting function is squared and added to a shifted version of itself, the resulting weighting is unity. An example of the coefficients of a possible weighting function for an eight-band TDAC transform is given in Table 8.1.

Figure 8.23 shows how the weighting function is combined with the overlapped transform process to provide perfect reconstruction of the original signal. Figure 8.23(a) shows the shape of the weighting function that is multiplied with the input

Table 8.1 Weighting function coefficients for an eight-band TDAC transform.

n	$W(n)$
0	0.00443
1	0.06153
2	0.24757
3	0.55031
4	0.82831
5	0.96674
6	0.99788
7	0.99999

signal before the transform is applied. Then, if the forward and inverse transform is applied to the weighting function only, Figure 8.23(b) shows the distorted weighting function that is produced by time-domain aliasing in the transform process. This distorted weighting function is then multiplied by the original weights to produce the final weighting for the output of the transform process as shown in Figure 8.23(c). This final weighting function now exhibits the desired properties for time-domain aliasing cancellation. Figure 8.23(d) shows that, if the output of each transform process is overlapped by half with the previous output, the original input signal is reconstructed perfectly.

The frequency analysis and synthesis used in the Dolby AC-3 audio coding algorithm is based on the transform process defined by Equations (8.28) and (8.29). The transform is performed on a single block of 512 audio samples or two blocks of 256 samples, and consequently the window function used has a length of 512 but has a similar shape to the function shown in Figure 8.23.

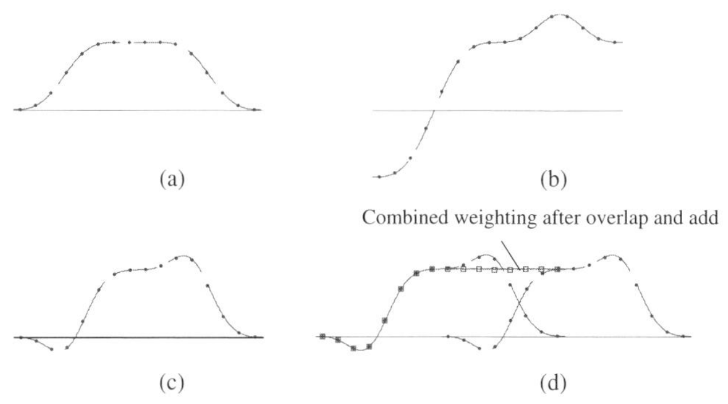

Figure 8.23 (a) The weighting function. (b) The distorted weighting function after the forward and inverse transform has been applied. (c) The final weighting function after multiplying the distorted function by the original weighting function. (d) If the final weighting function is overlapped by half and added, the result is a constant weighting of the reconstructed samples and a cancellation of the distortion introduced by the transform process.

8.7. SUMMARY

In this chapter, two different techniques for converting from time-domain samples to frequency-domain coefficients have been explained. In particular, the cosine-modulated filterbank approach used in the MPEG audio coding algorithms and the time-domain aliasing cancellation approach used in the Dolby AC-3 algorithm have been investigated. The following chapters explain how these frequency analysis and synthesis techniques have been incorporated into the design of the standard audio coding algorithms.

PROBLEMS

8.1 Briefly explain what is meant by the following terms:

- **(a)** Sampling
- **(b)** Aliasing
- **(c)** Nyquist sampling rate

8.2 If a signal has the magnitude spectrum given below:

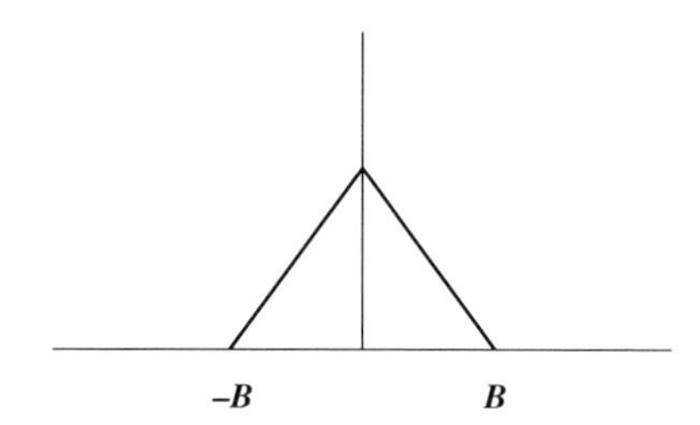

- **(a)** Sketch the spectrum of the sampled signal if the sampling frequency is $2B$.
- **(b)** Sketch the spectrum of the signal if it has been subsampled by a factor of 2.
- **(c)** Sketch the spectrum of the signal if it has been upsampled by a factor of 2.

8.3 Write the equations for the impulse response of the cosine-modulated analysis filters that would be used in a four-channel subband filterbank.

8.4 Write the equations for the impulse response of the cosine-modulated synthesis filters that would be used in a four-channel subband filterbank.

8.5 For the 32 PCM samples and the prototype filter impulse response given below, calculate the intermediate sum $y(j)$ that would be produced in a two-channel analysis filterbank using Rothweiler's method.

$x(n) = \{-4.02, 4.76, 6.73, 7.60, 15.30, 9.17, -6.86, -7.36, -3.91, -7.68, -4.02, 4.76, 6.73, 7.60, 15.30, 9.17, -6.86, -7.36, -3.91, -7.68, -4.02, 4.76, 6.73, 7.60, 15.30, 9.17, -6.86, -7.36, -3.91, -7.68, -4.02, 4.76\}$

$h(k) = \{0.00000, -0.00004, -0.00022, -0.00079, -0.00163, -0.00111, 0.00350, 0.01196, 0.01554, -0.00034, -0.03931, -0.07421, -0.05016, 0.07610, 0.28602, 0.48843, 0.57250, 0.48843, 0.28602, 0.07610, -0.05016, -0.07421, -0.03931, -0.00034, 0.01554, 0.01196, 0.00350, -0.00111, -0.00163, -0.00079, -0.00022, -0.00004\}$

8.6 For the intermediate sum $y(j)$ calculated in question 8.5, determine the output subband sample for the low-pass and high-pass filters that would be produced in the two-channel analysis filterbank. Verify that these subband samples are identical to the output samples that would be produced by a conventional implementation of the analysis filterbank.

8.7 If the value of the next low-pass and high-pass subband samples are -2.31 and 37.96, respectively, calculate the next four values for $v(j)$ that would be produced in a two-channel synthesis filterbank using Rothweiler's method.

8.8 For the following values of $v(j)$ calculate the values for $u(k)$ to be used to determine the next two output PCM samples of a synthesis filterbank using Rothweiler's method. Use the prototype filter given in question 8.5 to determine the next two reconstructed output samples that would be produced in the two-channel synthesis filter. Verify that these reconstructed PCM samples are identical to the samples that would be produced by a conventional implementation of the synthesis filterbank.

$v(j) = \{28.47, 0.00, -28.47, -35.65, 41.21, -0.00, -41.21, -120.83, 8.26, 0.00, -8.26, 63.83, -65.07, -0.00, 65.07, 42.28, -0.73, 0.00, 0.73, 5.50, 52.44, 0.00, -52.44, -48.48, 40.67, -0.00, -40.67, -121.14, 10.49, 0.00, -10.49, 63.38, -66.22, -0.00, 66.22, 42.68, -0.60, 0.00, 0.60, 5.84, 52.07, 0.00, -52.07, -48.25, 40.08, -0.00, -40.08, -119.09, 8.37, 0.00, -8.37, 66.81, -72.42, -0.00, 72.42, 55.31, -9.45, -0.00, 9.45, 24.97, 17.57, 0.00, -17.57, 33.72\}$

8.9 Modify the MATLAB code given in Section 8.5.1 to implement an efficient four-channel analysis filterbank using a prototype filter of length 64.

8.10 Modify the MATLAB code given in Section 8.5.2 to implement an efficient four-channel synthesis filterbank using a prototype filter of length 64.

8.11 For an efficient implementation of a four-channel cosine-modulated analysis filterbank using Rothweiler's method, how many additions and multiplications are required per output subband sample if the prototype filter impulse response has 64 sample values? How many additions and multiplications would be required for a conventional implementation of this filterbank?

8.12 Complete the following table of values for the function $\cos(2\pi nk/N)$ when $N = 4$:

k	n								
	0	1	2	3	4	5	6	7	8
0									
1									
2									
3									

Use these values to verify that the following equation holds for $N = 4$:

$$\sum_{k=0}^{N-1} \cos\left(\frac{2\pi k}{N} n\right) = \begin{cases} N, & \text{if } n = 0, N, 2N, \ldots \\ 0, & \text{otherwise} \end{cases}$$

8.13 Determine the output signal obtained from the DCT described in Equations (8.28) and (8.29) when the input signal is $x(n) = [0\ 0\ 4\ 4\ 4\ 4\ 0\ 0]$. Verify that this output signal is equal to the signal obtained using Equation (8.32).

8.14 Write the forward transform equation for a TDAC transform that would convert 16 input samples to eight coefficients.

8.15 Write the inverse transform equation for a TDAC transform that would convert eight coefficients into 16 output samples.

8.16 Calculate the output coefficients for the first two overlapping blocks of 16 samples from the following set of input sample values using the forward transform from question 16 and the following weighting function values.

$w(n) = \{0.00443,\ 0.06153,\ 0.24757,\ 0.55031,\ 0.82831,\ 0.96674,\ 0.99788,\ 0.99999,$
$0.99999, 0.99788, 0.96674, 0.82831, 0.55031, 0.24757, 0.06153, 0.00443\}$

8.17 Calculate the output samples for the two blocks of eight coefficients from question 8.16 using the inverse transform from question 17 and the following weighting function values.

8.18 Add the last eight samples of the first block of output samples from question 19 with the first eight samples of the second block and verify that the resulting output samples are identical to the corresponding input sample values given in question 8.16.

MATLAB EXERCISE 8.1

1. Modify the MATLAB code given in Section 8.5.1 to implement a 32-channel analysis filterbank. Use the MPEG-modified prototype filter given in Section 9.1.1 of Chapter 9 for the values of $c(k)$.
2. Construct a signal that is the sum of two sinusoids where the frequency of the first sinusoid is the center frequency of the first critical band of the ear and the frequency of the second sinusoid is 0.1 Bark higher than that of the first sinusoid.
3. Obtain the set of 32 subband samples that are produced by the analysis filterbank from part 1 (note that you should pass enough samples of the input signal through the filterbank so that the output subband samples are stable).
4. Repeat steps 2 and 3 keeping the frequency of the first sinusoid constant and increasing the frequency of the second sinusoid by 0.1 Bark until there is no overlap between the subband samples produced by the two sinusoids. Determine the difference in critical band rate required between the first and second sinusoids so that there is no overlap between the

subband samples produced by the two sinusoids and record this value in the following table:

Critical band no.	Selectivity Δz
1	
2	
3	
4	
5	
6	
7	
8	
9	
10	
11	
12	
13	

Critical band no.	Selectivity Δz
14	
15	
16	
17	
18	
19	
20	
21	
22	
23	
24	
25	

5. Repeat steps 2–4 for the remaining critical bands of the human auditory system.
6. Comment on the selectivity of this filterbank compared with the selectivity of the human auditory system across the range of audible frequencies.

MATLAB EXERCISE 8.2

1. Use Equation (8.28) given in Section 8.6 to implement a 256 point TDAC forward transform. Use the Dolby AC-3 weighting function given in Section 10.1.3 of Chapter 10.
2. Construct a signal that is the sum of two sinusoids where the frequency of the first sinusoid is the center frequency of the first critical band of the ear and the frequency of the second sinusoid is 0.1 Bark higher than that of the first sinusoid.
3. Obtain the set of 256 coefficients that are produced by the forward TDAC transform from part 1 (Note that you should pass enough samples of the input signal through the transform so that the output coefficients are stable).
4. Repeat steps 2 and 3 keeping the frequency of the first sinusoid constant and increasing the frequency of the second sinusoid by 0.1 Bark until the there is no overlap between the subband samples produced by the two sinusoids. Determine the difference in critical band rate required between the first and second sinusoids so that there is no overlap between the

coefficients produced by the two sinusoids and record this value in the following table:

Critical band no.	Selectivity Δz
1	
2	
3	
4	
5	
6	
7	
8	
9	
10	
11	
12	
13	

Critical band no.	Selectivity Δz
14	
15	
16	
17	
18	
19	
20	
21	
22	
23	
24	
25	

5. Repeat steps 2–4 for the remaining critical bands of the human auditory system.
6. Comment on the selectivity of this transform compared with the selectivity of the human auditory system across the range of audible frequencies.

REFERENCES

1. Rothweiler, J. H. Polyphase Quadrature Filters - A New Subband Coding Technique, *Proceedings of the IEEE International Conference on Acoustics Speech and Signal Processing*, 27.2, IEEE Press, Piscataway, N.J., 1983, pp. 1280–1283.
2. Princen, J., Bradley, A. Analysis/Synthesis filter bank design based on time domain aliasing cancellation, *IEEE Transactions on Acoustics, Speech and Signal Processing, Vol. 34, Issue 5*, Oct. 1986, pp. 1153–1161.

Chapter 9

MPEG Audio

The MPEG-1 audio standard describes an algorithm suitable for coding a monophonic or stereo audio signal with a bandwidth of up to 24 kHz. The algorithm is divided into three levels of coding complexity called *layers*. The Layer I algorithm is the simplest algorithm in terms of computational complexity and requires the highest bit rate to produce an output audio quality that is indistinguishable from the original. The Layers II and III algorithms include additional components that increase the computational complexity of the Layer I algorithm and consequently require lower bit rates to provide indistinguishable quality.

The MPEG-1 audio coding algorithm was finalized as an international standard in 1993 [1], and the algorithms used in the standard are based on two perceptual audio coding algorithms that were developed in the late 1980s. Layer I is a simplified version of the MUSICAM (masking-pattern universal subband integrated coding and multiplexing) algorithm. The MUSICAM algorithm was derived from an earlier subband coding scheme called MASCAM (masking pattern adapted subband coding and multiplexing) and has been adopted in Europe for use in digital audio broadcasting. Layer II is essentially identical to MUSICAM and Layer III is a combination of techniques from MUSICAM and the ASPEC (adaptive spectral perceptual entropy coding) transform-based algorithm.

The MPEG-2 audio standards extend the functionality of the MPEG-1 algorithm to allow lower input sampling rates and the coding of *multichannel* audio signals for surround-sound reproduction. Early psychoacoustic experiments showed that at least four channels are required to produce a realistic surround-sound field. But due to limitations in the technology available, two-channel stereo became the accepted standard for home sound reproduction. Today though, with the improvements in digital coding techniques and the increased capacity of storage devices, multichannel audio systems have become a viable alternative.

In particular, the ITU-R have recommended a five-channel audio system known as 3/2 stereo that consists of left and right channels (L and R), a center channel (C), and two rear surround channels (L_S and R_S) [2]. This system produces a realistic surround-sound field with a stable frontal sound image and a large listening area.

Digital Television, by John Arnold, Michael Frater and Mark Pickering.

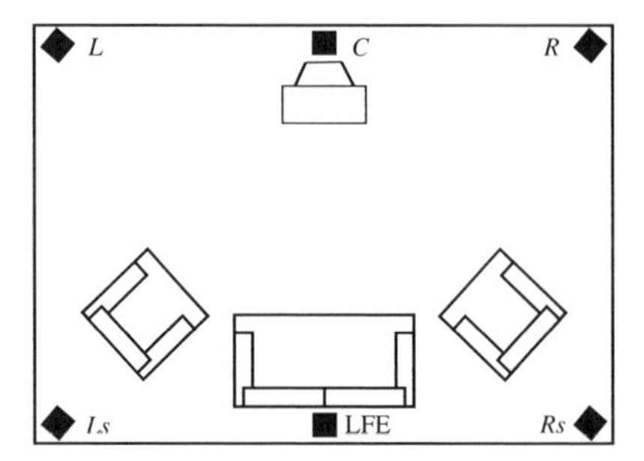

Figure 9.1 A typical speaker position for a 5.1 channel surround-sound listening environment.

An optional low-frequency enhancement channel (LFE) can be added to increase the level of frequency components between 15 and 120 Hz. A 3/2 stereo system combined with a LFE channel is usually referred to as a 5.1 channel system and a typical speaker configuration for such a system in a home listening environment is shown in Figure 9.1.

The MPEG-2 audio coding standard contains two completely separate algorithms. The first algorithm was designed to allow backward compatibility with MPEG-1 and is usually referred to as MPEG-2 BC (backward compatible). It was finalized as an international standard in 1995 with amendment 1 added in 1996 [3]. The second algorithm (known as MPEG-2 AAC (advanced audio coding)) was added as a new part to the MPEG-2 standard in 1997 [4].

The MPEG-2 BC standard uses essentially the same algorithms as the MPEG-1 encoder and decoder, but has added functionality to allow lower input sampling rates for mono and stereo inputs and multichannel audio at the MPEG-1 input sampling rates. The backward compatibility between MPEG-2 BC and MPEG-1 means that an MPEG-2 decoder can decode an MPEG-1 bit stream and an MPEG-1 decoder can derive a stereo signal from an MPEG-2 multichannel bit stream provided that an appropriate matrixing procedure is conducted before encoding. An MPEG-2 encoder may also produce a multichannel bit stream that is not backward compatible with an MPEG-1 decoder.

The MPEG-2 AAC algorithm was designed to include many of the latest audio coding compression techniques that could not be implemented while maintaining backward compatibility. These new compression techniques allow the MPEG-2 AAC coder to produce indistinguishable quality at significantly lower bit rates than MPEG-2 BC.

The DVB digital television standard specifies that an IRD (integrated receiver-decoder) must conform to the following guidelines for audio:

- MPEG-2 Layers I and II is supported by the IRD;
- The use of Layer II is recommended for the encoded bit stream;
- IRDs support single channel, dual channel, joint stereo, stereo, and the extraction of at least a stereo pair from MPEG-2 compatible multichannel audio;

- Sampling rates of 32, 44.1, and 48 kHz are supported by IRDs;
- The encoded bit stream does not use emphasis.

This chapter describes the MPEG-1 and MPEG-2 algorithms and bit-stream syntax associated with these minimum requirements of a DVB IRD.

9.1. MPEG-1 LAYER I,II ENCODERS

The algorithms used in the Layer I encoder and decoder are a subset of the algorithms used in layer II. This section describes the algorithms that are common to the Layers I and II encoders. The extra parts of the standard that are used only in the Layer II encoder are described in Section 9.2.

The MPEG-1 Layer I encoder is intended to be used to code wideband audio signals that have been converted to PCM (pulse code modulation) samples with sampling rates of 32, 44.1, or 48 kHz. The encoder can operate using one of the following four modes:

1. Single channel (a single audio signal)
2. Dual channel (two independent audio signals)
3. Stereo (a left and right stereo pair)
4. Joint stereo (a left and right stereo pair with the stereo redundancy used to reduce the output bit rate).

The Layer I, II encoders produce a digital bit stream with a constant output bit rate that is specified in the bit stream syntax. The maximum output bitrates are 448 and 384 kbits/s for Layers I and II, respectively.

The basic block diagram for a Layer I encoder is shown in Figure 9.2. The remainder of this section contains a very brief description of the algorithm and then

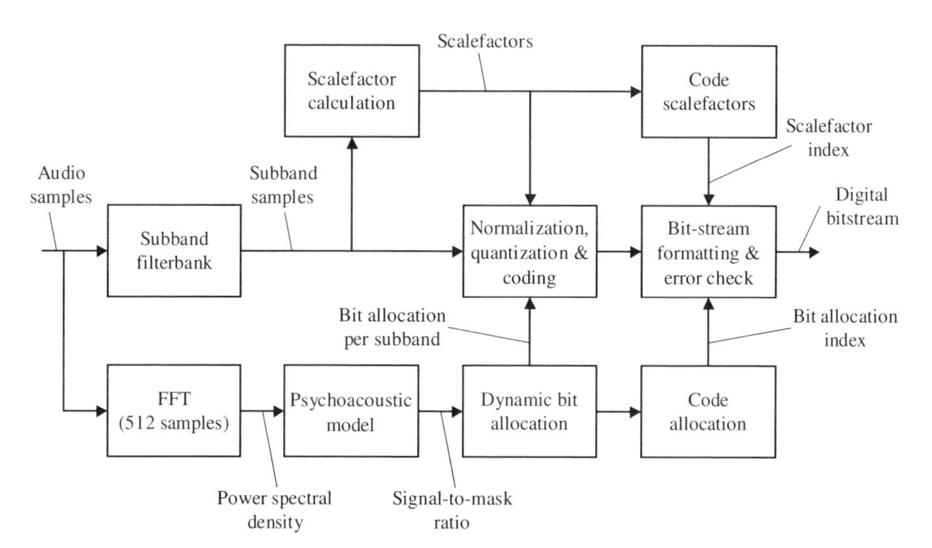

Figure 9.2 Basic block diagram of a Layer I encoder.

the individual components of the encoder are explained in more detail in the following sections.

For a Layer I encoder, the analysis filter bank converts 384 input PCM samples into a *block* of 12 sets of 32 subband samples. A *scalefactor* is then calculated for each subband and used to normalize the subband samples to a range of -1 to $+1$. The same scalefactor is used to normalize the 12 samples in each subband. In parallel to the filter bank, a psychoacoustic model is used to estimate the masking threshold associated with the current set of input samples. The masking threshold indicates the level of inaudible quantization noise that can be added to the subband samples during the quantization process. A dynamic bit allocation process then uses this masking threshold information to determine the number of bits required to transmit each subband sample. The block of normalized subband samples is then quantized using the quantizer step size that corresponds to the number of bits allocated for each subband. Then finally, the bits for the coded bit allocation, scalefactor information, and quantized subband samples are combined with header information and formatted into an audio *frame*.

9.1.1. Analysis Filterbank

The analysis filterbank used in the Layer I,II encoders is based on the efficient implementation of a cosine modulated filterbank proposed by Joseph Rothweiler in 1983. A detailed explanation of this filterbank technique is given in Section 8.5 of Chapter 8. The first 256 values of the modified prototype filter $c(k)$ are given in Table 9.1. The value of $c(257)$ is 0.03578097 and the remaining 255 values are found by time-reversing and negating the values with indexes 1 to 255 from Table 9.1.

The subband filterbank used in the Layer I,II encoders has the following minor limitations:

1. The individual frequency responses of adjacent subbands overlap, so a single frequency can affect two adjacent subband sample values.
2. The filterbank does not provide perfect reconstruction, so even if no quantization is performed there is still a very small error introduced into the output samples.
3. The equal width subbands do not correspond well to the unequal critical bandwidths of the human ear, especially at low frequencies.

9.1.2. Scalefactor Calculation

For a Layer I encoder, a *scalefactor* is calculated for each set of 12 subband samples. This scalefactor is used later in the calculation of the signal-to-mask ratio and also in the quantization and encoding of the subband samples. The scalefactor is calculated, for each subband, by finding the maximum of the absolute values of the 12 subband samples and then taking the lowest value from Table 9.2 that is greater than this maximum value.

The index in Table 9.2 is transmitted as a 6-bit binary number (MSB first) if a nonzero number of bits has been allocated to the subband in the bit allocation procedure.

Table 9.1 Values for the modified prototype filter used in the Layer I,II MPEG-1 audio coders.

	$w(8*i+j)$							
	$j=0$	$j=1$	$j=2$	$j=3$	$j=4$	$j=5$	$j=6$	$j=7$
$i=0$	0.000000000	−0.000000477	−0.000000477	−0.000000477	−0.000000477	−0.000000477	−0.000000477	−0.000000954
$i=1$	−0.000000954	−0.000000954	−0.000000954	−0.000001431	−0.000001431	−0.000001907	−0.000001907	−0.000002384
$i=2$	−0.000002384	−0.000002861	−0.000003338	−0.000003338	−0.000003815	−0.000004292	−0.000004768	−0.000005245
$i=3$	−0.000006199	−0.000006676	−0.000007629	−0.000008106	−0.000009060	−0.000010014	−0.000011444	−0.000012398
$i=4$	−0.000013828	−0.000014782	−0.000016689	−0.000018120	−0.000019550	−0.000021458	−0.000023365	−0.000025272
$i=5$	−0.000027657	−0.000030041	−0.000032425	−0.000034809	−0.000037670	−0.000040531	−0.000043392	−0.000046253
$i=6$	−0.000049591	−0.000052929	−0.000055790	−0.000059605	−0.000062943	−0.000066280	−0.000070095	−0.000073433
$i=7$	−0.000076771	−0.000080585	−0.000083923	−0.000087261	−0.000090599	−0.000093460	−0.000096321	−0.000099182
$i=8$	0.000101566	0.000103951	0.000105858	0.000107288	0.000108242	0.000108719	0.000108719	0.000108242
$i=9$	0.000106812	0.000105381	0.000102520	0.000099182	0.000095367	0.000090122	0.000084400	0.000077724
$i=10$	0.000069618	0.000060558	0.000050545	0.000039577	0.000027180	0.000013828	−0.000000954	−0.000017166
$i=11$	−0.000034332	−0.000052929	−0.000072956	−0.000093937	−0.000116348	−0.000140190	−0.000165462	−0.000191212
$i=12$	−0.000218868	−0.000247478	−0.000277042	−0.000307560	−0.000339031	−0.000371456	−0.000404358	−0.000438213
$i=13$	−0.000472546	−0.000507355	−0.000542164	−0.000576973	−0.000611782	−0.000646591	−0.000680923	−0.000714302
$i=14$	−0.000747204	−0.000779152	−0.000809669	−0.000838757	−0.000866413	−0.000891685	−0.000915051	−0.000935555
$i=15$	−0.000954151	−0.000968933	−0.000980854	−0.000989437	−0.000994205	−0.000995159	−0.000991821	−0.000983715
$i=16$	0.000971317	0.000953674	0.000930786	0.000902653	0.000868797	0.000829220	0.000783920	0.000731945
$i=17$	0.000674248	0.000610352	0.000539303	0.000462532	0.000378609	0.000288486	0.000191689	0.000088215
$i=18$	−0.000021458	−0.000137329	−0.000259876	−0.000388145	−0.000522137	−0.000661850	−0.000806808	−0.000956535
$i=19$	−0.001111031	−0.001269817	−0.001432419	−0.001597881	−0.001766682	−0.001937389	−0.002110004	−0.002283096
$i=20$	−0.002457142	−0.002630711	−0.002803326	−0.002974033	−0.003141880	−0.003306866	−0.003467083	−0.003622532
$i=21$	−0.003771782	−0.003914356	−0.004048824	−0.004174709	−0.004290581	−0.004395962	−0.004489899	−0.004570484

(continued)

Table 9.1 *(Continued)*

	$w(8*i+j)$							
	$j=0$	$j=1$	$j=2$	$j=3$	$j=4$	$j=5$	$j=6$	$j=7$
$i=22$	−0.004638195	−0.004691124	−0.004728317	−0.004748821	−0.004752159	−0.004737377	−0.004703045	−0.004649162
$i=23$	−0.004573822	−0.004477024	−0.004357815	−0.004215240	−0.004049301	−0.003858566	−0.003643036	−0.003401756
$i=24$	0.003134727	0.002841473	0.002521515	0.002174854	0.001800537	0.001399517	0.000971317	0.000515938
$i=25$	0.000033379	−0.000475883	−0.001011848	−0.001573563	−0.002161503	−0.002774239	−0.003411293	−0.004072189
$i=26$	−0.004756451	−0.005462170	−0.006189346	−0.006937027	−0.007703304	−0.008487225	−0.009287834	−0.010103703
$i=27$	−0.010933399	−0.011775017	−0.012627602	−0.013489246	−0.014358521	−0.015233517	−0.016112804	−0.016994476
$i=28$	−0.017876148	−0.018756866	−0.019634247	−0.020506859	−0.021372318	−0.022228718	−0.023074150	−0.023907185
$i=29$	−0.024725437	−0.025527000	−0.026310921	−0.027073860	−0.027815342	−0.028532982	−0.029224873	−0.029890060
$i=30$	−0.030526638	−0.031132698	−0.031706810	−0.032248020	−0.032754898	−0.033225536	−0.033659935	−0.034055710
$i=31$	−0.034412861	−0.034730434	−0.035007000	−0.035242081	−0.035435200	−0.035586357	−0.035694122	−0.035758972

Table 9.2 Layer I,II Scalefactors.

Index	Scalefactor	Index	Scalefactor
0	2.00000000000000	32	0.00123039165029
1	1.58740105196820	33	0.00097656250000
2	1.25992104989487	34	0.00077509816991
3	1.00000000000000	35	0.00061519582514
4	0.79370052598410	36	0.00048828125000
5	0.62996052494744	37	0.00038754908495
6	0.50000000000000	38	0.00030759791257
7	0.39685026299205	39	0.00024414062500
8	0.31498026247372	40	0.00019377454248
9	0.25000000000000	41	0.00015379895629
10	0.19842513149602	42	0.00012207031250
11	0.15749013123686	43	0.00009688727124
12	0.12500000000000	44	0.00007689947814
13	0.09921256574801	45	0.00006103515625
14	0.07874506561843	46	0.00004844363562
15	0.06250000000000	47	0.00003844973907
16	0.04960628287401	48	0.00003051757813
17	0.03937253280921	49	0.00002422181781
18	0.03125000000000	50	0.00001922486954
19	0.02480314143700	51	0.00001525878906
20	0.01968626640461	52	0.00001211090890
21	0.01562500000000	53	0.00000961243477
22	0.01240157071850	54	0.00000762939453
23	0.00984313320230	55	0.00000605545445
24	0.00781250000000	56	0.00000480621738
25	0.00620078535925	57	0.00000381469727
26	0.00492156660115	58	0.00000302772723
27	0.00390625000000	59	0.00000240310869
28	0.00310039267963	60	0.00000190734863
29	0.00246078330058	61	0.00000151386361
30	0.00195312500000	62	0.00000120155435
31	0.00155019633981		

9.1.3. Psychoacoustic Model 1

The details of how to implement a psychoacoustic model are *nonnormative*. This means that it is not necessary to define these implementation details in order to produce a bitstream that can be decoded by a decoder that complies with the standard. However, the MPEG-1 audio standard provides an *informative* annex that specifies the implementation details of two psychoacoustic models. It is suggested that model 1 be used for Layers I and II and model 2 be used in layer III. The psychoacoustic model is used to calculate the signal-to-mask ratio (SMR) for each subband. This information is then used to select the quantizer step size for each subband that guarantees that the quantization noise introduced is inaudible.

More specifically, the psychoacoustic model defines an algorithm to estimate the masking threshold for the current input audio signal. (For a more detailed explanation of auditory masking see Chapter 7.) This masking threshold is used to determine the number of bits required to transmit the subband samples, in a particular subband, with no audible quantization noise as shown in Figure 9.3. For each subband, the minimum level of the masking threshold indicates the worst-case level of masking. The SMR is then determined using this minimum masking level and the maximum signal level in that subband. Then, because the subband samples are normalized before they are quantized, there is a well-defined relationship between the number of bits used to transmit the subband sample and the signal-to-noise ratio (SNR) introduced by quantizing to this number of bits. To guarantee no quantization noise is audible in the decoded audio signal, it is simply a matter of allocating the number of bits to each subband that guarantees that the SNR is greater than the SMR as shown in Figure 9.3.

In psychoacoustic model 1 the SNR is calculated using the following nine steps:

1. Calculate the FFT of the input audio samples.
2. Determine the maximum sound pressure level in each subband.
3. Determine the threshold in quiet.
4. Find the tonal and nontonal components of the input audio signal.
5. Decimate the maskers to obtain only the relevant maskers.
6. Calculate the individual masking thresholds for each relevant masker.

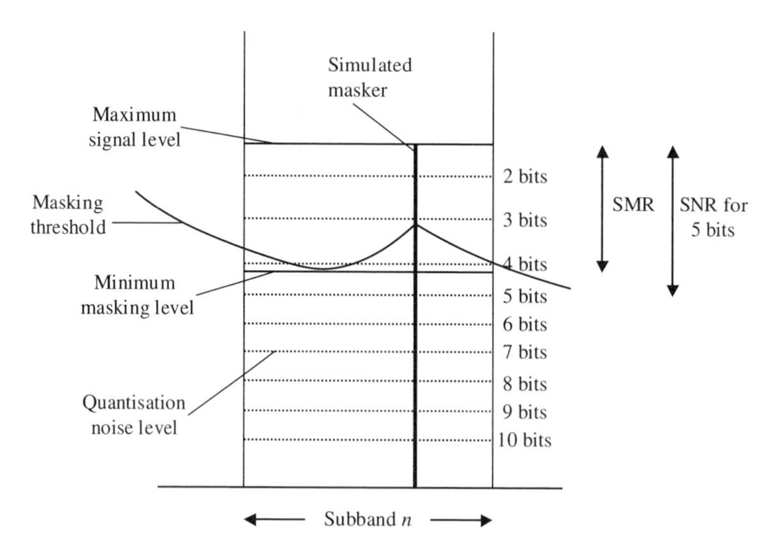

Figure 9.3 The masking threshold produced by the psychoacoustic model is used to determine the SMR for each subband.

7. Determine the global masking threshold.

8. Determine the minimum masking threshold in each subband.

9. Calculate the SMR ratio.

These nine steps are explained in more detail in the following sections.

9.1.3.1. *Calculating the FFT of the Input Audio Samples*

An estimate of the power spectral density of the audio samples is calculated using a 512-point FFT for Layer I or a 1024-point FFT for Layer II. In each case, the input samples are first multiplied by a Hann window function given by

$$w(n) = \sqrt{\tfrac{8}{12}}\left[1 - \cos\left(\frac{2n\pi}{N}\right)\right] \qquad \text{for } n = 0, \ldots, N-1 \tag{9.1}$$

where N is the number of points used to calculate the FFT. Because of the delay through the analysis filter, for the output of the FFT to coincide with the output subband samples of the analysis filter, a delay must be introduced at the input to the FFT. For Layer I, the input samples of the FFT should be delayed by 320 samples, and for Layer II the delay should be 192 samples.

Figure 9.4(a) shows the relative positions of the prototype analysis filter at the beginning and end of a Layer I frame, and Figure 9.4(b) shows the position of the Hann window used in the power spectral density calculation. Figure 9.4 shows that

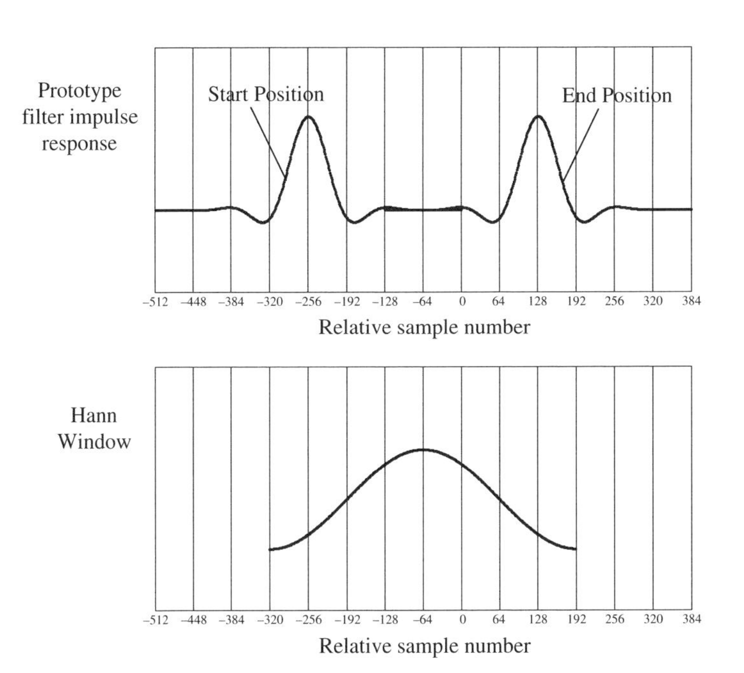

Figure 9.4 The relative position of the analysis filter and the Hann window function used in the FFT calculation.

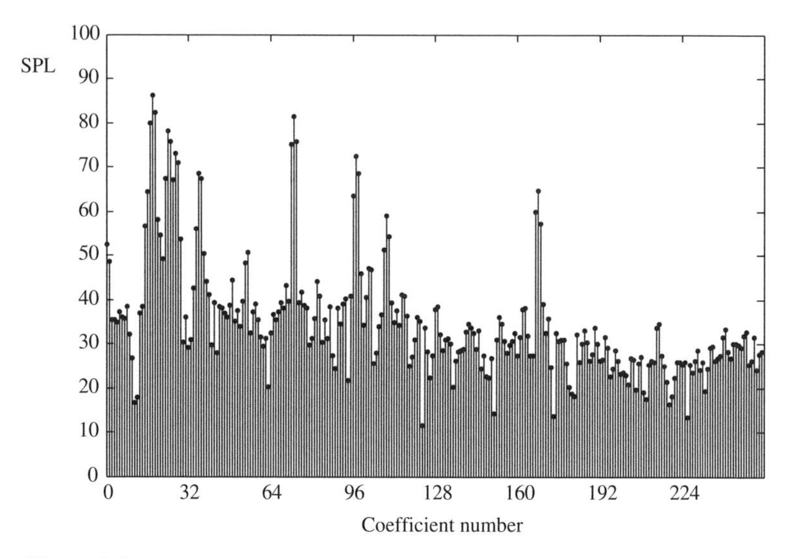

Figure 9.5 The power spectral density of a typical audio signal for a Layer 1 frame.

a delay of 320 samples is required so that the power spectral density calculation is operating on the same set of input samples as the analysis filter. Note that the Hann window does not extend to all input samples used in the analysis filter bank for the current frame. This disparity has a negligible effect because the samples lying outside the extent of the Hann window do not contribute significantly to the values of the output subband samples.

The power spectral density of the delayed input samples $x(n)$ is then given by

$$X(k) = 10\log_{10}\left(\left|\frac{1}{N}\sum_{n=0}^{N-1} w(n)x(n)e^{-j(2nk\pi/N)}\right|^2\right) \qquad \text{for } k = 0, \ldots, N/2-1 \quad (9.2)$$

To account for variations in the dynamic range of the input sample values, the power spectral density values should be normalized so that the maximum amplitude possible corresponds to a sound pressure level of 96 dB. Therefore, if the input samples represent a pure sinusoid with amplitude 2, the power spectral density values should be scaled so that the maximum value of $X(k)$ is 96 dB. Figure 9.5 shows an example of the power spectral density of a typical audio signal for a Layer I frame.

9.1.3.2. Determining the Maximum Sound Pressure Level in Each Subband

The power spectral density and the scalefactors for the frame are then used to determine the maximum sound pressure level in each subband. For Layer I, there

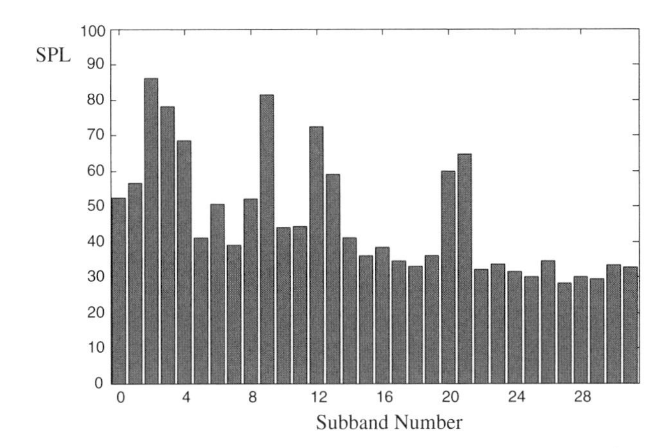

Figure 9.6 The maximum sound pressure level for each subband of the audio signal in Figure 9.5.

are eight spectral lines per subband and for Layer II there are 16 spectral lines per subband. The maximum sound pressure level of subband n is taken as the maximum of the spectral lines in the subband and the scalefactor for the subband converted to decibels. So the maximum sound pressure level, L_{sb}, of subband n is given by

$$L_{sb}(n) = \max_{k \text{ in subband } n} \left[X(k), 20\log_{10}\left(32768 \times \mathrm{scf}_{\max}(n)\right) - 10 \right] \tag{9.3}$$

In Layer I, the expression $\mathrm{scf}_{\max}(n)$ is simply the scalefactor for subband n, but in Layer II it denotes the maximum of the three scalefactors for subband n. Figure 9.6 shows the maximum sound pressure level for each subband of the audio signal from the previous example.

The following MATLAB code defines the algorithm used to determine the maximum sound pressure level in each subband. The array X contains the sound pressure level values for the FFT coefficients and scf_max contains the maximum scalefactor for each subband. The elements of the array Lsb are set to the maximum sound pressure level in each subband.

```
% step through the FFT coefficients corresponding to each subband
n = 1;
for k = 1:8:256
  % convert the maximum scalefactor for the nth subband to decibels
  scf_max_level(n) = 20*log10(32768*scf_max(n))-10;

  % find the maximum of the scalefactor level and the FFT coefficients
  % corresponding to subband n
  max_level = -inf;
  for m = 0:subband_width-1
    max_level = max([max_level X(k+m) scf_max_level(n)]);
  end
```

```
    % set the maximum sound pressure level for subband n to this
    % maximum level
    Lsb(n) = max_level;
    n = n+1;
end
```

9.1.3.3. Determining the Threshold in Quiet

The absolute threshold or the *threshold in quiet*, LT_q, is given in Tables A.1 and A.2 in Appendix for Layers I and II, respectively. Note that the values in these tables are given for a subset (indexed by i) of the original $N/2$ frequency values (indexed by k).

For Layer I, no subsampling is used for the first six subbands. For the next six subbands, every second frequency value is considered. Then for input sampling rates of 44.1 and 48 kHz, every fourth frequency value is considered up to a maximum of 20 kHz and for 32 kHz, every fourth frequency value is considered up to a maximum of 15 kHz.

For Layer II, no subsampling is used for the first three subbands and for the next three subbands, every second frequency value is considered. For the next six subbands, every fourth frequency value is considered. Then for input sampling rates of 44.1 and 48 kHz, every eighth frequency value is considered up to a maximum of 20 kHz and for 32 kHz, every eighth frequency value is considered up to a maximum of 15 kHz.

EXAMPLE 9.1

For a Layer I frame with an input sample rate of 44.1 kHz, the following equation describes the relationship between the index i of the subset of frequency values and the index k of the original frequency values:

$$i = \begin{cases} k, & \text{for } k = 1, 2, \ldots, 48 \\ \frac{k}{2} + 24, & \text{for } k = 50, 52, \ldots, 96 \\ \frac{k}{4} + 48, & \text{for } k = 100, 104, \ldots, 232 \end{cases}$$

■

9.1.3.4. Finding the Tonal and Nontonal Masking Components

The masking effect is different for tonal (sinusoidal) and nontonal maskers so it is necessary to determine the position of both tonal and nontonal masking components from the power spectral density of the input signal. The process for determining the location of these masking components is explained below.

Locating Local Maxima Because a tonal masker produces a large narrow spike in the power spectral density of the signal, the first step in this process is to determine the position of the local maxima.

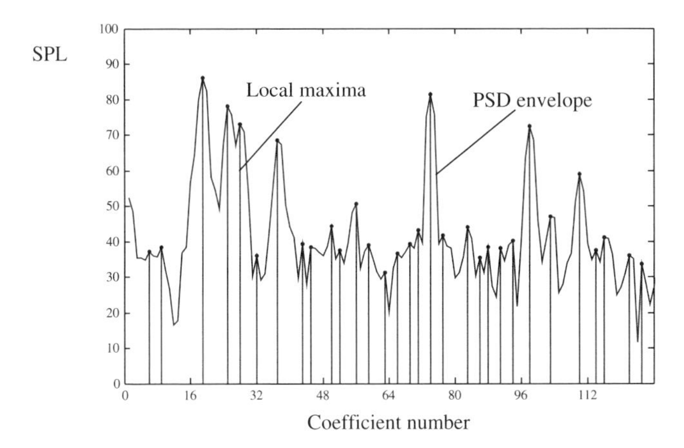

Figure 9.7 The envelope of the power spectral density and the local maxima.

A spectral line in $X(k)$ is considered to be a local maximum if the following condition holds:

$$X(k) > X(k-1) \quad \text{and} \quad X(k) > X(k+1) \tag{9.4}$$

Figure 9.7 shows the envelope for the first half of the power spectral density from the previous example with those components identified as local maxima using the above criteria.

The following MATLAB code defines the algorithm used to locate the local maxima. The elements of the array is_loc_max are set to 1 when an FFT coefficient is greater than its neighbors above and below in frequency.

```
is_loc_max = zeros(1,256);
for k = 2:255
  if ((X(k) > X(k-1)) && (X(k) > X(k+1)) )
    is_loc_max(k) = 1;
  else
    is_loc_max(k) = 0;
  end
end
```

Locating Tonal Maskers Not every local maximum represents a tonal masker. The next step requires the location of maxima that are significantly greater in amplitude than their neighboring spectral components. Because the human auditory system has a better frequency resolution at lower frequencies, tonal components at lower frequencies can be closer together than at higher frequencies and still be heard as two distinct tones. For this reason, a larger number of neighboring spectral components are considered as the frequency of the local maxima increases. So a local maximum is considered to be a tonal masker if the following equation holds

$$X(k) - X(k-j) \geq 7 \text{ dB} \tag{9.5}$$

where j is determined according to the value of k and is given by the following equations.
Layer I:

$$j = \begin{cases} -2, 2, & \text{for } 2 < k < 63 \\ -3, -2, 2, 3, & \text{for } 64 \le k < 127 \\ -6, \ldots, -2, 2, \ldots, 6, & \text{for } 128 \le k < 250 \end{cases}$$

Layer II:

$$j = \begin{cases} -2, 2, & \text{for } 2 < k < 63 \\ -3, -2, 2, 3, & \text{for } 64 \le k < 127 \\ -6, \ldots, -2, 2, \ldots, 6, & \text{for } 128 \le k < 255 \\ -12, \ldots, -2, 2, \ldots, 12, & \text{for } 256 \le k \le 500 \end{cases}$$

If a local maximum at frequency index k satisfies this equation then the sound pressure level of the tonal masker is given by

$$X_{tm}(k) = 10 \log_{10}\left(10^{\frac{X(k-1)}{10}} + 10^{\frac{X(k)}{10}} + 10^{\frac{X(k+1)}{10}}\right) \text{ dB} \tag{9.6}$$

Figure 9.8 shows the envelope for the first half of the power spectral density from the previous example together with those components identified as tonal maskers using the above criteria.

So the tonal maskers are not considered when calculating the sound pressure level of the nontonal maskers, all spectral components within the neighborhood of a tonal masker are set to $-\infty$ dB. The following MATLAB code defines the algorithm used to locate the tonal maskers. The elements of the array is_tonal_masker are set to 1 when an FFT coefficient is found to be a tonal masker. The elements of the array

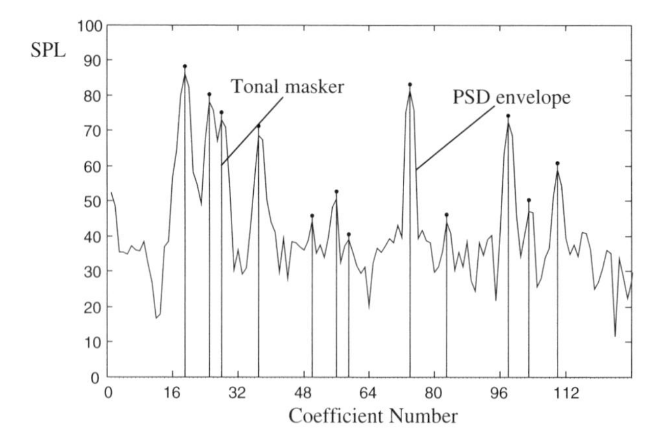

Figure 9.8 The envelope of the power spectral density and the tonal maskers.

Xtm are set to the simulated SPL value for a single tone masker at the location of the masker and minus infinity elsewhere.

```
is_tonal_masker = zeros(1,N);
Xtm = -inf*ones(1,256);

% step through the FFT coefficients one at a time
for k = 3:250
  % if the coefficient has previously been identified as a local maxima
  % check to determine if it is at a sufficient higher level compared to
  % its neighbours
  if (is_loc_max(k) == 1)
    % the frequency of the coefficient is used to determine the number
    % of coefficients to consider as neighbours
    if ((k >= 3) && (k < 64))
      neighbour_coeff = [−2 2];
      num_coeff = 2;
    elseif ((k >= 64) && (k < 128))
      neighbour_coeff = [−3 −2 2 3];
      num_coeff = 4;
    else
      neighbour_coeff = [−6 −5 −4 −3 −2 2 3 4 5 6];
      num_coeff = 10;
    end
    % if one of the coefficient's neighours is within 7 dB of the
    % coefficient then the tonal masker criterion is not satisfied
    satisfied = 1;
    for j = 1:num_coeff
      if (X(k) - X(k+neighbour_coeff(j)) < 7)
        satisfied = 0;
      end
    end
    % if the coefficient satisfies the criterion for a tonal masker
    if (satisfied)
      % calculate the level for a simulated single tone tonal masker
      masker_sum = 10^(X(k−1)/10) + 10^(X(k)/10) + 10^(X(k+1)/10);
      Xtm(k) = 10*log10(masker_sum);
      % set a boolean variable to locate the tonal maskers
      is_tonal_masker(k) = 1;
      % set coefficients in neighbourhood of masker to minus infinity
      % for the calculation of non-tonal maskers
      for j = 1:num_coeff
        X(k+neighbour_coeff(j)) = -inf;
      end
      X(k−1:k+1) = [-inf -inf -inf];
    end
  end
end
```

Locating Nontonal Maskers The SPL of the nontonal masking components are calculated from the spectral lines remaining after the components associated with the tonal maskers have been removed. The input frequency range is first divided up into a series of critical bands. Then one nontonal masking component is calculated for each critical band. The SPL of a nontonal masking component is calculated for each critical band by taking the sum of the power of the spectral components in the critical band (after the tonal components have been removed) and is given by

$$X_{nm}(k_{cb}) = 10\log_{10}\left(\sum_{k=k_l}^{k_u} 10^{(X(k)/10)}\right) \tag{9.7}$$

where k_l is the FFT index corresponding to the lowest frequency in the critical band and k_u is the FFT index corresponding to the highest frequency in the critical band. The location of the nontonal component for each critical band, k_{cb}, is taken as the index that is closest to the geometric mean of the indices of the critical band.

The following MATLAB code defines the algorithm used to locate the tonal maskers. The arrays lower, geom_mean, and upper contain the FFT index of the lower, geometric mean, and upper frequencies in each critical band and are given in Tables A.3 and A.4 in Appendix. The elements of the array is_non_tonal_masker are set to 1 for the index that is closest to the geometric mean of the indices of the critical band. The elements of the array Xnm are set to the simulated SPL value for a single tone masker at the location of the masker and minus infinity elsewhere.

```
is_non_tonal_masker = zeros(1,256);
Xnm = -inf*ones(1,256);

% step through each critical band
for crit_band = 1:num_critical_band
  % find the linear sum of the power of the SPL values in the critical
  % band after the tonal maskers have been removed
  linear_sum = 0;
  for k = lower(crit_band):upper(crit_band)
    linear_sum = linear_sum+10^(X(k)/10);
  end

  % set a boolean variable to locate the non-tonal maskers at the
  % coefficient index that is closest to the geometric mean of the indexes
  % in the critical band
  is_non_tonal_masker(geom_mean(crit_band)) = 1;
  % set the level for a simulated single tone tonal masker
  Xnm(geom_mean(crit_band)) = 10*log10(linear_sum);
  end
end
```

Figure 9.9 shows the envelope for the first half of the power spectral density from the previous example (after the components in the neighborhood of the tonal

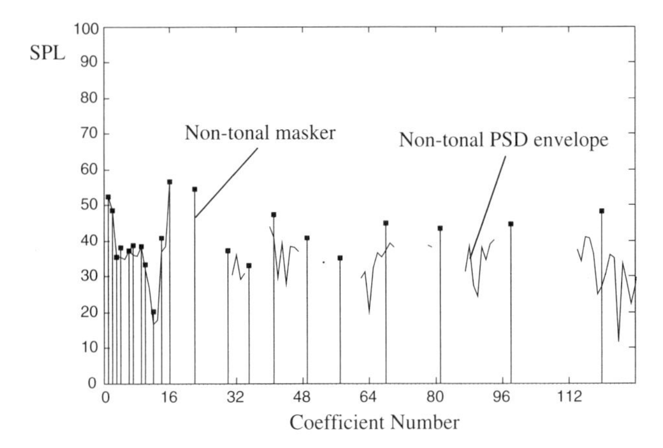

Figure 9.9 The envelope of the power spectral density after the components contributing to tonal masking have been removed and the nontonal maskers.

maskers have been removed) and the nontonal maskers for each critical band calculated using the above equations.

9.1.3.5. Decimating the Tonal and Nontonal Masking Components

Not all maskers from the previous step are used to calculate the global masking threshold. In this step, maskers can be removed if they do not meet certain requirements. Tonal or nontonal masking components are removed if their SPL is less than the absolute threshold, that is, tonal and nontonal masking components must satisfy the following conditions.

Tonal masking components:

$$X_{tm}(k) \geq LT_q(k) \tag{9.8}$$

Nontonal masking components:

$$X_{nm}(k) \geq LT_q(k) \tag{9.9}$$

The following MATLAB code defines the algorithm used to decimate the tonal and nontonal maskers that are below the threshold-in-quiet. The array LTq(k) contains the absolute threshold values at the frequency indicated by k. These absolute threshold values can be calculated for each input sampling rate from the values in Tables A.1 and A.2 of the Appendix.

```
% step through the FFT coefficients
for k = 1:256
  if (is_tonal_masker(k))
    % if the level of a tonal masker is below the threshold-in-quiet
    % then remove it
```

```
    if (Xtm(k) < LTq(k))
        is_tonal_masker(k) = 0;
      Xtm(k) = -inf;
    end
  elseif (is_non_tonal_masker(k))
    % if the level of a non-tonal masker is below the threshold-in-quiet
    % then remove it
    if (Xnm(k) < LTq(k))
      is_non_tonal_masker(k) = 0;
      Xnm(k) = -inf;
    end
  end
end
```

In addition, if two tonal masking components have critical band rates that are within 0.5 Bark of each other, the component with the lowest power is removed. The following MATLAB code defines the algorithm used to decimate the tonal maskers that are within 0.5 Bark of a larger tonal masker. The array band_rate(k) contains the critical band rate that corresponds to the frequency of the index k. The frequency to band rate conversion can be calculated using Equation (7.2) in Chapter 7.

```
% In each iteration find the smallest difference in critical band rate
% between two tonal maskers. If this difference is smaller than 0.5 then
% remove the tonal masker with the lowest power. Stop when no two tonal
% maskers are less than 0.5 bark apart.

% set the minimum difference in band rate to zero to start the while loop
min_dz = 0;

while min_dz < 0.5
  min_dz = inf;
  % step through the FFT coefficients
  for k = 1:256
    if is_tonal_masker(k)
      % if the coefficient is a tonal masker find its critical band
      % rate
      z1 = band_rate(k);

      % now step though the remaining coefficients
      for j = 1:256-k
        if is_tonal_masker(k+j)
          % if a second tonal masker is found find its critical
          % band rate
          z2 = band_rate(k+j);

          % find the difference in critical band rate for these
          % two tonal maskers
          dz = z2-z1;

          % check to see if this is the minimum difference
```

```
                % between two coefficients
                if dz < min_dz
                    % if this is the minimum difference update the
                    % minimum and remember the index of the coefficient
                    % with the lowest power
                    min_dz = dz;
                    if Xtm(k) < Xtm(k+j)
                        remove = k;
                    else
                        remove = k+j;
                    end
                end
            end
        end
    end
  end
  % if the minimum difference is less than 0.5 bark remove the
  % coefficient with the lowest power
  if min_dz < 0.5
    Xtm(remove) = -inf;
    is_tonal_masker(remove) = 0;
  end
end
```

Figure 9.10 shows the envelope for the first half of the power spectral density from the previous example together with the tonal and nontonal maskers that are used to calculate the individual masking thresholds in the next step.

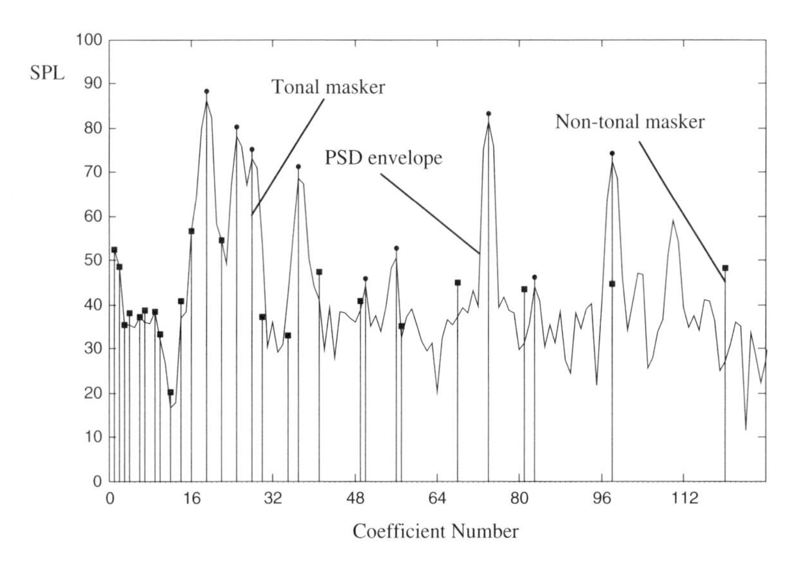

Figure 9.10 The envelope of the power spectral density and the tonal and nontonal maskers.

9.1.3.6. Calculating the Individual Masking Thresholds

The individual masking thresholds for the *i*th masking components at frequencies with index *j* are given by the following equations.

For tonal maskers:

$$LT_{tm}(i,j) = X_{tm}(k) + av_{tm}(k) + vf(k,j) \tag{9.10}$$

For non-tonal maskers:

$$LT_{nm}(i,j) = X_{nm}(k) + av_{nm}(k) + vf(k,j) \tag{9.11}$$

In these equations, k is the frequency index of the *i*th masking component, the terms $av_{tm}(k)$ and $av_{nm}(k)$ denote the *masking index* for tonal and nontonal maskers, respectively, and the term $vf(k,j)$ denotes the value of the *masking function* of the *i*th masking component at frequency index j. The masking index differs for tonal and nontonal maskers and is given by the following equations.

For tonal maskers:

$$av_{tm}(k) = -0.275\, z(k) - 6.025 \tag{9.12}$$

For nontonal maskers:

$$av_{nm}(k) = -0.175\, z(k) - 2.025 \tag{9.13}$$

In these equations, the term $z(k)$ denotes the critical band rate of the masker and can be found using the equation:

$$z(k) = \frac{28 f(k)}{f(k) + 2200} - 0.5 \text{ Bark} \tag{9.14}$$

where $f(k)$ is the frequency value corresponding to the index k.

The masking index sets a constant offset for the masking threshold of a masker. This offset is lower and also decreases more rapidly with critical band rate for tonal maskers than for nontonal maskers. The masking function defines the slope of the masking threshold for frequencies above and below the critical band rate of the masker. The masking function defines four different slopes for the masking threshold and is given by

$$vf(k,j) = \begin{cases} 17(dz+1) - (0.4X(k)+6), & \text{for } -3 \le dz < -1 \\ (0.4X(k)+6)dz, & \text{for } -1 \le dz < 0 \\ -17dz, & \text{for } 0 \le dz < 1 \\ -(dz-1)(17-0.15X(k)) - 17, & \text{for } 1 \le dz < 8 \end{cases} \tag{9.15}$$

where dz is the distance in Bark from the masker and is given by $dz = z(j) - z(k)$. The individual masking threshold for each masker is set to $-\infty$ dB for critical band rates less than 3 Bark below and greater than 8 Bark above the critical band rate of the masker. The following MATLAB code defines the algorithm used to calculate the individual masking thresholds. The elements of the array LTm(i,j) are set to the masking threshold values for the ith masking component at frequency index *j*.

```
LTm = -inf*ones(1,256);
i = 0;

% step through each FFT coefficient
for k = 1:256

  % if the coefficient is a masker
  if (is_tonal_masker(k) &Vert;&Vert; is_non_tonal_masker(k))

    % increment the index for the masker
    i = i+1;

    % find the critical band rate corresponding to the frequency of the
    % masker
    z = band_rate(k);

    % for this masker step through each frequency index
    for j = 1:256
      % find the difference in critical band rate between this
      % frequency index to the critical band rate of the masker
      dz = band_rate(j) - z;

      % calculate the individual masking threshold for this tonal
      % masker
      if (is_tonal_masker(k))
        avtm = −1.525 - 0.275*z - 4.5;
        vf = masking_function(dz,Xtm(k));
        LTm(i,j) = Xtm(k) + avtm + vf;
      end

      % calculate the individual masking threshold for this non-tonal
      % masker
      if (is_non_tonal_masker(k))
        avnm = −1.525 - 0.175*z - 0.5;
        vf = masking_function(dz,Xnm(k));
        LTm(i,j) = Xnm(k) + avnm + vf;
      end
    end
  end
end

function vf = masking_function(dz,X)
% calculate the masking function for the critical band rate difference dz
% and masker SPL value X

if (dz >= −3 && dz < −1)
  vf = 17*(dz + 1) - 0.4*X - 6;
```

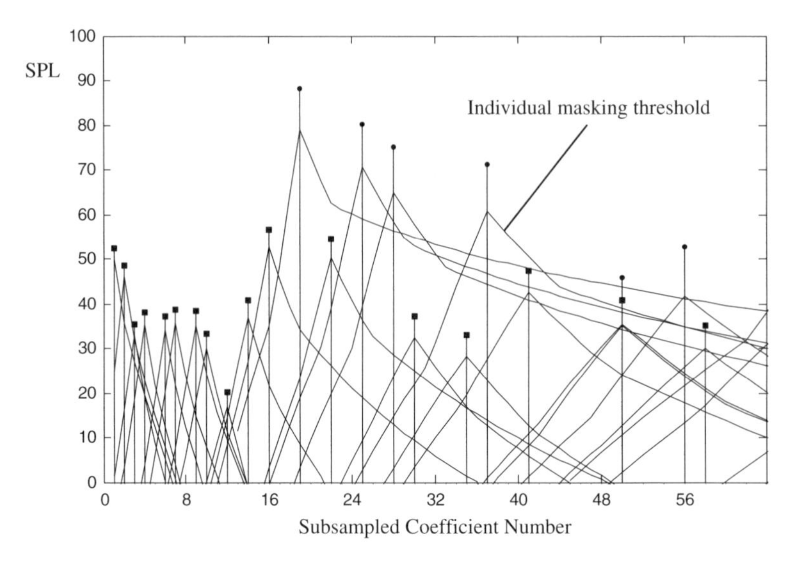

Figure 9.11 Individual masking thresholds for a subset of the tonal and nontonal maskers

```
elseif (dz >= −1 && dz < 0)
  vf = (0.4*X + 6)*dz;
elseif (dz >= 0 && dz < 1)
  vf = −17*dz;
elseif (dz >= 1 && dz < 8)
  vf = −(dz − 1)*(17 − 0.15*X) − 17;
else
  vf = -inf;
end
```

Figure 9.11 shows the individual masking thresholds for a subset of the tonal and nontonal maskers from the previous example.

9.1.3.7. Calculating the Global Masking Threshold

The global masking threshold is calculated for each frequency value by taking the sum of the individual masking thresholds, and the absolute threshold, and is given by

$$LT_g(k) = 10\log_{10}\left(10^{\frac{LT_q(k)}{10}} + \sum_{i=1}^{N_m} 10^{\frac{LT_m(i,k)}{10}}\right) \tag{9.16}$$

where N_m is the number maskers.

Figure 9.12 shows the global masking threshold calculated from the individual masking thresholds from the previous example.

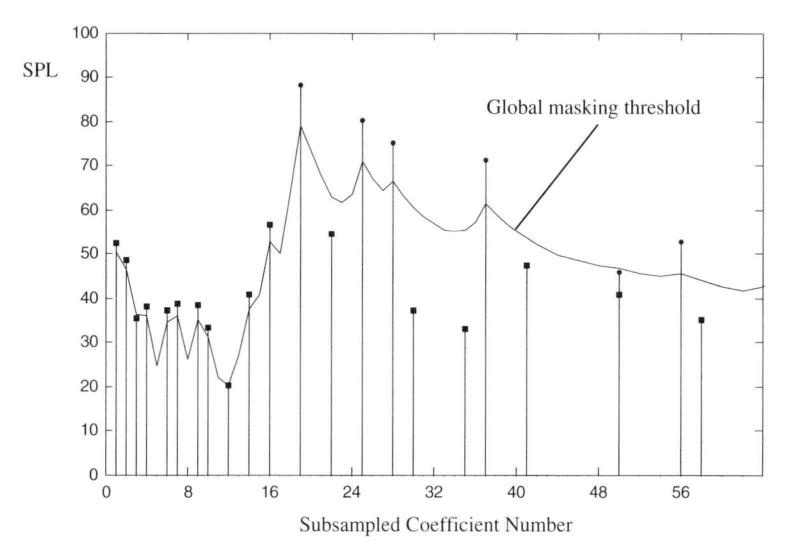

Figure 9.12 The global masking threshold for a subset of the tonal and nontonal maskers.

9.1.3.8. Calculating the Minimum Masking Threshold

The minimum masking threshold is calculated for each subband by finding the minimum of the global masking threshold over the subsampled frequency values contained in the subband and is given by

$$LT_{\min}(n) = \min_{k \text{ in subband } n}\left(LT_g(k)\right) \tag{9.17}$$

Figure 9.13 shows the minimum masking threshold for each subband calculated from the global masking threshold of the previous example. The dots on the plot of the global threshold indicate the frequency values at which the global threshold was calculated.

9.1.3.9. Calculating the Signal-to-Mask Ratio

The signal-to-mask ratio is calculated by subtracting the minimum masking threshold from the sound pressure level for each subband and is given by

$$SMR(n) = L_{sb}(n) - LT_{\min}(n) \tag{9.18}$$

Figure 9.14 shows the signal-to-mask ratio for each subband calculated using the above equation.

9.1.4. Dynamic Bit Allocation

The total number of bits available for each audio frame, *cb*, is determined from the number of 32-bit *slots* that are to be transmitted for that frame (see Section 9.1.7). Not all of the bits in each frame are available to transmit subband sample

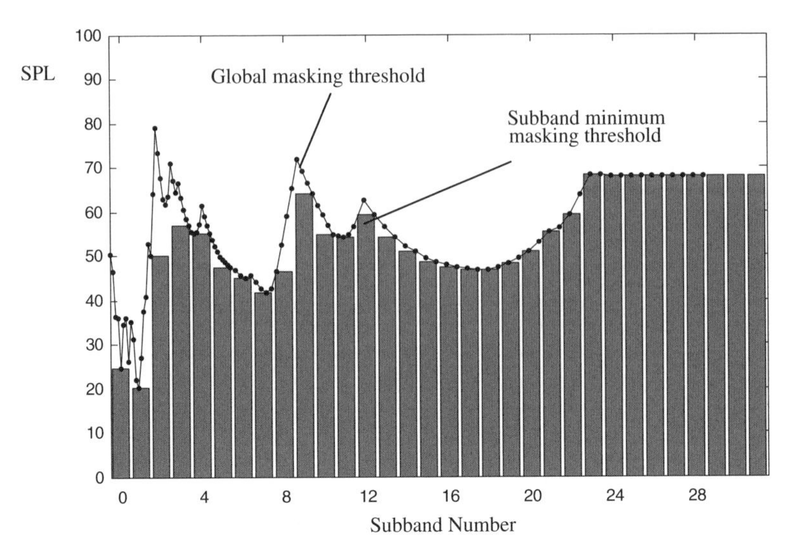

Figure 9.13 The minimum masking threshold for each subband and the global masking threshold.

and scalefactor information. The available number of bits can be determined by subtracting from the total number of bits available, *cb*, the number of bits required to transmit the header, *bhdr* (32 bits), the CRC check word if used, *bcrc*, (16 bits), the bit allocation for each subband, *bbal*, and any ancillary data, *banc* (see Section 9.4).

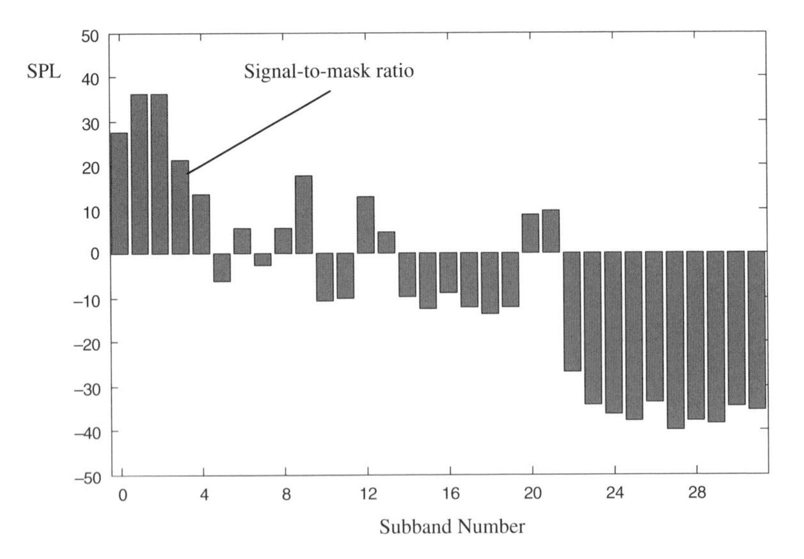

Figure 9.14 The signal-to-mask ratio for each subband.

So the number of bits available for subband sample and scalefactor information, *adb*, is

$$adb = cb - (bhdr + bcrc + bbal + banc) \quad (9.19)$$

The available bits are then allocated to the subbands using the principle of minimizing the total noise-to-mask ratio of the frame. The number of bits allocated to one sample can range from 0 to 15 bits with the exception of 1 bit per sample, which is not allowed.

The procedure to find the number of bits allocated to each subband begins by finding the *mask-to-noise-ratio* (MNR) for each subband. The MNR is calculated by subtracting the signal-to-mask ratio from the signal-to-noise ratio for each subband and is given by

$$\text{MNR} = \text{SNR} - \text{SMR} \quad (9.20)$$

The signal-to-mask ratio is the output of the psychoacoustic model and the signal-to-noise ratio depends on the number of bits allocated to the subband. The relationship between SNR and bits allocated to a subband is given in Table 9.3.

The allocation of bits to each subband is an iterative process that begins by allocating zero bits for the scalefactors and subband samples of each subband. The iterative process then continues using the following steps:

1. Use the number of bits allocated to each subband to determine the SNR for each subband and then calculate the MNR for each subband.

Table 9.3 SNR per number of bits allocated to each subband.

No. of bits	SNR (dB)	*A*	*B*
0	0.00		
2	7.00	0.750000000	−0.250000000
3	16.00	0.875000000	−0.125000000
4	25.28	0.937500000	−0.062500000
5	31.59	0.968750000	−0.031250000
6	37.75	0.984375000	−0.015625000
7	43.84	0.992187500	−0.007812500
8	49.89	0.996093750	−0.003906250
9	55.93	0.998046875	−0.001953125
10	61.96	0.999023438	−0.000976563
11	67.98	0.999511719	−0.000488281
12	74.01	0.999755859	−0.000244141
13	80.03	0.999877930	−0.000122070
14	86.05	0.999938965	−0.000061035
15	92.01	0.999969482	−0.000030518

2. Determine the subband that has the minimum MNR.
3. The number of bits allocated to this subband is then increased to the next highest allowable number of bits.
4. The number of bits required to transmit the samples, *bspl*, is updated to the number of bits required in step 2. In addition, if *bspl* has increased from zero to two then the number of bits required to transmit the scalefactor for this subband, *bscf* is increased from zero to six bits.
5. Recalculate the number of bits available for subband sample and scalefactor information, *adb*, using the following equation:

$$adb = cb - (bhdr + bcrc + bbal + bscf + bspl + banc) \qquad (9.21)$$

6. If *adb* is greater than any possible increase of *bscf* + *bspl* the iterative procedure continues from step 1.

9.1.5. Coding of Bit Allocation

The bit allocation information for subband **sb** of channel **ch** is transmitted using the 4 bit code word **allocation[ch][sb]**. The number of bits per sample that corresponds to each value of the **allocation[ch][sb]** code word is shown in Table 9.4.

Table 9.4 Bits per sample for each value of the allocation index.

allocation[ch][sb]	Bits per sample
"0000"	0
"0001"	2
"0010"	3
"0011"	4
"0100"	5
"0101"	6
"0110"	7
"0111"	8
"1000"	9
"1001"	10
"1010"	11
"1011"	12
"1100"	13
"1101"	14
"1110"	15
"1111"	forbidden

9.1.6. Quantization and Coding of Subband Samples

The subband samples are coded using a linear quantizer with a symmetric zero representation. This type of quantizer guarantees that small values close to zero are quantized to zero. The quantization process is defined by the following steps:

1. Normalize the subband sample by dividing its value by the scalefactor to obtain the normalized sample X.
2. Calculate the quantized value Q using the following formula:
$$Q = AX + B$$
where the values for the quantization coefficients A and B depend on the number of bits allocated to the subband and are given in Table 9.3.
3. Find Q_N by taking the N most significant bits of Q where N is the number of bits allocated in the previous bit allocation process and Q is stored as a two's complement binary fraction.
4. Invert the most significant bit of Q_N in order to avoid the code word containing all 1s.

EXAMPLE 9.2

To quantize a subband sample using three bits, the first step is to normalize the subband value by the scalefactor to give a value for X. Since the scalefactor must always be larger than the maximum sample value in each subband, the possible range of values for X is defined by $-1 < X < 1$. The value of X is then used to calculate a value for Q using the equation

$$Q = 0.875\,X - 0.125$$

and the possible range of values for Q is then defined by $-1 < Q < 0.75$. If Q is then stored as a two's complement binary fraction and Q_3 is taken as the first three bits of this binary fraction, there are only seven different possible values for Q_3. Table 9.5 shows the range of values for X and Q that correspond to each of these seven values of Q_3. The last column of Table 9.5 shows the actual code word that is transmitted after the most significant bit of Q_3 has been inverted.

Table 9.5 Ranges and code words for a 3-bit quantizer.

Range for X	Range for Q	Q_3	Code word
$-1 < X < \frac{-5}{7}$	$-1 < Q < -0.75$	100	000
$\frac{-5}{7} \leq X < \frac{-3}{7}$	$-0.75 \leq Q < -0.5$	101	001
$\frac{-3}{7} \leq X < \frac{-1}{7}$	$-0.5 \leq Q < -0.25$	110	010
$\frac{-1}{7} \leq X < \frac{1}{7}$	$-0.25 \leq Q < 0.0$	111	011
$\frac{1}{7} \leq X < \frac{3}{7}$	$0.0 \leq Q < 0.25$	000	100
$\frac{3}{7} \leq X < \frac{5}{7}$	$0.25 \leq Q < 0.5$	001	101
$\frac{5}{7} \leq X < 1$	$0.5 \leq Q < 0.75$	010	110

■

There is a provision in the MPEG-1 Audio standard for the inclusion of a variable amount of ancillary data provided by the user. The number of ancillary data bits is included in the calculation of the constant number of bits for each frame and reduces the number of bits available for transmitting audio data. Hence a large amount of ancillary data may result in a significant degradation of audio quality if the output bit rate is not increased to compensate for the inclusion of this extra data.

Care should also be taken that a group of bits in the ancillary data does not match the syncword that occurs at the beginning of each frame (see Section 9.4.3). If care is not taken to avoid such a match, recovering synchronization immediately after an error event may be made more difficult.

9.1.7. Formatting

The output of the audio encoder is transmitted in frames. Each frame contains the information required to decode 384 samples of the original audio signal. The Layer I and II encoders produce an output bit stream with a constant bitrate. The number of bits available to code a single frame is determined using this constant output bit rate.

Each frame must consist of an integer number of *slots*. In Layer I, a slot is defined as 32 bits and the number of slots in a frame, N_S, is calculated using the following formula:

$$N_S = \mathrm{int}\left[\frac{384R/F_S}{32}\right] \tag{9.22}$$

where R is the output bit rate and F_S is the input sampling frequency. The length of a frame is forced to be an integer number of slots so that every frame must start on a slot boundary. This alignment of a frame to a slot boundary is done to increase the speed of resynchronization after an error event has occurred. For some combinations of bit rate and input sampling frequency, if the number of slots is kept constant at N_S, the actual output bit rate gradually falls below the desired output bit rate. This means that, in practice, the number of slots in each frame is allowed to vary between N_S and $N_S + 1$ to maintain the desired output bit rate.

EXAMPLE 9.3

For a Layer I coder with an input sampling rate of 44.1 kHz and an output bit-rate of 192 kbit/s the number of slots per frame is given by

$$\begin{aligned} N_S &= \mathrm{int}\left[\frac{384\ \text{(samples/frame)}\ \times 192000\ \text{(bits/second)}/44100\ \text{(samples/second)}}{32\ \text{(bits/slot)}}\right] \\ &= 52 \quad \text{(slots/frame)} \end{aligned}$$

Then, if each frame is transmitted using 52 slots, after 100 frames the desired number of bits is

$$\frac{100 \times 384 \times 192,000}{44100} = 167184 \quad \text{(bits)}$$

but the actual number of bits produced by the decoder is

$$100 \times 52 \times 32 = 166,400 \text{ (bits)}$$

The actual number of output bits is 784 bits (or approximately 24 slots) less than the desired number of output bits. To compensate for this discrepancy, the encoder would need to transmit approximately every fourth frame using 53 slots instead of 52. ■

If a frame is coded using an extra slot, the value of the flag **padding_bit** is set to '1' in the header of the frame. The following MATLAB code defines the algorithm that can be used to determine if padding is necessary.

```
% for the first frame don't use an extra slot and set the variable rest to
% zero
if (frame == 0)
  rest = 0;
  padding_bit = 0;
else
  % for every other frame calculate the discrepancy between the desired
  % number of bits and the actual number of bits for a single frame
  if (layer==1)
    dif = mod(384*bitrate/sampling_frequency,32);
    bits_per_slot = 32;
  else
    dif = mod(1152*bitrate/sampling_frequency,8);
    bits_per_slot = 8;
  end
  % the variable rest is used to keep track of the difference between the
  % desired and actual number of bits
  rest = rest-dif;
  % when the value of rest is negative an extra slot is required to keep
  % the difference between the desired and actual number of bits used to
  % within the size of one slot
  if (rest < 0)
    % if an extra slot is required set the value of padding_bit to 1
    % and add the number of bits in a slot to the value of rest
    padding_bit = 1;
    rest = rest + bits_per_slot;
  else
    padding_bit = 0
  end
end
```

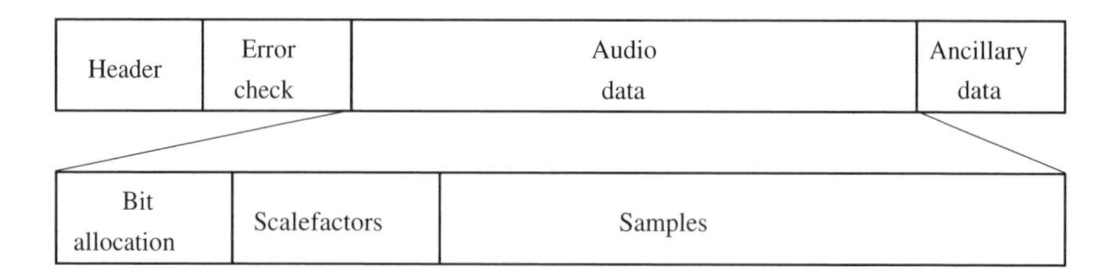

Figure 9.15 Format for a Layer I audio frame.

Figure 9.15 shows the format of a Layer I frame. The bit-stream syntax required for each audio frame is described later in Section 9.4.

9.2. LAYER II ENCODER

The major components that are changed or added to form a Layer II encoder are shown highlighted in Figure 9.16. The same analysis filter bank is used for both Layers I and II but in Layer II, three blocks of 12 × 32 subband samples are combined to form a *superblock*. This means a Layer II audio frame is effectively triple the size of a Layer I frame and contains the coded data for 1152 PCM samples compared with 384 samples for Layer I. The number of samples used in the FFT calculation is also increased to 1024 so that the power spectral density used in the psychoacoustic model is representative of the audio samples to be coded. A scalefactor is still calculated for each set of 12 samples in a subband so there are

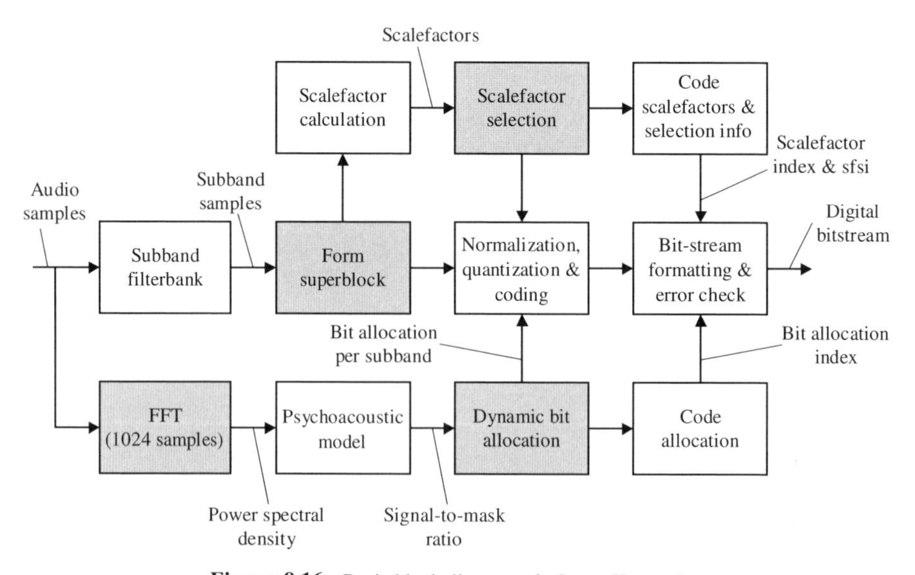

Figure 9.16 Basic block diagram of a Layer II encoder.

now three scalefactors for each subband in the superblock. Since there is often some redundancy between these three scalefactors, a more sophisticated method of scalefactor coding is adopted in Layer II. The final major modification occurs in the bit allocation procedure where the number of allowable quantizer step sizes is reduced for mid and high frequency subbands resulting in a saving in the bits used for the bit allocation code words.

9.2.1. Analysis Filterbank

The same analysis filterbank is used in both Layers I and II (see Section 9.1.1).

9.2.2. Scalefactor Calculation

As in Layer I, a *scalefactor* is calculated for each set of 12 subband samples by finding the maximum of the absolute values of the 12 subband samples and then taking the lowest value from Table 9.2 that is greater than this maximum value. The scalefactor index is then defined as the index from Table 9.2 that corresponds to this value.

9.2.3. Coding of Scalefactors

For a Layer II frame the audio data contains information from a *superblock* that consists of three consecutive blocks of 12 subband samples for each subband. This means that three scalefactors are calculated for each subband in the superblock. Often there is some redundancy between these three scalefactors and for this reason a Layer II encoder employs the following method to efficiently code the scalefactor information for a superblock.

The first step in coding the scalefactors is to calculate the two scalefactor differences, $\mathrm{d}scf_1$ and $\mathrm{d}scf_2$, using the following two equations:

$$\begin{aligned} dscf_1 &= scf_1 - scf_2 \\ dscf_2 &= scf_2 - scf_3 \end{aligned} \tag{9.23}$$

where scf_1, scf_2, and scf_3 are the successive scalefactor indices for the three blocks calculated as described in Section 9.2.2. The scalefactor differences are then categorized into one of five classes using the criteria described in Table 9.6.

The pair of scalefactor difference classes for each subband is then used to determine the method for transmitting the scalefactors from Table 9.7.

In Table 9.7, the third column titled "scalefactor used in encoder" indicates the scalefactors that are actually used in the later encoding procedures. The numbers in this column, indicate which scalefactor is used with 1, 2, and 3 indicating the scalefactors from the first, second, and third blocks in the superblock respectively and 4 indicating the maximum of the three scalefactor values (i.e. the scalefactor with the minimum scalefactor index).

Table 9.6 Scalefactor difference classes.

Class	d*scf*
1	$dscf \leq -3$
2	$-3 < dscf < 0$
3	$dscf = 0$
4	$0 < dscf < 3$
5	$dscf \geq 3$

Table 9.7 Scalefactor transmission patterns and selection information.

$Class_1$	$Class_2$	Scalefactors used in encoder	Transmission pattern	Selection information
1	1	1 2 3	1 2 3	0
1	2	1 2 2	1 2	3
1	3	1 2 2	1 2	3
1	4	1 3 3	1 3	3
1	5	1 2 3	1 2 3	0
2	1	1 1 3	1 3	1
2	2	1 1 1	1	2
2	3	1 1 1	1	2
2	4	4 4 4	4	2
2	5	1 1 3	1 3	1
3	1	1 1 1	1	2
3	2	1 1 1	1	2
3	3	1 1 1	1	2
3	4	3 3 3	3	2
3	5	1 1 3	1 3	1
4	1	2 2 2	2	2
4	2	2 2 2	2	2
4	3	2 2 2	2	2
4	4	3 3 3	3	2
4	5	1 2 3	1 2 3	0
5	1	1 2 3	1 2 3	0
5	2	1 2 2	1 2	3
5	3	1 2 2	1 2	3
5	4	1 3 3	1 3	3
5	5	1 2 3	1 2 3	0

Table 9.8 Meaning of the scalefactor selection information.

Selection information	Meaning
00	The first, second, and third scalefactor transmitted should be allocated to blocks 1, 2, and 3, respectively.
01	The first scalefactor transmitted should be allocated to blocks 1 and 2 and the second to block 3.
10	The scalefactor transmitted should be allocated to all three blocks.
11	The first scalefactor transmitted should be allocated to block 1 and the second to blocks 2 and 3.

The third column of Table 9.7 shows that there is often some duplication of the scalefactors across the three blocks and so it is not always necessary to transmit all three scalefactors for each subband in a frame. For this reason only the scalefactors indicated by the transmission pattern in the next column of Table 9.7 are actually transmitted. The next column of Table 9.7 contains the *scalefactor selection information* which is a number between 0 and 3 that is used to indicate how the transmitted scalefactor indices are to be allocated to the three blocks during the decoding process. The meaning of the four values of the scalefactor selection information is shown in Table 9.8.

The scalefactor selection information is transmitted as a two bit unsigned integer for those subbands that are allocated with more than zero bits in the following bit allocation procedure.

9.2.4. Dynamic Bit Allocation

The total number of bits available for each audio frame, *cb*, is determined from the number of 8-bit slots that are to be transmitted for that frame (see Section 9.2.8). Not all of the bits in each frame are available to transmit subband sample and scalefactor information. The available number of bits can be determined by subtracting from the total number of bits available, *cb*, the number of bits required to transmit the header, *bhdr*, (32 bits), the CRC check word if used, *bcrc*, (16 bits), the bit allocation for each subband, *bbal*, and any ancillary data, *banc*. So the number of bits available for subband sample and scalefactor information, *adb*, is given by the following equation:

$$adb = cb - (bhdr + bcrc + bbal + banc) \tag{9.24}$$

The available bits are then allocated to the subbands using the principle of minimizing the total noise-to-mask ratio of the frame. The number of bits allocated to one sample can range from 0 to 16 bits with the exception of 1 bit per sample, which is not allowed. For a Layer II encoder certain restrictions are made on the number of bits that can be allocated to each subband. These restrictions depend on the input sampling rate, the output bit rate and the subband. Tables and show the allowable number of bits for each subband for the various input sampling rates and output bit rates.

The procedure to find the number of bits allocated to each subband begins by finding the *mask-to-noise-ratio* (MNR) for each subband. The MNR is calculated by subtracting the signal-to-mask ratio from the signal-to-noise ratio for each subband and is given by

$$\text{MNR} = \text{SNR} - \text{SMR} \tag{9.25}$$

The signal-to-mask ratio is the output of the psychoacoustic model and the signal-to-noise ratio depends on the number of bits allocated to the subband. The relationship between SNR and bits allocated to a subband is given in Table 9.9. Note that the number of bits allocated in column 1 of Table 9.9 refers to the number of bits for the group of three subband samples in each granule (see Section 9.2.6).

Table 9.9 SNR per number of bits allocated to each subband for a Layer II encoder.

No. of bits	SNR (dB)	*A*	*B*
0	0.00		
5	7.00	0.750000000	−0.250000000
7	11.00	0.625000000	−0.375000000
9	16.00	0.875000000	−0.125000000
10	20.84	0.562500000	−0.437500000
12	25.28	0.937500000	−0.062500000
15	31.59	0.968750000	−0.031250000
18	37.75	0.984375000	−0.015625000
21	43.84	0.992187500	−0.007812500
24	49.89	0.996093750	−0.003906250
27	55.93	0.998046875	−0.001953125
30	61.96	0.999023438	−0.000976563
33	67.98	0.999511719	−0.000488281
36	74.01	0.999755859	−0.000244141
39	80.03	0.999877930	−0.000122070
42	86.05	0.999938965	−0.000061035
45	92.01	0.999969482	−0.000030518
48	98.01	0.999984741	−0.000015259

The allocation of bits to each subband is an iterative process that begins by allocating zero bits for the scalefactors and subband samples of each subband. The iterative process then continues using the following steps:

1. Use the number of bits allocated to each subband to determine the SNR for each subband and then calculate the MNR for each subband.
2. Determine the subband that has the minimum MNR.
3. The number of bits allocated to this subband is then increased to the next highest allowable number of bits.
4. The number of bits required to transmit the samples, *bspl*, is updated to the number of bits required in step 2. In addition, if *bspl* has increased from zero to five then the number of bits required to transmit the scalefactors, *bscf*, and the scalefactor selection information, *bsel*, is updated for this subband.
5. Recalculate the number of bits available for subband sample and scalefactor information, *adb*, using the following equation:

   $$adb = cb - (bhdr + bcrc + bbal + bsel + bscf + bspl + banc)$$

6. If *adb* is greater than any possible increase of *bsel* + *bscf* + *bspl* the iterative procedure continues from step 1.

9.2.5. Coding of Bit Allocation

The bit allocation information for each subband is coded as an unsigned integer, MSB first. The value of this integer is taken as the allocation index from Tables A.5, A.6, A.7 or A.8. The number of bits used to code the allocation index is indicated by the column labeled *nbal* in these tables. The number of bits per group of three subband samples that corresponds to each value of the allocation index is shown in the remaining columns.

9.2.6. Quantization and Coding of Subband Samples

The subband samples are coded using a linear quantizer with a symmetric zero representation. This type of quantizer guarantees that small values close to zeros are quantized to zero. The quantization process is defined by the following steps:

1. Normalize the subband sample by dividing its value by the scalefactor to obtain the normalize sample X.
2. Calculate the quantized value Q using the following formula:

   $$Q = AX + B$$

   where the values for the quantization coefficients A and B depend on the number of bits allocated to the subband and are given in Table 9.9.

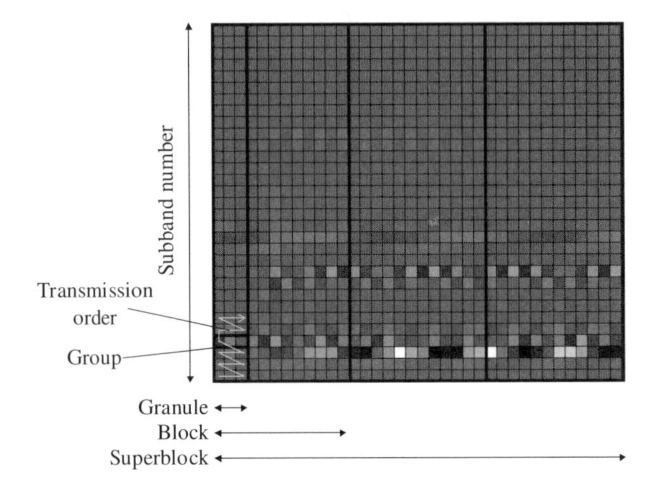

Figure 9.17 Subband sample transmission order for a single channel Layer II frame.

3. Find Q_N by taking the N most significant bits of Q, where N is the number of bits allocated in the previous bit allocation process and Q is stored as a two's complement binary fraction.
4. Invert the most significant bit of Q_N in order to avoid the code word containing all 1s.

For a Layer II coder, each block of subband samples is further subdivided into four *granules* as shown in Figure 9.17. Each shaded square in Figure 9.17 represents the value of a subband sample with gray indicating zero, white indicating a large positive value, and black indicating a large negative value. Each column of shaded squares represents one set of 32 subband sample outputs from the analysis filter and one granule consists of three of these sets of 32 subband samples. The Layer II coding algorithm specifies that subband samples be transmitted as sets of three samples for each subband in each granule. If two channels are to be coded, the two sets of three samples for each channel in each subband are transmitted consecutively.

If the number of bits allocated to the subband is 5, 7, or 10 then the code words corresponding to the set of three samples are treated as unsigned integers and *grouped* together into a single larger code word. Otherwise, each sample is transmitted as a separate code word.

If grouping is required, the three smaller code words denoted by w_a, w_b, and w_c are treated as unsigned integers and combined into a single larger code word using the following equations:

$$\begin{aligned} w_5 &= 9w_c + 3w_b + w_a \\ w_7 &= 25w_c + 5w_b + w_a \\ w_{10} &= 81w_c + 9w_b + w_a \end{aligned} \tag{9.26}$$

Then w_5, w_7, and w_{10} are transmitted as 5, 7, and 10-bit unsigned integers, respectively, MSB first.

EXAMPLE 9.4

To quantize three consecutive subband samples using five bits, the first step is to normalize the subband values by the scalefactor to give a value for X. Since the scalefactor must always be larger than the maximum sample value in each subband, the possible range of values for X is defined by $-1 < X < 1$. The value of X is then used to calculate a value for Q using the equation

$$Q = 0.75\,X - 0.25$$

and the possible range of values for Q is then defined by $-1 < Q < 0.5$.

If Q is then stored as a two's complement binary fraction and Q_2 is taken as the first two bits of this binary fraction, there are only three different possible values for Q_2. Table 9.10 shows the range of values for X and Q that correspond to each of these three values of Q_2. The last column of Table 9.10 shows the code word that is grouped together with two other code words to form a single larger code word after the most significant bit of Q_2 has been inverted.

Table 9.10 Ranges and code words for a 2-bit quantizer.

Range for X	Range for Q	Q_2	Code word
$-1 < X < \frac{-1}{3}$	$-1 < Q < -0.5$	10	00
$\frac{-1}{3} \leq X < \frac{1}{3}$	$-0.5 \leq Q < 0.0$	11	01
$\frac{1}{3} \leq X < 1$	$0.0 \leq Q < -0.5$	10	00

Then three consecutive code words produced in this way, w_a, w_b, and w_c, are treated as unsigned integers (i.e. with values of 0, 1, or 2) and grouped together to form a single code word using the equation

$$w_5 = 9w_c + 3w_b + w_a$$

This equation results in a value of w_5 between 0 and 26 that is transmitted as a 5-bit unsigned integer, MSB first. ■

9.2.7. Ancillary Data

See Section 9.1.6.

9.2.8. Formatting

The output of the audio encoder is transmitted in frames. In a Layer II bit stream, each frame contains the information required to decode 1152 samples of the original audio signal. Each frame must consist of an integer number of *slots*. In Layer II, a

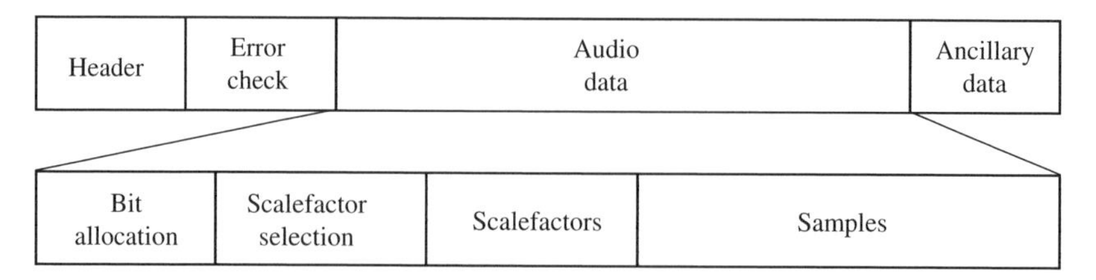

Figure 9.18 Format of a Layer II audio frame.

slot is defined as 8 bits and the number of slots in a frame, N_S, can be calculated using the following formula:

$$N_S = \text{int}\left[\frac{1152R/F_S}{8}\right] \tag{9.27}$$

where R is the output bit rate and F_S is the input sampling frequency. As for a Layer I frame, the number of slots in each frame may vary between N_S and $N_S + 1$. Figure 9.18 shows the general format of a Layer II frame. The bit stream syntax required for each audio frame is described later in Section 9.4.

9.3. JOINT STEREO CODING

The ability of the human auditory system to determine the location of sound sources depends on interaural level differences and interaural time differences. For frequencies above 2 kHz, it is the level differences between the envelope of the left and right signals that determines the perceived lateral position of the sound source. For Layer I and II encoders an optional joint stereo coding mode known as *intensity stereo coding* is available to exploit this limitation of the human auditory system.

In intensity stereo mode, the samples for the left and right channels for some subbands are added and transmitted as a single channel. The bit allocation, quantization and coding of the combined subband samples is performed as for a single channel but the scalefactor information is transmitted for both the left and right channels. In this way, the decoded left and right signals are identical except for the level. However, the ability of the human auditory system to perceive the location of the sound source is unaffected by this approximation.

The intensity stereo coding mode can be applied in one of four different configurations. The difference between the configurations is simply the lowest frequency subband that is coded using the intensity stereo method. The four possible choices for the lowest subband coded in the intensity stereo mode are subbands 4, 8, 12, or 16.

The first step in the intensity stereo process is to determine the number of subbands to be code in intensity stereo mode. For these subbands the left and right samples are added and the combined samples are scaled in the same way as for subbands from a single channel. It is important to note however that the scalefactor used in this step is not transmitted. Instead, the originally determined scalefactors for the left and right channels are transmitted. The combined and scaled samples are then quantized and coded as for a single channel using the higher of the bit allocations from the left and right channels.

Use of the intensity stereo mode generally results in a bit-rate saving of between 10 and 30 kbit/s for no perceivable degradation of audio quality and requires only a small increase in encoder and decoder complexity.

9.4. MPEG-1 SYNTAX

The syntax of the MPEG-1 audio bit sream is specified in this section. The order of arrival of the bits is defined using the C-like methodology described in Chapter 5 for the simple video coder syntax.

9.4.1. Audio Sequence Layer

The syntax specification for an MPEG-1 audio bit stream is given in Table 9.11. The bitstream consists of consecutive audio frames.

9.4.2. Audio Frame

The syntax specification for an MPEG-1 audio frame is given in Table 9.12.

Table 9.11 Syntax specification for an MPEG-1 audio bitstream.

	Syntax	Bits	Mnemonic
1	audio_sequence()		
2	{		
3	while (nextbits()==syncword){		
4	frame()		
5	}		
6	}		

Table 9.12 Syntax specification for an MPEG-1 audio frame.

	Syntax	Bits	Mnemonic
1	frame()		
2	{		
3	header()		
4	error_check()		
5	audio_data()		
6	ancillary_data()		
7	}		

9.4.3. Header

The header field contains information necessary for the decoder to maintain synchronization of the incoming bit stream and parameters that specify aspects of the audio service being transmitted. The syntax specification of the header field is given in Table 9.13

A description of the bit stream elements contained in a header field is given in Table 9.14.

The meaning of the **layer** code word is specified in Table 9.15.

The meaning of the **bitrate_index** code word is specified in Table 9.16. The specified bit rate is the total bit rate for all channels regardless of the coding mode. The "free" format is used to allow a fixed bit rate that is not specified in Table 9.16 to be used. In "free" format, the maximum bit rate allowed is 448 and 384 in Layers I and II, respectively. A continuously variable bit rate is not allowed for Layers I and II.

For Layer II, not all combinations of total bit rate and mode are allowed. The combinations that are allowed are shown in Table 9.17.

The meaning of the **sampling_frequency** code word is specified in Table 9.18.

The meaning of the **mode** code word is specified in Table 9.19.

Table 9.13 Syntax specification of the header field.

	Syntax	Bits	Mnemonic
1	header()		
2	{		
3	**syncword**	12	bslbf
4	**ID**	1	bslbf
5	**layer**	2	bslbf
6	**protection_bit**	1	bslbf
7	**bitrate_index**	4	bslbf
8	**sampling_frequency**	2	bslbf
9	**padding_bit**	1	bslbf
10	**private_bit**	1	bslbf
11	**mode**	2	bslbf
12	**mode_extension**	2	bslbf
13	**copyright**	1	bslbf
14	**original/copy**	1	bslbf
15	**emphasis**	2	bslbf
16	}		

Table 9.14 Description of the bit stream elements in a header field.

Element	Description
syncword	Used to detect the start of the header field. The value of the syncword is always 0×FFF.
ID	A value of "1" indicates the following bit stream was coded using MPEG-1. The value "0" is reserved.
layer	Indicates the layer used to produce the following bit sream (see Table 9.15).
protection_bit	A value of "1" indicates that the error_check field follows the header.
bitrate_index	Specifies the output bit rate of the coder (see Table 9.16).
sampling_frequency	Specifies the input sampling frequency (see Table 9.18).
padding_bit	A value of "1" indicates that the following audio frame contains an extra slot (see Section 9.1.7).
private_bit	Flag for private use.
mode	Specifies the coding mode (see Table 9.19).
mode_extension	Specifies the highest frequency subband that is not included in intensity stereo coding (see Table 9.20).
copyright	A value of "1" indicates that the following bit stream is protected by copyright.
original/copy	A value of "1" indicates that the following bit stream is an original bit stream.
emphasis	Specifies the type of emphasis/deemphasis used (see Table 9.21).

In intensity stereo mode the code word **mode_extension** specifies the highest frequency subband that is not included in intensity stereo coding. The meaning of the **mode_extension** code word is specified in Table 9.20.

The meaning of the **emphasis** code word is specified in Table 9.21.

Table 9.15 Meaning of Layer.

layer	Meaning
"11"	Layer I
"10"	Layer II
"01"	Layer III
"00"	Reserved

Table 9.16 Meaning of bitrate_index.

bitrate_index	Specified bitrate (kbits/s)	
	Layer I	Layer II
"0000"	free	free
"0001"	32	32
"0010"	64	48
"0011"	96	56
"0100"	128	64
"0101"	160	80
"0110"	192	96
"0111"	224	112
"1000"	256	128
"1001"	288	160
"1010"	320	192
"1011"	352	224
"1100"	384	256
"1101"	416	320
"1110"	448	384
"1111"	Forbidden	Forbidden

Table 9.17 Allowed modes for each bit rate of a Layer II encoder.

Bit rate (kbits/s)	Allowed modes
Free format	All modes
32	Single_channel
48	Single_channel
56	Single_channel
64	All modes
80	Single_channel
96	All modes
112	All modes
128	All modes
160	All modes
192	All modes
224	Stereo, intensity stereo, dual channel
256	Stereo, intensity stereo, dual channel
320	Stereo, intensity stereo, dual channel
384	Stereo, intensity stereo, dual channel

Table 9.18 Meaning of sampling_frequency.

sampling_frequency	Specified frequency (kHz)
"11"	44.1
"10"	48
"01"	32
"11"	Reserved

Table 9.19 Meaning of mode.

mode	Specified mode
"11"	Stereo
"10"	Intensity_stereo
"01"	Dual_channel
"11"	Single_channel

Table 9.20 Meaning of mode_extension.

mode_extension	Meaning
"00"	Subbands 4 – 31 in intensity_stereo, bound = 4
"10"	Subbands 8 – 31 in intensity_stereo, bound = 8
"01"	Subbands 12 – 31 in intensity_stereo, bound = 12
"11"	Subbands 16 – 31 in intensity_stereo, bound = 16

Table 9.21 Meaning of emphasis.

emphasis	Specified emphasis
"00"	None
"10"	50/15 μs
"01"	Reserved
"11"	CCITT J.17

Table 9.22 Syntax specification of the error_check field.

	Syntax	Bits	Mnemonic
1	error_check()		
2	{		
3	if (protection_bit==0)		
4	**crc_check**	16	rpchof
5	}		

9.4.4. Error Check

The syntax specification of the error_check field is given in Table 9.22.

The **crc_check** code word contains a 16 bit parity-check word. This code word is transmitted using the remainder polynomial coefficients, highest order first (rpchof) data type. Further details of the error checking procedure can be found in Ref. [1].

9.4.5. Audio Data, Layer I

The syntax specification of the audio_data field in Layer I is given in Table 9.23.

A description of the bit stream elements contained in an audio_data field for Layer I is given in Table 9.24.

9.4.6. Audio Data, Layer II

The syntax specification of the audio_data field in Layer II is given in Table 9.25.

A description of the bit stream elements contained in an audio_data field for Layer II is given in Table 9.26.

9.5. MPEG-1 LAYER I,II DECODERS

9.5.1. Bit Allocation Decoding

For a Layer I decoder, the 4-bit allocation parameter is read for all subbands. If the coding mode is joint stereo then only one bit allocation parameter is read for each subband above the subband indicated by the **mode_extension** parameter.

For a Layer II decoder, the number of bits used for the bit allocation parameter varies for different input sampling rates and output bit rates and can be found using tables in Appendix. The output bit rate used to determine the appropriate table is the

Table 9.23 Syntax specification of the audio_data field in Layer I.

	Syntax	Bits	Mnemonic
1	audio_data(){		
2	for (sb=0; sb<bound; sb++)		
3	for (ch=0; ch<nch; ch++)		
4	**allocation[ch][sb]**	4	uimsbf
5	for (sb=bound; sb<32; sb++){		
6	**allocation[0][sb]**	4	uimsbf
7	allocation[1][sb]=allocation[0][sb]		
8	}		
9	for (sb=0; sb<32; sb++)		
10	for (ch=0; ch<nch; ch++)		
11	if (allocation[ch][sb]!=0)		
12	**scalefactor[ch][sb]**	6	uimsbf
13	for (s=0; s<12; s++)		
14	for (sb=0; sb<bound; sb++)		
15	for (ch=0; ch<nch; ch++)		
16	if (allocation[ch][sb]!=0)		
17	**sample[ch][sb][s]**	2..15	uimsbf
18	for (sb=bound; sb<32; sb++)		
19	if (allocation[ch][sb]!=0)		
20	**sample[0][sb][s]**	2..15	uimsbf
21	}		
22	}		

bit rate for one channel so if the coding mode is not single channel then the output bit rate indicated in the frame header should be divided by two to find the bit rate per channel. The number of bits to read for each subband is indicated by the value of *nbal* in the appropriate table in Appendix. This allocation parameter is treated as

Table 9.24 Description of the bitstream elements in an audio_data field for Layer I.

Element	Description
allocation[ch][sb]	Specifies the bits per sample used to code the subband samples in subband **sb** of channel **ch** (see Section 9.1.4).
scalefactor[ch][sb]	Specifies the scalefactor used in subband **sb** of channel **ch** (see Section 9.1.2).
sample[ch][sb][s]	The coded value for sample **s** in subband **sb** of channel **ch** (see Section 9.1.6)

Table 9.25 Syntax specification of the audio_data field in Layer II.

	Syntax	Bits	Mnemonic
1	audio_data(){		
2	for (sb=0; sb<bound; sb++)		
3	for (ch=0; ch<nch; ch++)		
4	**allocation[ch][sb]**	2..4	uimsbf
5	for (sb=bound; sb<sblimit; sb++){		
6	**allocation[0][sb]**	2..4	uimsbf
7	allocation[1][sb]=allocation[0][sb]		
8	}		
9	for (sb=0; sb<sblimit; sb++)		
10	for (ch=0; ch<nch; ch++)		
11	if (allocation[ch][sb]!=0)		
12	**scfsi[ch][sb]**	2	uimsbf
13	for (sb=0; sb<32; sb++)		
14	for (ch=0; ch<nch; ch++)		
15	if (allocation[ch][sb]!=0)		
16	if (scfsi[ch][sb]==0){		
17	**scalefactor[ch][sb][0]**	6	uimsbf
18	**scalefactor[ch][sb][1]**	6	uimsbf
19	**scalefactor[ch][sb][2]**	6	uimsbf
20	}		
21	if ((scfsi[ch][sb]==1) || (scfsi[ch][sb]==3)){		
22	**scalefactor[ch][sb][0]**	6	uimsbf
23	**scalefactor[ch][sb][1]**	6	uimsbf
24	}		
25	if ((scfsi[ch][sb]==2)		
26	**scalefactor[ch][sb][0]**	6	uimsbf
27	}		
28	for (gr=0; gr<12; gr++){		
29	for (sb=0; sb<bound; sb++)		
30	for (ch=0; ch<nch; ch++)		
31	if (allocation[ch][sb]!=0){		
32	if (grouping[ch][sb])		
33	**samplecode[ch][sb][gr]**	5..10	uimsbf
34	else		
35	for (s=0; s< 3; s++)		
36	**sample[ch][sb][3*gr+s]**	3..16	uimsbf
37	}		
38	for (sb=bound; sb<sblimit; sb++)		
39	if (allocation[0][sb]!=0){		
40	if (grouping[0][sb])		
41	**samplecode[0][sb][gr]**	5..10	uimsbf
42	else		

Table 9.25 (*Continued*)

	Syntax	Bits	Mnemonic
43	for (s=0; s< 3; s++)		
44	**sample[0][sb][3*gr+s]**	3..16	uimsbf
45	}		
46	}		
47	}		

an unsigned integer and used as an index into the remaining columns of the table. The value indicated by this index represents the number of bits transmitted in the corresponding subband. This number of bits can then in turn be used as an index into Table 9.27 that shows the number of bits used to code the quantized samples and the requantization coefficients C and D.

9.5.2. Scalefactor Selection Information Decoding

For a layer II decoder the 2-bit scalefactor selection information (*scfsi*) code word is read for all subbands that have a nonzero bit allocation. The meaning of the *scfsi* code word can be found in Table 9.8.

9.5.3. Scalefactor Decoding

For a Layer I decoder, a 6-bit scalefactor code word for each subband with a nonzero bit allocation is read from the bit stream and used as an index into Table 9.2 to find the scalefactor for the corresponding subband.

Table 9.26 Description of the bit stream elements in an audio_data field for Layer II.

Element	Description
allocation[ch][sb]	Specifies the bits per sample used to code the subband samples in subband **sb** of channel **ch** (see Section 9.2.4).
scfsi[ch][sb]	Specifies the number of scalefactors transferred for subband **sb** of channel **ch** (see Section 9.2.3).
scalefactor[ch][sb][p]	Specifies the scalefactor used subband **sb** of channel **ch** for part **p** of the frame (see Section 9.2.3)
samplecode[ch][sb][gr]	The code word for the three consecutive sample values in granule **gr** in subband **sb** of channel **ch** if grouping is used (see Section 9.2.6).
sample[ch][sb][s]	The coded value for sample **s** in subband **sb** of channel **ch** (see Section 9.2.6).

Table 9.27 Classes of quantization for a Layer II decoder.

No. of bits	C	D
5	1.33333333333	0.50000000000
7	1.60000000000	0.50000000000
9	1.14285714286	0.25000000000
10	1.77777777777	0.50000000000
12	1.06666666666	0.12500000000
15	1.03225806452	0.06250000000
18	1.01587301587	0.03125000000
21	1.00787401575	0.01562500000
24	1.00392156863	0.00781250000
27	1.00195694716	0.00390625000
30	1.00097751711	0.00195312500
33	1.00048851979	0.00097656250
36	1.00024420024	0.00048828125
39	1.00012208522	0.00024414063
42	1.00006103888	0.00012207031
45	1.00003051851	0.00006103516
48	1.00001525902	0.00003051758

For a Layer II decoder, the number of 6-bit scalefactor code words corresponding to each subband with a nonzero bit allocation is indicated by the scalefactor selection information. These scalefactor code words are read from the bit stream and used as an index into Table 9.2 to determine the three scalefactors for the corresponding subband.

9.5.4. Requantization of Subband Samples

9.5.4.1. *Layer I*

The number of bits indicated by the bit allocation parameter is read from the bit stream. This N-bit code word is then used to find the value of the requantized and rescaled subband sample using the following steps:

1. Invert the most significant bit of the N-bit code word to obtain the 2's complement fractional number R_N.

2. Calculate the requantized value R using the following formula:

$$R = \frac{2^N}{2^N - 1}\left(R_N + 2^{1-N}\right)$$

3. Rescale the subband sample by multiplying R by the scalefactor for the corresponding subband to obtain the rescaled subband sample S.

9.5.4.2. Layer II

If the number of bits allocated to the subband is 5, 7, or 10 then the code words corresponding to the set of three consecutive samples are grouped together into a single larger code word. The algorithm used to convert this group code word back into three individual code words is defined by the following MATLAB code. The variable nlevels has the value 3, 5, or 9 if the number of bits allocated to the subband is 5, 7, or 10, respectively.

```
for i = 1:3
  code word(i) = mod(group_code,nlevels)
  group_code = floor(group_code/nlevels);
end
```

Then, for each individual N-bit code word, the following steps are used to find the value of the requantized and rescaled subband sample:

1. Invert the most significant bit of the N-bit code word to obtain the 2's complement fractional number R_N.
2. Calculate the requantized value R using the following formula:

$$R = C(R_N + D)$$

   where the values for the requantization coefficients C and D depend on the number of bits allocated to the subband and are given in Table 9.27.

3. Rescale the subband sample by multiplying R by the scalefactor for the corresponding subband to obtain the rescaled subband sample S.

9.5.5. Synthesis Filterbank

The synthesis filterbank used in the Layer I,II encoders is based on the efficient implementation of a cosine modulated filterbank proposed by Joseph Rothweiler in 1983. A detailed explanation of this filter bank technique is given in Section 8.5 of Chapter 8.

9.6. MPEG-2

The DVB guidelines state that an IRD need only be able to decode a stereo pair from an MPEG-2 BC bit stream. So this section describes the frame formatting and matrixing procedures required to produce an MPEG-1 compatible frame within an MPEG-2 bit stream.

9.6.1. Backwards-Compatible MPEG-2 Frame Formatting

If a 5.1 channel signal is to be transmitted using an MPEG-2 encoder and backwards compatibility with an MPEG-1 decoder is required, the first step is to down-mix

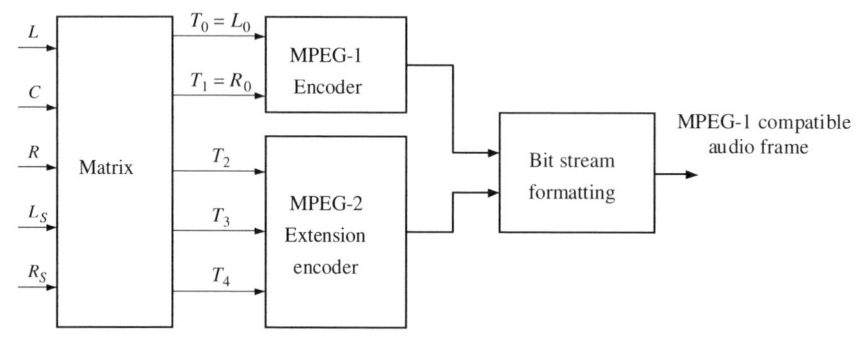

Figure 9.19 The down-mixed stereo pair is encoded using an MPEG-1 encoder and the remaining multichannel data is encoded using an MPEG-2 extension encoder.

the five input channels into a basic stereo pair. This down-mixing process is usually referred to as *matrixing* and the resulting stereo pair is denoted by L_0 and R_0. This stereo pair is then allocated to *transmission channels* T_0 and T_1 and encoded using an MPEG-1 encoder. Transmission channels T_2, T_3, and T_4 contain the audio data required to reconstruct the remaining multichannel signal and are coded using an MPEG-2 *extension encoder* as shown in Figure 9.19.

Backwards compatibility with an MPEG-1 decoder is achieved by placing the data for transmission channels T_0 and T_1 in the audio data field of an MPEG-1 frame and all the multichannel extension data in the ancillary data field of the frame as shown in Figure 9.20. An MPEG-1 decoder simply ignores the multichannel data in the ancillary data field and decodes the stereo pair L_0/ R_0. An MPEG-2 decoder decodes both the stereo pair and the multichannel data and performs a dematrixing procedure to obtain the decoded 5.1 multichannel signals.

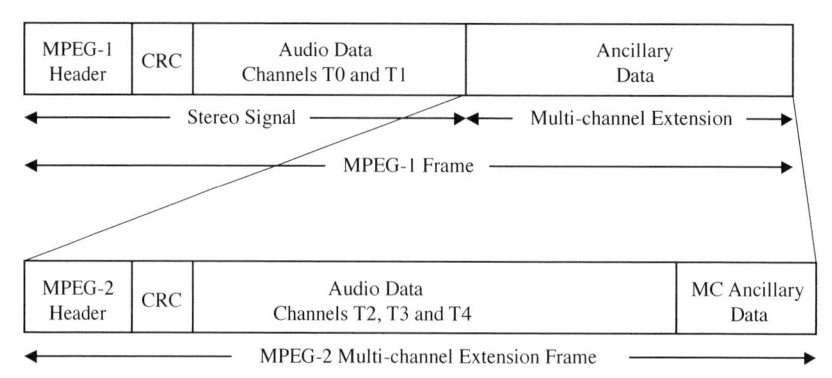

Figure 9.20 Data format of an MPEG-2 audio frame that is backwards compatible with an MPEG-1 decoder.

Table 9.28 Filter coefficients for the filter used in matrixing procedure 2.

Sampling frequency (kHz)	a_0	b_0	b_1	b_2
32	486	2048	−471	370
44.1	295	2048	−1394	521
48	294	2048	−1388	520

9.6.2. Matrixing Procedures for Backwards Compatibility

The MPEG-2 standard defines four possible matrixing procedures that can be used to obtain the L_0/R_0 stereo pair from a 5.1 channel signal. The equations used for procedures 0, 1, and 3 are

$$L_0 = \alpha\,(L + \beta C + \gamma L_S) \tag{9.28}$$

$$R_0 = \alpha\,(R + \beta C + \gamma R_S) \tag{9.29}$$

Matrixing procedure 2 can be used to produce a L_0/R_0 stereo pair that is compatible with a Dolby Pro Logic decoder and the equations used for this procedure are

$$L_0 = \alpha\,(L + \beta C + \gamma jS) \tag{9.30}$$

$$R_0 = \alpha\,(R + \beta C + \gamma jS) \tag{9.31}$$

To obtain the signal jS used in Equations (9.30) and (9.31), half Dolby B-type encoding is applied to the signals L_S and R_S followed by a 90° phase shift. The monophonic component of the resulting signals is then calculated and limited in bandwidth to the range 100–7 kHz to produce the surround signal jS. The transfer function of the bandwidth-limiting filter is given in Equation (9.32) and the coefficients to be used in this equation are given in Table 9.28 for each input sampling frequency.

$$H(z) = \frac{a_0\left(1 + 2z^{-1} + z^{-2}\right)}{b_0 + b_1 z^{-1} + b_2 z^{-2}} \tag{9.32}$$

The attenuation factors, α, β, and γ, differ for each procedure and are given in Table 9.29.

9.7. SUMMARY

The DVB digital television standard prescribes the use of an MPEG-1 or backwards compatible MPEG-2 bit stream to transmit audio information. The MPEG-1 audio standard defines a multilayer coding algorithm with each additional layer providing increased complexity and coding efficiency. The MPEG-2 BC audio standard defines a mechanism for transmitting a multichannel audio signal that can be decoded by an MPEG-1 decoder. The encoding algorithm, syntax, and decoding algorithm for the first two layers of the MPEG-1 standard and the matrixing operations required to allow an MPEG-1 decoder to decode an MPEG-2 bit stream are described in this chapter.

Table 9.29 Attenuation factors for each matrixing procedure.

Matrixing procedure	α	β	γ
0	$\frac{1}{1+\sqrt{2}}$	$\frac{1}{\sqrt{2}}$	$\frac{1}{\sqrt{2}}$
1	$\frac{1}{1.5+0.5\sqrt{2}}$	$\frac{1}{\sqrt{2}}$	$\frac{1}{2}$
2	$\frac{1}{1+\sqrt{2}}$	$\frac{1}{\sqrt{2}}$	$\frac{1}{\sqrt{2}}$
3	1	0	0

PROBLEMS

9.1 Draw a simple block diagram of an MPEG-1 encoder that includes the following system components and briefly explain the function of each of these components:

(a) Analysis filterbank

(b) Psychoacoustic model

(c) Dynamic bit allocation

(d) Normalization quantization and coding

(e) Bitstream formatting

9.2 Briefly explain the four audio coding modes that are possible with an MPEG-1 audio coder.

9.3 Explain how a scalefactor is calculated for each of the subbands in an audio block.

9.4 Explain why a psychoacoustic model is used to calculate a signal-to-mask ratio value for each subband.

9.5 Explain what is meant by the following terms when used in the context of Psychoacoustic model 1 in the MPEG-1 encoder:

(a) Maximum sound pressure level in each subband

(b) Threshold-in-quiet

(c) Tonal and Nontonal components

(d) Individual masking threshold

(e) Global masking threshold

(f) Signal-to-mask ratio

9.6 Explain how the global masking threshold curve is calculated from the individual masking curves for each masker.

9.7 Explain how the minimum masking threshold for each subband is calculated from the global masking curve.

9.8 Describe the process for allocating the number of bits to be used to quantize the subband samples in each subband for a Layer I coder.

9.9 Describe the process for quantizing and coding subband samples in each subband for a Layer I coder.

9.10 Complete the following table by determining the actual bits to be transmitted to the decoder for the given normalized subband sample and *bspl* values.

Subband sample	*bspl*	Code word
0.2311	3	
−0.6068	5	
0.4860	7	
−0.8913	9	

9.11 Calculate the number of slots per frame for a Layer I coder with an input sampling rate of 48 kHz and output bitrate of 128 kbit/s.

9.12 With the aid of a diagram describe the format of an MPEG-1 Layer I audio frame.

9.13 Briefly explain the major differences between the MPEG-1 Layer 1 encoding algorithm and the MPEG-1 layer II encoding algorithm.

9.14 Complete the following table by determining the pair of scalefactor difference classes, the scalefactors to be used in the encoder, the transmission pattern and the scalefactor selection information for each set of three scalefactor indices.

Scalefactor indices	Class_1	Class_2	Scalefactors used in encoder	Transmission pattern	Selection information
5,10,20					
32,34,36					
20,25,26					
16,18,17					
48,32,34					
36,30,24					

9.15 Describe the process for allocating the number of bits to be used to quantize the subband samples in each subband for a Layer II coder.

9.16 Describe the process for quantizing and coding subband samples in each subband for a Layer II coder.

9.17 With the aid of a diagram, explain what is meant by the following terms when used in the context of quantizing and coding subband samples in an MPEG-1 layer II encoder:

(a) Superblock

(b) Block

(c) Granule

(d) Group

9.18 Complete the following table by determining the actual bits to be transmitted to the decoder for the given sets of three normalized subband sample values.

Subband samples	*bspl*	Code word
−0.1106, 0.8436, −0.1886	5	
0.2309, 0.4764, 0.8709	7	
0.5839, −0.6475, 0.8338	9	
−0.1795, −0.2943, −0.7222	10	

9.19 Calculate the number of slots per frame for a Layer II coder with an input sampling rate of 48 kHz and output bitrate of 128 kbit/s.

9.20 With the aid of a diagram describe the format of an MPEG-1 Layer II audio frame.

9.21 Determine the actual bits required in the header of an MPEG-1 audio frame for the audio stream with the following configuration:

Property	Value
ID	ISO/IEC 11172-3
Layer	I
Error protection redundancy	None
Bitrate	128 kbit/s
Sampling frequency	44.1 kHz
Padding slot required	No
Mode	Single channel
Copyright	Protected
Original	Yes
Emphasis	None

9.22 Determine the actual bits required in the header of an MPEG-1 audio frame for the audio stream with the following configuration:

Property	Value
ID	ISO/IEC 11172-3
Layer	II
Error protection redundancy	None
Bitrate	256 kbit/s
Sampling frequency	48 kHz
Padding slot required	No
Mode	Intensity_stereo
Subbands in intensity_stereo	4–31
Copyright	Protected
Original	Yes
Emphasis	None

9.23 Decode the following bit stream that represents a single audio frame from an MPEG-1 Layer II bit stream.

9.24 With the aid of diagrams, explain how a multichannel audio signal can be encoded with an MPEG-2 encoder to produce an MPEG-1 compatible audio frame.

MATLAB EXERCISE 9.1

The aim of this exercise is to implement the Psychoacoustic model used in the MPEG-1 Layer I/II encoder.

1. Write a MATLAB function to determine the scalefactor for each subband given a block of 12 sets of 32 subband samples and using the scalefactor values defined in Section 9.1.2.
2. Write a MATLAB function to calculate the FFT of a set of 512 input audio samples using the equation defined in Section 9.1.3.1.

3. Write a MATLAB function to determine the maximum sound pressure level in each subband given a set of 256 FFT coefficients and the scalefactors for each subband of the corresponding subband samples using the equation defined in Section 9.1.3.2.
4. Write a MATLAB function to determine the threshold in quiet at the subsampled frequency values for each input sampling frequency.
5. Write a MATLAB function to locate the tonal and nontonal masking components for a set of 256 FFT coefficients using the code given in Section 9.1.3.4.
6. Write a MATLAB function to decimate the tonal and nontonal masking components for a set of 256 FFT coefficients using the code given in Section 9.1.3.5.
7. Write a MATLAB function to calculate the individual masking threshold for a set of tonal and nontonal masking components using the code given in Section 9.1.3.6.
8. Write a MATLAB function to calculate the global masking threshold for a set of individual masking thresholds using the equation given in Section 9.1.3.7.
9. Write a MATLAB function to calculate the minimum masking threshold in each subband for a given global masking thresholds using the equation given in Section 9.1.3.8.
10. Write a MATLAB function to calculate the signal-to-mask ratio for a set of minimum masking threshold and sound pressure level values using the pseudocode given in Section 9.1.3.9.
11. Combine these functions to produce a MATLAB program that calculates the signal-to-mask-ratio for each subband given a block of 12 sets of 32 subband samples.
12. Write a MATLAB function to calculate the number of bits allocated to each subband given a set of signal-to-mask ratio values for each subband.
13. Combine the functions from 11 and 12 and determine the number of bits allocated to each subband for a block of 12 sets of 32 subband samples.

MATLAB EXERCISE 9.2

The aim of this exercise is to implement the quantization and inverse quantization algorithms used in the MPEG Layer I encoder.

1. Write a MATLAB function to quantize a block of subband sample values with the number of bits specified by a set of bit allocation values for each subband using the quantization algorithm given in Section 9.1.6.
2. Write a MATLAB function to produce a block of subband samples values from a bit stream containing the code words corresponding to a set of quantized subband sample values.

MATLAB EXERCISE 9.3

The aim of this exercise is to implement the scalefactor coding and decoding algorithms used in the MPEG-1 Layer II encoder and decoder.

1. Write a MATLAB function to determine the two scalefactor difference classes for a set of three scalefactor indices using the algorithm described in Section 9.2.2.
2. Write a MATLAB function to determine the scalefactor to be used in the encoder and the scalefactor selection information given a set of three scalefactor indices and their scalefactor difference classes.
3. Write a MATLAB function to produce a set of three scalefactor indices given a bit stream containing the scalefactors and scalefactor selection information for an MPEG-1 Layer II audio frame.

REFERENCES

1. ISO/IEC 11172-3: 1993, Information technology—Coding of moving pictures and associated audio for digital storage media at up to about 1.5 Mbit/s, Part 3: Audio.
2. ITU-R Recommendation BS-775-1, Multichannel stereophonic sound system with and without accompanying picture.
3. ISO/IEC 13818-3: 1995/Amd. 1: 1996, Information technology—Generic coding of moving pictures and associated audio information, Part 3: Audio.
4. ISO/IEC 13818-7: 1997/Cor. 1: 1998, Information technology—Generic coding of moving pictures and associated audio information, Part 7: Advanced audio coding.

Chapter 10

Dolby AC-3 Audio

The Dolby AC-3 standard describes an algorithm suitable for coding an audio signal with channel formats ranging from monophonic to 5.1 channels into a serial bit stream with data rates ranging from 32 to 640 kbits/s. The AC-3 algorithm is based on the transform coding technology developed for the AC-2 coding algorithm in 1989. The AC-2 algorithm operated on stereo signals only and relied on a 4-2-4 multichannel matrix system to transmit a four-channel surround-sound signal. By 1990 it was realized that the data compression performed by the 4-2-4 matrix system could be better performed as part of the coding algorithm and the development of a multichannel audio coder was initiated. The first implementation of such a multichannel coder was used for providing a 5.1 channel digital soundtrack in cinemas in 1991. This early implementation of the AC-3 algorithm (or Dolby digital as it is usually called) converted a 5.1 channel input signal into a serial bit stream with a single output bit rate of 320 kbits/s. This rate was determined by the maximum amount of data that could be reliably placed and extracted from the area between the sprocket holes on one side of the 35 mm film.

In February 1992, the U.S. Advanced Television Systems Committee formally recommended the use of a 5.1 channel audio signal for the U.S. HDTV service. The AC-3 coder was considered as a possible technique to provide this audio signal. However, the requirements placed for an audio coder providing such a service are more diverse than for a coder intended solely for digital cinema applications. So the basic AC-3 algorithm was improved to provide extra functionality. These extra features included support for a range of output bit rates, the ability to downmix the 5.1 channel output to fewer channels, and the ability to reproduce the output signal with a restricted dynamic range. In late 1993, the AC-3 coding algorithm was formally evaluated using subjective testing and was subsequently recommended for use in the U.S. HDTV system. The standard was initially published in November 1994 and a revised edition containing a backwards-compatible alternate bit-stream syntax was published in August 2001 [1]. (The latest version of the standard can be found at www.atsc.org.)

A block diagram of the AC-3 encoder is shown in Figure 10.1. The first step in the encoding process is to convert a sequence of audio input samples into a block of

Digital Television, by John Arnold, Michael Frater and Mark Pickering.

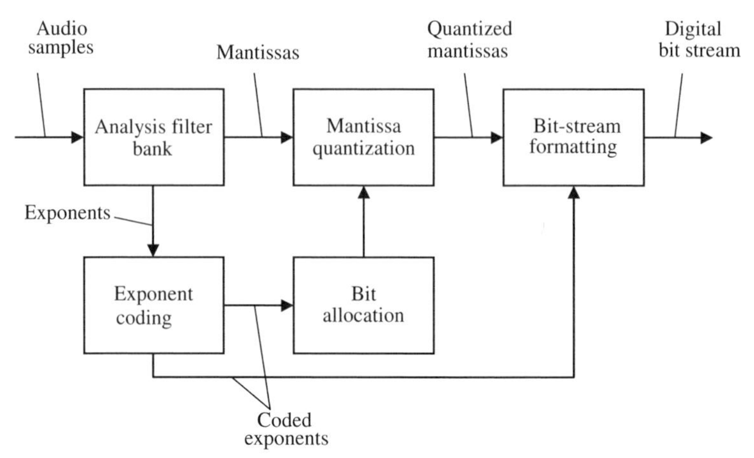

Figure 10.1 Block diagram of the AC-3 encoder. © Advanced Television Standards Committee Inc. 2001. A copy of this standard is available at http://www.atsc.org.

frequency coefficients. This is achieved using a time-domain aliasing cancellation (TDAC) transform approach as discussed in Section 8.6 of Chapter 8. Then each frequency coefficient in the block is represented in floating point format as an exponent and a mantissa. The exponent acts as a scalefactor for the mantissa and indicates the number of leading zeros in the binary integer representation of the coefficient. Exponent values can range from 0 to 24 and are fixed at 24 if the coefficient has more leading zeros. The set of exponents is also used as an approximation of the power spectral density of the signal and is referred to as the *spectral envelope* of the signal. This spectral envelope is then used in the bit-allocation process to determine the quantizer step size for the mantissas for each coefficient. The bit-allocation process uses a model of frequency masking to determine the precision required for each mantissa.

An AC-3 bit stream consists of a sequence of synchronization frames as shown in Figure 10.2. Each synchronization frame includes six *audio blocks*, each of which contains the coded data representing 256 new input samples. Each synchronization frame begins with a synchronization information (SI) header, which contains information necessary to acquire and maintain synchronization, and a bit-stream information (BSI) header that contains parameters describing the coded audio service. The synchronization frame ends with an error check field that contains a CRC code word used for error detection, and an auxiliary (Aux) data field may also be included after the coded audio blocks.

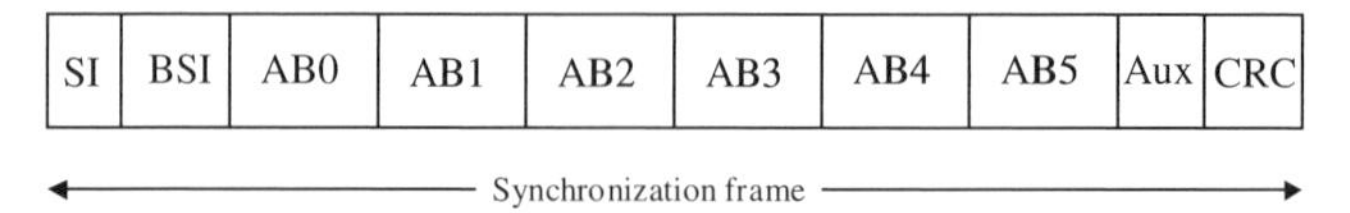

Figure 10.2 Format of an AC-3 synchronization frame.

The algorithms and bit-stream syntax associated with a basic implementation of the AC-3 encoder and decoder are described in more detail in the following sections.

10.1. ENCODER

A more detailed flow diagram of the encoding process is shown in Figure 10.3, and the individual components of this diagram are explained in more detail in the following section.

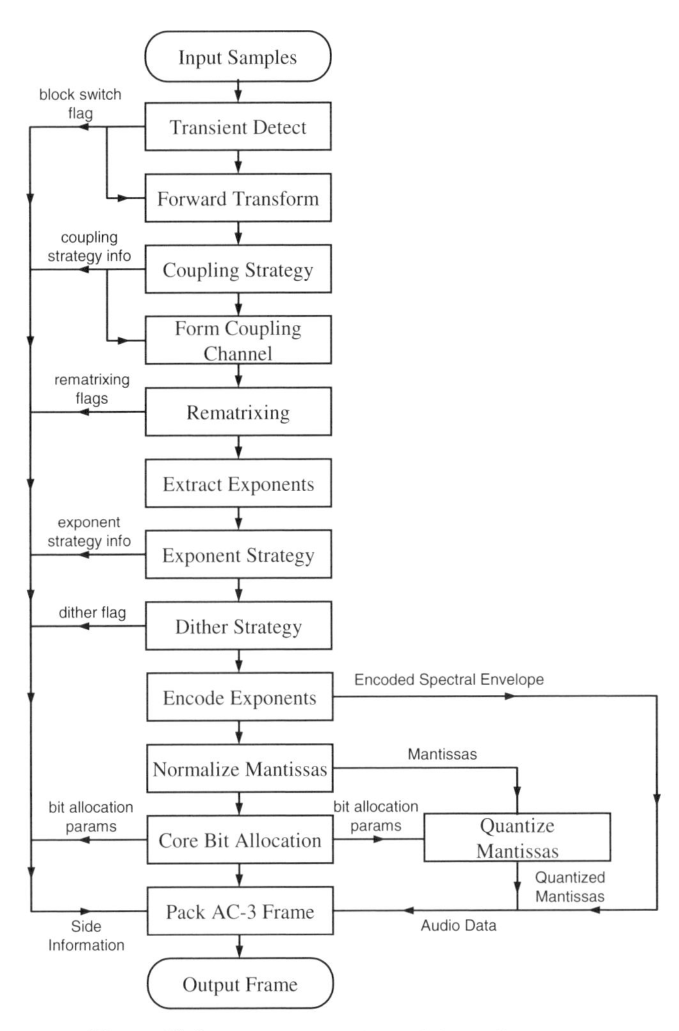

Figure 10.3 Flow diagram of the AC-3 encoding process.

Table 10.1 Sample rate code words.

fscod	Sampling rate (kHz)
"00"	48
"01"	44.1
"10"	32
"11"	Reserved

10.1.1. Audio Input Format

The AC-3 encoder accepts audio samples with a precision of up to 24 bits. The sampling rate of the input signal is directly proportional to the output. The input sample rate must be chosen such that each AC-3 audio frame consists of the audio data for 1536 audio samples. If the input audio data is sampled at a rate other than that required, sample rate conversion must be preformed to achieve the desired input sample rate. The input sample rate is indicated by the 2-bit code word **fscod**. The sampling rate indicated by each value of this code word is shown in Table 10.1.

The input audio signal may have one of the eight possible channel configurations (usually referred to as *audio coding modes*) as shown in Table 10.2. The p/q notation is typically used in multichannel systems to indicate p front and q back channels. These channels are referred to as *full-bandwidth* channels as opposed to the low-frequency effects channel that typically has a much smaller bandwidth. So if the low-frequency effects channel is present, the total number of channels to be transmitted is equal to the number of full-bandwidth channels plus one. The 1+1 notation is used to indicate two completely independent input channels.

Individual input channels may also be filtered to remove the DC components. This allows more efficient coding of the input signal. A typical encoder would filter the

Table 10.2 Audio coding modes.

acmod	Audio coding mode	nfchans	Channel array ordering
"000"	1+1	2	Ch1, Ch2
"001"	1/0	1	C
"010"	2/0	2	L, R
"011"	3/0	3	L, C, R
"100"	2/1	3	L, R, S
"101"	3/1	4	L, C, R, S
"110"	2/2	4	L, R, Ls, Rs
"111"	3/2	5	L, C, R, Ls, Rs

input signals using a high-pass filter with a cutoff frequency of 3 Hz. The LFE channel may also be filtered using a low-pass filter with a cutoff frequency of 120 Hz.

10.1.2. Transient Detection

One common problem that is encountered by most digital audio encoders is the occurrence of *preecho.* This problem is associated with transients in the audio signal that take the form of a period of silence followed by the rapid onset of a loud percussive sound such as a cymbal clash or a drum beat. If these transients occur in the middle of an audio block, the quantization strategy for that block is determined based on the loud sound rather than the period of silence, and this generally results in relatively large quantizer step sizes being chosen for the entire block. Because the frequency-domain coefficients translate back to sinusoidal waveforms that extend over the entire reconstructed audio signal, the large amount of distortion associated with the coarse quantization strategy is present in the period of silence at the beginning of the block and is not hidden by either frequency masking or backward temporal masking. Hence, the noise at the beginning of the block is usually clearly audible and is referred to as preecho.

One solution to this problem is to use a shorter window length for the transform so that backward temporal masking can be used to mask the preechoes. However, shorter window lengths result in an increase in block overhead information and an increase in the overlap of subbands in the frequency domain. Therefore, a typical approach in audio coding is to switch between long and short transform windows depending on the transient behavior of the input audio signal.

In the AC-3 encoder, transient detection is used to determine when to switch between a single 512-sample window and two 256-sample windows. The transient detection algorithm is divided into four basic steps:

1. high-pass filtering,
2. segmentation of the block into subblocks,
3. peak amplitude detection within each subblock, and
4. threshold comparison.

The high-pass filter used in the first step should have a cutoff frequency of 8 kHz. Each block of 256 high-pass filtered samples is then segmented into a hierarchical tree of subblocks. Level 1 in the tree corresponds to the entire 256-sample block, Level 2 consists of two 128-sample subblocks, and Level 3 consists of four 64-sample subblocks. The peak detection algorithm then determines the sample with the largest magnitude in each subblock in each level of the hierarchical tree. The peak magnitude in subblock k of Level j is denoted by $P_{j,k}$.

The first step in the threshold comparison process is to determine if the maximum signal level in the block is above a *silence threshold.* This is achieved by comparing $P_{1,1}$ with the silence threshold value of 100/32768. If the value of $P_{1,1}$ is below the silence threshold, the remaining threshold comparison is not performed and a transient is not detected in the current 256-sample block. If $P_{1,1}$ is above the silence

Table 10.3 Threshold values of the transient detection algorithm.

Level (j)	T_j
1	0.1
2	0.075
3	0.05

threshold, the next step in the threshold comparison process is to compare the ratio of the peak values of adjacent subblocks at each level with a predefined threshold T_j. The threshold values for each level are shown in Table 10.3.

If any of these ratios is above the threshold, a transient is detected in the current 256-sample block. Note that for the first subblock of each level, the ratio of peak values is calculated using the last subblock of the same level from the previously calculated tree. For example, $P_{3,1}$ of the current tree and $P_{3,4}$ of the preceding tree would be used to calculate the ratio for the first subblock in Level 3 of the current tree.

If a transient is detected in the second half of the 512-sample block, this is indicated by a value of 1 for the **blksw[ch]** flag, and the following forward transform operation is performed using two 256-point transform operations.

10.1.3. Forward Transform

The TDAC transform process requires that the input audio samples be multiplied by a window function. The values for the first half of this window function are given in Table 10.4. These values are mirrored and repeated to form a 512-point symmetrical window function.

The forward transform is then performed using the following equation:

$$X(k)=\frac{-2}{N}\sum_{n=0}^{N-1} w(n)x(n)\cos\left(\frac{2\pi}{4N}(2n+1)(2k+1)+\frac{\pi}{4}(2k+1)(1+\alpha)\right)$$
$$\text{for } k=0,\ldots,N/2-1 \tag{10.1}$$

Where $x(n)$ is the set of input samples, $w(n)$ is the window function with the values shown in Table 10.4, and $X(k)$ is the set of transform coefficients. The variable α is used to control the time offset for the transform basis functions and is given by

$$\alpha=\begin{cases}-1, & \text{for the first short transform}\\ 0, & \text{for the long transform}\\ +1, & \text{for the second short transform}\end{cases} \tag{10.2}$$

Figure 10.4 shows an example of a set of 512 samples of a typical audio waveform, and Figure 10.5 shows the set of 256 output coefficients for these input samples when the forward transform is applied using a single long window.

Table 10.4 Window function for the TDAC transform.

	$w(i+j)$									
	$j=0$	$j=1$	$j=2$	$j=3$	$j=4$	$j=5$	$j=6$	$j=7$	$j=8$	$j=9$
$i=0$	0.00014	0.00024	0.00037	0.00051	0.00067	0.00086	0.00107	0.00130	0.00157	0.00187
$i=10$	0.00220	0.00256	0.00297	0.00341	0.00390	0.00443	0.00501	0.00564	0.00632	0.00706
$i=20$	0.00785	0.00871	0.00962	0.01061	0.01166	0.01279	0.01399	0.01526	0.01662	0.01806
$i=30$	0.01959	0.02121	0.02292	0.02472	0.02662	0.02863	0.03073	0.03294	0.03527	0.03770
$i=40$	0.04025	0.04292	0.04571	0.04862	0.05165	0.05481	0.05810	0.06153	0.06508	0.06878
$i=50$	0.07261	0.07658	0.08069	0.08495	0.08935	0.09389	0.09859	0.10343	0.10842	0.11356
$i=60$	0.11885	0.12429	0.12988	0.13563	0.14152	0.14757	0.15376	0.16011	0.16661	0.17325
$i=70$	0.18005	0.18699	0.19407	0.20130	0.20867	0.21618	0.22382	0.23161	0.23952	0.24757
$i=80$	0.25574	0.26404	0.27246	0.28100	0.28965	0.29841	0.30729	0.31626	0.32533	0.33450
$i=90$	0.34376	0.35311	0.36253	0.37204	0.38161	0.39126	0.40096	0.41072	0.42054	0.43040
$i=100$	0.44030	0.45023	0.46020	0.47019	0.48020	0.49022	0.50025	0.51028	0.52031	0.53033
$i=110$	0.54033	0.55031	0.56026	0.57019	0.58007	0.58991	0.59970	0.60944	0.61912	0.62873
$i=120$	0.63827	0.64774	0.65713	0.66643	0.67564	0.68476	0.69377	0.70269	0.71150	0.72019
$i=130$	0.72877	0.73723	0.74557	0.75378	0.76186	0.76981	0.77762	0.78530	0.79283	0.80022
$i=140$	0.80747	0.81457	0.82151	0.82831	0.83496	0.84145	0.84779	0.85398	0.86001	0.86588
$i=150$	0.87160	0.87716	0.88257	0.88782	0.89291	0.89785	0.90264	0.90728	0.91176	0.91610
$i=160$	0.92028	0.92432	0.92822	0.93197	0.93558	0.93906	0.94240	0.94560	0.94867	0.95162
$i=170$	0.95444	0.95713	0.95971	0.96217	0.96451	0.96674	0.96887	0.97089	0.97281	0.97463
$i=180$	0.97635	0.97799	0.97953	0.98099	0.98236	0.98366	0.98488	0.98602	0.98710	0.98811
$i=190$	0.98905	0.98994	0.99076	0.99153	0.99225	0.99291	0.99353	0.99411	0.99464	0.99513
$i=200$	0.99558	0.99600	0.99639	0.99674	0.99706	0.99736	0.99763	0.99788	0.99811	0.99831
$i=210$	0.99850	0.99867	0.99882	0.99895	0.99908	0.99919	0.99929	0.99938	0.99946	0.99953
$i=220$	0.99959	0.99965	0.99969	0.99974	0.99978	0.99981	0.99984	0.99986	0.99988	0.99990
$i=230$	0.99992	0.99993	0.99994	0.99995	0.99996	0.99997	0.99998	0.99998	0.99998	0.99999
$i=240$	0.99999	0.99999	0.99999	1.00000	1.00000	1.00000	1.00000	1.00000	1.00000	1.00000
$i=250$	1.00000	1.00000	1.00000	1.00000	1.00000	1.00000				

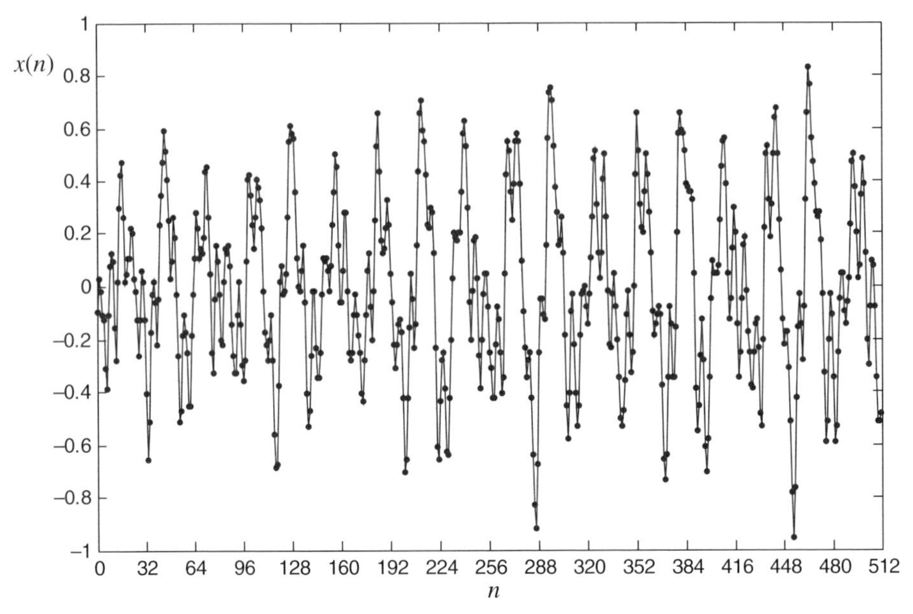

Figure 10.4 An example of a set of 512 input samples of an audio waveform.

Each frequency coefficient is then converted to an exponent and a mantissa. The exponent indicates the number of leading zeros in the binary integer representation of the coefficient. The exponents are integer values ranging from 0 to 24 and are fixed at 24 if the coefficient has more leading zeros. For example, a coefficient with a value of 0.1 would be represented as 0.8×2^{-3}, where 0.8 is the mantissa value and

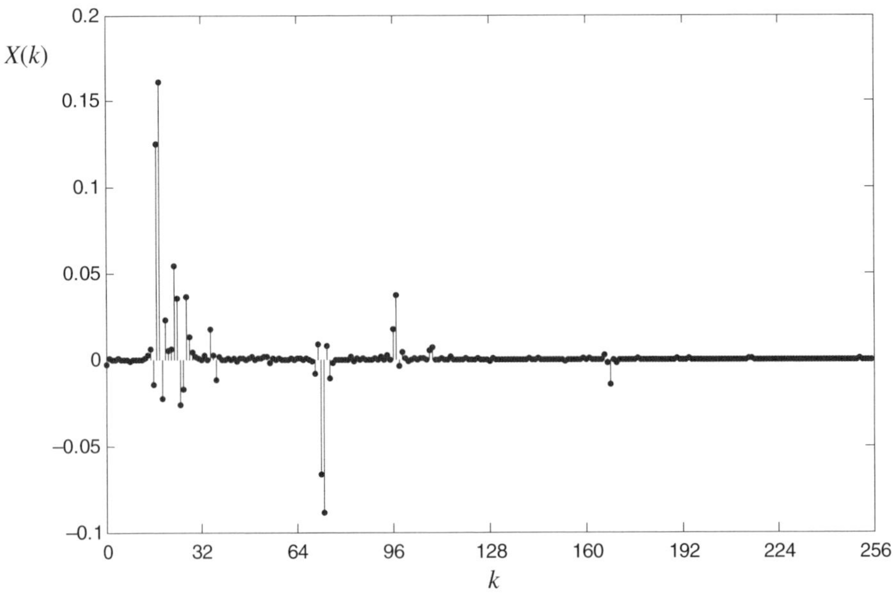

Figure 10.5 Output coefficients of the forward transform for the input samples shown in Figure 10.4.

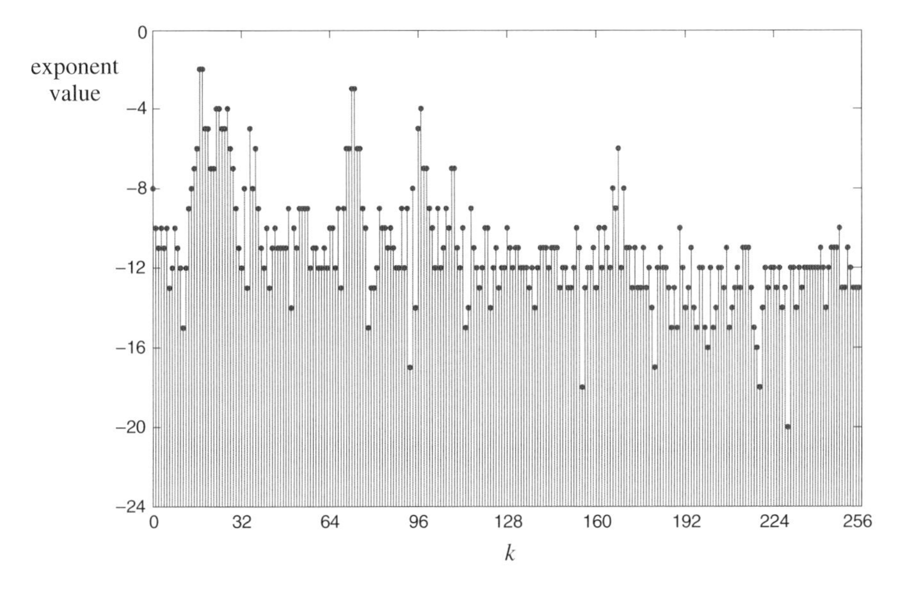

Figure 10.6 Exponent values for the output coefficients shown in Figure 10.5.

3 is the exponent value. Figure 10.6 shows the exponent values for the output coefficients shown in Figure 10.5. Note that the exponent values are always negative, so only the magnitude is required to be transmitted to the decoder. The method for coding the exponent and mantissa values is discussed in Sections 10.1.7 and 10.1.9.

10.1.4. Channel Coupling

For multichannel audio signals, there is often a significant amount of correlation between the different channels in the audio signal. *Channel coupling* is a means of exploiting this correlation to reduce the output bit rate of the AC-3 coder. The basis of the technique is to combine the high-frequency portion of several channels to form a single *coupling channel*. The coefficients of this coupling channel are transmitted as a set of mantissas and exponents in the same manner as the uncoupled coefficients of each channel. Then, in order to reconstruct an approximation of the original channels at the decoder, the ratio of the magnitude of the original coefficients in each channel to the coefficients in the coupling channel is also calculated. This ratio is called a *coupling coordinate* and is not calculated for each coefficient but is instead calculated for groups of coefficients called *coupling bands*.

The justification for reducing the bit rate in this way stems from the way the human ear perceives the directionality of an audio source. At frequencies above about 2 kHz, the ear determines the direction of a source using the level difference and interaural time delay of the envelope of the signal. So, for multichannel signals, as long as approximately the same signal envelope for the high-frequency components is maintained, the ear still perceives the correct directional information from the decoded sound source.

Table 10.5 Coupling subbands.

Subband No.	Start coefficient	End coefficient
0	37	48
1	49	60
2	61	72
3	73	84
4	85	96
5	97	108
6	109	120
7	121	132
8	133	144
9	145	156
10	157	168
11	169	180
12	181	192
13	193	204
14	205	216
15	217	228
16	229	240
17	241	252

Coupling may be performed on frequency coefficients 37 to 252 only. These coefficients are grouped into 18 *coupling subbands*, with 12 coefficients in each subband. The coefficient numbers corresponding to the beginning and end of each coupling subband are given in Table 10.5.

The parameters **cplbegf** and **cplendf** indicate the first and last coupling subbands, respectively, that are included in the coupling process. The parameter **cplendf** is transmitted as a 4-bit unsigned integer and, hence, has a maximum value of 15, so the decoder adds 2 to the value of **cplendf** to obtain the last coupling band used.

The coupling subbands may be combined to form coupling bands (coupling coordinates are transmitted for each coupling band in each channel included in the coupling process). The parameter **cplbndstrc[sbnd]** is used to indicate which subbands have been combined and which subbands are treated independently. **cplbndstrc[sbnd]** is treated as an array of single bit elements with one element for each subband included in the coupling process, except for the first subband. No element is transmitted in **cplbndstrc[sbnd]** for the first subband included in the coupling process. A value of 1 for a bit in **cplbndstrc[sbnd]** indicates that the corresponding subband should be combined with the subband immediately below it in frequency, and a value of 0 indicates that the subband should start a new coupling band. For example, a basic encoder coupling strategy would produce the following coupling parameters:

cplbegf = 6
cplendf = 12
cplbndstrc = 00110111

These parameters indicate that the coupling is performed using subbands 6 to 14 and that four coupling bands are formed from subbands 6, 7, 8–10, and 11–14.

The coefficients of the coupling channel are calculated by averaging the coefficients of the individual channels included in the coupling process. Only those coefficients that correspond to the coupling bands are included in this process. However, simply averaging the coefficients may lead to an inaccurate estimate of the combined signal power when the coefficients to be combined represent signals that are close to 180° out-of-phase with each other. If this is the case, then the coefficients in the out-of-phase bands may be negated before the coupling channel is determined. This does not usually result in any perceived difference in the directionality of the sound source since the human auditory system does not use the interaural phase shift to determine directionality at these frequencies.

For the special case of 2/0 mode (left and right side channels only) it is possible to restore the relative phase of the two channels at the decoder. This is achieved by sending phase restoration information to the decoder using the parameter **phsflg[bnd]**. This parameter is treated as an array of single bit elements with one element for each coupling band. A value of 1 for a bit in **phsflg[bnd]** indicates that the coefficients of the right channel in the corresponding coupling band have been negated before the coupling channel was formed.

EXAMPLE 10.1

Figure 10.7 shows an example of the left and right audio signals for a stereo pair. Note that the amplitude of the high-frequency components is much larger in the right-hand signal than in the left.

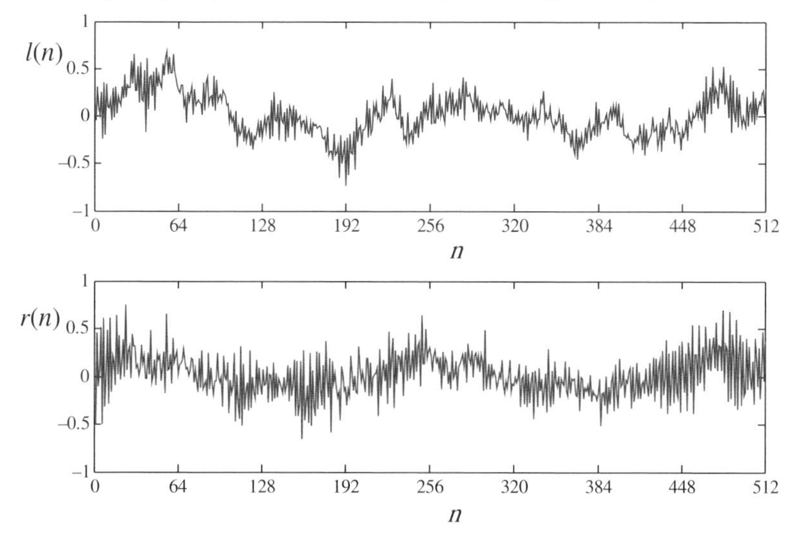

Figure 10.7 Audio signals for the left and right side channels of a stereo pair.

If coupling is to be performed on these two channels using subbands 6 to 14, Figure 10.8 shows the frequency coefficients that would be used for the two channels.

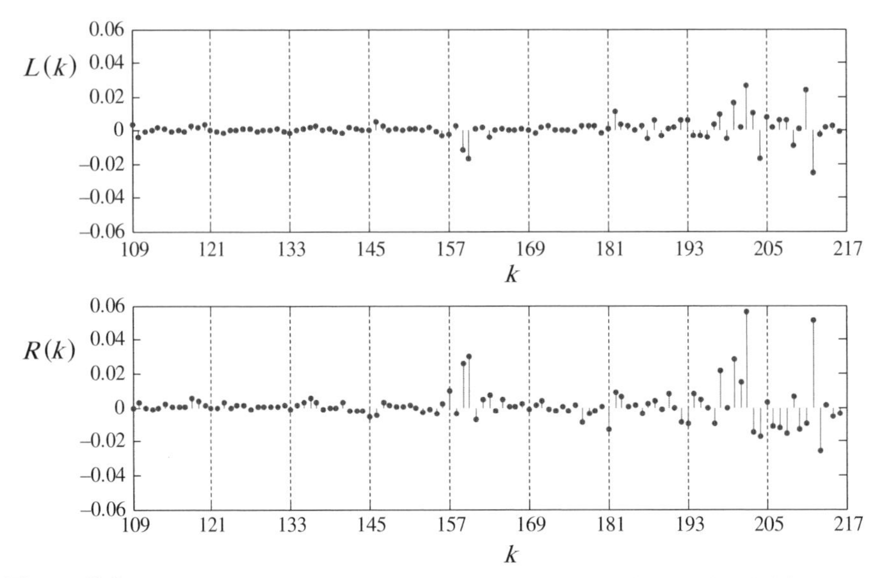

Figure 10.8 Frequency coefficients corresponding to subbands 6 to 14 for the left and right channels shown in Figure 10.7.

The next step in the coupling process is to adjust the phase of the right channel to avoid phase cancellation and combine subbands into bands. One suitable coupling strategy for these channels would be to form six coupling bands from subbands 6, 7, 8, 9–11, 12–13 and 14. For this strategy, the coupling parameters to be transmitted to the decoder are

cplbegf = 6
cplendf = 12
cplbndstrc = 00011010
phsflg = 010101

Now the coupling channel can be formed by averaging the coefficients from the left channel and the phase-adjusted right channel. The resulting coupling channel coefficients are shown in Figure 10.9.

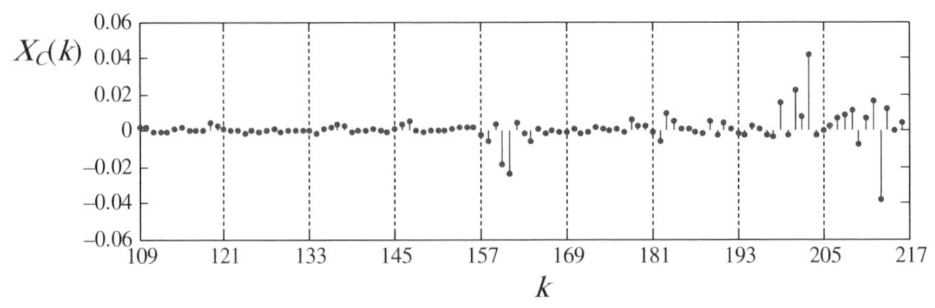

Figure 10.9 Coupling channel coefficients. ■

The next step in the coupling process is to calculate coupling coordinates for each band in each coupled channel. These coupling coordinates are determined by dividing the average magnitude of the original coefficients in a coupled band by the average magnitude of the coefficients in the same band of the coupling channel. The coupling coordinates for each band are transmitted using a 4-bit mantissa **cplcomant[ch][bnd]** and a 4-bit exponent **cplcoexp[ch][bnd]**.

To guarantee that the coupling coordinate values used to generate the mantissa and exponent are always in the range 0.0 to 1.0, coupling coordinate values should be limited to the range 0.0 to 8.0. Then, before converting the coordinate values to a floating-point representation, they are scaled to the range 0.0 to 1.0 by dividing by a factor of 8.0.

The coupling exponents indicate the number of leading zeros in the binary representation of the fractional coupling coordinates. Except for the case when the exponent value is 15, the mantissa values are always positive values in the range 0.5–1.0. Hence when the exponent value is less than 15, the most significant bit of the mantissa is always 1 and is not transmitted. So when the exponent value is less than 15, the 4-bit value of **cplcomant[ch][bnd]** is generated by multiplying the mantissa value by 32 and then subtracting 16 from the result. If the exponent value is equal to 15, **cplcomant[ch][bnd]** is calculated by multiplying the mantissa value by 16.

The dynamic range of the coupling coordinates can be increased by transmitting a 2-bit master coupling coordinate **mstrcplco[ch]** for any of the coupled channels. The exponent values for all coupling coordinates in the corresponding channel are increased three times the value of **mstrcplco[ch]**.

EXAMPLE 10.2

Figure 10.10 shows the frequency coefficient values and the average coefficient magnitudes for each coupling band of the left and right side channels shown in Figure 10.7, and Figure 10.11

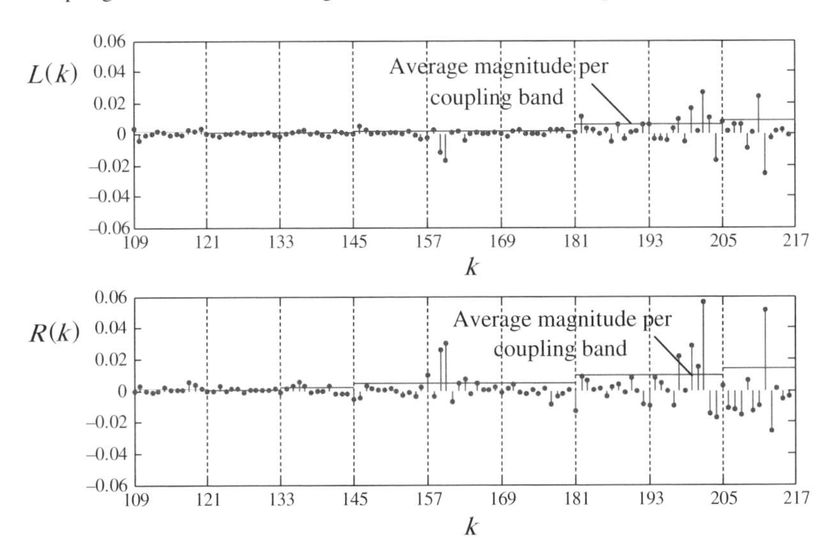

Figure 10.10 Frequency coefficient values and the average coefficient magnitudes for each coupling band of the left and right side channels shown in Figure 10.7.

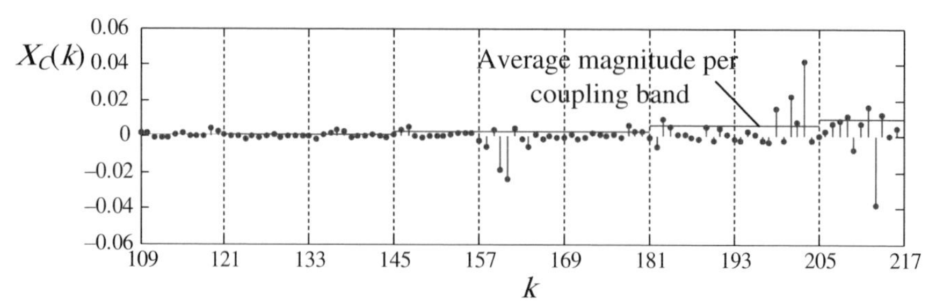

Figure 10.11 Frequency coefficient values and the average coefficient magnitudes for each band of the coupling channel.

shows the average coefficient magnitudes for each band of the coupling channel generated in the previous example.

Now, using these average magnitude values, the coupling coordinates for the six bands in each channel are generated. The values of the average coefficient magnitudes for each band are shown in Table 10.6 along with the corresponding coupling coordinates for each band of the left and right side channels.

Table 10.6 Average coefficient magnitudes and the corresponding coupling coordinates for each coupling band.

	Average magnitude ($\times 10^{-3}$)			Coupling coordinates	
Coupling band	Left	Right	Coupling	Left	Right
0	1.68	1.60	1.28	1.3131	1.2522
1	0.80	0.82	0.54	1.4803	1.5296
2	1.02	2.00	1.09	0.9353	1.8273
3	2.08	4.36	2.89	0.7216	1.5112
4	5.58	9.44	5.94	0.9393	1.5908
5	8.65	14.34	9.52	0.9078	1.5056

Then, after converting these coupling coordinates into a floating point representation, the values to be transmitted for **cplcomant[ch][bnd]** and **cplcoexp[ch][bnd]** are shown in

Table 10.7 Transmitted values for **cplcomant[ch][bnd]** and **cplcoexp[ch][bnd]** and the corresponding decoded coupling coordinates.

cplcomant		**cplcoexp**		Decoded coordinates	
Left	Right	Left	Right	Left	Right
5	4	2	2	1.3125	1.25
7	8	2	2	1.4375	1.5
13	13	3	2	0.90625	1.8125
7	8	3	2	0.71875	1.5
14	9	3	2	0.9375	1.5625
13	8	3	2	0.90625	1.5

Table 10.7 along with corresponding values for the coupling coordinates that would be generated at the decoder.

At the decoder, the values of each element of **phsflg[bnd]** and the decoded coupling coordinates would be used to reconstruct the high-frequency portions of the original left and right side channels. Figure 10.12 shows the original and reconstructed frequency coefficients that would be obtained for the left and right side channels shown in Figure 10.7. Figure 10.13 shows the reconstructed audio signals obtained by performing the inverse TDAC transform on the original low-frequency coefficients and the reconstructed high-frequency coefficients. Figures 10.7 and 10.13 show that the coupling process has maintained the relationship between the high-frequency envelopes of the two channels.

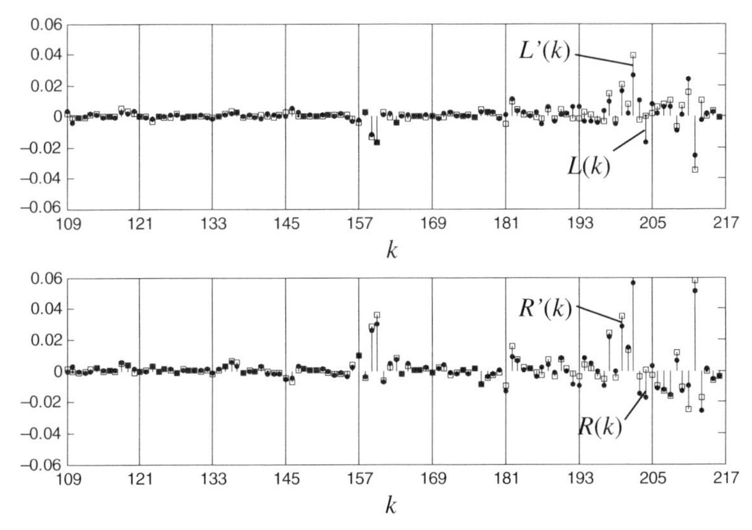

Figure 10.12 Original and reconstructed frequency coefficients for the left and right side channels shown in Figure 10.7.

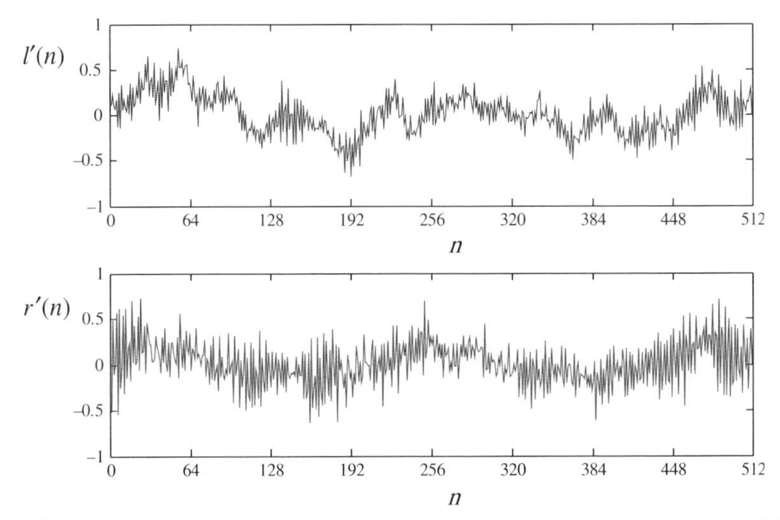

Figure 10.13 Reconstructed audio signals for the left and right side channels shown in Figure 10.7. ■

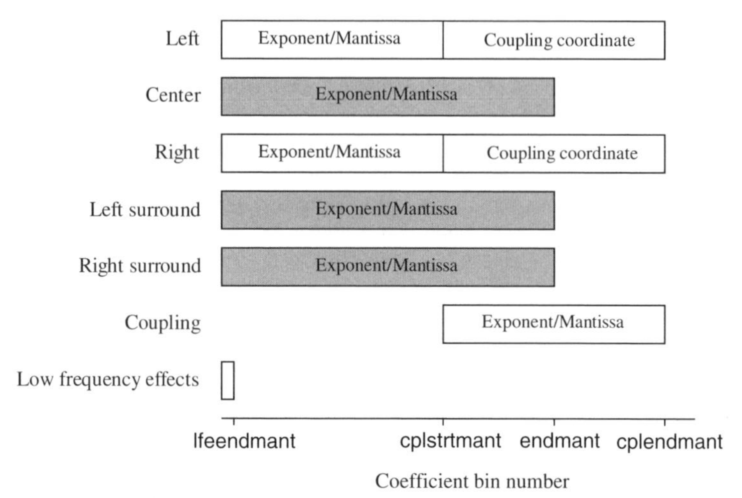

Figure 10.14 Structure of the channels to be coded when coupling is in use.

So, after coupling has been performed, there are a number of different types of channels that are transmitted to the receiver. Consider the example shown in Figure 10.14 where only the left and right side channels are included in the coupling process. The coefficients of the *coupled* channels (the left and right side channel in this example) are coded individually as exponent and mantissa values up to the coefficient bin number indicated by the parameter cplstrtmant. Then above this bin number, the coefficients are coded using coupling parameters up to the bin number indicated by cplendmant. The coefficients of the *coupling* channel are only required for bin numbers between cplstrtmant and cplendmant and are coded individually as exponent and mantissa values. The remaining full bandwidth channels that are not included in the coupling process are referred to as *independent* channels, and these are shown as shaded parts in Figure 10.14. The coefficients of the independent channels are coded individually as exponent and mantissa values up to the bin number indicated by fbwendmant[ch]. Note that the value of fbwendmant[ch] may be different for each independent channel. Finally the coefficients in the low-frequency effects channel are coded individually as exponent and mantissa values up to the bin number indicated by lfeendmant. The starting bin numbers for the full bandwidth and LFE channels are indicated by fbwstrtmant[ch] and lfestrtmant, respectively.

10.1.5. Rematrixing

Rematrixing is an additional channel combination technique that is employed in the 2/0 audio coding mode. When rematrixing is in use, the coder is able to transmit a $L+R$ signal and a $L-R$ signal instead of the original L and R stereo signals. If coupling is also in use, rematrixing is only performed on those frequencies below the coupling frequency.

The use of rematrixing allows a significant reduction in the bit rate when the original left and right side signals are highly correlated. Rematrixing is also important for preserving compatibility with Dolby surround algorithms. Consider the case where the input left and right side signals are transmitted without rematrixing. The received left and right side signals, denoted by L_r and R_r, are

$$\begin{aligned} L_r &= L + N_1 \\ R_r &= R + N_2 \end{aligned} \tag{10.3}$$

where N_1 and N_2 denote the uncorrelated quantization noise associated with the encoding process. The Dolby Pro Logic decoder takes these received signals and constructs a center and surround channel, denoted by C and S, respectively, which are as follows:

$$\begin{aligned} C &= 0.5\left(L + R + \left(N_1 + N_2\right)\right) \\ S &= 0.5\left(L - R + \left(N_1 - N_2\right)\right) \end{aligned} \tag{10.4}$$

So, for the decoded center channel, the noise signals add, but the resulting noise remains masked by $L + R$ since this signal is of similar magnitude to the original left and right side signals. However, for the decoded surround channel, because the noise signals are uncorrelated they also add, but in this case the $L - R$ signal is much smaller in magnitude than the original left and right side signals and may not mask the decoded noise signal.

Now consider the case when rematrixing is in use. The received left and right side signals are

$$\begin{aligned} L_r &= 0.5\left(L + R\right) + N_1 \\ R_r &= 0.5\left(L - R\right) + N_2 \end{aligned} \tag{10.5}$$

Note that, in this case, the encoding process attempts to mask all quantization noise in the $L - R$ signal, and hence N_2 is typically much smaller in magnitude than N_1. The Dolby Pro Logic decoder takes these received signals and constructs the center and surround channel as follows:

$$\begin{aligned} C &= L_r \\ S &= R_r \end{aligned} \tag{10.6}$$

Hence the quantization noise remains masked for the decoded center and surround signals. The decoded left and right side signals, denoted by L_d and R_d, are

$$\begin{aligned} L_d &= L_r + R_r = L + \left(N_1 + N_2\right) \\ R_d &= L_r - R_r = R + \left(N_1 - N_2\right) \end{aligned} \tag{10.7}$$

So for both these signals the uncorrelated noise signals add but are dominated by the much larger N_1. Hence most, if not all, of the decoded noise remains masked by the L and R signals, which are of similar magnitude to the $0.5(L + R)$ signal that was used in the encoding process.

10.1.5.1. Rematrixing Frequency Bands

The rematrixing process is conducted using independent frequency bands. The boundary locations for these bands are defined using the coefficient numbers and depend on the number of coefficients that are included in the coupling process. Tables 10.8 to 10.11 show the four rematrixing band boundaries to be used for different values of **cplbegf**.

Table 10.8 Rematrixing bands, coupling not in use.

Band No.	Start Coeff.	End Coeff.
0	13	24
1	25	36
2	37	60
3	61	252

Table 10.9 Rematrixing bands, **cplbegf** > 2.

Band No.	Start Coeff.	End Coeff.
0	13	24
1	25	36
2	37	60
3	61	36 + **cplbegf** × 12

Table 10.10 Rematrixing bands, $2 \geq$ **cplbegf** > 0.

Band No.	Start Coeff.	End Coeff.
0	13	24
1	25	36
2	37	36 + **cplbegf** × 12

Table 10.11 Rematrixing bands, **cplbegf** = 0.

Band No.	Start Coeff.	End Coeff.
0	13	24
1	25	36

10.1.5.2. Encoding Technique

In the 2/0 audio coding mode, rematrixing is always in use. However, for each rematrixing band it is still possible to transmit the original L and R channels rather than rematrixing them into sum and difference channels. The decision to rematrix or not is made using the sum of the squares of the coefficients in the band. This sum is calculated for the L and R channels and for the $L+R$ and $L-R$ combinations. If the minimum of these four sums for a rematrixing band corresponds to the L or R channels, then L and R channels are transmitted for that band. Alternatively, if the minimum sum corresponds to the $L+R$ or the $L-R$ combination, then the rematrixed sum and difference channels $0.5(L+R)$ and $0.5(L-R)$ are transmitted.

The type of channel to be transmitted for each band is indicated by the parameter **rematflg[rbnd]**. This parameter is treated as an array of single bit elements with one element for each rematrixing band. A value of 1 for a bit in **rematflg[rbnd]** indicates that the coefficients in the band have been rematrixed into sum and difference channels.

10.1.6. Extract Exponents

After the coupling (and rematrixing if in 2/0 mode) strategies have been determined, the coefficients that are to be transmitted to the decoder are known. These coefficients are transmitted in floating point form as a mantissa and an exponent value.

The exponent information is coded differentially across the transmitted frequency range for the channel. The first exponent of the channel is sent as a 4-bit absolute value with a range of 0–15. Then successive exponent values are transmitted as differential values. These differential values are referred to as *differential exponents* and are limited to one of the following five values: −2, −1, 0, 1, or 2. The decoded differential exponents can be used to form an approximation of the power spectral density of the audio signal. This approximation is commonly referred to as the *coded spectral envelope* of the signal.

The differential exponents can be calculated using three different methods that are referred to as *exponent strategies*. The difference between the three methods is simply the number of original exponents corresponding to each differential exponent. The first method is called the D-15 exponent strategy, and in this method a differential exponent is transmitted for every original exponent value. The D-15 strategy results in the highest resolution spectral envelope and also requires the most number of bits to transmit. For this reason the D-15 strategy is typically used for signals with a spectral envelope that remains relatively stable over time. For this type of signal, the coded spectral envelope is usually transmitted once at the beginning of each audio frame and is reused for each of the six audio blocks in the frame.

EXAMPLE 10.3

Figure 10.15 shows the original exponent values for the first 36 coefficients of Figure 10.6 together with the differentially coded spectral envelope for the signal when the D-15 exponent strategy is in use. Figure 10.16 shows the differential exponents that would be transmitted to the decoder for the spectral envelope shown in Figure 10.15.

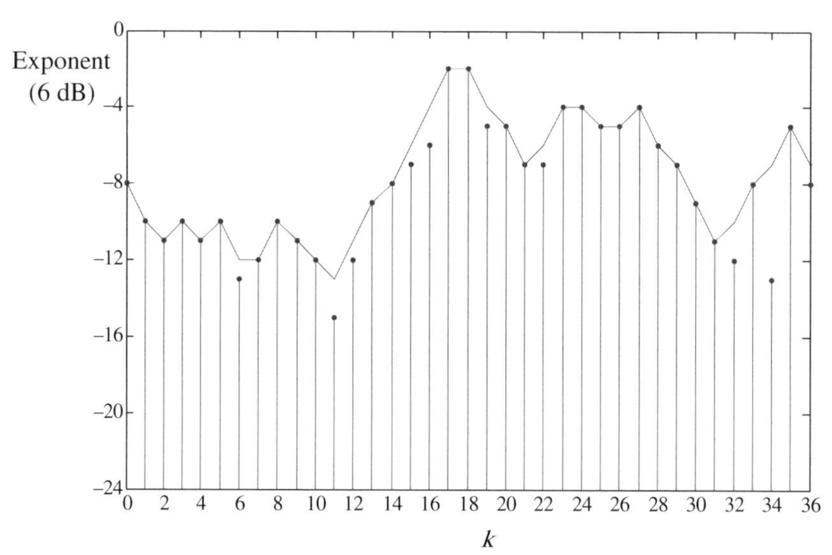

Figure 10.15 Original exponent values and the corresponding differentially coded spectral envelope for the first 36 coefficients shown in Figure 10.6.

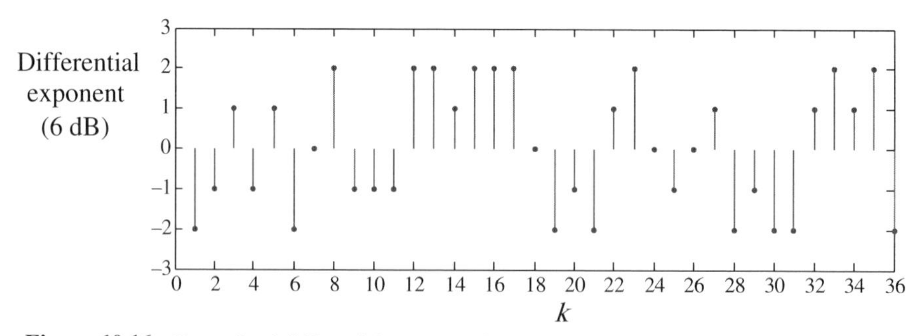

Figure 10.16 Transmitted differential exponents for the spectral envelope shown in Figure 10.15.

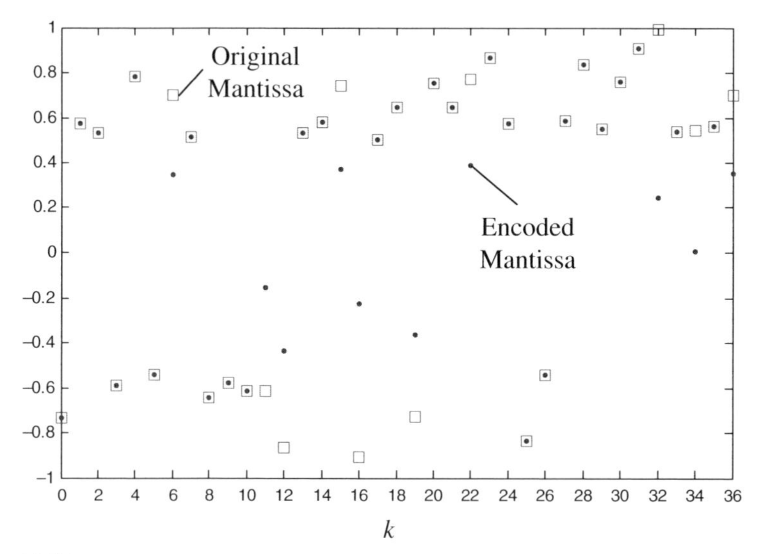

Figure 10.17 Original mantissa values and the mantissa values to be quantized and transmitted for the first 36 coefficients of Figure 10.5.

Figure 10.15 shows that the limited range of possible differential exponent values means that the coded spectral envelope may not always exactly match the original exponent values. However, if the coded spectral envelope is always greater than or equal to the actual exponent values, then the discrepancy simply means that the mantissa value to be quantized and transmitted is smaller than the original mantissa value. To illustrate this point, Figure 10.17 shows the original mantissa values for the first 36 coefficients of Figure 10.5 together with the mantissa values obtained by dividing the same 36 coefficients by the coded spectral envelope shown in Figure 10.15. ■

The second method is called the D-25 exponent strategy, and in this method a differential exponent is transmitted for every two original exponent values. The D-25 strategy results in a medium-resolution spectral envelope and is typically used for signals with a spectral envelope that remains relatively stable for two to three audio blocks. Hence, if the D-25 exponent strategy is in use, the coded spectral envelope is usually transmitted twice for each audio frame and the same spectral envelope is used for three consecutive audio blocks in the frame.

The final method is called the D-45 exponent strategy, and in this method a differential exponent is transmitted for every four original exponent values. The D-45 strategy results in a low-resolution spectral envelope and is typically used for transient signals where the spectral envelope changes for each block but is usually relatively flat in nature. Consequently, a coded spectral envelope is usually transmitted for every audio block if the D-45 strategy is in use.

EXAMPLE 10.4

Figure 10.19 shows the coded spectral envelope produced using the D-45 exponent strategy for the coefficients of the transient signal shown in Figure 10.18.

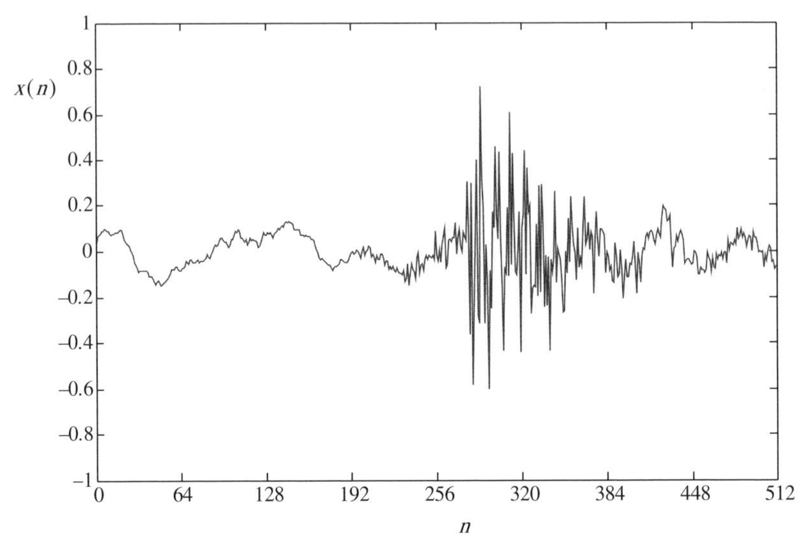

Figure 10.18 Transient audio signal produced by a castanet "click."

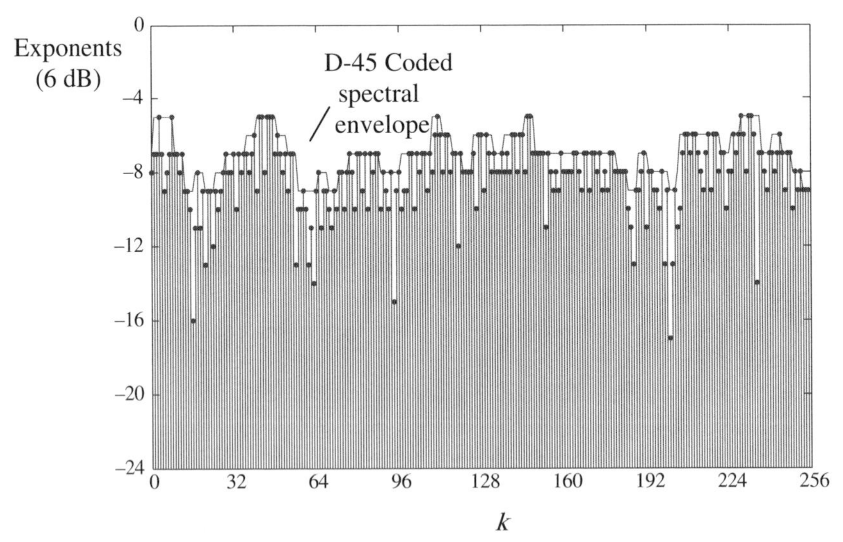

Figure 10.19 Original exponent values and the D-45 coded spectral envelope for the coefficients of the transient signal shown in Figure 10.18. ■

Table 10.12 Exponent strategy code words.

chexpstr[ch], cplexpstr	Exponent strategy
"00"	Reuse prior exponents
"01"	D15
"10"	D25
"11"	D45

The exponent strategy in use for a full bandwidth channel is transmitted to the decoder using a 2-bit code word for each channel designated by **chexpstr[ch]**. If the coupling channel is present, the 2-bit code word **cplexpstr** is used. The exponent strategy corresponding to each value of these code words is shown in Table 10.12 for full bandwidth and coupling channels.

If the low-frequency effects channel is enabled, the exponent strategy is transmitted to the decoder using a single bit code word designated by **lfeexpstr**. A value of 1 for **lfeexpstr** indicates that a new spectral envelope coded using the D-15 exponent strategy will be transmitted for the current audio block, otherwise the spectral envelope from the previous block should be reused.

If a full bandwidth channel is not included in the coupling process, it is often not necessary to transmit the entire set of coefficients for the channel. In such cases, the last coefficient to be transmitted is indicated by a *channel bandwidth code*, designated by **chbwcod[ch]**. The channel bandwidth code is transmitted to the decoder as a 6-bit unsigned integer and is calculated from the end mantissa bin number, endmant[ch], using the following formula:

$$\mathbf{chbwcod[ch]} = ((\text{endmant[ch]} - 37)/3) + 12 \tag{10.8}$$

Valid values for **chbwcod[ch]** lie in the range 0–60.

10.1.7. Encode Exponents

The spectral envelope for a channel is transmitted using a 4-bit absolute value for the first exponent of the channel followed by a set of differential exponents. For full bandwidth and low-frequency effects channels, the initial absolute value is limited to the range 0 to 15, and exponent values larger than 15 are truncated. For the coupling channel, the absolute exponent is limited to even values and is divided by a factor of 2 before transmission. The initial absolute exponent value is transmitted to the decoder using a 4-bit unsigned integer designated by **cplabsexp, exps[ch][0]**, and **lfeexps[0]** for the coupling, full bandwidth, and low-frequency effects channels, respectively.

Each differential exponent is first mapped to a positive integer, called a *mapped value*, by adding a factor of 2. Then each set of three adjacent mapped values is grouped into a 7-bit *grouped value* using the formula:

$$G = 25M_1 + 5M_2 + M_3 \quad (10.9)$$

where G denotes the 7-bit grouped value, and M_1, M_2, and M_3 are three adjacent mapped values. The grouped values are transmitted to the decoder using 7-bit unsigned integers designated by **cplexps[grp]**, **exps[ch][grp]**, and **lfeexps[grp]** for the coupling, full bandwidth, and low frequency effects channels, respectively.

10.1.8. Bit Allocation

The bit-allocation strategy used in the AC-3 algorithm consists of two separate procedures; a standard *core* bit-allocation procedure and an optional *delta* bit-allocation procedure. The core bit-allocation algorithm uses the coded spectral envelope as an approximation to the power spectral density of the transmitted signal. A standard psychoacoustic model is then used to determine a masking threshold for the signal. This threshold can be used to determine the number of bits to be allocated to each mantissa for a given level of perceived distortion at the receiver. The psychoacoustic model used in the core bit-allocation procedure is defined as part of the standard. The approach used is to iteratively decrease the level of noise introduced by the coding process until the required bit rate is reached and then transmit the final setup of the model to the decoder via a number of parameters in the bit stream. The decoder can then reproduce the identical bit-allocation values for each coefficient and correctly decode the transmitted coefficients.

The advantage of this approach is that it is not necessary to transmit the actual bit allocation for each coefficient in the bit stream. The decoder has the coded spectral envelope and the parameters of the psychoacoustic model that were used in the encoding process, so it can repeat the calculations and arrive at identical bit-allocation values. The disadvantage of this approach is that the psychoacoustic model is fixed and cannot be improved or changed for different input signals. To overcome this problem, the syntax allows for delta bit-allocation parameters to be transmitted so that the bit-allocation values obtained by the standard psychoacoustic model can be adjusted to match the output of an alternative model. These adjustments should only be minor because the standard psychoacoustic model usually provides a good approximation to the required acoustic model.

10.1.8.1. Initialization

The first step in the core bit-allocation process is to initialize the strtmant and endmant parameters for each of the channels. These parameters are determined using **chincpl[ch]** for each full bandwidth channel, **cplbegf** and **cplendf** for the coupled and coupling channels, and **chbwcod[ch]** for the independent channels. The following MATLAB code defines the algorithm used to initialize these parameters. The value of the parameter nfchans is equal to the number of full bandwidth channels and is determined from **acmod** as shown in Table 10.2.

```
for ch = 0:nfchans-1
  fbwstrtmant(ch+1) = 0;
  if chincpl(ch+1)
    fbwendmant(ch+1) = 37 + (12*cplbegf);
  else
    fbwendmant(ch+1) = 37 + (3*(chbwcod(ch+1) + 12));
  end
end

cplstrtmant = 37 + (12*cplbegf);
cplendmant = 37 + (12*(cplendf + 3));

lfestrtmant = 0;
lfeendmant = 7;
```

Then, in the following procedures, the parameters strtmant and endmant are equal to fbwstrtmant[ch] and fbwendmant[ch] for the full bandwidth channels, cplstrtmant and cplendmant for the coupling channel, and lfestrtmant and lfeendmant for the LFE channel.

10.1.8.2. Exponent to PSD Conversion

The next step in the core bit-allocation procedure is to convert the coded spectral envelope to a power spectral density function. The conversion process consists of simply shifting the range of coded spectral envelope values from $[-24, 0]$ to $[0, 3027]$. The algorithm used is defined by the following MATLAB code:

```
for bin = strtmant:endmant
      psd(bin+1) = 3072 - exponent(bin+1)*128;
end
```

10.1.8.3. PSD Integration

The next step in the process is to convert the domain of the PSD values from a fine-resolution linear frequency scale into a scale that approximates the Bark scale. To perform this conversion, the frequency scale is subdivided into 50 nonuniform frequency *bands*. The power spectral density value for each band, designated by bndpsd[band], is found by calculating the integral of the PSD values in the band. The width of the bands is nonuniform and is derived from the critical bandwidths at the corresponding frequency. The starting bin number for each band is defined using the array bndtab[band], and the width of each band is defined using the array bndsz[band]. The values contained in these two arrays are shown in Table 10.13.

The conversion from a linear frequency scale into bands corresponding to the Bark scale requires the integration of the linear values corresponding to the logarithmic PSD values. So a direct computation of bndpsd[band] would require the

Table 10.13 Bark scale banding structure.

band	bndtab[band]	bndsz[band]	band	bndtab[band]	bndsz[band]
0	0	1	25	25	1
1	1	1	26	26	1
2	2	1	27	27	1
3	3	1	28	28	3
4	4	1	29	31	3
5	5	1	30	34	3
6	6	1	31	37	3
7	7	1	32	40	3
8	8	1	33	43	3
9	9	1	34	46	3
10	10	1	35	49	6
11	11	1	36	55	6
12	12	1	37	61	6
13	13	1	38	67	6
14	14	1	39	73	6
15	15	1	40	79	6
16	16	1	41	85	12
17	17	1	42	97	12
18	18	1	43	109	12
19	19	1	44	121	12
20	20	1	45	133	24
21	21	1	46	157	24
22	22	1	47	181	24
23	23	1	48	205	24
24	24	1	49	229	24

conversion of the logarithmic PSD values into linear values, then a summation, and finally a conversion back to a logarithmic scale for the banded PSD values. However, because the logarithmic PSD values are available, this process of *log addition* can be performed more efficiently if the following relationship is used:

$$\log_2(a+b) = \max(\log_2(a), \log_2(b)) + \log_2\left(1 + 2^{-|\log_2(a) - \log_2(b)|}\right) \tag{10.10}$$

Equation (10.10) shows that if a and b are dissimilar and hence $|\log(a) - \log(b)|$ is large, then the second term on the right-hand side becomes insignificant. Consequently, $\log_2(a+b)$ is dominated by the larger values of a and b and can be approximated by $\max(\log_2(a), \log_2(b))$. Alternatively, if a and b are similar then the second term on the right-hand side is significant and provides the required adjustment to max($\log_2(a)$,$\log_2(b)$). The maximum adjustment is required when a and b are equal, in which case the second term becomes $\log_2(1 + 1) = 1$.

In the AC-3 encoder, the log addition is performed using a maximum operation for the first term on the right-hand side of Equation (10.10) and a subtraction, absolute value, and table lookup for the second term. The output of the table lookup operation is the integer part of 64 $\log_2(1 + 2^{-d/32})$, if d is the index to the table. So the algorithm used to calculate the banded PSD values is defined by the following MATLAB code. The values in the lookup table, latab[], used in the log-addition process are shown in Table 10.14, and the table used to convert bin numbers to band numbers, masktab[ ], is shown in Table 10.15.

Table 10.14 Log addition lookup table values, latab[i+j].

	$j=0$	$j=1$	$j=2$	$j=3$	$j=4$	$j=5$	$j=6$	$j=7$	$j=8$	$j=9$
$i=0$	64	63	62	61	60	59	58	57	56	55
$i=10$	54	53	52	52	51	50	49	48	47	47
$i=20$	46	45	44	44	43	42	41	41	40	39
$i=30$	38	38	37	36	36	35	35	34	33	33
$i=40$	32	32	31	30	30	29	29	28	28	27
$i=50$	27	26	26	25	25	24	24	23	23	22
$i=60$	22	21	21	21	20	20	19	19	19	18
$i=70$	18	18	17	17	17	16	16	16	15	15
$i=80$	15	14	14	14	13	13	13	13	12	12
$i=90$	12	12	11	11	11	11	10	10	10	10
$i=100$	10	9	9	9	9	9	8	8	8	8
$i=110$	8	8	7	7	7	7	7	7	6	6
$i=120$	6	6	6	6	6	6	5	5	5	5
$i=130$	5	5	5	5	4	4	4	4	4	4
$i=140$	4	4	4	4	4	3	3	3	3	3
$i=150$	3	3	3	3	3	3	3	3	3	2
$i=160$	2	2	2	2	2	2	2	2	2	2
$i=170$	2	2	2	2	2	2	2	2	1	1
$i=180$	1	1	1	1	1	1	1	1	1	1
$i=190$	1	1	1	1	1	1	1	1	1	1
$i=200$	1	1	1	1	1	1	1	1	1	1
$i=210$	0	0	0	0	0	0	0	0	0	0
$i=220$	0	0	0	0	0	0	0	0	0	0
$i=230$	0	0	0	0	0	0	0	0	0	0
$i=240$	0	0	0	0	0	0	0	0	0	0
$i=250$	0	0	0	0	0	0	0	0	0	0

Table 10.15 Bin number to band number conversion table, masktab[i+j].

	$j=0$	$j=1$	$j=2$	$j=3$	$j=4$	$j=5$	$j=6$	$j=7$	$j=8$	$j=9$
$i=0$	0	1	2	3	4	5	6	7	8	9
$i=10$	10	11	12	13	14	15	16	17	18	19
$i=20$	20	21	22	23	24	25	26	27	28	28
$i=30$	28	29	29	29	30	30	30	31	31	31
$i=40$	32	32	32	33	33	33	34	34	34	35
$i=50$	35	35	35	35	35	36	36	36	36	36
$i=60$	36	37	37	37	37	37	37	38	38	38
$i=70$	38	38	38	39	39	39	39	39	39	40
$i=80$	40	40	40	40	40	41	41	41	41	41
$i=90$	41	41	41	41	41	41	41	42	42	42
$i=100$	42	42	42	42	42	42	42	42	42	43
$i=110$	43	43	43	43	43	43	43	43	43	43
$i=120$	43	44	44	44	44	44	44	44	44	44
$i=130$	44	44	44	45	45	45	45	45	45	45
$i=140$	45	45	45	45	45	45	45	45	45	45
$i=150$	45	45	45	45	45	45	45	46	46	46
$i=160$	46	46	46	46	46	46	46	46	46	46
$i=170$	46	46	46	46	46	46	46	46	46	46
$i=180$	46	47	47	47	47	47	47	47	47	47
$i=190$	47	47	47	47	47	47	47	47	47	47
$i=200$	47	47	47	47	47	48	48	48	48	48
$i=210$	48	48	48	48	48	48	48	48	48	48
$i=220$	48	48	48	48	48	48	48	48	48	49
$i=230$	49	49	49	49	49	49	49	49	49	49
$i=240$	49	49	49	49	49	49	49	49	49	49
$i=250$	49	49	49	0	0	0				


```
bin = strtmant;
band = masktab(strtmant11);
lastbin = bin;
while (lastbin < endmant)

    % calculate the last coefficient bin number in the current critical
    % band
    lastbin = min(bndtab(band+1) + bndsz(band+1), endmant);

    % find the PSD values for the current band using the process of log
    % addition
    bndpsd(band+1) = psd(bin+1);
    bin = bin+1;
```

```
    for i = bin:lastbin-1
        bndpsd(band+1) = logadd(bndpsd(band+1),psd(bin+1));
        bin = bin+1;
    end
    band = band+1;
end

function x = logadd(a,b)

c = a - b;

% find the address into the look up table latab using the difference in the
% PSD values
d = min(floor(abs(c)/2),255);

if (c >= 0)
    x = a + latab(d+1);
else
    x = b + latab(d+1);
end
```

10.1.8.4. The Spreading Function

The banded PSD values are then used to estimate the masking threshold curve by convolving them with a *spreading function*. The spreading function is approximated by a fast-decaying upward masking curve and a slow-decaying upward masking curve as shown in Figure 10.20. Downward masking is ignored in this algorithm to reduce computational complexity at the expense of calculating a more conservative masking threshold.

A simplified convolution is then performed using the two masking curves and the banded PSD values. The calculations for the convolution operation are carried out using the logarithmic amplitude values available in the **bndpsd[band]** array. This simplifies the process since, in the log domain, multiplication can be

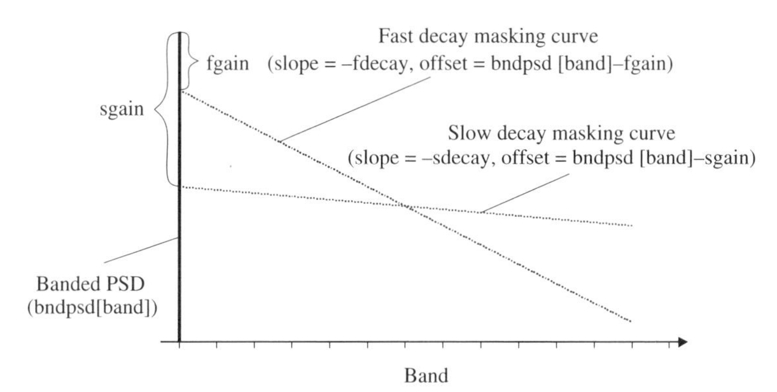

Figure 10.20 Linear masking curves used to approximate the upward masking spreading function.

performed by addition, and log additions can be approximated by a maximum value operation.

To explain this simplification further, consider the case of performing a linear convolution of a signal x_n and a filter with exponentially decaying coefficients [1,1/2,1/4,1/8].

In the linear domain, the output of this convolution is

$$y_n = \frac{x_{n-3}}{8} + \frac{x_{n-2}}{4} + \frac{x_{n-1}}{2} + x_n \tag{10.11}$$

However, if the convolution is performed in the log domain and log addition is replaced with the maximum operator, the output of the convolution is

$$\log_2(y_n) = \max\left(\log_2(x_{n-3}) - 3, \log_2(x_{n-2}) - 2, \log_2(x_{n-1}) - 1, \log_2(x_n)\right) \tag{10.12}$$

For the special case of a filter with exponentially decaying coefficients, the process can be simplified further by performing the maximum operation in a recursive manner. The output of the convolution can now be written as a function of the previous output and is

$$\log_2(y_n) = \max\left(\log_2(y_{n-1}) - 1, \log_2(x_n)\right) \tag{10.13}$$

To see why this recursive simplification is possible, consider the previous convolution output given by

$$\log_2(y_{n-1}) = \max\left(\log_2(x_{n-3}) - 2, \log_2(x_{n-2}) - 1, \log_2(x_{n-1})\right) \tag{10.14}$$

Then, if both sides of this equation are decayed, the result is

$$\begin{aligned} \log_2(y_{n-1}) - 1 &= \max\left(\log_2(x_{n-3}) - 2, \log_2(x_{n-2}) - 1, \log_2(x_{n-1})\right) - 1 \\ &= \max\left(\log_2(x_{n-3}) - 3, \log_2(x_{n-2}) - 2, \log_2(x_{n-1}) - 1\right) \end{aligned} \tag{10.15}$$

Hence, the output of the convolution process in the log domain can be obtained by simply finding the maximum of the current input and a decayed version of the previous output.

An example illustrating this simplification process is shown in Figure 10.21. Figure 10.21 shows that, when $n = 3$, the maximum of the decayed values of the previous inputs ($\log_2(x_0) - 3$) is equal to the decayed previous output value ($\log_2(y_2) - 1$). So the output of the convolution for $n = 3$ is given by the maximum of $\log_2(y_2) - 1$ and $\log_2(x_3)$.

This simplified convolution process is conducted for the two linear masking curves, and the final output is referred to as the *excitation function* for the current set

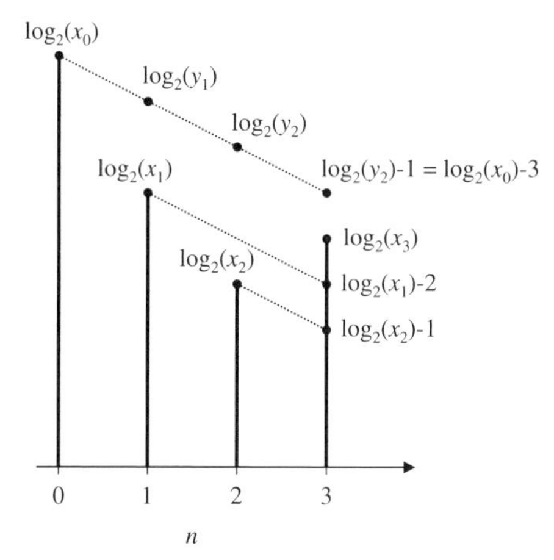

Figure 10.21 Illustration of the simplification process for performing convolution in the log domain with exponentially decaying filter coefficients.

of banded PSD values. The convolution process to determine the excitation function is illustrated in Figure 10.22. The output of the convolution of the fast-decaying masking curve with the banded PSD values is stored in the fastleak parameter as shown in Figure 10.22(a). The output of the convolution of the slow-decaying masking curve

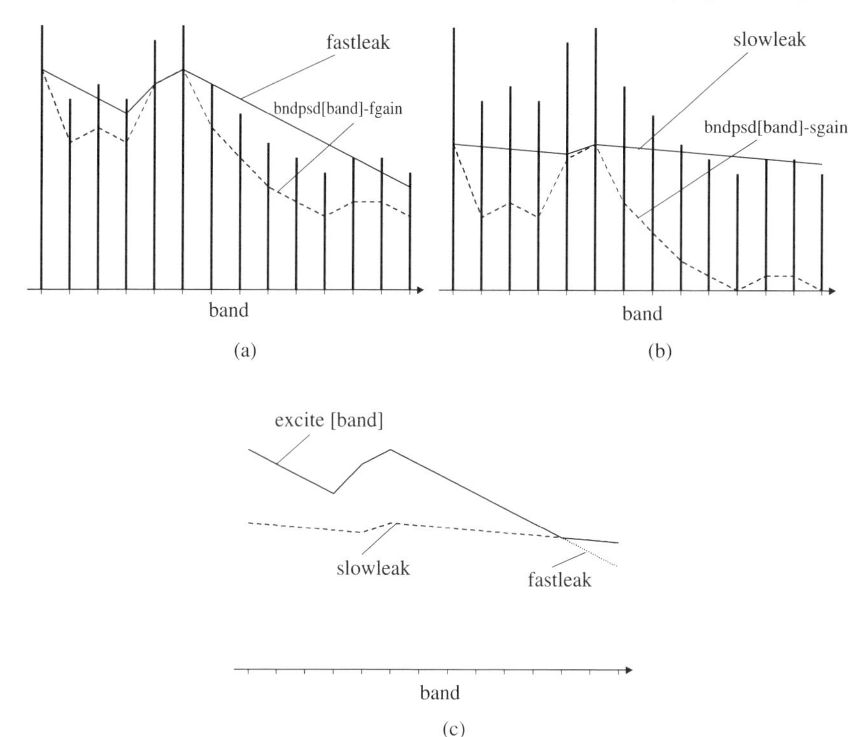

Figure 10.22 An example illustrating the convolution process used to form the excitation function.

with the banded PSD values is stored in the slowleak parameter as shown in Figure 10.22(b). The final excitation function, designated by excite[band], is calculated by taking the maximum of the fastleak and slowleak values for each band as shown in Figure 10.22(c).

10.1.8.5. Compensation for Decoder Selectivity

A problem exists for the lower frequency coefficients of the decoder filterbank. At these low frequencies there is a significant amount of overlap between the subbands of the decoder filterbank. So if quantization distortion is introduced into one band at the encoder, then a significant amount of this distortion appears in neighboring frequencies at the output of the decoder filter bank. The problem occurs predominantly at low frequencies where the slope of the masking curve can exceed the roll-off of the filter frequency response.

In the AC-3 encoder a compensation for this poor decoder frequency selectivity is incorporated into the calculation of the excitation function. The compensation is only applied to frequencies between 0 and approximately 700 Hz, and this range is further subdivided into two smaller frequency ranges. For frequencies below about 200 Hz, no upward masking is assumed when calculating the excitation function. For frequencies between 200 and 700 Hz, upward masking is only enabled for frequencies above the first significant spectral component.

In addition to these limitations on upward masking, the compensating factor, lowcomp, is included in the calculation of the excitation function for frequencies between 0 and about 2.3 kHz. The value for lowcomp is calculated using the slope between consecutive banded PSD values using the algorithm defined by the following MATLAB code.

```
% for frequencies below about 700 Hz
if band < 7
  % if two consecutive PSD values have a difference of 256 (+12 dB)
  % the value of lowcomp is set to 384 (an 18 dB adjustment)
  if bandpsd0 + 256 == bandpsd1
    lowcomp = 384;
  % if two consecutive PSD values have a difference of −128 or −256 (−6
  % or −12 dB) the value for lowcomp is taken as the larger of: the
  % current value of lowcomp minus 64 and zero
  elseif bandpsd0 > bandpsd1
    lowcomp = max(0,lowcomp − 64);
  end
  % if two consecutive PSD values have a difference of 128 or 0 (+6 or 0
  % dB) the value for lowcomp remains unchanged
% for frequencies between about 700 Hz and 1.8 kHz the algorithm is the
% same except that the value of lowcomp is set to 320 (a 12 dB adjustment)
elseif band < 20
```

```
  if bandpsd0 + 256 == bandpsd1
    lowcomp = 320;
  elseif bandpsd0 > bandpsd1
    lowcomp = max(0,lowcomp - 64);
  end
% for frequencies between about 1.8 kHz and 2.3 kHz, the value for lowcomp
% is taken as the larger of: the current value of lowcomp minus 128, and
% zero
else
  lowcomp 5 max(0,lowcomp - 128);
end
```

For frequencies below about 700 Hz, if any two consecutive PSD values have a difference of 256 (i.e. a positive slope of 12 dB), the value of lowcomp is set to 384 (i.e., an 18-dB adjustment). If two consecutive PSD values have a difference of 128 or 0 (i.e., a positive slope of 6 or 0 dB), the value for lowcomp remains unchanged. If two consecutive PSD values have a difference of -128 or -256 (i.e., a slope of -6 or -12 dB), the value for lowcomp is taken as the larger of the current value of lowcomp minus 64 and zero.

For frequencies between about 700 Hz and 1.8 kHz, the algorithm is the same except that the value of lowcomp is set to 320 for a slope of 12 dB and, for frequencies between about 1.8 and 2.3 kHz, the value for lowcomp is taken as the larger of the current value of lowcomp minus 128 and zero.

So the full algorithm to compute the excitation function, including low-frequency selectivity compensation, is defined by the following MATLAB code:

```
bndstrt = masktab(strtmant+1);
bndend = masktab(endmant) + 1;
% for fbw and lfe channels
if (bndstrt == 0)
  % for frequencies below about 200 Hz, no upward masking is assumed when
  % calculating the excitation function.
  lowcomp = calc_lowcomp(lowcomp,bndpsd(1),bndpsd(2),0);
  excite(1) = bndpsd(1) - fgain - lowcomp;
  lowcomp = calc_lowcomp(lowcomp,bndpsd(2),bndpsd(3),1);
  excite(2) = bndpsd(2) - fgain - lowcomp;
  begin = 7;

  % for frequencies between 200 and 700 Hz, upward masking is only
  % enabled for frequencies above the first significant spectral
  % component.
  for band = 2:6
    % skip for the last band of the lfe channel
      if ((bndend ~= 7) || (band ~= 6))
```

```
        lowcomp = calc_lowcomp(lowcomp,bndpsd(band+1),bndpsd(band+2),band);
      end
      % calculate the starting values for the convolution
      fastleak = bndpsd(band+1) - fgain;
      slowleak = bndpsd(band+1) - sgain;
      % but don't perform masking until a significant spectral component
      % is found
      excite(band+1) = bndpsd(band+1) - fgain - lowcomp;
      % skip for the last band of the lfe channel
      if ((bndend ~= 7) || (band ~= 6))
        % if a significant spectral component is found begin upward
        % masking
        if (bndpsd(band+1) <= bndpsd(band+2))
          begin = band + 1;
          break;
        end
      end
end
for band = begin:min(bndend,22)-1
      % skip for the last band of the lfe channel
      if ((bndend ~= 7) || (band ~= 6))
        lowcomp = calc_lowcomp(lowcomp,bndpsd(band+1),bndpsd(band+2),band);
      end
      % perform the convolution in the log domain for the slow decaying
      % masking curve
      fastleak = fastleak - fdecay;
      fastleak = max(fastleak, bndpsd(band+1) - fgain);
      % perform the convolution in the log domain for the fast decaying
      % masking curve
      slowleak = slowleak - sdecay;
      slowleak = max(slowleak, bndpsd(band+1) - sgain);
      % the excitation function is taken as the maximum of the fast
      % decaying and slow decaying masking curves
      excite(band+1) = max(fastleak - lowcomp, slowleak);
   end
   begin = 22;
% for the coupling channel
else
   begin = bndstrt ;
end
for band = begin:bndend-1
   fastleak = fastleak - fdecay;
```

```
    fastleak = max(fastleak, bndpsd(band+1) - fgain);
    slowleak = slowleak - sdecay;
    slowleak = max(slowleak, bndpsd(band+1) - sgain);
    excite(band+1) = max(fastleak, slowleak);
end
```

where strtmant is equal to fbwstrtmant[ch], cplstrtmant, or lfestrtmant and endmant is equal to fbwendmant[ch], cplendmant, or lfeendmant for the full bandwidth, coupling, and LFE channels, respectively.

The values for the parameters sdecay, fdecay, and sgain are constant for all channels and are transmitted to the decoder using the code words **sdcycod**, **fdcycod**, and **sgaincod**. These code words are treated as addresses into the corresponding lookup tables shown in Table 10.16.

The value for the parameter fgain may vary with the type of channel and is transmitted to the decoder using the code words **fgaincod[ch]**, **cplfgaincod**, and **lfefgaincod** for the full bandwidth, coupling, and LFE channels, respectively. These code words are treated as addresses into the lookup table shown in Table 10.17.

For the coupling channel, initial values for the fastleak and slowleak parameter other than zero may be used. These nonzero values are transmitted to the decoder using the 3-bit unsigned integers, **cplfleak** and **cplsleak**. The values for these parameters are calculated from the initial values of fastleak and slowleak using the following equations:

$$\textbf{cplfleak} = (\text{fastleak} - 768) >> 8 \tag{10.16}$$

$$\textbf{cplsleak} = (\text{slowleak} - 768) >> 8 \tag{10.17}$$

10.1.8.6. Masking Curve Computation

Once the excitation function has been determined, the next step in the bit-allocation process is to compute the masking threshold curve. This curve is calculated using the absolute threshold of hearing, the excitation function, and the parameter dbknee. This last parameter is used to adjust the masking curve according to the absolute level of the masking signal. The results of experimental

Table 10.16 Bit allocation tables slowdec[ ], fastdec[ ], and slowgain[ ].

sdcycod	slowdec[ ]	**fdcycod**	fastdec[ ]	**sgaincod**	slowgain[ ]
“00”	0x0f	“00”	0x3f	“00”	0x540
“01”	0x11	“01”	0x53	“01”	0x4d8
“10”	0x13	“10”	0x67	“10”	0x478
“11”	0x15	“11”	0x7b	“11”	0x410

Table 10.17 Bit allocation table fastgain[ ].

fgaincod	fastgain[ ]
"000"	0x080
"001"	0x100
"010"	0x180
"011"	0x200
"100"	0x280
"101"	0x300
"110"	0x380
"111"	0x400

listening tests have shown that the masking threshold increases by about 5 dB for each 20 dB reduction in the absolute level of the masking tone. The parameter dbknee is used to indicate the level below which the masking threshold is adjusted upward. If the masking signal is less than dbknee, the masking threshold is adjusted upward by an amount proportional to the difference between the masking signal and dbknee. If the masking signal is greater than dbknee, the masking curve is not adjusted. The value of dbknee can be set to zero so that no adjustment for masking signal level is included in the calculation. The algorithm used to determine the masking threshold curve is defined by the following MATLAB code:

```
bndstrt = masktab(strtmant+1)
bndend = masktab(endmant)
for band = bndstrt:bndend
  % if the masking signal is less than dbknee, the masking threshold is
  % adjusted upwards by an amount proportional to the difference between
  % the masking signal and dbknee.
  if bndpsd(band+1) < dbknee
    excite(band+1) = excite(band+1) + (dbknee - bndpsd(band+1))/4;
  end
  % if the masking signal is greater than dbknee, the masking curve is
  % not adjusted.
  mask(band+1) = max(excite(band+1), hth(fscod+1,band+1))
end
```

The array hth[**fscod**][band] contains values for the absolute threshold for each band at each input sampling rate and is shown in Table 10.18.

The value for the parameters dbknee is constant for all channels and is transmitted to the decoder using the code word **dbpbcod**. This code word is treated as an address in the lookup table shown in Table 10.19.

Table 10.18 Absolute threshold table, hth[**fscod**][band].

	fs (kHz)/**fscod**				fs (kHz)/**fscod**		
Band	48/0	44.1/1	32/2	Band	48/0	44.1/1	32/2
0	1232	1264	1408	25	832	848	896
1	1232	1264	1408	26	816	832	896
2	1088	1120	1200	27	800	832	880
3	1024	1040	1104	28	784	800	864
4	992	992	1056	29	768	784	848
5	960	976	1008	30	752	768	832
6	944	960	992	31	752	752	816
7	944	944	976	32	752	752	800
8	928	944	960	33	752	752	784
9	928	928	944	34	768	752	768
10	928	928	944	35	784	768	752
11	928	928	944	36	832	800	752
12	928	928	928	37	912	848	752
13	912	928	928	38	992	912	768
14	912	912	928	39	1056	992	784
15	912	912	928	40	1120	1056	816
16	896	912	928	41	1168	1104	848
17	896	896	928	42	1184	1184	960
18	880	896	928	43	1120	1168	1040
19	880	896	928	44	1088	1120	1136
20	864	880	912	45	1088	1088	1184
21	864	880	912	46	1312	1152	1120
22	848	864	912	47	2048	1584	1088
23	848	864	912	48	2112	2112	1104
24	832	848	896	49	2112	2112	1248

Table 10.19 Bit-allocation table dbpbtab[ ].

dbpbcod	dbpbtab[ ]
"00"	0x000
"01"	0x700
"10"	0x900
"11"	0xb00

10.1.8.7. Delta Bit Allocation

The AC-3 coder provides a means of adjusting the masking curve produced by applying the default masking model. This adjustment might be necessary for special types of signals or if a more sophisticated masking model is to be used. The adjustment to the masking curve is achieved by transmitting a set of *delta bit-allocation* parameters as side information. These parameters describe the necessary adjustments to transform the default masking curve into the alternate curve.

The existence of delta bit-allocation parameters in the bit stream is indicated by the single bit code word **deltbaie**. A value of 1 for **deltbaie** indicates that some delta bit-allocation information follows in the bit stream. A value of 0 for **deltbaie** in blocks 2 to 6 indicates that the delta bit-allocation information from the previous block is to be reused. However, if **deltbaie** is 0 in the first block of the audio frame, this indicates that no delta bit-allocation information is to be used in the frame.

The delta bit-allocation strategy to be used in the current block is indicated by the 2-bit code word **deltbae[ch]** for the full bandwidth channels or **cpldeltbae** for the coupling channel as shown in Table 10.20.

The delta bit-allocation parameters describe a set of variable length and constant amplitude *segments* where the amplitude of each segment is the difference between the default masking curve and the alternate curve. The properties of these segments are transmitted as follows.

The number of segments is indicated by a 3-bit unsigned integer **deltnseg[ch]**. The actual number of segments ranges from 1 to 8 and is calculated by adding 1 to the transmitted 3-bit value. For the first segment transmitted (seg = 0), the first band to be modified by that segment is indicated by a 5-bit unsigned integer **deltoffst[ch][seg]**. For subsequent segments, this parameter indicates the offset from the last band of the previous segment to the first band of the current segment. The number of bands spanned by each segment is indicated by the 4-bit unsigned integer **deltlen[ch][seg]**. The required adjustment to the default masking curve for the current segment is indicated by the 3-bit code word **deltba[ch][seg]** as shown in Table 10.21.

Table 10.20 Delta bit-allocation strategy.

cpldeltbae/deltbae	Strategy
"00"	Reuse previous state
"01"	New info follows
"10"	Perform no delta allocation
"11"	Reserved

Table 10.21 Delta bit allocation code words and adjustments.

cpldeltba/deltba	Adjustment (dB)
"000"	−24
"001"	−18
"010"	−12
"011"	−6
"100"	6
"101"	12
"110"	18
"111"	24

The corresponding information, is transmitted for the coupling channel using the parameters **cpldeltnseg**, **cpldeltoffst[seg]**, **cpldeltlen[seg]**, and **cpldeltba[seg]**.

So if the default masking curve is to be modified using the delta bit-allocation parameters, the following pseudocode defines the algorithm used to make these adjustments for a single channel. The parameters **deltbae**, **deltnseg**, **deltoffst[seg]**, **deltlen[seg]**, and **deltba[seg]** can be replaced by the parameters associated with the channel being coded. The following MATLAB code defines the algorithm used to perform the delta bit-allocation procedure.

```
if ((deltbae == 0) || (deltbae == 1))
  band = 0 ;
  % the number of segments is indicated by deltnseg.
  for seg = 0:deltnseg
    % deltoffst indicates the offset from the last band of the previous
    % segment to the first band of the current segment.
    band = band + deltoffst(seg+1);
    % the required adjustment to the default masking curve for the
    % current segment is indicated by deltba
    if (deltba(seg+1) >= 4)
      delta = (deltba(seg+1) - 3)*128;
    else
      delta = (deltba(seg+1) - 4)*128;
    end
    % The number of bands spanned by each segment is indicated by
    % deltlen
    for k = 0:deltlen(seg+1)−1
        mask(band+1) = mask(band+1) + delta ;
```

```
            band = band + 1;
        end
    end
end
```

10.1.8.8. Compute Bit Allocation

In the final step of the bit-allocation process, the signal-to-mask ratio is obtained by subtracting the masking threshold values from the fine resolution PSD values. This signal-to-mask ratio is converted to a 6-bit number and used as an index in a lookup table to obtain the number of bits allocated to each coefficient mantissa. The total number of bits allocated to coding the mantissa values is constrained to be less than or equal to a fixed number of bits available for the current audio frame. The encoder controls the number of bits allocated by adjusting a constant offset, snroffset, and a lower bound, snrfloor, for the masking threshold values in the array mask[band]. The adjustments are made to snroffset in an iterative manner until the desired number of bits is produced. The value for snroffset at each iteration is calculated using a global coarse offset value, **csnroffst**, and a fine offset value for each full bandwidth channel, **fsnroffst[ch]**, the coupling channel, **cplfsnroffst**, and the low frequency channel, **lfefsnroffst**. The value for snroffset for each channel is then calculated from these parameters using Equation (10.18) with the term fsnroffst replaced by the fine offset parameter associated with the channel being coded.

$$\text{snroffset} = ((\mathbf{csnroffst} - 15) \times 16 + \text{fsnroffst}) \times 4 \tag{10.18}$$

Once the required number of bits has been allocated to each channel, the coarse offset parameter is transmitted to the decoder as a 6-bit unsigned integer and the fine offset parameters for each channel are transmitted using 4-bit unsigned integers. The following MATLAB code defines the algorithm used to calculate the bit-allocation pointer, designated by bap[bin], for each mantissa value.

```
bin = strtmant;
band = masktab(strtmant+1);
lastbin = bin;

while (endmant > lastbin)
  lastbin = min(bndtab(band+1) + bndsz(band+1),endmant);

  % the encoder controls the number of bits allocated by adjusting a
  % constant offset, snroffset, and a lower bound, snrfloor, for the
  % masking threshold values.

  mask(band+1) = mask(band+1) - snroffset;
  mask(band+1) = mask(band+1) - snrfloor;
```

```
  if (mask(band+1) < 0)
    mask(band+1) = 0;
  end

  mask(band+1) = bitand(mask(band+1),8160);
  mask(band+1) = mask(band+1) + snrfloor;

  for k = bin:lastbin-1

    % the signal-to-mask ratio is obtained by subtracting the masking
    % threshold values from the fine resolution PSD values.

    address = floor((psd(bin+1) - mask(band+1))/32);

    % This signal to mask ratio is converted to a 6-bit number and used
    % as an index in to a lookup table to obtain the number of bits
    % allocated to each coefficient mantissa.

    address = min(63, max(0, address)) ;
    bap(bin+1) = baptab(address+1);

    bin = bin + 1;

  end

  band = band+1;
end
```

The lookup table, baptab[smr], to translate the signal-to-mask ratio values into bit-allocation pointers is shown in Table 10.22.

The value for the parameter snrfloor is constant for all channels and is transmitted to the decoder using the code word **floorcod**. This code word is treated as an address in the lookup table shown in Table 10.23.

10.1.9. Quantize Mantissas

Now that the bits to be used for each coefficient number have been allocated, the corresponding mantissa can be quantized with the appropriate level of precision. This quantization is performed using a symmetric uniform quantizer for bit-allocation pointer (bap) values less than 5 and an asymmetric quantizer for bap values greater than 5.

Table 10.24 shows the number of quantizer levels for each bap value and also the number of bits required to code the mantissa. For bap values of 1 and 2, three quantized values are grouped together and coded as 5- and 7-bit code words, respectively, and for the bap value of 4, two quantized values are grouped together and coded as a 7-bit code word.

Before quantization is performed, each mantissa is *normalized* by shifting its binary representation to the left by the number of positions indicated in the corresponding coded exponent value. Then either symmetric or asymmetric quantization is performed on the normalized mantissa values as given in the following section.

Table 10.22 Bit-allocation pointer table.

smr	baptab[smr]	smr	baptab[smr]
0	0	32	10
1	1	33	10
2	1	34	10
3	1	35	11
4	1	36	11
5	1	37	11
6	2	38	11
7	2	39	12
8	3	40	12
9	3	41	12
10	3	42	12
11	4	43	13
12	4	44	13
13	5	45	13
14	5	46	13
15	6	47	14
16	6	48	14
17	6	49	14
18	6	50	14
19	7	51	14
20	7	52	14
21	7	53	14
22	7	54	14
23	8	55	15
24	8	56	15
25	8	57	15
26	8	58	15
27	9	59	15
28	9	60	15
29	9	61	15
30	9	62	15
31	10	63	15

10.1.9.1. *Symmetric Quantization (1 $\leq$ bap $\leq$ 5)*

For bap values of 1–5, the quantized mantissa value is represented by the code word mantissa_code assigned using a table lookup. The decimal value of the code word assigned to each quantized mantissa value is shown in Tables 10.25–10.29.

Table 10.23 Bit-allocation table floortab[ ].

floorcod	floortab[ ]
"000"	0x2f0
"001"	0x2b0
"010"	0x270
"011"	0x230
"100"	0x1f0
"101"	0x170
"110"	0x0f0
"111"	0xf800

Table 10.24 Number of quantizer levels and mantissa bits.

bap	Quantizer levels	Mantissa bits (qntztab[bap])
0	0	0
1	3	5/3
2	5	7/3
3	7	3
4	11	7/2
5	15	4
6	32	5
7	64	6
8	128	7
9	256	8
10	512	9
11	1024	10
12	2048	11
13	4096	12
14	16384	14
15	65536	16

Table 10.25 Mantissa code words for symmetric quantization (bap = 1).

mantissa_code	mantissa value
0	−2/3
1	0
2	2/3

Table 10.26 Mantissa code words for symmetric quantization (bap = 2).

mantissa_code	mantissa value
0	−4/5
1	−2/5
2	0
3	2/5
4	4/5

For bap values of 1 and 2, three mantissa_code values are treated as unsigned integers and grouped together to form a single larger code word using the following equations:

bap = 1:

$$\text{group\_code} = 9 \times \text{mantissa\_code}[a] + 3 \times \text{mantissa\_code}[b] + \text{mantissa\_code}[c] \tag{10.19}$$

bap = 2:

$$\text{group\_code} = 25 \times \text{mantissa\_code}[a] + 5 \times \text{mantissa\_code}[b] + \text{mantissa\_code}[c] \tag{10.20}$$

For a bap value of 4, two mantissa_code values are treated as unsigned integers and grouped together to form a single larger code word using the following equation:

bap = 4:

$$\text{group\_code} = 11 \times \text{mantissa\_code}[a] + \text{mantissa\_code}[b] \tag{10.21}$$

Table 10.27 Mantissa code words for symmetric quantization (bap = 3).

mantissa_code	mantissa value
0	−6/7
1	−4/7
2	−2/7
3	0
4	2/7
5	4/7
6	6/7

Table 10.28 Mantissa code words for symmetric quantization (bap = 4).

mantissa_code	mantissa value
0	−10/11
1	−8/11
2	−6/11
3	−4/11
4	−2/11
5	0
6	2/11
7	4/11
8	6/11
9	8/11
10	10/11

Table 10.29 Mantissa code words for symmetric quantization (bap = 5).

mantissa_code	mantissa value
0	−14/15
1	−12/15
2	−10/15
3	−8/15
4	−6/15
5	−4/15
6	−2/15
7	0
8	2/15
9	4/15
10	6/15
11	8/15
12	10/15
13	12/15
14	14/15

where index a refers to the mantissa_code value with the lowest frequency in the group and indexes b and c refer to values with successively higher frequencies.

Note that these mantissa_code values are not required to be immediately adjacent to each other in frequency. The group_code value is constructed from the next available mantissa_code values in order of ascending frequency. If the number of mantissa_code values in a channel is not a multiple of the group size, then the mantissa_code value from the next channel to be processed is used to complete the group. If the number of mantissa_code values in an audio block is not a multiple of the group size, then the last group in the block is completed by padding with dummy mantissa values.

10.1.9.2. Asymmetric Quantization ($6 \leq$ bap ≤ 15)

For bap values of 6 to 15, the quantized mantissa value is defined as the fractional 2's complement representation of the mantissa value. So, if the number of bits allocated to each mantissa is designated by qntztab[bap] and the decimal point is considered to be to the left of the MSB, the 2's complement form of the normalized mantissa can represent values in the range -1.0 to $(1.0 - 2^{-(\mathrm{qntztab[bap]}-1)})$.

10.1.9.3. Dither for Zero Bit Mantissas

Dither is the term commonly used for the substitution of zero values with random noise. In the AC-3 decoder, those coefficient mantissas that have been allocated a bap of zero may be reconstructed using random noise. The use of dither is controlled by the single bit parameter **dithflag**. A value of 1 for **dithflag** indicates that dither should be used, and a value of 0 indicates that the coefficient should be reconstructed as a true zero value. A **dithflag** bit is transmitted for each channel, and the dither noise is added after individual channel coefficients have been extracted from the coupling channel. The dither noise is added in this way so that the noise in each channel remains uncorrelated.

The random noise sequence used to generate the dither values should have a uniform distribution and ideally has values in the range -0.707 to 0.707. However, this range can be approximated to -0.75 to 0.75 or even -0.5 to 0.5 to reduce computational complexity.

10.1.10. Dialog Normalization

In some cases there may be a significant loudness difference for normal spoken dialog between different portions of a television broadcast. For example, the loudness level of dialog in a commercial may be different from that in the movie being broadcast at the time. The **dialnorm** parameter in the AC-3 bit stream provides a mechanism for alleviating this problem and allowing normal dialog from all parts of the broadcast signal to be reproduced at the same loudness level. The **dialnorm** parameter is interpreted as a 5-bit unsigned integer that indicates the subjective level of spoken dialog relative to the maximum possible reproduced sound level. For example a **dialnorm** value of 25 indicates that the level of spoken dialog in the current transmission is at a level of 25 dB below the maximum output level of the decoder.

The **dialnorm** parameter is not intended to be used by the decoder itself but instead can be used in the section of the sound reproduction system that controls the output sound level. This section of the system typically uses **dialnorm** to adjust the output sound level so that spoken dialog is always reproduced at a constant volume level while allowing the maximum volume level to vary for different parts of the broadcast signal.

EXAMPLE 10.5

Consider the situation where the soundtrack of a movie is currently being reproduced with a **dialnorm** value of 20 and the listener has adjusted the desired output volume to 65 dB. The volume control section of the sound reproduction system adjusts the output volume so that the maximum decoded signal level is reproduced at 85 dB (i.e., 20 dB above the level for spoken dialog) and, hence, normal spoken dialog is reproduced at the desired 65 dB.

Now consider what happens if a commercial message is inserted into the broadcast and is transmitted with a **dialnorm** value of 10. The volume control section can use this value to automatically adjust the output volume so that the maximum decoded signal level is reproduced at 75 dB. Consequently, the spoken dialog in the commercial message is automatically normalized to 65 dB (10 dB below the maximum level) and is reproduced at the same output volume as the dialog in the movie. ■

10.1.11. Dynamic Range Compression

There is often a requirement for different listeners to be able to reproduce the same audio signal with different amounts of *dynamic range*. The original dynamic range for some audio signals can be very wide. For example a soundtrack for a feature film may be produced with loud sounds, such as explosions, that are 20 dB louder than normal dialog and quiet sounds, such as leaves rustling, that are 50 dB quieter than the level for dialog.

There are many listening situations where this amount of dynamic range is unacceptable, and it is desirable to reduce the volume of loud sounds and also increase the volume of quiet sounds. These limitations on the reproduced volume of signals are collectively known as *dynamic range compression*. However, there also situations where the listener would like to receive the signal with the original dynamic range.

To satisfy the requirements of both these situations, the AC-3 bit stream may contain a sequence of dynamic range control values. These control values are transmitted as 8-bit code words designated by **dynrng**. A **dynrng** code word may be transmitted at the start of each audio block and is used to indicate the alteration to the original sound level required to implement dynamic range compression.

EXAMPLE 10.6

Consider the case of a soundtrack for a feature film that has an original dialog level of 25 dB less than the maximum level. In this soundtrack, very loud sounds can reach the maximum level of 25 dB greater than the dialog level and quiet sounds can reach 50 dB below the dialog level.

A typical set of **dynrng** parameters would indicate a reduction in level for those sounds above the dialog level and an increase in level for those sounds below the dialog level. The loudest sounds might be reduced by 15 dB, and the quietest sounds might be increased by 20 dB. Hence, these **dynrng** parameters provide a well-defined mechanism for compressing the reproduced dynamic range of the transmitted audio signal from 75 to 40 dB. Use of the **dynrng** parameters also allows the reproduction system to maintain a constant volume level for spoken dialog while reducing the volume of loud sounds, such as explosions, and increasing the volume of quiet sounds that would otherwise be inaudible. ■

The default action for the AC-3 decoder is to implement the gain control indicated by the **dynrng** parameters. However, users may optionally decide to fully or partially disable the dynamic range compression in order to reproduce the signal with some or all of its original dynamic range.

10.1.11.1. Detailed Implementation

The 8-bit **dynrng** parameter may be transmitted with any audio block. The presence of this parameter is indicated with the 1 bit parameter **dynrnge**, with a value of 1 for **dynrnge** indicating that a **dynrng** parameter follows in the bit stream. A value of 0 for **dynrnge** in blocks 2 to 6 of an audio frame indicates that the **dynrng** parameter from the previous block should be used. For the first block in the audio frame, a value of 0 indicates that a **dynrng** code word of "00000000" should be used.

The first three bits of the **dynrng** code word are used to indicate a gain change that can be implemented using a simple left or right side shift. Table 10.30 shows the number of shifts indicated by the first three bits of the **dynrng** code word and the corresponding gain adjustment in dB.

The remaining 5 bits of the **dynrng** code word are interpreted as the five least significant bits of a 6-bit unsigned binary fraction with a leading value of 1. So the fractional values that can be represented in this way range from 0.111111_2 (63/64 or −0.14 dB) to 0.100000_2 (1/2 or −6.02 dB). These fractional gain adjustments can then be implemented using a 6-bit multiply operation.

The gain adjustments indicated by the two segments of the **dynrng** code word are consecutively implemented and, hence, allow for gain changes of $-18.06 - 6.02 = -24.08$ dB to $24.08 - 0.14 = 23.94$ dB in steps of 0.14 dB.

Table 10.30 Interpretation of the three most significant bits of **dynrng**.

Three most significant bits of **dynrng**	Number of shifts	Equivalent gain adjustment (dB)
011	4 left	24.08
010	3 left	18.06
001	2 left	12.04
000	1 left	6.02
111	none	0
110	1 right	−6.02
101	2 right	−12.04
100	3 right	−18.06

10.1.12. Heavy Compression

The AC-3 syntax provides a second means of transmitting dynamic range compression information in the BSI header of the audio frame. The 8-bit code word **compr** can be used to indicate a larger reduction in dynamic range than the **dynrng** code words in each audio block. The gain adjustment indicated by **compr** can also be used to guarantee that a monophonic downmix of the transmitted audio signal does not exceed a certain peak level.

The large dynamic range reduction (or *heavy compression*) provided by the **compr** control signal may be useful when the audio signal is to be delivered in a quiet environment such as a hotel room or an airline seat. The gain adjustment indicated by **compr** has twice the range of that indicated by the **dynrng** parameters and half the resolution (i.e., −48 to 48 dB rather than −24 to 24 dB in steps of 0.28 rather than 0.14 dB) but only has a time resolution of an audio frame rather than a block (i.e., 32 to 48 ms rather than 5.33 to 8 ms).

The peak downmix level limitation may be required when a monophonic downmix is to be used to modulate an RF signal and overmodulation is to be avoided. For example, a digital set-top decoder may be required to modulate video and audio signals onto an RF channel in order for the broadcast signal to be received by an analog television receiver.

User implementations that require the peak downmix level to be constrained should use the **compr** code words rather than the **dynrng** control signal. However, it may only be necessary to transmit **compr** parameters when the compression provided by **dynrng** does not adequately restrict the peak level. In such cases, if the decoder has been instructed to use **compr**, and the compr code word is not present in the bit stream, then the decoder should use the **dynrng** code words for peak level limitation.

10.1.12.1. Detailed Implementation

The 8-bit **compr** parameter may be transmitted with any audio frame. The presence of this parameter is indicated by the 1 bit parameter **compre**. A value of 1 for **compre** indicates that a **compr** parameter follows in the bit stream.

The first four bits of the compr code word are used to indicate a gain change that can be implemented using a simple left or right shift. Table 10.31 shows the number of shifts indicated by the first three bits of the **compr** code word and the corresponding gain adjustment in dB.

The remaining four bits of the **compr** code word are interpreted as the four least significant bits of a 5-bit unsigned binary fraction with a leading value of 1. So the fractional values that can be represented in this way range from 0.11111_2 (31/32 or -0.28 dB) to 0.10000_2 (1/2 or -6.02 dB). These fractional gain adjustments can then be implemented using a 5-bit multiply operation.

The gain adjustments indicated by the two segments of the **compr** code word are consecutively implemented and hence allow for gain changes of $-42.14 -6.02 = -48.16$ dB to $48.16 - 0.28 = 47.88$ dB in steps of 0.28 dB.

10.1.13. Downmixing

If the number of speakers in the reproduction system does not match the number of channels provided in the transmitted audio signal, then some form of *downmixing*

Table 10.31 Interpretation of the four most significant bits of **compr**.

Four most significant bits of **compr**	Number of shifts	Equivalent gain adjustment (dB)
0111	8 left	48.16
0110	7 left	42.14
0101	6 left	36.12
0100	5 left	30.10
0011	4 left	24.08
0010	3 left	18.06
0001	2 left	12.04
0000	1 left	6.02
1111	none	0
1110	1 right	−6.02
1101	2 right	−12.04
1100	3 right	−18.06
1011	4 right	−24.08
1010	5 right	−30.10
1001	6 right	−36.12
1000	7 right	−42.14

is required. The downmixing procedure for an AC-3 decoder is standardized so that audio content providers can be certain of the final audio quality of systems with various loudspeaker configurations.

10.1.13.1. General Downmix Procedure.

The unnormalized downmix coefficients for a range of channel and speaker configurations can be determined using the following pseudocode. Once these coefficients have been determined, it may be necessary to apply a normalization procedure to prevent arithmetic overflow. This normalization is achieved by attenuating all the downmix coefficients equally so that the sum of the coefficients used to create any single output channel does not exceed 1.0.

Pseudocode

```
downmix()
{
  if (acmod == 0) /* 1+1 mode, dual independent mono channels present */
  {
    if (output_nfront == 1) /* 1 front loudspeaker (center) */
    {
      if (dualmode == Chan 1) /* Ch1 output requested */
      {
        route left into center ;
      }
      else if (dualmode == Chan 2) /* Ch2 output requested */
      {
        route right into center ;
      }
      else
      {
        mix left into center with −6 dB gain ;
        mix right into center with −6 dB gain ;
      }
    }
    else if (output_nfront == 2) /* 2 front loudspeakers (left, right) */
    {
      if (dualmode == Stereo)
      /* output of both mono channels requested */
      {
        route left into left ;
        route right into right ;
      }
      else if (dualmode == Chan 1)
      {
        mix left into left with −3 dB gain ;
        mix left into right with −3 dB gain ;
      }
```

```
        else if (dualmode == Chan 2)
        {
            mix right into left with −3 dB gain ;
            mix right into right with −3 dB gain ;
        }
        else /* mono sum of both mono channels requested */
        {
            mix left into left with −6 dB gain ;
            mix right into left with −6 dB gain ;
            mix left into right with −6 dB gain ;
            mix right into right with −6 dB gain ;
        }
    }
    else /* output_nfront == 3 */
    {
        if (dualmode == Stereo)
        {
            route left into left ;
            route right into right ;
        }
        else if (dualmode == Chan 1)
        {
            route left into center ;
        }
        else if (dualmode == Chan 2)
        {
            route right into center ;
        }
        else
        {
            mix left into center with −6 dB gain ;
            mix right into center with −6 dB gain ;
        }
    }
}
else /* acmod > 0 */
{
    for i = { left, center, right, leftsur/monosur, rightsur }
    {
        if (exists(input_chan[i])) and (exists(output_chan[i]))
        {
            route input_chan[i] into output_chan[i] ;
        }
    }
    if (output_mode == 2/0 Dolby Surround compatible)
    /* 2 ch matrix encoded output requested */
    {
        if (input_nfront != 2)
        {
```

```
        mix center into left with −3 dB gain ;
        mix center into right with −3 dB gain ;
    }
    if (input_nrear == 1)
    {
        mix -mono surround into left with −3 dB gain ;
        mix mono surround into right with −3 dB gain ;
    }
    else if (input_nrear == 2)
    {
        mix -left surround into left with −3 dB gain ;
        mix -right surround into left with −3 dB gain ;
        mix left surround into right with −3 dB gain ;
        mix right surround into right with −3 dB gain ;
    }
}
else if (output_mode == 1/0) /* center only */
{
    if (input_nfront != 1)
    {
        mix left into center with −3 dB gain ;
        mix right into center with −3 dB gain ;
    }
    if (input_nfront == 3)
    {
    mix center into center using clev and +3 dB gain ;
    }
    if (input_nrear == 1)
    {
        mix mono surround into center using slev;
    }
    else if (input_nrear == 2)
    {
        mix left surround into center using slev and −3 dB gain ;
        mix right surround into center using slev and −3 dB gain ;
    }
}
else /* more than center output requested */
{
if (output_nfront == 2)
{
    if (input_nfront == 1)
    {
        mix center into left with −3 dB gain ;
        mix center into right with −3 dB gain ;
    }
    else if (input_nfront == 3)
    {
        mix center into left using clev ;
```

```
            mix center into right using clev ;
        }
    }
    if (input_nrear == 1) /* single surround channel coded */
    {
        if (output_nrear == 0) /* no surround loudspeakers */
        {
            mix mono surround into left with slev and −3 dB gain ;
            mix mono surround into right with slev and −3 dB gain ;
        }
        else if (output_nrear == 2) /* two surround loudspeaker channels */
        {
            mix mono srnd into left surround with −3 dB gain ;
            mix mono srnd into right surround with −3 dB gain ;
        }
    }
    else if (input_nrear == 2) /* two surround channels encoded */
    {
        if (output_nrear == 0)
        {
            mix left surround into left using slev ;
            mix right surround into right using slev ;
        }
        else if (output_nrear == 1).
        {
            mix left srnd into mono surround with −3 dB gain ;
            mix right srnd into mono surround with −3 dB gain ;
        }
    }
  }
}
```


The downmix coefficients clev and slev are indicated by the 2-bit code words **cmixlev** and **surmixlev**, respectively. These code words are transmitted in the BSI part of the audio frame, and the coefficient value corresponding to each code word is shown in Table 10.32. If the reserved code is received for **cmixlev**, the decoder should use the intermediate value of 0.596. If the reserved code is received for **surmixlev**, the decoder should use the intermediate value of 0.5.

10.1.13.2. Downmixing to Stereo

If surround channels are transmitted and the output required is a stereo pair, then two types of downmix are required. The first is a downmix to a conventional stereo

Table 10.32 Center and surround mix levels.

cmixlev	clev	**surmixlev**	slev
00	0.707	00	0.707
01	0.596	01	0.5
10	0.5	10	0
11	Reserved	11	Reserved

pair, L_o and R_o, and the second is a downmix to a matrix surround encoded stereo pair, L_t and R_t.

If all five channels are transmitted (3/2 mode) and the output required is a conventional stereo pair, then the following downmix equations are to be used.

$$L_o = L + \text{clev} \times C + \text{slev} \times L_s \tag{10.22}$$

$$R_o = R + \text{clev} \times C + \text{slev} \times R_s \tag{10.23}$$

If the output required is a matrix surround encoded stereo pair, then the following downmix equations are to be used.

$$L_t = L + 0.707 \times (C - L_s - R_s) \tag{10.24}$$

$$R_t = R + 0.707 \times (C + L_s + R_s) \tag{10.25}$$

If the center channel is not transmitted (2/2 mode), the same equations can be used without the C term.

If only a single surround channel is transmitted (3/1 mode) and the output required is a conventional stereo pair, then the following downmix equations are to be used.

$$L_o = L + \text{clev} \times C + 0.707 \times \text{slev} \times S \tag{10.26}$$

$$R_o = R + \text{clev} \times C + 0.707 \times \text{slev} \times S \tag{10.27}$$

If the output required is a matrix surround encoded stereo pair, then the following downmix equations are to be used.

$$L_t = L + 0.707 \times (C - S) \tag{10.28}$$

$$R_t = R + 0.707 \times (C + S) \tag{10.29}$$

If the center channel is not transmitted (2/1 mode), the same equations can be used without the C term. If the surround channels are not transmitted (3/0 mode), the same equations can be used without the L_s, R_s, or S terms.

10.1.13.3. Downmixing to Mono

If the output required is a monophonic signal, the downmixed stereo signals may be further downmixed to mono using a simple summation of the two channels. However, if the matrix surround encoded pair is used, the surround information is lost, so it is desirable to use the conventional stereo pair.

If the monophonic signal is to be calculated directly and all channels are transmitted, the following equation is to be used.

$$M = L + R + 2.0 \times \text{clev} \times C + \text{slev} \times (L_s + R_s) \qquad (10.30)$$

If the center channel is not transmitted (2/2 mode), the same equation can be used without the C term. If only a single surround channel is transmitted (3/1 mode), the following equation is to be used.

$$M = L + R + 2.0 \times \text{clev} \times C + 1.4 \times \text{slev} \times S \qquad (10.31)$$

If the center or surround channels are not transmitted (2/1, 3/0, or 2/0 mode), the same equation may be used without the C or S terms.

10.1.13.4. Normalizing Downmixing Coefficients

If all the channels to be included in a downmixed signal have their maximum value at the same time, the values in the signal will exceed the maximum available wordlength. This problem is known as *arithmetic overflow* and can be avoided if the coefficients used in the downmixing equations described in the previous section are scaled downwards. A different scaling could be applied for all possible combinations of input and output channels, but this would require a very large number of scaled coefficients to be defined. Instead, the worst-case situation is used to determine the maximum scaling factor required when clev and slev are both equal to 0.707.

For a conventional stereo output, if unscaled coefficients are used, the maximum downmixed signal has a value of $1 + 0.707 + 0.707 = 2.414$. So the maximum scaling required for the coefficients is $1/2.414 = 0.4143$. Table 10.33 shows the unscaled and scaled versions of all the possible downmixing coefficients required to produce a conventional stereo pair. The downmixing equation is typically implemented using

Table 10.33 Scaled downmix coefficients for a conventional stereo signal.

Unscaled coefficient	Scaled coefficient	6-bit Quantized coefficient
1.0	0.414	26/64
0.707	0.293	18/64
0.596	0.247	15/64
0.5	0.207	13/64
0.354	0.147	9/64

Table 10.34 Scaled downmix coefficients for a matrix surround encoded stereo signal.

Unscaled coefficient	Scaled coefficient	6-bit Quantized coefficient
1.0	0.3204	26/64
0.707	0.293	18/64

fixed length binary arithmetic, and if this is the case the 6-bit quantized coefficients shown in Table 10.33 produce a downmixed signal with the required accuracy. If a mono signal is required, the coefficients need to be scaled by a further factor of 0.5.

For a matrix surround encoded stereo output, if unscaled coefficients are used, the maximum downmixed signal has a value of $1 + 0.707 + 0.707 + 0.707 = 3.121$. So the maximum scaling required for the coefficients is $1/3.121 = 0.3204$. Table 10.34 shows the unscaled, scaled, and 6-bit quantized versions of all the possible downmixing coefficients required to produce a matrix surround encoded stereo pair.

10.2. SYNTAX

The syntax of the AC-3 bit stream is specified in this section. The order of arrival of the bits is defined using the methodology described in Chapter 5. If the bit-stream element to be transmitted is larger than 1 bit, it is transmitted in one of two ways: (a) if the element specifies a numerical value, it is transmitted with the most significant bit first and (b) if the element is a bit field, it is transmitted with the left bit first. In some cases, the meaning for a possible value of a bit-stream element is "reserved." If the decoder erroneously receives a reserved value for an element, there are two possible responses for the decoder: (a) the decoder cannot decode any value and (b) the decoder can use a default value for the element.

10.2.1. Syntax Specification

The structure of an AC-3 synchronization frame is shown in Figure 10.2. The synchronization frame begins with a syncinfo (SI) header. The bit-stream information (BSI) header is transmitted next, and this is followed by six audio blocks (AB). Each audio block contains the information required to decode 256 audio samples. The synchronization frame may then contain an optional auxiliary data field (Aux) and ends with an error check field that contains a cyclic redundancy code word (CRC) used for error detection. The syntax specification for an AC-3 bit stream is given in Table 10.35. The bit stream consists of consecutive synchronization frames.

Table 10.35 Syntax specification for an AC-3 bit stream.

Syntax	Bits

```
AC-3_bit stream()
{
  while(true)
  {
    syncframe();
  }
}/* end of AC-3 bit stream */
```


The syntax specification for a syncframe is given in Table 10.36.

Each field of the syncframe is described in the following sections.

10.2.1.1. Synchronization Information (syncinfo)

The syncinfo field contains information necessary for the decoder to maintain synchronization of the incoming bit stream. The syntax specification of the syncinfo field is given in Table 10.37.

A description of the bit stream elements contained in the syncinfo field is given in Table 10.38.

10.2.1.2. Bit-Stream Information (bsi)

The BSI field contains parameters that specify aspects of the audio service being transmitted. The syntax specification of the BSI field is given in Table 10.39.

Table 10.36 Syntax specification for a syncframe.

Syntax	Bits

```
syncframe()
{
  syncinfo();
  bsi();
  for( blk = 0; blk < 6; blk++)
  {
    audblk();
  }
  auxdata();
  errorcheck();
}/* end of syncframe */
```

Table 10.37 Syntax specification for a syncinfo field.

Syntax	Bits
syncinfo()	
{	
syncword;	16
crc1;	16
fscod;	2
frmsizecod;	6
}/* end of syncinfo */	

A description of the bit stream elements contained in a BSI field is given in Table 10.40.

10.2.1.3. Audio Block (audblk)

Each audio block contains the information required to decode 256 audio samples. The syntax specification of the audblk field is given in Table 10.41.

A description of the bit stream elements contained in an audblk field is given in Table 10.42.

Table 10.38 Description of the bit stream elements in a syncinfo field.

Element	Description
syncword	Used to locate the start of the syncframe (the value of the syncword is always 0x0B77)
crc1	Error check bits
fscod	Specifies the input sampling rate (see Table 10.1)
frmsizecod	Specifies the number of 16-bit words in the current syncframe (see Ref. [1])

Table 10.39 Syntax specification of the BSI field.

Syntax	Bits
bsi()	
{	
bsid	5
bsmod	3
acmod	3
if((acmod & 0x1) && (acmod != 0x1))/* if 3 front channels */ **{cmixlev}**	2
if(acmod & 0x4) /* if a surround channel exists */ **{surmixlev}**	2

(*continued*)

Table 10.39 (*Continued*)

Syntax	Bits
if(acmod == 0x2) /* if in 2/0 mode */ **{dsurmod}**	2
lfeon	1
dialnorm	5
compre	1
if(compre) {**compr**}	8
langcode	1
if(langcode) **{langcod}**	8
audprodie	1
if(audprodie)	
{	
mixlevel	5
roomtyp	2
}	
if(acmod == 0)/* if 1+1 mode (dual mono, so some items need a second value)*/	
{	
dialnorm2	5
compr2e	1
if(compr2e) **{compr2}**	8
lngcod2e	1
if(langcod2e) **{langcod2}**	8
audprodi2e	1
if(audprodi2e)	
{	
mixlevel2	5
roomtyp2	2
}	
}	
copyrightb	1
origbs	1
timecod1e	1
if(timecod1e) **{timecod1}**	14
timecod2e	1
if(timecod2e) **{timecod2}**	14
addbsie	1
if(addbsie)	
{	
addbsil	6
addbsi (addbsil+1)	8
}	
} /* end of bsi */	

Table 10.40 Description of the bit stream elements in a BSI field.

Element	Description
bsid	Specifies the version of the standard (currently 8).
bsmod	Specifies the type of service that this audio bit stream is providing (see Ref. [1]).
acmod	Specifies the channel configuration in use (see Table 10.2).
cmixlev	Specifies the downmix level for the center channel relative to the left and right side channels (see Ref. [1]).
surmixlev	Specifies the downmix level for the surround channels relative to the left and right side channels (see Ref. [1]).
dsurmod	Specifies whether the program has been encoded using Dolby surround mode (see Ref. [1]).
lfeon	A value of "1" indicates that the low-frequency effects (LFE) channel is in use.
dialnorm	Specifies how far the average dialog level is below the maximum level (see Section 10.1.10).
compre	A value of "1" indicates that a compression gain word follows in the bit stream.
compr	The compression gain word (see Section 10.1.12).
langcode	A value of "1" indicates that a language code follows in the bit stream.
langcod	The language code word (see Ref. [1]).
audprodie	A value of "1" indicates that audio production information follows in the bit stream.
mixlevel	Specifies the absolute sound pressure level of a channel (see Ref. [1]).
roomtyp	Specifies the type of room used in the final mixing process (see Ref. [1]).
dialnorm2	Used for channel 2 in dual mono mode.
compr2e	Used for channel 2 in dual mono mode.
compr2	Used for channel 2 in dual mono mode.
langcod2e	Used for channel 2 in dual mono mode.
langcod2	Used for channel 2 in dual mono mode.
audprodi2e	Used for channel 2 in dual mono mode.
mixlevel2	Used for channel 2 in dual mono mode.
roomtyp2	Used for channel 2 in dual mono mode.
copyrightb	A value of "1" indicates that the following bit stream is protected by copyright.
origbs	A value of "1" indicates that the following bit stream is an original bit stream.
timecod1e	A value of "1" indicates that the first half of a time code follows in the bit stream.
timecod1	Specifies a time with resolution of 8 s and full scale of 24 h (see Ref. [1]).
timecod2e	A value of "1" indicates that the second half of a time code follows in the bit stream.
timecod2	Specifies a time with resolution of 1/64th of a frame and full scale of 8 s (see Ref. [1]).
addbsie	A value of "1" indicates that additional information follows in the bit stream.
addbsil	Specifies the length in bytes of the additional information that follows.
addbsi	Additional information.

Table 10.41 Syntax specification of the audblk field.

Syntax	Bits
audblk()	
{	
/* These fields for block switch and dither flags */	
for(ch = 0; ch < nfchans; ch++) {**blksw[ch]**}	1
for(ch = 0; ch < nfchans; ch++) {**dithflag[ch]**}	1
/* These fields for dynamic range control */	
dynrnge	1
if(dynrnge) {**dynrng**}	8
if(acmod == 0) /* if 1+1 mode */	
{	
dynrng2e	1
if(dynrng2e) {**dynrng2**}	8
}	
/* These fields for coupling strategy information */	
cplstre	1
if(cplstre)	
{	
cplinu	1
if(cplinu)	
{	
for(ch = 0; ch < nfchans; ch++) {**chincpl[ch]**}	1
if(acmod == 0x2) {**phsflginu**} /* if in 2/0 mode */	1
cplbegf	4
cplendf	4
/* ncplsubnd = 3 + cplendf - cplbegf */	
for(bnd = 1; bnd < ncplsubnd; bnd++) {**cplbndstrc[bnd]**}	1
}	
}	
/* These fields for coupling coordinates, phase flags */	
if(cplinu)	
{	
for(ch = 0; ch < nfchans; ch++)	
{	
if(chincpl[ch])	
{	
cplcoe[ch]	1
if(cplcoe[ch])	
{	
mstrcplco[ch]	2
/* ncplbnd derived from ncplsubnd, and cplbndstrc */	
for(bnd = 0; bnd < ncplbnd; bnd++)	
{	
cplcoexp[ch][bnd]	4

Table 10.41 (*Continued*)

```
Syntax                                                                Bits
                cplcomant[ch][bnd]                                    4
              }
            }
          }
        }
        if((acmod==0x2)&&phsflginu&&(cplcoe[0]||cplcoe[1]))
        {
          for(bnd = 0; bnd < ncplbnd; bnd++) {phsflg[bnd]}            1
        }
      }
      /* These fields for rematrixing operation in the 2/0 mode */
      if(acmod == 0x2) /* if in 2/0 mode */
      {
        rematstr                                                      1
        if(rematstr)
        {
        if((cplbegf > 2) || (cplinu == 0))
        {
          for(rbnd = 0; rbnd < 4; rbnd++) {rematflg[rbnd]}            1
          if((2 >= cplbegf > 0) && cplinu)
          {
            for(rbnd = 0; rbnd < 3; rbnd++) {rematflg[rbnd]}          1
          }
          if((cplbegf == 0) && cplinu)
          {
            for(rbnd = 0; rbnd < 2; rbnd++) {rematflg[rbnd]}          1
          }
        }
      }
      /* These fields for exponent strategy */
      if(cplinu) {cplexpstr}                                          2
      for(ch = 0; ch < nfchans; ch++) {chexpstr[ch]}                  2
      if(lfeon) {lfeexpstr}                                           1
      for(ch = 0; ch < nfchans; ch++)
      {
        if(chexpstr[ch] != reuse)
        {
          if(!chincpl[ch]) {chbwcod[ch]}                              6
        }
      }
      /* These fields for exponents */
      if(cplinu) /* exponents for the coupling channel */
      {
```

(*continued*)

Table 10.41 (*Continued*)

Syntax	Bits
if(cplexpstr != reuse)	
}	
cplabsexp	4
/* ncplgrps derived from ncplsubnd, cplexpstr */	
for(grp = 0; grp< ncplgrps; grp++) **{cplexps[grp]}**	7
}	
}	
for(ch = 0; ch < nfchans; ch++) /* exponents for full bandwidth channels */	
{	
if(chexpstr[ch] != reuse)	
{	
exps[ch][0]	4
/* nchgrps derived from chexpstr[ch], and cplbegf or chbwcod[ch] */	
for(grp = 1; grp <= nchgrps[ch]; grp++) **{exps[ch][grp]}**	7
gainrng[ch]	2
}	
}	
if(lfeon) /* exponents for the low-frequency effects channel */	
{	
if(lfeexpstr != reuse)	
{	
lfeexps[0]	4
/* nlfegrps = 2 */	
for(grp = 1; grp <= nlfegrps; grp++) **{lfeexps[grp]}**	7
}	
}	
/* These fields for bit-allocation parametric information */	
baie	1
if(baie)	
{	
sdcycod	2
fdcycod	2
sgaincod	2
dbpbcod	2
floorcod	3
}	
snroffste	1
if(snroffste)	
{	
csnroffst	6
if(cplinu)	
{	
cplfsnroffst	4
cplfgaincod	3

Table 10.41 (*Continued*)

Syntax	Bits
}	
for(ch = 0; ch < nfchans; ch++)	
}	
fsnroffst[ch]	4
fgaincod[ch]	3
}	
if(lfeon)	
{	
lfefsnroffst	4
lfefgaincod	3
}	
}	
if(cplinu)	
{	
cplleake	1
if(cplleake)	
{	
cplfleak	3
cplsleak	3
}	
}	
/* These fields for delta bit-allocation information */	
deltbaie	1
if(deltbaie)	
{	
if(cplinu) **{cpldeltbae}**	2
for(ch = 0; ch < nfchans; ch++) **{deltbae[ch]}**	2
if(cplinu)	
{	
if(cpldeltbae==new info follows)	
{	
cpldeltnseg	3
for(seg = 0; seg <= cpldeltnseg; seg++)	
{	
cpldeltoffst[seg]	5
cpldeltlen[seg]	4
cpldeltba[seg]	3
}	
}	
}	
for(ch = 0; ch < nfchans; ch++)	
{	

(*continued*)

Table 10.41 (*Continued*)

Syntax	Bits
if(deltbae[ch]==new info follows)	
{	
deltnseg[ch]	3
for (seg= 0; seg <= deltnseg[ch]; seg ++	
{	
deltoffst[ch][seg]	5
deltlen[ch][seg]	4
deltba[ch][seg]	3
}	
}	
}	
}	
/* These fields for inclusion of unused dummy data */	
skiple	1
if(skiple)	
{	
skipl	9
skipfld	skipl × 8
}	
/* These fields for quantized mantissa values */	
got_cplchan = 0	
for (ch = 0; ch < nfchans; ch++)	
{	
for (bin = 0; bin < nchmant[ch]; bin++) **{chmant[ch][bin]}**	(0 – 16)
if (cplinu && chincpl[ch] && !got_cplchan)	
{	
for (bin = 0; bin < ncplmant; bin++) **{cplmant[bin]}**	(0 – 16)
got_cplchan = 1	
}	
}	
if(lfeon) /* mantissas of low-frequency effects channel */	
{	
for (bin = 0; bin < nlfemant; bin++) **{lfemant[bin]}**	(0 – 16)
}	
} /* end of audblk */	

Table 10.42 Description of the bit stream elements in an audblk field.

Element	Description
blksw[ch]	A value of "1" indicates that the current audio block was coded using two blocks in channel **ch** (see Section 10.1.2).
dithflag[ch]	A value of "1" indicates that the decoder should use dither in channel **ch** (see Section 10.1.9.3).
dynrnge	A value of "1" indicates that a dynamic range gain word follows in the bit stream.
dynrng	The dynamic range gain word (see Section 10.1.11).
dynrng2e	Used for channel 2 in dual mono mode.
dynrng2	Used for channel 2 in dual mono mode.
cplstre	A value of "1" indicates that new coupling information follows.
cplinu	A value of "1" indicates that coupling is in use.
chincpl[ch]	A value of "1" indicates that channel **ch** is coupled.
phsflginu	A value of "1" indicates that phase flags are included with the coupling information.
cplbegf	Specifies the lowest frequency subband included in coupling (see Section 10.1.4).
cplendf	Specifies the highest frequency subband included in coupling (see Section 10.1.4).
cplbndstrc[sbnd]	A value of "1" indicates that the corresponding subband should be combined with the subband immediately below it in frequency (see Section 10.1.4).
cplcoe[ch]	A value of "1" indicates that coupling coordinates for channel **ch** follow (see Section 10.1.4).
mstrcplco[ch]	Specifies the master coupling coordinate gain factor for channel **ch** (see Section 10.1.4).
cplcoexp[ch][bnd]	The coupling coordinate exponent (see Section 10.1.4).
cplcomant[ch][bnd]	The coupling coordinate mantissa (see Section 10.1.4).
phsflg[bnd]	A value of "1" indicates that the coefficients of the right channel in coupling band **bnd** were negated before the coupling channel was formed (see Section 10.1.4).
rematstr	A value of "1" indicates that new rematrix flags follow.
rematflg[rbnd]	A value of "1" indicates that the coefficients in rematrixing band **rbnd** have been rematrixed (see Section 10.1.5).
cplexpstr	Specifies the method used to code exponents in the coupling channel (see Table 10.12).
chexpstr[ch]	Specifies the method used to code exponents in the full bandwidth channel **ch** (see Table 10.12).
lfeexpstr	Specifies the method used to code exponents in the LFE channel (see Section 10.1.6).
chbwcod[ch]	Specifies the highest frequency band included in the full bandwidth channel **ch** (see Section 10.1.6).
cplabsexp	The coupling absolute exponent (see Section 10.1.7).

(*continued*)

Table 10.42 *(Continued)*

Element	Description
cplexps[grp]	Specifies the value of differentially coded exponents for the coupling exponent group **grp** (see Section 10.1.7).
exps[ch][grp]	Specifies the encoded exponents for channel **ch** in group **grp** (see Section 10.1.7).
gainrng[ch]	Specifies an increase in dynamic range for channel **ch** (see Ref. [1]).
lfeexps[grp]	Specifies the encoded exponents for the LFE channel in group **grp** (see Section 10.1.7).
baie	A value of "1" indicates that bit-allocation information follows.
sdcycod	Specifies the slow delay parameter sdecay (see Section 10.1.8).
fdcycod	Specifies the fast-delay parameter fdecay (see Section 10.1.8).
sgaincod	Specifies the slow-gain parameter sgain (see Section 10.1.8).
dbpbcod	Specifies the parameter dbknee (see Section 10.1.8).
floorcod	Specifies the parameter floor (see Section 10.1.8).
snroffste	A value of "1" indicates that SNR offset parameters follow.
csnroffst	Specifies the coarse SNR offset (see Section 10.1.8).
cplfsnroffst	Specifies the fine SNR offset for the coupling channel (see Section 10.1.8).
cplfgaincod	Specifies the fast gain parameter fgain for the coupling channel (see Section 10.1.8).
fsnroffst[ch]	Specifies the fine SNR offset for the channel **ch** (see Section 10.1.8).
fgaincod[ch]	Specifies the fast gain parameter fgain for the channel **ch** (see Section 10.1.8).
lfefsnroffst	Specifies the fine SNR offset for the LFE channel (see Section 10.1.8).
lfefgaincod	Specifies the fast gain parameter fgain for the LFE channel (see Section 10.1.8).
cplleake	A value of "1" indicates that leak initialization parameters follow (see Section 10.1.8).
cplfleak	Specifies the parameter fastleak for the coupling channel (see Section 10.1.8).
cplsleak	Specifies the parameter slowleak for the coupling channel (see Section 10.1.8).
deltbaie	A value of "1" indicates that delta bit allocation follows.
cpldeltbae	Specifies the delta bit-allocation mode for the coupling channel (see Section 10.1.8.7).
cpldeltnseg	Specifies the number of delta bit-allocation segments in the coupling channel (see Section 10.1.8.7).
cpldeltoffst[seg]	Specifies the offset from the last band of the previous segment to the first band of the segment **seg** in the coupling channel (see Section 10.1.8.7).

Table 10.42 *(Continued)*

Element	Description
cpldeltlen[seg]	Specifies the number of bands in the segment **seg** in the coupling channel (see Section 10.1.8.7).
cpldeltba[seg]	Specifies the adjustment to the default masking curve in the segment **seg** in the coupling channel (see Section 10.1.8.7).
deltbae[ch]	Specifies the delta bit-allocation mode for channel **ch** (see Section 10.1.8.7).
deltnseg[ch]	Specifies the number of delta bit-allocation segments in channel **ch** (see Section 10.1.8.7).
deltoffst[ch][seg]	Specifies the offset from the last band of the previous segment to the first band of the segment **seg** in channel **ch** (see Section 10.1.8.7).
deltlen[ch][seg]	Specifies the number of bands in the segment **seg** in channel **ch** (see Section 10.1.8.7).
deltba[ch][seg]	Specifies the adjustment to the default masking curve in the segment **seg** in channel **ch** (see Section 10.1.8.7).
skiple	A value of "1" indicates that padding bytes follow.
skipl	Specifies the number of padding bytes that follow.
skipfld	The padding bytes.
chmant[ch][bin]	The quantized mantissa value for frequency index **bin** in channel **ch** (see Section 10.1.9).
cplmant[bin]	The quantized mantissa value for frequency index **bin** in the coupling channel (see Section 10.1.9).
lfemant[bin]	The quantized mantissa value for frequency index **bin** in the LFE channel (see Section 10.1.9).

10.2.1.4. Auxiliary Data (auxdata)

User information may be placed at the end of a synchronization frame in the auxiliary data field. The syntax specification of the auxdata field is given in Table 10.43.

The auxiliary data is placed in the auxdata field so that an auxiliary data decoder can begin decoding from the end of the field. The auxiliary data decoder first examines the last bit of the auxdata field. This bit contains the flag **auxdatae**. If the value of **auxdatae** is "1," the data decoder examines the previous 14 bits that contain the **auxdatal** code word. This code word specifies the length in bits of the user data contained in the auxdata field. The decoder can then back up **auxdatal** bits from the end of the **auxbits** field and begin decoding the auxiliary data. Note that the length of the auxbits field, nauxbits, may be greater than **auxdatal**. The auxiliary data field is structured in this way so that the auxiliary data can be decoded independently without having to first decode the audio data contained in the synchronization frame.

Table 10.43 Syntax specification of the auxdata field.

Syntax	Bits
auxdata()	
{	
auxbits	nauxbits
if(auxdatae)	
{	
auxdatal	14
}	
auxdatae	1
} /* end of auxdata */	

10.2.1.5. Frame Error Detection Field (Error Check)

The syntax specification of the error check field is given in Table 10.44.

The one bit code word **crcrsv** is reserved for use in certain situations to guarantee that **crc2** is not equal to the syncword. The details of the error checking procedure using the 16 bit CRC code word **crc2** can be found in Ref. [1].

10.3. DECODER

10.3.1. Decode Exponents

The number of exponents to decode for each channel is determined from the bit-stream parameters **chbwcod[ch]** if the channel is not included in coupling and **cplbegf** and **cplendf** for the coupled and coupling channels. The starting bin number for the full bandwidth channels is always 0, that is fbwstrtmant[ch] = 0. The starting bin number for the coupling channel is given by

$$\text{cplstrtmant} = (\mathbf{cplbegf} \times 12) + 37 \tag{10.32}$$

Table 10.44 Syntax specification of the error check field.

Syntax	Bits
errorcheck()	
{	
crcrsv	1
crc2	16
} /* end of errorcheck */	

The last mantissa bin number for the full bandwidth channels is defined by

$$\text{fbwendmant[ch]} = ((\textbf{chbwcod[ch]} + 12) \times 3) + 37 \tag{10.33}$$

and for the coupling channel is defined by

$$\text{cplendmant} = ((\textbf{cplendf} + 3) \times 12) + 37 \tag{10.34}$$

For the low-frequency effects channel the start and end mantissa numbers are always given by lfestrtmant = 0 and lfeendmant = 7, respectively.

The number of grouped exponents to decode depends on the exponent strategy and (for independent and coupled channels) is given by the following equations.

D-15 mode:

$$\text{nchgrps[ch]} = \text{truncate}((\text{endmant[ch]} - 1)/3) \tag{10.35}$$

D-25 mode:

$$\text{nchgrps[ch]} = \text{truncate}((\text{endmant[ch]} - 1 + 3)/6) \tag{10.36}$$

D-45 mode:

$$\text{nchgrps[ch]} = \text{truncate}((\text{endmant[ch]} - 1 + 9)/12) \tag{10.37}$$

For the coupling channel, the number of grouped exponents to decode is given by the following equations.

D-15 mode:

$$\text{ncplgrps} = (\text{cplendmant} - \text{cplstrtmant})/3 \tag{10.38}$$

D-25 mode:

$$\text{ncplgrps} = (\text{cplendmant} - \text{cplstrtmant})/6 \tag{10.39}$$

D-45 mode:

$$\text{ncplgrps} = (\text{cplendmant} - \text{cplstrtmant})/12 \tag{10.40}$$

For the low-frequency effects channel, the number of groups is always given by nlfegrps = 2.

Then to obtain the set of 5-bit absolute exponents for each coefficient bin number the following procedure is used:

1. Each 7-bit grouped value, G, is decoded into three mapped values using the following equations (where % denotes the modulus operation):

$$\begin{aligned} M_1 &= \text{truncate}(G/25) \\ M_2 &= \text{truncate}((G \,\%\, 25)/5) \\ M_3 &= (G \,\%\, 25) \,\%\, 5 \end{aligned} \tag{10.41}$$

2. Each mapped value is converted to a differential exponent by subtracting a factor of 2.
3. The set of differential exponents is converted to a set of absolute exponents by adding each differential exponent to the previous absolute exponent value.
4. For the D-25 and D-45 modes, each absolute exponent is copied to the corresponding two or four coefficient bins.

10.3.2. Bit Allocation

The bit-allocation procedure is the same as that defined in Section 10.1.8., except that the value for snroffset is calculated using the transmitted values of the global coarse offset parameter, **csnroffst**, and the fine offset parameter for each full bandwidth channel, **fsnroffst[ch]**, the coupling channel, **cplfsnroffst**, and the low frequency channel, **lfefsnroffst**.

The setup of the psychoacoustic model is determined from the parameters transmitted in the bit stream so the following MATLAB code defines the initialization of the model.

For all channels:

```
sdecay = slowdec(sdcycod+1);
fdecay = fastdec(fdcycod+1);
sgain = slowgain(sgaincod+1);
dbknee = dbpbtab(dbpbcod+1);
snrfloor = floortab(floorcod+1);
```

For the full bandwidth channels:

```
strtmant = fbwstrtmant(ch);
endmant = fbwendmant(ch);
lowcomp = 0 ;
fgain = fastgain(fgaincod(ch)+1);
snroffset(ch) = (((csnroffst - 15)*16) + fsnroffst(ch))*4;
```

For the coupling channel:

```
startmant = cplstrtmant;
endmant = cplendmant;
fgain = fastgain(cplfgaincod+1);
snroffset = (((csnroffst - 15)*16) + cplfsnroffst)*4;
fastleak = (cplfleak*256) + 768 ;
slowleak = (cplsleak*256) + 768 ;
```

For the LFE channel:

```
strtmant = lfestrtmant;
endmant = lfeendmant;
lowcomp = 0;
fgain = fastgain(lfefgaincod+1);
snroffset = (((csnroffst - 15)*16) + lfefsnroffst)*4;
```

10.3.3. Decode Coefficients

For each coefficient that is not included in coupling, a fixed point value is obtained by right shifting the mantissa value by its exponent as follows:

$$\text{coefficient[bin]} = \text{mantissa[bin]} >> \text{exponent[bin]} \tag{10.42}$$

The value for mantissa[bin] depends on the type of quantization used and is determined using one of the following procedures.

10.3.3.1. Symmetric Quantization (1 $\leq$ bap $\leq$ 5)

For bap[bin] values of 1–5, the value for mantissa[bin] is determined using the code word mantissa_code[bin] as follows:

$$\text{mantissa[bin]} = \text{mantissa\_value[mantissa\_code[bin]]} \tag{10.43}$$

where the mantissa_value corresponding to each mantissa_code[bin] value can be found in Tables 10.25 – 10.29.

For bap[bin] values of 3 or 5, the value of mantissa_code[bin] is obtained directly from the corresponding 3 or 4 bits of the bit stream, respectively.

For a bap[bin] value of 1, the corresponding 5 bits of the bit stream represent a group_code value for the next three mantissa_code values to be processed that have been allocated a bap[bin] value of 1. Similarly, for a bap[bin] value of 2, the corresponding 7 bits of the bit stream represents a group_code value for three mantissa_code values, and for a bap[bin] value of 4, the corresponding 7 bits of the bit stream represents a group_code values for two mantissa_code values.

The mantissa_code values corresponding to each group_code value are extracted using the following equations:

bap = 1:

$$\begin{aligned}
\text{mantissa_code[a]} &= \text{truncate(group_code/9)} \\
\text{mantissa_code[b]} &= \text{truncate((group_code\%9)/3))} \\
\text{mantissa_code[c]} &= \text{(group_code\%9)\%3}
\end{aligned} \tag{10.44}$$

bap = 2:

$$\begin{aligned}
\text{mantissa_code[a]} &= \text{truncate(group_code/25)} \\
\text{mantissa_code[b]} &= \text{truncate((group_code\%25)/5))} \\
\text{mantissa_code[c]} &= \text{(group_code\%25)\%5}
\end{aligned} \tag{10.45}$$

bap = 4:

$$\begin{aligned}
\text{mantissa_code[a]} &= \text{truncate(group_code/11)} \\
\text{mantissa_code[b]} &= \text{group_code\%11}
\end{aligned} \tag{10.46}$$

10.3.3.2. Asymmetric Quantization ($6 \leq$ bap ≤ 15)

For bap[bin] values of 6 to 15, the value for mantissa[bin] is obtained directly from the corresponding qntztab[bap[bin]] bits of the bit stream.

10.3.4. Decoupling

If coupling is in use, the high-frequency coefficients of the coupled channels must be reconstructed from the coupling channel coefficients and the coupling coordinates of the coupled channels. The following MATLAB code defines the algorithm used to reconstruct the coupling coordinates for each coupling band of the coupled channels from the coupling coordinate mantissa and exponent values.

```
if (cplcoexp(ch+1,bnd+1) == 15)
   cplco_temp(ch+1,bnd+1) = cplcomant(ch+1,bnd+1)/16;
else
   cplco_temp(ch+1,bnd+1) = (cplcomant(ch+1,bnd+1)+16)/32;
end
cplexp = (cplcoexp(ch+1,bnd+1)+3*mstrcplco(ch+1));
cplco(ch+1,bnd+1) = cplco_temp(ch+1,bnd+1)*(2^(-cplexp));
```

Then using the **cplbndstrc** array, the coupling coordinates for bands that span multiple coupling subbands are repeated to reconstruct the coupling coordinates for each subband.

The mantissa values for each coupled coefficient can then be determined from these coupling coordinates using the algorithm defined by the following MATLAB code.

```
for sbnd = cplbegf:cplendf+2
   for bin = 0:11
      s = sbnd*12+bin+37;
      chmant(ch+1,s+1) = cplmant(s+1)*cplco(ch+1,sbnd+1)*8;
   end
end
```

Then for each coefficient that is included in coupling, a fixed point value is obtained by right shifting this mantissa value by the corresponding exponent from the coupling channel.

10.3.5. Inverse Transform

Once the transform coefficients have been recovered from the decoded mantissa and exponent values, the inverse transform is performed to obtain the reconstructed sample values. The inverse transform is performed using the following equation:

$$y(n) = -w(n) \sum_{k=0}^{N/2-1} X(k) \cos\left(\frac{2\pi}{4N}(2n+1)(2k+1) + \frac{\pi}{4}(2k+1)(1+\alpha)\right) \quad (10.47)$$
$$\text{for } n = 0, \ldots, N-1$$

where $X(k)$ is the set of decoded transform coeffcients, $w(n)$ is the window function with the values shown in Table 10.4, and $y(n)$ is the set of windowed output samples. The variable α is used to control the time offset for the transform basis functions and is given by

$$\alpha = \begin{cases} -1, & \text{for the first short transform} \\ 0, & \text{for the long transform} \\ +1, & \text{for the second short transform} \end{cases} \quad (10.48)$$

10.3.6. Overlap and Add

The final step in the TDAC transform is to overlap the first half of the set of windowed output samples with the second half of the previous block. The following MATLAB code defines this overlap-and-add process.

```
for n=0:N/2-1
   output(n+1) = 2*(y(n+1)+delay(n+1));
   delay(n+1) = y(N/2+n+1);
end
```

10.4. SUMMARY

The ATSC digital television standard prescribes the use of an AC-3 bit stream to transmit audio information. The AC-3 audio standard defines a multichannel coding algorithm that is capable of transmitting up to five full bandwidth channels and one low-frequency effects channel in one of the eight possible channel configurations. In this chapter, the encoding algorithm, syntax, and decoding algorithm for the AC-3 standard are explained.

PROBLEMS

10.1 Draw a simple block diagram of an AC-3 encoder that includes the following system components, and briefly explain the function of each of these components:

(a) Frequency analysis

(b) Exponent coding

(c) Bit allocation

(d) Mantissa quantization

(e) Bit-stream formatting

10.2 With the aid of a diagram describe the format of an AC-3 synchronization frame.

10.3 List the eight audio coding modes that are possible with an AC-3 coder and specify the order in which the channels are coded in each mode.

10.4 Explain what is meant by the term *preecho* in digital audio coding and describe the system used in the AC-3 coder to overcome this problem.

10.5 Explain how the output coefficients of the TDAC transform are converted into an exponent/mantissa representation and the reason for transmitting the coefficients in this manner.

10.6 Explain why channel coupling is used in the AC-3 encoder and discuss the psychoacoustic effect that allows this type of data rate reduction to be performed without introducing audible distortion in the reconstructed audio signal.

10.7 Explain what is meant by the following terms when used in the context of channel coding in the AC-3 encoder:

(a) Coupling subband

(b) Coupling band

(c) Coupling coordinates

(d) Phase flag

(e) Master coupling coordinate

(f) Coupling channel

(g) Coupled channel

(h) Independent channel

10.8 Explain why rematrixing is used in the 2/0 audio coding mode.

10.9 Explain how the coefficient exponent values are transmitted to the decoder using differential exponents.

10.10 Explain the difference between the three exponent strategies D15, D25, and D45, and discuss the types of audio signals that would be best coded using each strategy.

10.11 Determine the absolute exponent value for the first exponent and the differential exponent values that would be transmitted to the decoder for the original exponent values given below. Assume that the D-15 exponent coding strategy is in use.

10.12 Determine the absolute exponent value for the first exponent and the differential exponent values that would be transmitted to the decoder for the original exponent values given below. Assume that the D-45 exponent coding strategy is in use.

10.13 Explain how differential exponents are transmitted to the decoder using *mapped values* and *grouped values*.

10.14 Complete the following table by calculating the actual bits to be transmitted to the decoder for each set of three differential exponent values.

Differential exponent values	Grouped value code word
−2,1,0	
−1,−1,−1	
1,0,−1	

10.15 Discuss the advantages and disadvantages of the AC-3 bit-allocation philosophy of using a standard core bit-allocation procedure that can be modified by delta bit-allocation parameters.

10.16 Explain what is meant by the following terms when used in the context of bit allocation in the AC-3 encoder:

(a) PSD integration

(b) Bark scale band

(c) Log addition

(d) Spreading function

(e) Fast decay masking curve

(f) Slow decay masking curve

(g) Excitation function

10.17 Explain why it is necessary to compensate for the selectivity of the TDAC transform at low frequencies.

10.18 Explain how the masking threshold curve is calculated from the excitation function and the absolute threshold of hearing.

10.19 Describe the process for adjusting the masking curve produced by the core bit-allocation procedure using delta bit-allocation parameters.

10.20 Describe the process for determining the bit-allocation pointer values for each Bark scale band.

10.21 Explain the difference between the symmetric and asymmetric quantization procedures used to quantize the mantissa values.

10.22 Complete the following table by determining the actual bits to be transmitted to the decoder for the given mantissa and bap values.

Mantissa	bap	Code word
0.2311	3	
0.6068	5	
0.4860	7	
0.8913	9	

10.23 Explain the concept of dialog normalization and give examples of when it may be beneficial to use this procedure.

10.24 Explain the concept of dynamic range compression and give examples of when it may be beneficial to use this procedure.

10.25 Complete the following table by determining the normalized standard downmix equations that should be used in the audio reproduction system for the following input/output channel configurations:

Input channels	Output channels	Equations
L,C,R,S	L_O,R_O	
L,C,R,S	L_t,R_t	
L,C,R,S	C	
L,C,R,Ls,Rs	L_O,R_O	
L,C,R,Ls,Rs	L_t,R_t	
L,C,R,S	L,C,R,Ls,Rs	

10.26 Determine the actual bits required in the BSI header of an AC-3 syncframe for the audio stream with the following configuration:

Property	Value
Version	8
Type of service	C
Audio coding mode	1/0
Low-frequency effects channel	Off
Average dialog level	−25 dB
Heavy compression	Off
Copyright	Protected
Original	Yes
Time code (h:min:s:frame)	1:45:15:0

10.27 Determine the actual bits required in the BSI header of an AC-3 syncframe for the audio stream with the following configuration:

Property	Value
Version	8
Type of service	CM
Audio coding mode	3/2
Center mix level	−4.5 dB
Surround mix level	−6 dB
Low-frequency effects channel	on
Average dialogue level	−30 dB
Heavy compression	off
Peak mixing level	100 dB
Room type	Large room, X curve monitor
Copyright	Protected
Original	Yes
Time code (h:min:s:frame)	2:05:10:0

MATLAB EXERCISE 10.1

The aim of this exercise is to implement the spectral envelope coding and decoding algorithms used in the AC-3 encoder and decoder.

1. Write a MATLAB function to calculate the mantissa and exponent values for a set of transform coefficients using the procedure described in Section 10.1.6.
2. Write a MATLAB function to calculate the differential exponent values for a set of absolute exponent values using the procedure described in Section 10.1.7. Your function should have the exponent coding strategy as an input and be able to produce differential exponent values for the three exponent coding strategies.
3. Write a MATLAB function to convert a set of differential exponent values into mapped values and then take three mapped values and produce the bits corresponding to a grouped value.
4. Write a MATLAB function to produce a set of coefficient exponent values from a bit stream containing the code words corresponding to a set of coded group values.
5. Discuss techniques that may be used to automatically choose the most appropriate exponent coding strategy.

MATLAB EXERCISE 10.2

The aim of this exercise is to implement the bit-allocation algorithm used in the AC-3 encoder.

1. Write a MATLAB function to initialize the strtmant and endmant parameters using the code given in Section 10.1.8.1.
2. Write a MATLAB function to convert the exponent values specified by the coded spectral envelope into power spectral density values using the code given in Section 10.1.8.2.
3. Write a MATLAB function to calculate the power spectral density values for the set of 50 Bark scale bands given in Table 10.13 using the code given in Section 10.1.8.3.
4. Write a MATLAB function to calculate the excitation function values for the set of 50 Bark scale bands given in Table 10.13 using the code given in Section 10.1.8.4.
5. Write a MATLAB function to calculate the masking curve values for the set of 50 Bark scale bands given in Table 10.13, using the pseudocode given in Section 10.1.8.6.
6. Write a MATLAB function to calculate the bit-allocation pointer values for the set of transform coefficients, using the pseudocode given in Section 10.1.8.8.

7. Now combine these functions to produce a MATLAB function that determines bit-allocation pointer values for a given set of input coefficients. Determine the required snroffset values to produce an audio block with the following output bitrates:

   (a) 32 kbit/s

   (b) 128 kbit/s

MATLAB EXERCISE 10.3

The aim of this exercise is to implement the quantization and inverse quantization algorithms used in the AC-3 encoder and decoder.

1. Write a MATLAB function to quantize a set of coefficient mantissa values with the number of bits specified by a set of bit-allocation values for each coefficient using the quantization algorithm given in Section 10.1.9.
2. Write a MATLAB function to produce a set of coefficient mantissa values from a bit stream containing the code words corresponding to a set of quantized mantissa values.

REFERENCES

1. ATSC Standard: Digital Audio Compression (AC-3), Revision A, Doc. A/52A, 20 August 2001.

Chapter 11

MPEG-2 Systems

11.1. INTRODUCTION

The video and audio parts of the MPEG-2 standard define the bit-stream syntaxes and the signal processing required for the decoding of video and audio bit streams. Each of these bit streams, known as an *elementary stream*, is generated by an elementary stream encoder (either video or audio) and must be stored or transported across a communications system before being passed to an elementary stream decoder. The systems part of the MPEG-2 standard [1] defines a number of services that are required for the delivery of these elementary streams. These services are as follows:

- *Multiplexing.* Multiplexing is used to interleave two or more elementary streams into a single stream that can be carried on a communications channel. This is achieved using a packet-based multiplex. Multiplexing is discussed in Section 11.3.
- *Timing.* The major purpose of the timing function is to provide synchronization. First, it is necessary during decoding to synchronize the operation of a decoder to the timing of the encoder that generated the bit stream being decoded. Second, synchronization between compressed streams on decoding and playback must be provided. Without this type of synchronization, lip-sync cannot be provided. Timing is discussed in Section 11.4.
- *Buffer management.* A decoder stores a received bit stream until it is ready to decode it. Buffer management is required to ensure that the decoder's internal buffers do not overflow, resulting in the loss of parts of an elementary stream. Buffer management is discussed in Section 11.5.
- *Transmission of control data.* A variety of control data must be transmitted to carry configuration information about elementary streams (allowing them to be reassembled into audio–visual programs), the network carrying the multiplexed stream, and the conditional access system used. This data takes the form of a number of tables and is discussed in Section 11.6.

This chapter describes the structure of MPEG-2 systems and its use in the *digital video broadcast* (*DVB*) and *advanced television standards committee* (*ATSC*) suites

Digital Television, by John Arnold, Michael Frater and Mark Pickering.

of standards. The aim is to provide the reader with an understanding of the operation of MPEG-2 systems but not to describe fully its syntax. A complete description of the syntax can be found in the MPEG-2 systems standard. A number of extensions to MPEG-2 systems have been defined by DVB and ATSC. These extensions both increase the functionality of the systems layer and ease the task of the decoder in some circumstances. They are discussed separately in Chapter 12.

11.2. SERVICE OVERVIEW

The structure of a simple MPEG-2 encoder [2] encoding a single program is shown in Figure 11.1. A video signal is converted into a video bit stream by the video encoder. An audio signal is converted into an audio bit stream by the audio encoder. These two elementary streams are conveyed to the multiplexer, which multiplexes them together and inserts control information to produce a single systems bit stream.

The systems bit stream, illustrated in Figure 11.2, contains interleaved video, audio, and control data. The interleaving is achieved by placing data from the elementary streams in packets. This allows much greater flexibility than would be possible with a periodic time-division multiplexing, particularly for streams that have vastly different data rates and streams whose data rate changes over time.

The various components making up an MPEG-2 systems bit stream are distinguished by the contents of the packet header, in which each component is allocated a unique value of the payload identifier (PID) field.

A television program encoded as an MPEG-2 bit stream contains, therefore, a number of multiplexed elementary streams, each of which contains data that can be decoded by either a video or audio decoder. The MPEG-2 decoder (Figure 11.3) [3] is responsible for decoding of the received systems bit stream and the reassembly of the video and audio components of the program. The systems decoder performs the first stage of decoding, extracting the elementary streams and passing them to the elementary stream decoders, that is, the video and audio decoders. The elementary stream decoders decode the elementary streams and perform the signal processing required to reconstruct the video and audio signals.

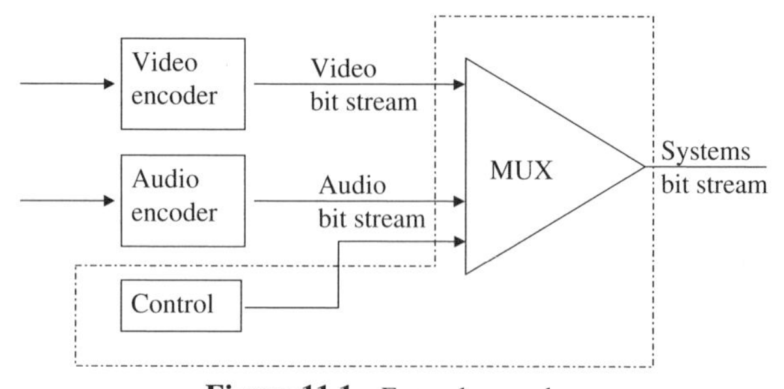

Figure 11.1 Example encoder.

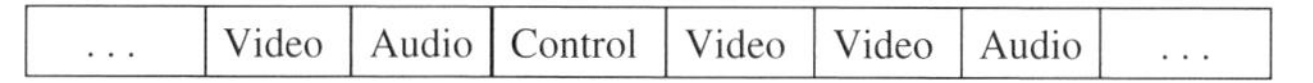

Figure 11.2 Example systems bit stream.

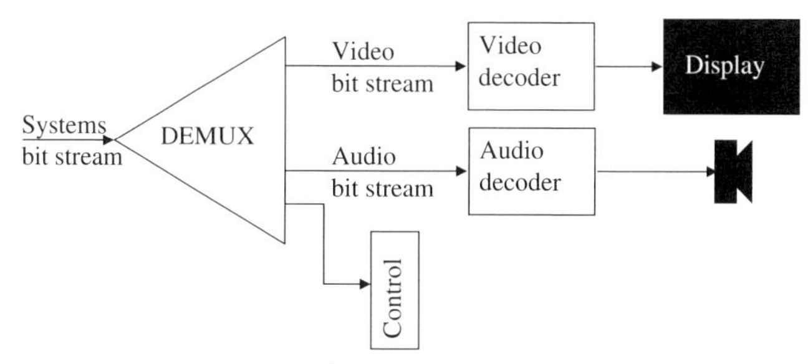

Figure 11.3 Example decoder.

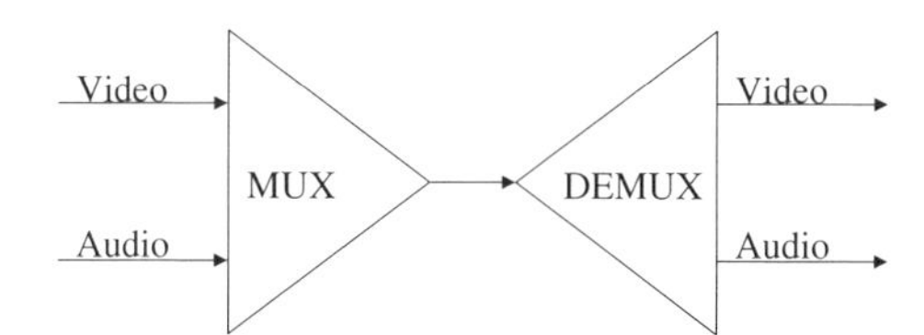

Figure 11.4 Single-program operation.

The systems encoding and decoding processes for a single program can be represented schematically as shown in Figure 11.4. The elementary stream encoders and decoders are not shown.

For a systems bit stream containing multiple programs, the systems encoder and decoder can be represented as shown in Figure 11.5. Elementary streams from the multiple programs are passed into the systems encoder. The resulting systems bit stream is then transmitted to the decoder over a single channel. The systems decoder extracts the elementary streams that make up the multiple programs, passing the streams corresponding to one program to the elementary stream decoders and discarding all other elementary streams.

The sharing of elementary streams between programs is also possible. In the example shown in Figure 11.6, a video stream is shared between two programs that have distinct audio streams. This approach might be used for the broadcast of a

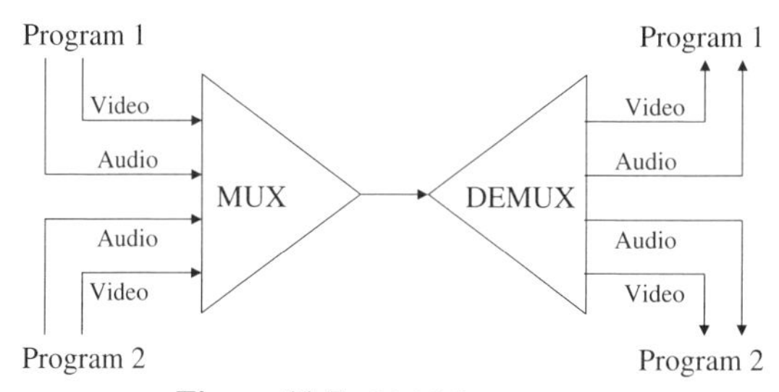

Figure 11.5 Multiple programs.

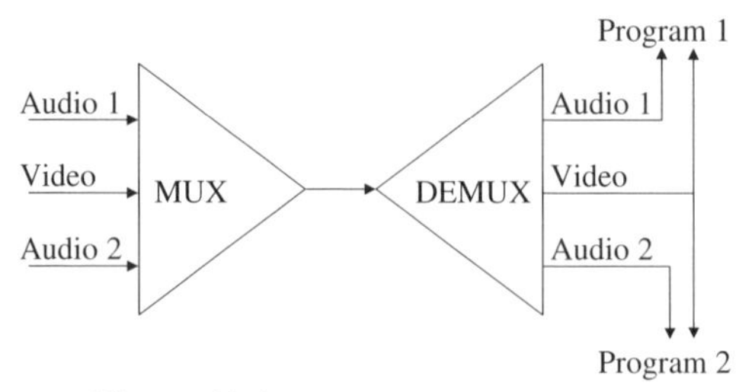

Figure 11.6 Shared elementary streams.

movie dubbed into a number of languages or for a sports broadcast to be viewed in several languages. In the latter case, the soundtracks may not only be in different languages but may also be generated independently, expressing national biases in the outcome of the event.

Although the primary types of data carried by the MPEG-2 systems layer are MPEG-2 video and MPEG-2 audio, it is often necessary to carry other types of data not standardized by MPEG-2. Such data, known as private data, may be used for the transfer of key management data for conditional access systems, the use of non-MPEG-2 audio standards (such as Dolby AC-3), or carrying information associated with a program such as subtitles. The combined use of video, audio, and private data elementary streams is illustrated in Figure 11.7. Both DVB and ATSC make extensive use of private data.

In order to allow synchronization between elementary streams, two types of timing information are carried in an MPEG-2 systems bit stream. Clock references are used to pass the current time to the decoder; time stamps provide information on the time at which specific actions are to occur.

The term *access unit* is used to describe the basic unit of data that is presented to the viewer by the decoder. For a video elementary stream, the access unit corresponds to a picture. For an audio elementary stream, the access unit corresponds to an audio frame.

The control information carried by the MPEG-2 systems layer provides information about timing, buffer management, and the structure of the bit stream, including the *program-specific information* (PSI) that allows a decoder to group elementary streams into programs and to decode one program from a systems bit stream containing several programs. The PSI is organized into a number of tables. Two of these tables effectively provide a table of contents for the bit stream: the

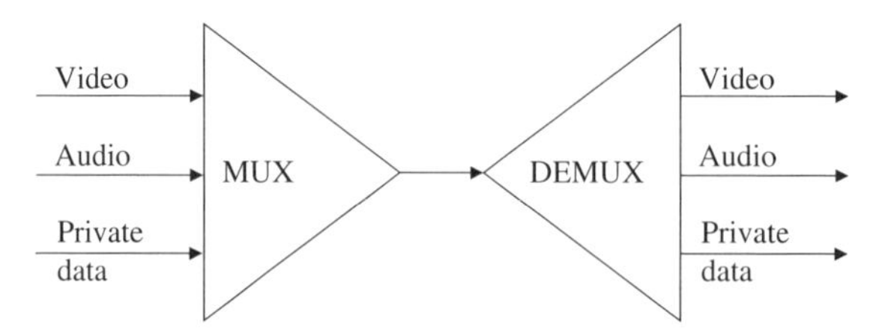

Figure 11.7 Use of private data.

program association table and the program map table. Each program has its own associated program map table, which contains a list of PIDs for the packets carrying the elementary streams making up the program. The program association table carries a list of PIDs for the packets carrying the program map tables of all program carried in the system bit stream, providing a single entry point for the system bit stream.

11.3. MULTIPLEXER STRUCTURE

The MPEG-2 systems multiplexer is a two-layer, packet-oriented multiplex. The upper sublayer, the *packetized elementary stream* (PES) sublayer, breaks the video and audio bit streams into packets and attaches a packet header that contains packetization, timing, and control information. The PES packets are passed to one of the two multiplexing sublayers. The *transport stream* (TS) is the more commonly used of the multiplexing sublayers and is designed to provide robust transmission on unreliable channels. The *program stream* is provided as an alternative to the transport stream and can be used to provide backward compatibility with MPEG-1. The DVB and ATSC standards for digital television use only the transport stream.

In this section, the properties of the MPEG-2 multiplexer are discussed.

11.3.1. PES Sublayer

The PES sublayer packetizes bit streams from elementary stream encoders. Each packet consists of a header, containing configuration information about the packet, timing information and information about the type of the payload, and a payload containing the elementary stream data.

The PES sublayer can be thought of as an adaptation layer between the continuous bit streams generated by the video and audio encoders and the multiplex of the transport stream or program stream. This adaptation layer captures a number of important types of information about the elementary stream including timing and control information. PES packets are passed to either the transport stream or program stream for multiplexing.

11.3.1.1. PES Packet Header

Each PES packet header begins with a 32-bit MPEG start code. The first 24 bits have value 0x000001. The remaining 8 bits are the **stream_id** of the data in the packet, which specifies the type of data carried by the PES packet. This **stream_id** is used extensively by both DVB and ATSC digital television systems.

The remainder of the PES packet header contains the following types of information: packetization information, which provides information about the structure of the packet; timing information, which is used by the decoder for synchronization; and control information, which provides other information about the elementary stream.

The major fields of the PES packet header are shown in Table 11.1. Fields carrying data that only describes the structure of the PES packet header have been omitted. All other fields are included. Even though the lengths of many of the fields in the PES packet header are not multiples of 8 bits, the syntax guarantees that the length of the header is always an integer number of bytes.

The length of a PES packet is specified in the 16-bit **PES_packet_length** field of the PES header. The maximum length that can be specified in this field is $2^{16}-1$ bits, which is the maximum length allowed for all PES packets except for those containing video elementary streams. In the case of video elementary streams, a value of 0 may be placed in this field, indicating that the length is unbounded. Where the PES length is unbounded, a decoder must determine the length from the transmitted bit stream. The reason that the unbounded length option is permitted for video elementary streams is that start-code emulation is prevented in these streams. (Chapter 5 provides an explanation of start-code emulation.) Because audio elementary streams

Table 11.1 Fields of PES packet header.

Field	Size (bits)	Use
packet_start_code_prefix	24	0x000001
stream_id	8	Identifies the type of the elementary stream, for example, video, audio, or private data
PES_packet_length	16	Number of bytes in the PES packet following this field
PES_scrambling_control	2	Indicates whether or not the payload of the PES packet is scrambled
PES_priority	1	"1" indicates higher priority that "0"
data_alignment	1	"1" indicates that the PES header is immediately followed by a video start code or audio sync word
Copyright	1	Indicates that the payload is protected by copyright
PTS	33	Presentation time stamp
DTS	33	Decoding time stamp
ESCR	42	Elementary stream clock reference
trick_mode_control	3	Signals the use of pause, fast forward, and fast reverse modes—primarily for video-on-demand applications
previous_PES_packet_CRC	16	Intended for network maintenance—only calculated on payload because header can be modified in transport
stuffing_byte	8	Up to 32 bytes of stuffing may be added to a PES header

do not use the same start-code structure, start-code emulation is therefore not prevented in these streams.

The use of scrambling on the body of the PES packet can be indicated using the **PES_scrambling_control** field. The binary value 00 indicates no scrambling; the remaining three values indicate that scrambling is used. Differences in the meanings of these three values are not defined by MPEG-2 but are left to vendors of conditional access systems.

The primary type of timing information carried by the PES packet header is time stamps. The structure of these fields is discussed in Section 11.4. Two types of time stamps may be used: PTS and DTS. The PTS tells the decoder the time at which decoded information should be presented to the viewer and may be used with video, audio, and private data. The decoding time stamp is for use where a picture must be decoded before it is displayed and tells the decoder when decoding should begin. (More information on clock references is provided in Section 11.4.2.) Clock references can also be carried in the PES header using the elementary stream clock reference (ESCR). In principle, this means that each elementary stream can be synchronized to an independent clock. In digital television, all streams making up a program are usually required to be synchronized to a single clock, and this is not used.

The **stuffing_byte** may be used to pad a header. One use of this field is to allow the length of the PES header to remain unchanged when a field is deleted during transmission. This might occur when a time stamp is deleted during transcoding.

Alignment between the boundaries of PES packets and the video bit stream offers a number of advantages, especially for error resilience. Byte alignment is always preserved, because a PES packet always contains a whole number of bytes of the video bit stream. Byte alignment of start codes in the video bit stream is achieved by inserting 0 bits at the end of a video bit stream until its length in bits is a multiple of 8. Setting the **data_alignment** field of the PES header to 1 indicates that the PES header is immediately followed by a video start code or audio sync word. For video, this means that the PES packet is aligned with the slice layer, picture layer, GOP layer, or sequence layer. Further information on alignment may be carried in the PSI referring to this stream as discussed in Section 11.6.2.2.

11.3.1.2. Overheads Due to the PES Layer

The basic length of the PES header used with video and audio elementary streams is 8 bytes. A major component of this first part of the header is a group of flags that indicate the presence or absence of other parts of the PES header. Where a time stamp is present in the PES header, an additional overhead of 5 bytes is added to the basic overhead. Time stamps are required no more often than once per picture and are often required much less often. Their overhead amounts to much less than 0.1% of the total rate. Where the ESCR is present, a further 6 bytes of overhead is added. A further 2 bytes is added if the optional CRC is transmitted, which does not usually occur in normal transmission. Further variable overheads are added by the use of trick modes, or inclusion of private data or stuffing bytes in the header.

The average overhead of the PES layer is therefore approximately 8 bytes per PES packet, with the additional data added by the use of optional fields not contributing significantly. The total overhead (O) therefore depends on the number of PES packets per second (P) and the data rate in bytes per second (R):

$$O = 8\frac{P}{R}$$

EXAMPLE 11.1—PES Layer Overheads

Calculate the percentage overhead for a 4-Mbit/s, Rec.601-resolution video sequence (25 pictures per second, with each picture containing 576 lines) with one slice per row of macroblocks and one PES packet per slice. Repeat the calculation for HDTV (1088 lines per picture, with a frame rate of 25 frames per second) at 20 Mbit/s.

In 576 lines, there are 36 rows of macroblocks. With one slice per row of macroblocks and one PES packet per slice, we have $25 \times 36 = 900$ PES packets per second or 7200 byte/s. For a video elementary stream at Rec.601 resolution and 4 Mbit/s (500 Kbyte/s) with one PES packet per slice, the overhead is therefore 1.4%.

In 1088 lines, we have 68 rows of macroblocks. With one slice per row of macroblocks and one PES packet per slice, we have $25 \times 68 = 1700$ PES packets per second or 13.6 Kbyte/s. For HDTV at 20 Mbit/s (2.5 Mbyte/s) with one PES packet per slice, the overhead is approximately 0.54%. The use of larger PES packets reduces these overheads to even lower levels. ■

11.3.2. Transport Stream Sublayer

The purpose of the transport stream sublayer is to provide a robust multiplex for PES packets from a number of elementary streams. All data is carried in packets with length 188 bytes. This packet length is chosen to allow an MPEG-2 transport stream packet to fit into four asynchronous transfer mode (ATM) cells, allowing for the overhead of the ATM adaptation layer. The first four bytes of each packet are the packet header. The remaining bytes may optionally contain an adaptation field, payload bytes, or an adaptation field followed by payload bytes.

Each section of each PSI table is carried as the payload of one or more transport stream packets. Multiple sections with a common PID may be carried in a single transport stream packet.

11.3.2.1. *Transport Stream Packet Header*

The allocation of the contents of the transport stream packet header is shown in Table 11.2. The first byte of the transport stream packet header is the **sync_byte**. This field always has the value 0x47 and is used for synchronization. When a decoder begins the decoding of a new transport stream, it first searches the stream for the value 0x47. Because there is no protection against emulation of the **sync_byte** (unlike the start codes used in MPEG-2 video), the decoder must ensure that

Table 11.2 Transport stream packet header. Fields are transmitted in the order indicated. Each field is transmitted with its most significant bit first.

Field	Size (bits)	Use
sync_byte	8	Always 0100 0111 (0x47)—used for Synchronization.
transport_error_indicator	1	Set if network's transport layer detects an error.
payload_unit_start_indicator	1	Packet begins with new PES packet or PSI section.
transport_priority	1	"1" indicates higher priority than "0."
PID	13	Payload ID; indicates number and type of elementary stream.
transport_scrambling_control	2	Indicates use of encryption.
adaptation_field_control	2	Indicates presence or absence of adaptation field and payload.
continuity_counter	4	Increases with each transport stream packet with same PID. Can be used to detect lost packets.

a number of correct **sync_byte** fields are received spaced exactly 188 bytes apart. The **transport_error_indicator** is usually set to 0, but may be set to 1 where the transmission network's transport layer detects an error that is passed uncorrected to the MPEG-2 decoder.

The **payload_unit_start_indicator** field is used to indicate whether the transport stream packet contains the first byte of a new PES packet or a new section of a PSI table. If the transport stream carries PES data, setting the **payload_unit_start_indicator** to 1 indicates that the payload of the transport stream packet begins with a new PES packet, that is, the PES start code consisting of 0x000001 followed by the 8-bit **stream_id**; a value of 0 indicates that no PES packet starts in this transport stream packet. For transport stream packets carrying PSI data, setting the **payload_unit_start_indicator** to 1 means that the first byte of the payload of the transport stream packet points to the start of the first new PSI section to start in this transport stream packet. The possible uses of **payload_unit_start_indicator** are illustrated in Figure 11.8.

The **transport_priority** field may be used by an encoder to indicate to the transport layer of the transmission network that some transport stream packets have higher priority than others. This may be useful in distribution systems, such as those based on ATM switching, that are able to provide higher levels of protection against error and loss to some packets. The **PID** identifies the elementary stream with which the transport stream packet is associated. The 13 bits assigned to this field permit up to 8192 different streams to be uniquely identified. Some values of this field are preassigned, such as 0x0000 being used for the program map table. A **PID** value of 0x47 should not be used, as this may be confused by a decoder with the **sync_byte**. The **transport_scrambling_**

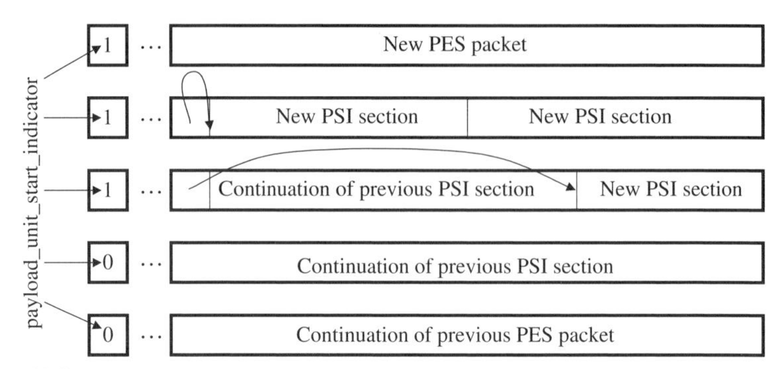

Figure 11.8 Uses of payload_unit_start_indicator in the MPEG-2 transport stream packet header.

control field is used to indicate if the contents of the transport stream packet are scrambled for conditional access. A value of 0 indicates no scrambling. All other values indicate the use of conditional access. Definition of any difference in meaning between the available nonzero values is left to the implementor. The **adaptation_field_control** indicates whether or not the adaptation field (described below) and payload are present in the transport stream packet. Two bits are required to signal one of the three possible packet configurations: adaptation field only, payload only, or both adaptation field and payload. The **continuity_counter** is increased by one for each successive packet having the same value of **PID**.

11.3.2.2. Adaptation Field

Table 11.3 shows a summary of the contents of the adaptation field, listing fields whose value impacts on the remainder of the transport stream but omitting flags and field lengths that affect only the contents of the adaptation field.

The purposes of the adaptation field are to carry the clock references used by the decoder to recover information about the timing of the stream (described in Section 11.4) and to provide information to assist buffer management, including notification of discontinuities in the clock reference, splicing points, and the latest valid arrival time for a transport stream packet. Without these clock references, a decoder would not be able to reliably decode a received bit stream and to regenerate the video frame and audio sample clocks.

Splicing points provide support for the editing of MPEG-2 systems bit streams. Editing without the use of splicing points requires that the system time clock run continuously, that is, without break. The major impact of this is that when two MPEG-2 bit streams are spliced together, the timing of one must be altered so that the system time clock is continuous across the point where they are joined. This usually requires that the transport stream of at least one of the sequences is recoded, although it may be possible to avoid recoding of video and audio material. The VBV requirements (see Chapter 6) of video, however, must also be met. Meeting this requirement may sometimes require recoding of video.

Table 11.3 Summary of adaptation field contents.

Field	Size (bits)	Use
discontinuity_indicator	1	Indicates discontinuity in system time clock—the next program clock reference in a transport stream packet with same PID begins a new system time clock.
random_access_indicator	1	Current transport stream packet contains information to aid random access—a video sequence header for a video elementary stream or the start of an audio frame for an audio elementary stream.
elementary_stream_priority_indicator	1	Indicates priority of transport stream packet within elementary stream.
program_clock_reference	42	PCR—this clock reference is used by the decoder to reconstruct the STC.
original_program_clock_reference	42	Original PCR—used to transmit the original value of the PCR when it has been modified during remultiplexing or transcoding.
splice_countdown	8	Number of transport stream packets of same PID until a splicing point.
transport_private_data_length	8	Number of bytes of private data contained in this transport stream packet.
ltw_offset	15	Legal time window in which transport stream packet should arrive to prevent illegal buffer conditions.
piecewise_rate	22	Notional rate used to define end of legal time window.
splice_type	4	Assists the decoder to prevent buffer overflow at the next splice.
DTS_next_AU	33	Decoding time stamp for the next access unit, which begins in the transport stream packet after **splice_countdown** reaches 0.

The purpose of splicing points in MPEG-2 systems bit streams is to avoid the need for recoding of either systems bit streams or elementary streams during editing. Where a splice occurs in an MPEG-2 systems bit stream, the **discontinuity_indicator** field of the transport stream packet after the splice is set to 1. This means that the **continuity_counter** field of the header of the first transport stream after the splice may have an arbitrary value. The **discontinuity_indicator** field of the transport stream packet after the splice carrying the program clock reference (PCR) is also set to 1. This means that the decoder must resynchronize to a new system time clock. The **splice_countdown** field of the adaptation field of transport stream packets preceding the splice is used to inform the decoder of the precise location of the splice. The **splice_type** field in the adaptation field of transport stream packets leading up to the splice is used to provide information to the decoder on the peak data rates that can be expected as a result of the splice, assisting the decoder with its buffer management.

Stuffing bytes, whose value is 0xFF, are available in the adaptation field. These are used to fill up transport stream packets that are only partially filled by the payload and adaptation fields. This might happen in the last transport stream packet of a PES packet.

Null transport stream packets, whose PID is alway 0x1FFF, are allowed. These transport stream packets can be inserted and deleted when transport streams are multiplexed to provide fine adjustment of bit rate. They can also be inserted in place of transport stream packets from a program that is deleted from a transport stream. The payload bytes of a null transport stream packet may take any value.

Private data may also be carried in the adaptation field. This private data may be used to extend the functionality of the adaptation field but is obviously available only to proprietary decoders.

An adaptation field containing a single clock reference is 8 bytes in length.

11.3.2.3. Transport Stream Payload

The payload of a transport stream packet consists of elementary stream data or PSI data. Each transport stream packet may carry data from only one elementary stream. Alternatively, a transport stream packet may carry data from one or more PSI sections. Multiple PSI sections may be carried in one transport stream payload, subject to the constraint that these PSI sections have the same associated PID. This often prevents carrying sections from multiple tables in a single transport stream packet.

11.3.2.4. Alignment between Transport Stream and PES

Byte alignment of the PES packet is always preserved by the transport stream. PES packets may only begin at the beginning of a transport stream packet. The last transport stream packet carrying a PES packet or PSI section may be only partially full; the remaining bytes of this transport stream packet have the value 0xFF.

11.3.2.5. Overhead Due to Transport Stream

The overhead due to the transport stream packet header (O_1) is 4 bytes per packet (total length 188 bytes), giving $O_1 = 2.1\%$, which is mostly associated with synchronization. Additional overheads come from the last transport stream packet carrying a PES packet being only partially full and the transmission of the adaptation field.

The amount of overhead due to partially full packets depends on the alignment between elementary stream (video, audio, or private) data and the PES layer and the proportion of the last transport stream packet that is empty. We assume that this last packet is on average half full. Then the overhead due to partially full transport stream packets (O_2) depends on the number of transport stream packets per second (T) and the number of PES packets per second (P):

$$O_2 = \frac{0.5P}{T}$$

The overhead due to the adaptation field depends on how often this field is transmitted. In a transport stream, it is necessary to transmit the clock at least once every 100 ms, that is, 10 times per second (see Section 11.4). The clock reference is transmitted more regularly where there are discontinuities in the clock reference due to, for example, splicing. Based on the minimum 8 bytes required for an adaptation field containing a single clock reference, an overhead of 80 bytes per second is required for each program. This is insignificant for digital television applications.

The significant sources of overhead are therefore due to the transport stream packet header and the use of partially filled transport stream packets due to alignment between the PES packets and elementary streams. The total overhead O is therefore the sum of O_1 and O_2:

$$O = \frac{4}{188} + \frac{0.5P}{T}$$

EXAMPLE 11.2—Overhead Due to Transport Stream

Calculate the overheads due to the transport stream for Rec.601 video (25 pictures per second, with each picture containing 576 lines) at 4 Mbit/s and HDTV (1088 lines per picture, with a frame rate of 25 frames per second) at 20 Mbit/s, assuming alignment between the PES and video layers at either the slice or picture layer. Assume that each slice contains one complete row of macroblocks.

For all cases, the overhead due to the transport stream packet header is approximately 2.1%.

For Rec.601 video, there are 576/16 = 36 rows of macroblocks per picture, or 900 rows of macroblocks per second, corresponding to 900 PES packets per second if the alignment between PES and video layers is at the slice layer. For each of these PES packets, there is an average of one transport streams packet that is approximately half full. At 4 Mbps (500 kbyte/s), there are 500,000/184 = 2717 transport stream packets per second. The overhead due to partially full transport stream packets is therefore approximately 16.6%, giving a total overhead of approximately 19%.

If the alignment between PES and video layers is at the picture layer, there are 25 PES packets per second. The overhead for this case is therefore approximately 0.5%, giving a total overhead of approximately 2.6%.

For HDTV, with alignment between video and PES at the slice layer, there are $25 \times 68 = 1700$ PES packets per second. On average, each of these leads to one transport stream packet that is half full. For HDTV at 20 Mbps (13,587 transport stream packets per second), the overhead due to partially full packets is approximately 6.3%. With alignment between video and PES at the picture layer, the overhead due to partially full transport stream

Table 11.4 Summary of transport stream overheads for video elementary streams in digital television applications.

	Rec.601 at 4 Mbps	HDTV at 20 Mbps
One slice per PES packet	19%	8%
One picture per PES packet	2.6%	2.2%

packets is approximately 0.1%. (This latter case would almost certainly require the use of the unbounded option on the **PES_packet_length** field.)

These overheads are summarized in Table 11.4. ■

11.3.3. Program Stream Sublayer

The program stream is provided for backward compatibility with MPEG-1 systems. It supports variable length packets, formed by grouping PES packets into *packs*. Each pack consists of a pack header possibly containing an optional system header, and zero or more PES packets. Although the program stream is used for DVDs, it is not used in either DVB or ATSC and is not discussed further here.

11.4. TIMING

The MPEG-2 timing functions are required to provide synchronization of the decoder to the encoder and between the elementary streams making up a program. Without synchronization of the decoder to the encoder, the decoder's video frame clock may run slower than the encoder's. Even if the difference in clock speed is very small, video frames will eventually have to be dropped when the internal buffers of the decoder overflow. On the contrary, if the decoder's video frame clock runs faster than the encoder's, extra video frames will have to be inserted by playing back some frames twice. In addition, if synchronization between elementary streams is not possible, the decoder is unable to preserve features such as lip synchronization in corresponding audio and video streams.

Synchronization of the decoder to the encoder requires that the two devices have the same concept of time. Synchronization between elementary streams requires that the encoder can pass information to the decoder on the times at which events associated with an elementary stream should occur. The common concept of time used by the encoder and decoder is the *system time clock* (STC). Inclusion of regular *clock references* in an MPEG-2 systems bit stream allows the transfer of the STC from encoder to decoder, allowing synchronization of decoder to encoder. The use of *time stamps* associated with events in individual elementary streams permits synchronization between these elementary streams. The concepts of the STC, clock references, and time stamps (illustrated in Fig. 11.9) are discussed in the following sections.

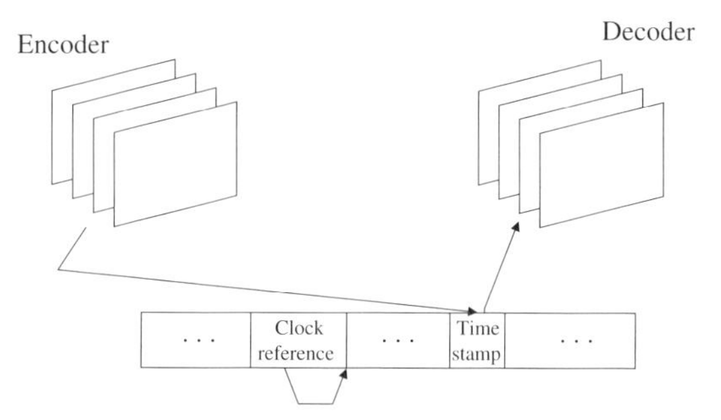

Figure 11.9 Time stamps and clock references.

11.4.1. System Time Clock

All timing in an MPEG-2 encoder or decoder is defined with respect to the STC. This 27-MHz clock is notionally generated by all encoders and must be reconstructed by decoders from clock references transmitted in the bit stream. A decoder can use the STC to manage its internal buffers, to generate the video frame and audio sample clocks, and to identify the precise time at which presentation units should be displayed. The relationship between the 27-MHz STC and the notional frequencies of the various video and audio clocks is shown in Table 11.5.

Each program in an MPEG-2 transport stream may have its own STC. There is no requirement for any timing relationship between these different STCs. Even within one program, discontinuities in the STC are permitted. These are commonly introduced at splicing points, where material from different sources is joined.

11.4.2. Clock References and Reconstruction of the STC

Clock references transmitted in an MPEG-2 systems bit stream are used by a decoder to reconstruct the STC. In simple terms, this means that clock references are used by the decoder to find out what the current time is.

Table 11.5 Divisors required to generate notional video and audio clocks from 27-MHz STC.

Clock	Divisor
25-Hz video frame	1,080,000
50-Hz video field	540,000
30-Hz video frame	900,000
60-Hz video field	450,000
44.1-kHz audio	30,000/49
48-kHz audio	1,125/2

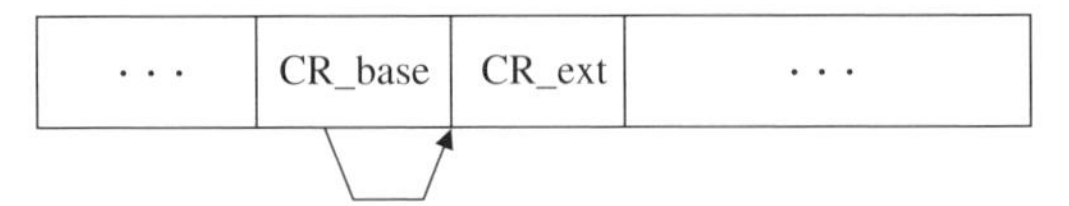

Figure 11.10 Timing of clock reference.

A clock reference is transmitted in the bit stream in two parts whose length totals 42 bits. The first is the **CR_base** (33 bits) and the second is the **CR_ext** (9 bits). The **CR_base** field specifies the time with respect to a 90-kHz clock derived by dividing the STC by 300. The **CR_ext** field specifies the remainder from this division. The full clock reference is used to reconstruct the STC. Time stamps are specified with respect to the 90-kHz clock of the **CR_base**.

The value of the clock reference is specified with respect to the location of the last bit of **CR_base** as illustrated in Figure 11.10.

The decoder uses the received clock references to regenerate the STC. This may be achieved by the use of a phase-locked loop (Fig. 11.11), which is used to adjust the operating frequency of a 27-MHz oscillator so as to keep it synchronized to a corresponding oscillator in the encoder. The phase-locked loop works by comparing received clock references to its internally generated STC. The differences are passed into a low-pass filter, whose output is used to speed up or slow down the oscillator from which the internal STC is derived. The use of a well-designed low-pass filter prevents jumps in the value of the STC and short-term, large variations in the operating frequency of the oscillator.

Simpler decoders may use a free-running oscillator and simply update the current value of the STC every time a clock reference is received, reducing complexity by removing the need for a phase-locked loop. Such decoders are likely to be particularly susceptible to delay jitter in the distribution system, which can cause larger jumps in the value of the STC.

Where STC discontinuities are signaled in the systems stream, the MPEG-2 decoder discards the memory in its clock recovery circuit, allowing the new STC to be generated. In the case of the structure shown in Figure 11.11, this requires setting to 0 all the values stored inside the low-pass filter. In digital television applications, the interval between clock references transmitted by an encoder must never exceed 100 ms. STC discontinuities may cause additional clock references to be transmitted.

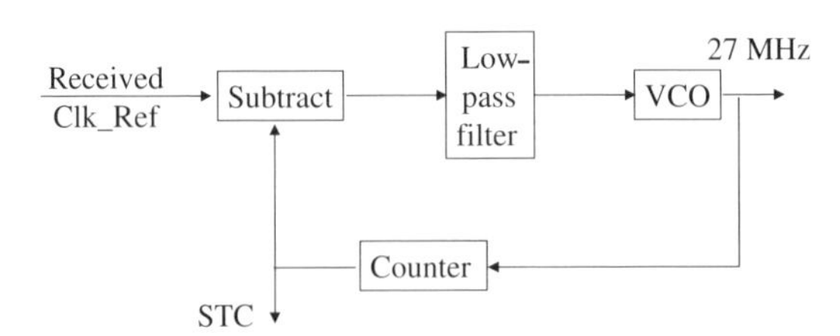

Figure 11.11 Outline structure of clock recovery circuit.

EXAMPLE 11.3—Clock Reference

How is the STC value 4,505,342 represented?

For an STC value of 4,505,342, the value of **CR_base** is the integer part of 4,505,342/300 = 15,017 (represented in hexadecimal as 0x000003AA9). The value of **CR_ext** is the remainder from the division 4,505,342/300, which is 242 (0xF2 in hexadecimal). ■

11.4.3. Time Stamps

Time stamps tell a decoder when an event associated with an elementary stream should occur. There are two types of time stamp: the *presentation time stamp* (PTS) and the *decoding time stamp* (DTS). The PTS is used to determine the display time of, for example, a video picture. This allows synchronization between elementary streams and generation of accurate video frame and audio sample clocks. The DTS may be used by a decoder to determine when it is necessary to start decoding the data for an access unit. The specification of DTS is based on the *system target decoder* (STD), which is a notional decoder architecture specified by MPEG-2. For decoders with different architectures, the DTS can be translated to a value appropriate to that architecture. The DTS is used only where the decoding time differs from the presentation time. One example is in a video sequence incorporating B-pictures, where I and P-pictures immediately following a B-picture (in presentation order) are transmitted before the B-picture.

Each time stamp is encoded as a 33-bit field. These 33 bits specify the time at which an event (presentation or decoding) associated with an elementary stream occur. The value of the time stamp provides the time with respect to the 90-kHz clock derived by dividing the STC by 300 and encoded as the base part of the clock reference.

11.5. BUFFER MANAGEMENT

The basic structure of an end-to-end MPEG-2 system is shown in Figure 11.12. Video and audio arrive at the left side of their respective elementary stream encoder, pass though the system and emerge at the right side with all timing relationships preserved. For the usual case of a constant video picture rate and a constant audio sample rate, this means that the video picture rates at either end of the system are the same, as are the audio sample rates. Any timing relationship between video and audio is also preserved.

In a simple system, the timing relationships may be preserved at all points through the chain from encoder to decoder. Because this would imply, for example, that all macroblocks of video were coded with the same number of bits, this situation never arises in practice. The usual situation is that the output of the video encoder has constant picture rate but variable bit rate. The buffer converts this output into a constant rate bit stream. Because the number of bits used to code each picture is not constant, pictures occupy different amounts of time in the output bit stream of the buffer.

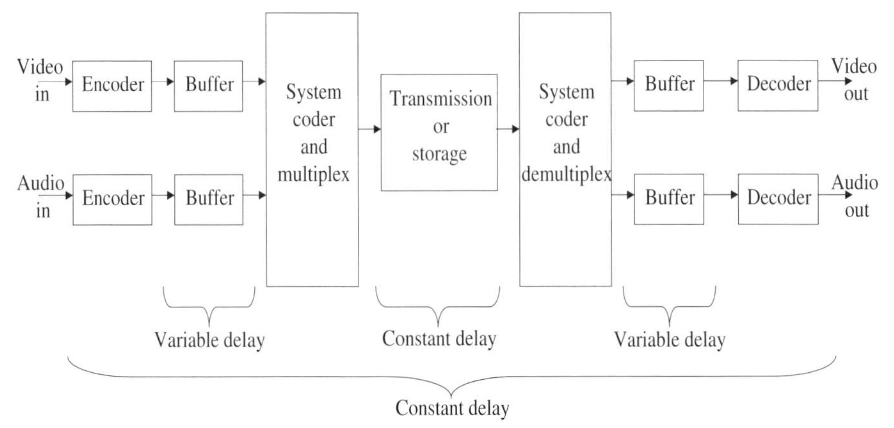

Figure 11.12 Basic structure of end-to-end MPEG-2 system.

A smaller amount of variability in the delay is introduced by multiplexing services at different rates. If, for example, there are 10 packets of video generated for each packet of audio, each of the 10 consecutive video packets experiences a different delay in multiplexing, which depends on how close it lies to an audio packet. If there were two audio packets for every nine video packets, then the audio packets experience variable multiplexing delay as well.

The bit stream that travels through the channel can be thought of as being constant bit rate but variable sample rate (i.e., variable picture rate for video). At the decoder, the buffer converts this signal back into a constant sample rate, variable bit-rate stream that is passed to the elementary stream decoder.

Practical decoders have finite-length buffers. Providing a high-quality service requires that the bit stream is such that these buffers do not overflow. This guarantee is provided by placing constraints on the minimum-length buffers that a decoder provides and on the amount of variability in sample rate that can be introduced into a bit stream by the encoder buffering.

An important consequence of the requirement to preserve timing relationships through encoding, multiplexing, demultiplexing, and decoding is that the end-to-end delay is constant. This means that an audio sample and a video picture arriving at the encoder at the same time will also be the output from the decoder at the same time.

The STD is defined by MPEG-2 as a reference structure against which parameters such as timing and buffer sizes can be defined. The STD defines the interconnections, transfer rates, and buffer sizes for a notional decoder, in which it is assumed that the elementary stream decoders operate instantaneously. There is no requirement, however, that a manufacturer implement this architecture in a real decoder. The purpose of the STD is to provide a framework allowing manufacturers to build decoders that are capable of decoding all compliant bit streams without loss of data due to buffer overflows. In other words, specifying buffer sizes with respect to the STD provides enough information so that a manufacturer can determine the buffer sizes required for a different decoder architecture.

11.6. PROGRAM-SPECIFIC INFORMATION

The MPEG-2 PSI provides information to the decoder on the structure of the MPEG-2 systems bit stream. This information includes the structure of the transport stream itself, the grouping of elementary streams to form programs, the types of these elementary streams (i.e., video, audio, or private data), the conditional access system (if any) used, and details of the distribution network. This information is contained in four tables: the *program association table* (PAT), the *program map table* (PMT), the *conditional access table* (CAT), and the *network information table* (NIT).

The program map table and conditional access table provide information about the *programs* carried in the transport stream. An MPEG-2 program can be thought of as comprising a set of elementary streams that are decoded and displayed simultaneously. Programs may share elementary streams. Elementary streams can exist for which there is no PSI information, which are therefore not part of any program and would not normally be decoded or displayed. PSI information may also be provided for programs whose elementary streams are not contained in the transport stream. The program map table and conditional access table carry most of their data in descriptors. Depending on where they are placed, a descriptor may refer to a single elementary stream, to all the elementary streams making up a program, or to the conditional access system.

The specific purposes of the tables are as follows:

- *Program association table.* The program association table is used to inform the decoder of the PIDs in which the program map table is carried for each program as well as the PID used for the network information table (if present).
- *Program map table.* Each program has its own program map table section, which specifies the PIDs of the elementary streams that make up the program and the type of each of these streams. Descriptors in the program map table may be associated with the whole program or with one elementary stream.
- *Conditional access table.* Where encryption is used to provide conditional access, the conditional access table is used to inform the decoder of the type of encryption used. A descriptor placed in the conditional access table provides system-wide information on conditional access.
- *Network information table.* A decoder requires a variety of information about the transmission network in order to receive and decode the transport streams carried by it. This type of information may be carried by the network information table.

11.6.1. MPEG-2 Descriptors

Most of the data in the PSI is carried in the descriptors. This section describes the generic syntax of MPEG-2 descriptors, before describing the structure of each individual descriptor.

Table 11.6 Generic syntax of descriptors used in MPEG-2 PSI tables.

Syntax	Number of bits	Mnemonic
*type*_descriptor() {		
descriptor_tag	8	uimsbf
descriptor_length	8	uimsbf
...		
}		

11.6.1.1. Generic Descriptor Syntax

Because of the great variability in the types of information specified in the program map table and conditional access table, specific fields are not defined for most types of data carried in these tables. The basic idea of a descriptor is that it carries information on both the type of the data and its value. This is in contrast to most data in MPEG-2, where the type is not explicitly transmitted. The generic syntax of a descriptor contained in an MPEG-2 PSI table is shown in Table 11.6. (This pseudo-C syntax is explained in Chapter 5.)

The **descriptor_tag** field uniquely identifies the descriptor. For example, the value 0x02 identifies the descriptor as a video_stream_descriptor. The meanings of the possible values of the **descriptor_tag** field are shown in Table 11.7. The descriptor type is independent of the table in which the descriptor is transmitted.

Table 11.7 Descriptor names and purposes associated with values of the descriptor_tag.

descriptor_tag	Descriptor	Purpose
0	Reserved.	Not currently used.
1	Reserved.	Not currently used.
2	Video stream descriptor.	Provides basic information about a video elementary stream, such as profile and level, frame rate, and chroma format.
3	Audio stream descriptor.	Provides basic information about an audio elementary stream, including the type (e.g., layer number) of the audio stream.
4	Hierarchy descriptor.	Identifies program elements containing components of hierarchically coded audio and video, including spatial, temporal, and SNR scalability.
5	Registration descriptor.	Identifies formats for private data.
6	Data stream alignment descriptor.	Used to indicate the type of alignment for an elementary stream with data alignment == 1 in the PES header. For example, in video, alignment may be at the slice, picture, GOP, or sequence layer.

Table 11.7 (*Continued*)

descriptor_tag	Descriptor	Purpose
7	Target background grid descriptor.	Specifies the size of the display area where a decoded video stream is not to occupy the whole display area.
8	Video window descriptor.	Specifies the offset from the top left corner of the display area where a decoded video stream is not to occupy the whole display area.
9	CA descriptor.	Used in the PMT to indicate that conditional access is used to scramble a program. The type of CA used is specified. It may be used to carry (as private data) system-wide information (in the CAT) or elementary-stream-specific information (in the PMT)
10	ISO 639 language descriptor.	Specifies the language of the associated program (e.g., French) and the audio type (e.g., hearing impaired or visual impaired commentary).
11	System clock descriptor.	Specifies the source and accuracy of the clock used to generate clock references.
12	Multiplex buffer utilization descriptor.	Provides bounds on the occupancy of system target decoder multiplex buffers. Primarily intended for use in remultiplexing.
13	Copyright descriptor.	Identifies the owner of copyright on a program element and carries private data specific to the copyright owner.
14	Maximum bit-rate descriptor.	Indicates the maximum bit rate in a program.
15	Private data indicator descriptor.	Contains 32 bits of private data.
16	Smoothing buffer descriptor.	Specifies the size and leak rate of the smoothing buffer associated with a program element. This descriptor can be used by a decoder to assist with buffer management.
17	STD descriptor.	Used in buffer management for video streams.
18	IBP descriptor.	Contains information on the sequence of picture types in a video sequence.
19–63	Reserved.	Reserved for future ISO use.
64–255	User private.	Available for private definition.

At least in principle, this means that any descriptor can be transmitted in either the conditional access table or program map table. The **descriptor_length** field specifies the number of bytes of data contained in the descriptor following the **descriptor_length** field. The use of the **descriptor_length** field allows a decoder that is unable to decode a particular descriptor to skip over its data and begin decoding the data that immediately follows. The remaining bits making up a descriptor carry the value of the descriptor. The syntax of these bits is defined separately for each descriptor.

11.6.1.2. MPEG-2 Audio Stream Descriptor

The MPEG-2 audio stream descriptor is used to provide information on an MPEG-2 audio elementary stream as shown in Table 11.8. The **free_format_flag** is set to 1 where one or more audio frames may have a value of 0 in the **bitrate_index** field in the audio elementary stream. The **ID** and **layer** fields have the same meanings as the corresponding fields in the associated audio elementary stream (Chapter 9). The **variable_rate_audio_indicator** is set where the bit rate may vary between consecutive audio frames.

A simplified representation of the structure of the MPEG-2 audio stream descriptor is shown in Figure 11.13, leaving out fields that are only used in the decoding of the descriptor. In this representation, an "*" next to a field represents a field that is optional or may be repeated, whereas a brace indicates that a group of fields is optional or may be repeated. This simplified representation is used throughout the remainder of this chapter and also in Chapter 12. Readers requiring more detailed information on the syntax of the various descriptors are referred to the MPEG-2 systems standard [1].

Table 11.8 Syntax of the MPEG-2 audio stream descriptor.

Pseudo-C syntax	Number of bits	Data type	Description
audio_stream_descriptor() {			Function header—does not contribute bits to the stream.
descriptor_tag	8	uimsbf	Standard descriptor header—see Table 11.6
descriptor_length	8	uimsbf	
free_format_flag	1	bslbf	Data from MPEG-2 audio header describing the audio elementary stream—specific to the audio stream descriptor.
ID	1	bslbf	
Layer	2	bslbf	
variable_rate_audio_indicator	1	bslbf	
Reserved	3	bslbf	Pad to byte boundary.
}			Function close—contributes no bits to the stream

audio_stream_descriptor	
free_format_flag	flag
ID	flag
layer	{1, 2, 3}
variable_rate_audio_indicator	flag

Figure 11.13 MPEG-2 audio stream descriptor.

EXAMPLE 11.4—Audio Stream Descriptor

Using the simplified representation, what are the contents of the audio stream descriptor describing a constant bit-rate audio stream with the **bitrate_index** *fields taking values other than 0000 and* **ID** *field set to 1 and* **Layer** *2?*

The audio stream descriptor carrying this information is shown in Figure 11.14.

audio_stream_descriptor	
free_format_flag	0
ID	1
layer	2
variable_rate_audio_indicator	0

Figure 11.14 Example MPEG-2 audio stream descriptor. ■

11.6.1.3. MPEG-2 Video Stream Descriptor

The purpose of the video stream descriptor (Fig. 11.15) is to provide information about one video elementary stream. This descriptor is usually carried in the program map table.

The video frame clock associated with a video elementary stream may be fixed or variable rate, as indicated by the **multiple_frame_rate_flag**. For streams where the frame rate is fixed, it may be signaled in the **frame_rate_code** as one of 23.976, 24.0, 25.0, 29.97, 30.0, 50.0, 59.94, or 60.0 Hz. Twenty five hertz is commonly used for European standard-definition television, with 29.97 and 30.0 Hz used for U.S. standard-definition television. Twenty four hertz may be used for movies shot on film. In 25-Hz television systems, such movies have usually been sped up to a frame rate of 25 Hz for broadcast.

video_stream_descriptor	
multiple_frame_rate_flag	0
frame_rate_code	29.97 Hz
MPEG_1_only_flag	0
constrained_parameter_flag	0
still_picture_flag	0
if (MPEG_1_only_flag == 0)	
profile_and_level_indication	Main profile at main level
chroma_format	4:2:0
frame_rate_extension_flag	0

Figure 11.15 Structure of the MPEG-2 video stream descriptor.

Video sequences that are compliant with MPEG-1 are indicated by setting the **MPEG_1_only_flag**, with the **constrained_parameter_flag** set to 1 for MPEG-1 sequences that comply with the MPEG-1 constrained parameters. Both of these flags are set to 0 for MPEG-2 video sequences. The **still_picture_flag** is set only in those sequences that consist only of still pictures. For MPEG-2 video sequences, the profile and level indication (following the format of MPEG-2 video discussed in Chapter 6) and chroma format (4:2:0, 4:2:2, or 4:4:4) are signaled. The use of nonzero values in the frame rate extension fields in the video sequence header (see Chapter 6) is indicated by setting the **frame_rate_extension_flag**.

EXAMPLE 11.5—Video Stream Descriptor

Using the simplified representation, show the contents of a video stream descriptor for an MPEG-2, main profile at main level, video sequence with a frame rate of 29.97 Hz, which includes pictures other than still pictures. The chroma format is 4:2:0.

The video stream descriptor carrying this data is shown in Figure 11.16.

video_stream_descriptor

multiple_frame_rate_flag	0
frame_rate_code	29.97 Hz
MPEG_1_only_flag	0
constrained_parameter_flag	1
still_picture_flag	0
if (MPEG_1_only_flag == 0)	
profile_and_level_indication	Main profile at main level
chroma_format	4:2:0
frame_rate_extension_flag	0

Figure 11.16 Example of MPEG-2 video stream descriptor. ■

11.6.1.4. MPEG-2 Hierarchy Descriptor

The MPEG-2 hierarchy descriptor (Fig. 11.17) carries information on scalable hierarchies in video, audio, and private data elementary streams. For MPEG-2 video, the **hierarchy_type** may be used to identify an elementary stream as the base layer or an enhancement layer for spatial, SNR or temporal scalability, or data partitioning. An MPEG-2 audio elementary stream may be indicated as the base layer or an MPEG-2 audio extension bit stream. The unique index of a particular elementary stream in a scalable hierarchy is indicated by the **hierarchy_layer_index**. The index of the layer in the hierarchy immediately below the current layer is specified by

hierarchy_descriptor

hierarchy_type	{MPEG-2 video spatial/SNR/temporal/ data partitioning, MPEG-2 audio extension
hierarchy_layer_index	Unsigned integer
hierarchy_embedded_layer_index	Unsigned integer
hierarchy_channel	Unsigned integer

Figure 11.17 Structure of the MPEG-2 hierarchy descriptor.

registration_descriptor
format_identifier Unsigned integer
<< additional identification info >>

Figure 11.18 Structure of the MPEG-2 registration descriptor.

the **hierarchy_embedded_layer_index**, which has no meaning for a base layer. The **hierarchy_channel** is used to indicate an ordering of channels in a hierarchy, with the most robust of these channels having the lowest value.

Because scalable hierarchies are not supported by either ATSC or DVB, we do not consider the use of this descriptor further.

11.6.1.5. MPEG-2 Registration Descriptor

MPEG-2 provides several opportunities for private data to be included in an MPEG-2 systems bit stream, a feature that is exploited by both DVB and ATSC. The registration descriptor (Fig. 11.18) signals the type of any private data in the bit stream using the **format_identifier**, which is a unique number issued by a registration authority.

11.6.1.6. MPEG-2 Data Stream Alignment Descriptor

The MPEG-2 data stream alignment descriptor, illustrated in Figure 11.19, indicates the type of alignment between the associated video elementary stream and the PES sublayer. The specified alignment applies to those PES packets in which the **data_alignment_indicator** is set to 1. The allowed values of **alignment_type** are "slice or picture," indicating alignment at the slice or picture layer, "picture," indicating alignment at the picture layer, "GOP or SEQ," indicating alignment at the GOP or sequence layer, and "SEQ," indicating alignment at the sequence layer. Specifying alignment at the slice layer is useful for error robustness because it allows a decoder to resynchronize and begin decoding at the first slice header after an error has occurred.

data_stream_alignment_descriptor
alignment_type {slice or picture, picture, GOP or SEQ, SEQ}

Figure 11.19 Structure of the MPEG-2 data stream alignment descriptor.

EXAMPLE 11.6—Data Stream Alignment Indicator

Using the simplified representation, show the contents of a data stream alignment descriptor that indicates that each PES packet with **data_alignment_indicator** *set to 1 begins with a picture header.*

This descriptor is shown in Figure 11.20.

data_stream_alignment_descriptor
alignment_type picture

Figure 11.20 Example of data stream alignment descriptor. ■

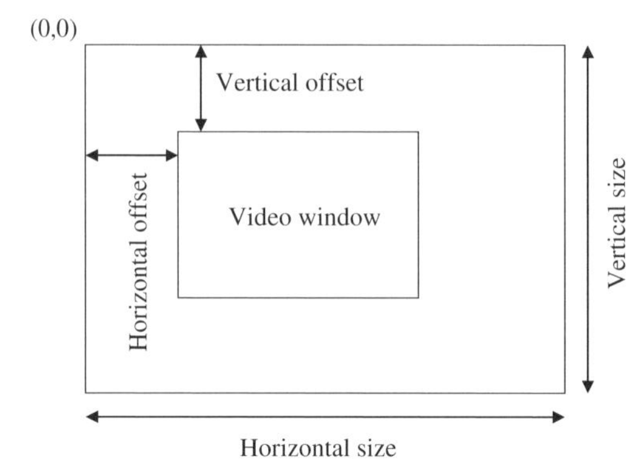

Figure 11.21 Video window filling only part of a display.

11.6.1.7. Descriptors for Video Windows

An MPEG-2 video elementary stream may carry a video sequence that is designed to fill only part of the available display window as illustrated in Figure 11.21. This may arise where a mosaic of video windows, with each window coded in its own elementary stream, is to be displayed. MPEG-2 provides two descriptors to convey information on the window occupied by a video sequence: the target background grid descriptor, which describes the background grid, and the video window descriptor, which specifies the offset in the background grid for one video sequence.

MPEG-2 Target Background Grid Descriptor The MPEG-2 target background grid descriptor specifies a background grid for the associated video elementary stream as shown in Figure 11.22. It specifies the number of pixels vertically and horizontally that make up the grid, which is intended to cover the whole display. The aspect ratio for the grid is also specified, either for the display as a whole or for the individual pixels making up the display. The values of the **horizontal_size** and **vertical_size** fields are calculated as shown in Figure 11.21.

EXAMPLE 11.7—Target Background Grid Descriptor

Using the simplified representation, show the contents of a target background grid descriptor whose target background grid has 720 pixels horizontally and 576 pixels vertically, with a display aspect ratio of 3:4.

target_background_grid_descriptor

horizontal_size	Unsigned integer
vertical_size	Unsigned integer
aspect_ratio_information	{1:1, 3:4, 9:16, 1:2.21}

Figure 11.22 Structure of the MPEG-2 target background grid descriptor.

This descriptor is shown in Figure 11.23.

target_background_grid_descriptor	
horizontal_size	720
vertical_size	576
aspect_ratio_information	3:4

Figure 11.23 Example of MPEG-2 target background grid descriptor. ■

video_window_descriptor	
horizontal_offset	Unsigned integer
vertical_offset	Unsigned integer
window_priority	Unsigned integer

Figure 11.24 MPEG-2 video window descriptor.

MPEG-2 Video Window Descriptor The MPEG-2 video window descriptor (Fig. 11.24) specifies the offset (horizontal and vertical) in the grid specified by the target background grid descriptor for an associated video elementary stream as well as a priority value. The available priority values are between 0 and 7, and higher priority windows should be displayed on top of lower priority windows. The values of the **horizontal_offset** and **vertical_offset** fields are calculated as shown in Figure 11.21.

EXAMPLE 11.8—Video Window Descriptor

Using the simplified representation, show a video window descriptor for which a horizontal offset of 134 pixels from the left side of the display and a vertical offset of 0 pixel form the top of the display are specified. This particular window has priority seven, which indicates that it should appear in front of windows with priority between zero and six but behind windows with priority eight or higher.

The descriptor is shown in Figure 11.25.

video_window_descriptor	
horizontal_offset	134
vertical_offset	0
window_priority	7

Figure 11.25 Example of MPEG-2 video window descriptor. ■

11.6.1.8. MPEG-2 CA Descriptor

The MPEG-2 CA descriptor carries a registered identifier for the conditional access system associated with a program or elementary stream using the structure shown in Figure 11.26. The registered identifier for the conditional access system is carried in the **CA_system_ID**, whereas the PID of the transport stream packets containing data associated with the CA system is specified by **CA_PID**. Additional

CA_descriptor

CA_system_ID	Unsigned integer
CA_PID	Unsigned integer
<< private data bytes >>	

Figure 11.26 Structure of the MPEG-2 CA descriptor.

private data associated with the particular conditional access system may be carried in this descriptor. Any decoder can skip over this private data by using the **descriptor_length** field.

EXAMPLE 11.9—CA Descriptor

Using the simplified representation, show an MPEG-2 CA descriptor indicating that the assigned identifier for the CA used in the associated elementary stream or program is 0x0234 and that information associated with this CA is carried in transport stream packets with PID 0x0934. The remaining data in the CA descriptor is private data with value 0x238F2426.

The descriptor is shown in Figure 11.27.

CA_descriptor

CA_system_ID	0x0234
CA_PID	0x0934
private_data_byte	0x238F2426

Figure 11.27 Example of MPEG-2 CA descriptor. ■

11.6.1.9. MPEG-2 ISO 639 Language Descriptor

The languages associated with an elementary stream are specified by the ISO 639 language descriptor (see Fig. 11.28), using the format of ISO 639 [4]. This descriptor can also indicate other properties that the audio may possess: "clean effects," which means that the audio contains no language; "hearing impaired," which indicates a stream specially designed for the hearing impaired that emphasizes speech over background sounds and effects; and "visual impaired commentary," which is an audio stream containing additional commentary to assist the visually impaired.

ISO_639_language_descriptor

* {	ISO_639_language_code	ISO 639 language code
	audio_type	{Undefined, clean effects, hearing impaired, visual impaired commentary}

Figure 11.28 Structure of the MPEG-2 ISO 639 language descriptor.

EXAMPLE 11.10—ISO 639 Language Descriptor

Using the simplified representation, show an ISO 639 language descriptor for an English audio service with no special properties.

The descriptor is shown in Figure 11.29.

ISO_639_language_descriptor

ISO_639_language_code	English
audio_type	Undefined

Figure 11.29 Example of MPEG-2 ISO 639 language descriptor. ■

11.6.1.10. MPEG-2 System Clock Descriptor

The purpose of the MPEG-2 system clock descriptor is to specify the accuracy of the system time clock generated by the encoder as shown in Figure 11.30. The accuracy is specified in parts per million (PPM) using the **clock_accuracy_integer** and **clock_accuracy_exponent** fields and is calculated from $clock\_accuracy\_integer \times 10^{-clock\_accuracy\_exponent}$ ppm. Where an external source that may be available to a decoder is used by an encoder to generate the system clock, the **external_clock_reference_indicator** may be set to 1.

system_clock_descriptor

external_clock_reference_indicator	flag
clock_accuracy_integer	Unsigned integer
clock_accuracy_exponent	Unsigned integer

Figure 11.30 Structure of the MPEG-2 system clock descriptor.

EXAMPLE 11.11—System Clock Descriptor

Using the simplified representation, show a system clock descriptor describing a system clock based on an external time reference that may be available to the decoder and with an accuracy of 5.4 ppm.

The descriptor is shown in Figure 11.31.

system_clock_descriptor

external_clock_reference_indicator	1
clock_accuracy_integer	540
clock_accuracy_exponent	2

Figure 11.31 Example of MPEG-2 system clock descriptor. ■

multiplex_buffer_utilization_descriptor	
bound_valid_flag	flag
LTW_offset_lower_bound	Unsigned integer
LTW_offset_upper_bound	Unsigned integer

Figure 11.32 Structure of the MPEG-2 multiplex buffer utilization descriptor.

11.6.1.11. MPEG-2 Multiplex Buffer Utilization Descriptor

The MPEG-2 multiplex buffer descriptor provides upper and lower bounds on the occupancy of the multiplex buffer in the system target decoder using the structure shown in Figure 11.32. This information is intended for use by a remultiplexer. The **bound_valid_flag** is used to signal that the other information in the descriptor is valid. The remaining two fields in the descriptor specify the upper and lower bounds, each specified as a number of 90-kHz clock cycles, that is, a multiple of 11.1 μs.

EXAMPLE 11.12—Multiplex Buffer Utilization Descriptor

Using the simplified representation, show a multiplex buffer utilization descriptor containing valid bounds with a lower bound of 0.389 ms and an upper bound of 193.244 ms.

The descriptor is shown in Figure 11.33.

multiplex_buffer_ut ilization_descriptor	
bound_valid_flag	1
LTW_offset_lower_bound	0x0023
LTW_offset_upper_bound	0x43F0

Figure 11.33 Example of MPEG-2 multiplex buffer utilization descriptor. ■

11.6.1.12. MPEG-2 Copyright Descriptor

The MPEG-2 copyright descriptor (Fig. 11.34) is used to enable identification of rights associated with audio–visual material. The value of the **copyright_identifier** is assigned by a registration authority. A particular value may indicate that rights are owned by a particular organization. Additional information may also be attached, but the format of this information is not standardized by MPEG-2.

11.6.1.13. MPEG-2 Maximum Bit-Rate Descriptor

The MPEG-2 maximum bit-rate descriptor, illustrated in Figure 11.35, tells a decoder the maximum bitrate (including transport overhead) that is associated with

copyright_descriptor	
copyright_identifier	Unsigned integer
<< additional_copyright_info >>	

Figure 11.34 Structure of the MPEG-2 copyright descriptor.

maximum_bitrate_descriptor	
maximum_bitrate	Unsigned integer

Figure 11.35 Structure of the MPEG-2 maximum bit-rate descriptor.

a program or elementary stream in multiples of 400 bit/s (50 byte/s). It is carried in the program map table.

EXAMPLE 11.13—Maximum Bit-Rate Descriptor

Using the simplified representation, show a maximum bit-rate descriptor specifying a maximum bitrate of 4 Mbit/s (500 kbyte/s).

The descriptor is shown in Figure 11.36.

maximum_bitrate_descriptor	
maximum_bitrate	10,000

Figure 11.36 Example of MPEG-2 maximum bit-rate descriptor. ■

11.6.1.14. MPEG-2 Private Data Indicator Descriptor

The MPEG-2 private data indicator descriptor (Fig. 11.37) carries one field to provide information about private data. The meaning of this field is not defined by MPEG-2.

11.6.1.15. MPEG-2 Smoothing Buffer Descriptor

The MPEG-2 smoothing buffer descriptor provides information on the size and leak rate of the smoothing buffer of the associated elementary stream. Its structure is shown in Figure 11.38. The leak rate is specified in units of 400 bit/s. The buffer size is specified in units of 1 byte.

private_data_indicator_descriptor	
private_data_indicator	Unsigned integer

Figure 11.37 Structure of the MPEG-2 private data indicator descriptor.

smoothing_buffer_descriptor	
sb_leak_rate	Unsigned integer
sb_size	Unsigned integer

Figure 11.38 Structure of the MPEG-2 smoothing buffer descriptor.

EXAMPLE 11.14—Smoothing Buffer Descriptor

Using the simplified representation, show an MPEG-2 smoothing buffer descriptor, specifying a leaking rate of 4Mbit/s and a buffer size of 20,000 bytes.

The descriptor is shown in Figure 11.39.

smoothing_buffer_descriptor	
sb_leak_rate	10,000
sb_size	20,000

Figure 11.39 Example of MPEG-2 smoothing buffer descriptor. ■

STD_descriptor	
leak_valid_flag	flag

Figure 11.40 Structure of the MPEG-2 STD descriptor.

11.6.1.16. MPEG-2 STD Descriptor

The MPEG-2 (system target decoder) STD descriptor (Fig. 11.40) indicates whether the leak method (**leak_valid_flag** = 1) or vbv_delay method (**leak_valid_flag** = 0) is used in the system target decoder.

11.6.1.17. MPEG-2 IBP Descriptor

The MPEG-2 IBP descriptor, illustrated in Figure 11.41, provides information on the GOP structure in an MPEG-2 video elementary stream. The **closed_gop_flag** is set to 1 for elementary streams in which a GOP header immediately precedes each I-picture. Elementary streams in which all GOPs have the same structure (i.e., the same sequence of B- and P-pictures between I-pictures) have the **identical_gop_flag** set to 1. The maximum number of frames that may be transmitted in one GOP is signaled in the **max_gop_length**.

IBP_descriptor	
closed_gop_flag	flag
identical_gop_flag	flag
max_gop-length	Unsigned integer

Figure 11.41 Structure of the MPEG-2 IBP descriptor.

EXAMPLE 11.15—IBP Descriptor

Using the simplified syntax, show an MPEG-2 IBP descriptor, in which a GOP header does not always appear before an I-picture, all GOPs are of the same structure, and the maximum GOP length is 15 pictures.

The descriptor is shown in Figure 11.42.

IBP_descriptor	
closed_gop_flag	0
identical_gop_flag	1
max_gop-length	15

Figure 11.42 Example of MPEG-2 IBP descriptor. ■

11.6.2. MPEG-2 Tables

The MPEG-2 descriptors described in Section 11.6.1 above are carried within one of the three PSI tables (program association table, program map table, or conditional access table) whose syntax is defined by MPEG-2. In this section, the generic syntax of these tables is presented, followed by a discussion of the structure of each table.

11.6.2.1. Generic Table Syntax

Each table is transmitted as one or more sections. Each section may be up to 1024 bytes in length. The generic syntax of the MPEG-2 PSI section is shown in Table 11.9.

The type of table, that is, program association table, program map table, conditional access table, or network information table is specified by the **table_id** field. The values 0x00, 0x01, and 0x02 are assigned to the program association table, conditional access table, and program map table, respectively. Values between 0x40 and 0xFE are available for tables defined outside the scope of MPEG-2 (including the network information table). Other values are not allowed. The identification of table type is followed by the 1-bit **section_syntax_indicator** field whose value is always 1. The reserved bits always take the value 00 and are reserved for future use by the ISO.

The **section_length** field specifies the number of bytes in the current table section following the **section_length** field (i.e., three less than the total number of bytes in the section). This field allows a decoder to identify the location of the end of the section, even if it does not understand the internal syntax of a particular table. For example, an MPEG-2 systems decoder is able to skip over the non-MPEG-2 tables defined by DVB without being able to decode the internal contents of these tables.

Table 11.9 Generic syntax of MPEG-2 PSI tables.

Syntax	Number of bits	Mnemonic
TS_table_section() {		
table_id	8	uimsbf
section_syntax_indicator	1	bslbf
"0"	1	bslbf
reserved	2	bslbf
section_length	12	uimsbf
—		
reserved	2	bslbf
version_number	5	uimsbf
current_next_indicator	1	bslbf
section_number	8	uimsbf
last_section_number	8	uimsbf
—		
CRC_32	32	rpchof

MPEG-2 allows tables longer than 1024 bytes to be transmitted as multiple sections, although many encoders and decoders do not support multisection tables. The various sections of these multisection tables are distinguished by their different values in the **section_number** field.

Each time the contents of a table are changed, the **version_number** field of the table is increased by 1. This allows the new and old contents to be distinguished from one another. An updated table may be transmitted before the new information is current. If this transmission occurs a number of times, it provides a powerful protection against transmission errors introduced in the channel. The **current_next_indicator** is used to indicate whether the table contents being transmitted refer to the current value or the next future value.

The **last_section_number** field specifies the largest value of the **section_number** field used in the table. This enables a decoder to determine the total number of sections in a table when any one of those sections is received.

Each section of each table includes a 32-bit cyclic redundancy code (CRC) (**CRC_32**) so that a decoder can determine confidently whether or not it has been correctly received. This CRC is used on PSI tables but not for other data transmitted in MPEG-2 bit streams because of the catastrophic effect on the quality of the decoded programs of using a corrupted table. Errors introduced into other parts of the bit stream will have effects that are neither as longlasting nor as catastrophic. The mnemonic "rpchof" indicates a field whose value can be used to detect transmission errors.

11.6.2.2. Program Association Table

When it begins decoding an MPEG-2 systems bit stream, the decoder first searches for the program association table, which can be readily identified because it is always transmitted with a PID of 0x00. The purpose of the program association table is to associate a program number with the PID of the program map table that contains the details of the elementary streams making up that program. The program association table also specifies the PID that is used to transmit the network information table. The program association table must be present in every transport stream and is not scrambled.

The structure of the program map table, shown in Figure 11.43, consists of two parts. The first (**transport_stream_id**) assigns a unique label that can be used to

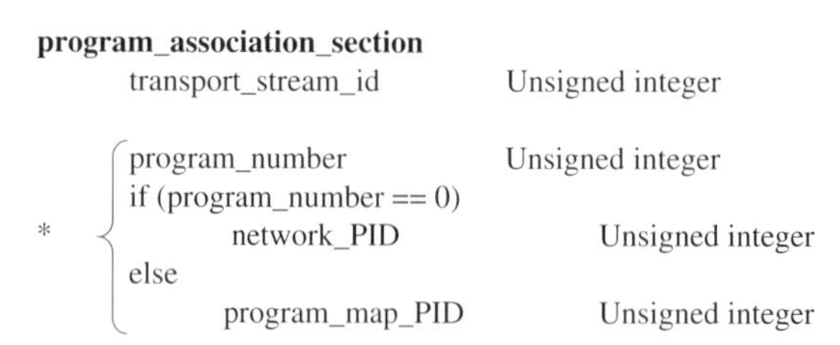

Figure 11.43 Structure of the program association table.

identify this transport stream. The second is a list of one or more programs, for each of which the PID of the transport stream packets carrying the program map table is specified. Program number 0 is used to indicate that the PID used to carry the network information table follows. The program association table carries no descriptors.

A decoder that is able to switch between a number of systems bit streams may monitor all of these bit streams for updates in the program association table to reduce the time required to switch between streams. Without this, there may be long delays associated with a change of systems bit stream.

EXAMPLE 11.16—Program Association Table

Using the simplified representation, show a program association section for a transport_stream_id of 0x1200. The section is to signal that a network information table is carried with PID 0x0010 and that programs 1–3 are carried with PIDs 0x0103, 0x0538, and 0x0456, respectively.

A program association section meeting this specification is shown in Figure 11.44. The order of the programs is arbitrary: In this case we have chosen the sequence 0, 1, 3, 2.

```
program_association_section
    transport_stream_id     0x1200

    program_number          0
    network_PID             0x0010

    program_number          1
    program_map_PID         0x0103

    program_number          3
    program_map_PID         0x0456

    program_number          2
    program_map_PID         0x0538
```

Figure 11.44 Example of program association section. ■

11.6.2.3. Program Map Table

Each section of the program map table provides information about one program. The PID used to carry the program map table section for each program is specified in the program association table. The program map table specifies the PIDs for the elementary streams that make up a program and the PID containing the clock reference for the program. An encoder may assign a separate PID that is used only to carry the clock reference, or the clock reference PID may be shared with one of the elementary stream PIDs. For each elementary stream making up the program, the type of the elementary stream is specified. Normally, the elementary streams of this program are carried in the same transport stream as the section of the program map

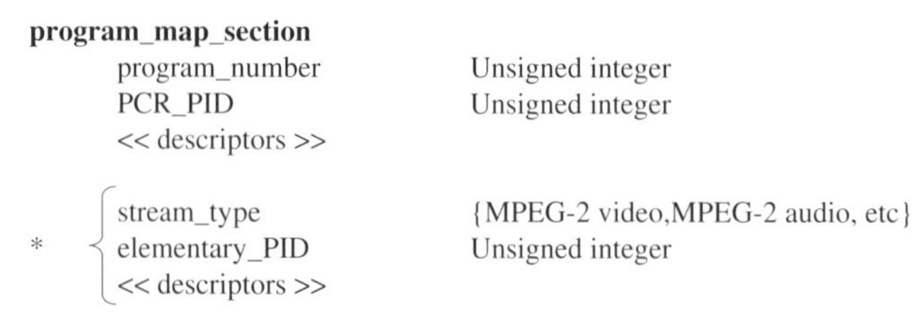

Figure 11.45 Structure of the program map table.

table referring to the program. Program-map-table data must be transmitted for each program carried in a transport stream. The program map table is not scrambled.

The basic structure of the program map table is illustrated in Figure 11.45. The main part of the table contains the program number, which is used by both DVB and ATSC as a means of identifying the group of elementary streams making up the program, and the PID of the transport stream packets carrying the clock reference for this program (**PCR_PID**). Descriptors carried immediately following the **PCR_PID** field refer to the whole program.

For each elementary stream making up the program, the program map table carries the stream type and the PID of the transport stream packets carrying the elementary stream. The stream type specifies the elementary stream decoder to which arriving packets are sent at the decoder. Descriptors following these fields refer only to one elementary stream. For video elementary streams, the video stream descriptor identifies basic coding parameters of the elementary stream, including frame rate, profile and level, and chroma format. For MPEG-2 audio elementary streams, the audio stream descriptor identifies the audio coder used including layer information.

EXAMPLE 11.17—Program Map Table

Using the simplified representation, show a program map table containing one video and one audio elementary stream. The PCR is carried in transport stream packets with the same PID as the video and the maximum bit rate for the program, including all elementary streams and systems overhead, is 5 Mbit/s.

The program map table is shown in Figure 11.46.

The video elementary stream is carried in transport stream packets with PID 0x0308 and is described by two descriptors: the video stream descriptor and data_stream_alignment descriptor. The video stream descriptor shows that the frame rate of the video is 25 Hz, that it is MPEG-2 main profile at main level, and that the chroma format is 4:2:0. The data stream alignment descriptor tells a decoder that the elementary stream is aligned to the PES layer at the slice layer.

The audio elementary stream is MPEG-2, Layer-2 audio. The ISO 639 language descriptor shows that the language of the audio stream is English. The value "undefined" for **audio_type** means that the audio stream does not have special properties such as being designed for hearing impaired.

program_map_section

program_number	2	
PCR_PID	0x0308	
maximum_bitrate_descriptor		
	maximum_bitrate	5 Mbit/s
stream_type	MPEG-2 video	
elementary_PID	0x0308	
video_stream_descriptor		
	multiple_frame_rate_flag	0
	frame_rate_code	25 Hz
	MPEG_1_only_flag	0
	constrained_parameter_flag	0
	still_picture_flag	0
	profile_and_level_indication	Main profile at main level
	chroma format	4:2:0
	frame_rate_extension_flag	0
data_stream_alignment_descriptor		
	alignment_type	Slice layer
stream_type	MPEG-2 BC audio	
elementary_PID	0x0309	
audio_stream_descriptor		
	free_format_flag	0
	ID	1
	layer	2
	variable_rate_audio_indicator	0
ISO_639_language_descriptor		
	ISO_639_language_code	English
	audio_type	Undefined

Figure 11.46 Example of program map table. ■

11.6.2.4. Conditional Access Table

The conditional access table carries information on conditional access schemes that may be used to control access to programs by scrambling. It is only transmitted where conditional access is used. If present, the conditional access table is transmitted with PID 0x01. Its major purposes are to flag the use of conditional access, to identify the conditional access system in use, and to specify the locations of conditional access data in a transport stream. The conditional access table may carry descriptors providing system-wide configuration management information for the conditional access system. This might include the delivery of keys to be used to decrypt encrypted programs.

All data in the conditional access table is carried in descriptors as shown in the structure of Figure 11.47.

The CA descriptor is usually the only descriptor carried in the conditional access table. Its contents point to system wide access management or control information, in contrast to the information associated with particular programs that is carried by this descriptor when it appears in the program map table.

conditional_access_section
<< descriptors >>

Figure 11.47 Structure of the conditional access table.

EXAMPLE 11.18—Conditional Access Table

Using the simplified representation, show a conditional access table that carries one CA descriptor that identifies the CA system in use and PID in which information associated with the CA system it carried.

The conditional access section is shown in Figure 11.48.

```
conditional_access_section
    CA_descriptor
        CA_system_ID    0x0234
        CA_PID          0x0934
```

Figure 11.48 Example of conditional access table. ■

11.6.2.5. Network Information Table

The purpose of the network information table is to carry information on the physical characteristics of the network being used to deliver the transport stream and possibly on other networks being used to deliver other transport streams. Its contents are not defined by MPEG-2. For a satellite delivery network, the types of information that might be carried in the network information table include the transmission frequency, modulation characteristics, and polarization. The network information table is transmitted in its own PID as specified in the program association table.

11.6.3. Overheads Due to PSI

An MPEG-2 decoder cannot extract the elementary streams making up a program from a systems bit stream without first receiving the PSI data describing the structure of the multiplex. Typically, random access points in digital television occur at GOP boundaries. An approximate length of the PSI can be obtained by assuming that the PSI is transmitted once per GOP, that each table occupies exactly one transport stream packet, and that each program map section (describing one program) is transmitted in its own transport stream packet.

EXAMPLE 11.19—PSI Overheads

For typical standard-definition television, random access is usually limited by the GOP structure of the video to approximately every half second (15 frames for 30 frames per second and 12 frames for 25 frames per second). For a systems bit stream at 4 Mbit/s containing one program, the overhead if PSI data is transmitted twice per second is likely to be approximately 6 kbyte/s, which is equivalent to 1.2% of the systems bit-stream capacity.

11.7. MPEG-2 DECODER OPERATION

When beginning to decode a transport stream, an MPEG-2 decoder undertakes three major tasks: synchronization to the transport stream, acquisition of PSI data, and program selection and decoding. Each task must be completed before the following task can begin.

11.7.1. Synchronization to Transport Stream

The sequence of operations required to begin decoding a program contained in a transport stream is as follows. First, the decoder must acquire synchronization with the transport stream, identifying the locations of the boundaries of the transport stream packets. Because there is no guarantee that the **sync_byte** value (0x47) cannot appear elsewhere in a transport stream, this synchronization process involves testing the **sync_byte** value for several successive packets. For some delivery systems, such as an ATM network, the underlying transport layer may assist the decoder in transport stream synchronization.

11.7.2. PSI Decoding

Following synchronization, the decoder scans the transport stream to find the program association table, which always uses a PID of 0x00. From the program association table, the decoder obtains one or more PIDs that are used to carry the program map table, which contains lists of PIDS of elementary streams making up the programs contained in the transport stream. The viewer is now able to select a program for decoding, after which the systems decoder extracts the elementary stream data for that program and discards all other data in the transport stream. Each of the elementary streams making up the program is passed to the appropriate elementary stream decoder. After decoding, the contents of each elementary stream are displayed.

Random access within transport streams is required where a decoder is able to switch between two or more streams. This may happen in terrestrial broadcast when changing frequencies. It may also occur when a decoder connected to two or more types of distribution network (e.g., cable and satellite) changes from one network to another. Random access can be facilitated by the regular retransmission of tables within each transport stream. MPEG-2 does not specify how often tables should be retransmitted. This approach is used in DVB. Random access may also be facilitated by placing additional restrictions on the transport stream. For example, the PIDs of the elementary streams making up a program may be required to be contiguous. This approach is used in ATSC.

11.7.3. Program Reassembly

An MPEG-2 decoder is only required to be capable of decoding one program, even though a transport stream may contain elementary streams for many programs.

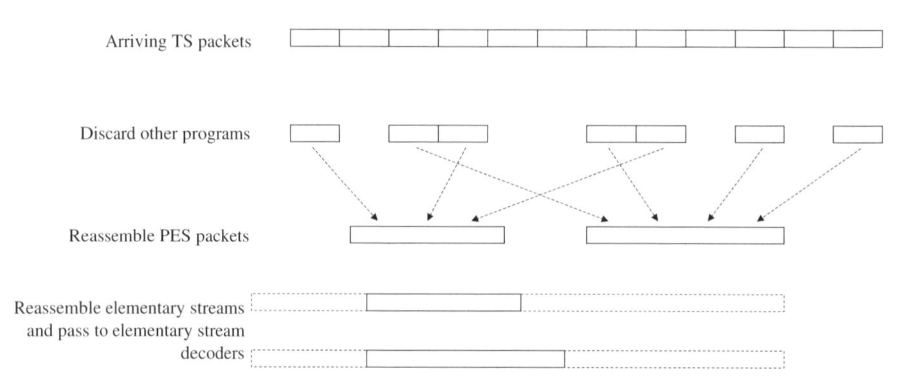

Figure 11.49 Program reassembly in MPEG-2 decoder.

The transport stream packets for a particular elementary stream are identified by their unique PID. Program reassembly in a decoder begins with breaking up of the MPEG-2 systems bit stream into transport stream packets. The decoder scans the PIDs in the transport stream packet headers and discards all those transport stream packets whose PID is not associated with an elementary stream associated with the current program. PES packets are then reassembled for each elementary stream that forms part of the current program, from which the elementary streams are extracted and passed to an elementary stream decoder. This process is illustrated in Figure 11.49.

EXAMPLE 11.20—MPEG-2 Transport Stream Decoding

An example of five packets from an MPEG-2 transport stream is shown in Table 11.10. *Each table entry shows the contents of 1 byte of the transport stream. The first 16 bytes are shown in the top row, the next 16 on the second row, and so on. For this example stream, identify the contents of the header of each transport stream packet.*

In this example, the first byte has value 0x47 and may be a transport packet **sync_byte**. This can be confirmed by observing that the value 188 byte later (0x0B, 0xC) has the same value as do all the entries at higher multiples of 188 bytes. In this example, all sync bytes are shown in italics.

The PID of the first transport stream packet is 0x0233. Unless the decoder has already read the program association table and program map table for this transport stream, this packet cannot be decoded. The 13-bit PID of the second transport stream packet is 0x0000, indicating the presence of the program association table. The **payload_unit_start_indicator** is also set to 1, showing that a new program association section starts in this transport stream packet. The first payload byte of the transport packet (at location (0x0C,0)) has value 5, indicating an offset of 5 bytes to the start of the program association section (at location (0x0C,6) where a value of zero corresponds to **table_id** = 0, which is consistent with the program association section.

Table 11.10 Example of MPEG-2 transport stream.

	0	1	2	3	4	5	6	7	8	9	A	B	C	D	E	F
00	*47*	02	33	15	E4	C3	74	04	D2	71	9D	CA	EB	BC	2D	67
01	EF	EA	69	E4	0E	5A	D0	02	23	33	32	9A	45	32	03	BF
02	71	EE	77	6B	D8	86	33	AC	D6	05	AE	61	D4	80	B5	6D
03	4D	30	31	AE	4D	8A	26	B2	60	DC	DA	97	7F	E6	D2	A5
04	D1	A9	57	4A	57	88	BA	4F	D6	91	5E	B3	8B	71	B1	9F
05	CB	F4	85	E1	2C	FA	45	40	E0	BC	22	03	E4	32	4C	A9
06	48	78	10	FD	95	6C	83	55	6E	39	94	C2	87	A3	35	61
07	C8	AE	76	91	CB	0F	9A	0C	6A	4E	DF	03	C4	F8	FD	C9
08	70	7F	36	A4	51	F5	BA	69	BE	44	70	EE	AE	36	D6	A0
09	22	35	9B	A1	5E	93	73	0B	06	50	03	62	AE	17	09	9C
0A	9B	04	04	30	96	0E	5E	A1	B7	B1	15	74	71	5A	27	AC
0B	B2	BA	7A	8E	1E	73	B7	E4	45	41	DD	3B	*47*	40	00	15
0C	05	14	A4	30	D8	2C	00	00	70	57	50	5D	64	97	1E	09
0D	75	DE	EF	43	29	DF	3C	A5	F7	AA	DE	02	23	D1	6E	E3
0E	BC	AF	58	2A	27	30	6C	DB	7D	D0	75	75	73	69	E6	01
0F	4C	0C	B1	A6	FB	8D	66	32	A0	BB	60	02	6B	C0	CB	EB
10	D8	5E	9E	BB	31	E7	91	A1	3C	8C	EE	55	A7	64	A0	B2
11	65	69	A7	D6	5F	6C	98	90	B7	82	C6	7D	2F	B3	FB	CE
12	B4	7C	1D	AA	5D	23	91	D2	AC	FF	F6	0F	5C	8C	43	98
13	0C	92	B3	F6	C0	BD	6E	A2	CD	15	F2	EA	9A	40	DF	83
14	BB	6C	F6	12	8D	4A	DB	55	AE	0D	5B	7F	6F	8F	9D	1D
15	E5	C1	CA	D0	AB	33	45	A0	89	0F	16	45	68	79	E8	98
16	54	7A	98	29	D4	F4	98	07	CF	9C	B3	17	6C	60	2A	D5
17	D6	73	F4	25	DE	C4	71	9E	*47*	02	33	16	30	7D	68	76
18	9C	12	50	9B	2C	9E	3E	96	81	76	8A	F1	57	66	4E	69
19	49	64	80	B8	4E	1C	71	77	03	A9	B9	48	43	B5	C8	FC
1A	79	E7	73	CD	D4	2A	64	85	B7	91	75	71	16	71	5D	4D
1B	DA	C2	F3	8E	03	98	D0	FA	38	B4	85	EE	B6	3A	73	2C
1C	F8	5B	0C	C1	E5	49	40	EE	21	F0	B3	D9	35	74	14	D9

(continued)

Table 11.10 *(Continued)*

	0	1	2	3	4	5	6	7	8	9	A	B	C	D	E	F
1D	8F	51	5F	DE	5F	12	33	0C	91	1F	85	1D	C5	60	D2	0B
1E	99	F2	49	E3	1A	10	3B	EE	10	43	FF	36	7F	4A	AC	F5
1F	C4	AA	21	18	03	49	D1	FC	04	D1	9F	8F	3E	D2	43	C0
20	A8	36	9A	9A	A8	2E	A2	2B	8A	9F	AF	AD	E0	03	4F	C7
21	4E	ED	AD	13	12	03	3A	84	75	B4	95	82	13	31	61	46
22	C5	50	A3	FC	80	F2	D3	EA	1C	CF	E8	28	1F	C3	B8	A6
23	C1	A9	E2	45	47	00	00	16	D9	57	77	E9	3A	DC	A8	E4
24	7C	FE	5F	88	2E	80	6C	A9	AC	F5	31	1C	90	F8	06	DE
25	06	84	31	B7	40	EF	23	85	E5	F1	55	6F	78	26	22	88
26	B9	66	5B	49	DE	A0	3D	FA	A3	3A	AE	AA	22	05	43	1D
27	11	DA	2E	08	BB	89	46	5E	03	E3	DD	41	91	28	98	54
28	A8	DD	91	FB	CA	27	D5	31	A3	AB	C5	61	71	7B	9B	2D
29	00	CA	83	36	1A	28	68	68	0D	F1	26	62	4F	2B	E5	52
2A	BB	69	66	81	2B	86	A4	04	D6	CD	B2	76	15	D2	31	72
2B	03	4F	E0	D5	55	E1	7A	8F	9D	A9	9D	AF	82	B6	83	9B
2C	F7	D2	51	96	21	41	CD	AA	03	8F	74	E7	48	10	7A	FB
2D	EC	8F	A6	C5	1B	00	8A	01	73	32	C9	9E	03	E4	C2	E8
2E	C2	61	54	81	90	C4	C7	7B	CD	78	33	94	AA	AD	F1	C5
2F	47	02	33	17	A1	CA	72	86	2B	21	38	1B	24	74	C9	47
30	39	E8	01	96	8A	A7	50	3B	6A	4C	AC	F0	57	90	1E	2B
31	47	8E	7C	F3	3B	7A	86	CA	31	E8	EC	03	C4	F2	D0	EC
32	32	AC	ED	58	98	9D	00	FB	E6	B1	70	B3	9C	4C	DB	1C
33	4A	18	65	55	F1	D6	42	0A	01	93	BE	CE	A3	40	24	A6
34	F2	D0	EE	4F	44	89	29	36	37	A6	0D	3A	AA	4F	4E	B8
35	F4	21	11	20	2A	E9	22	9D	44	38	B6	8C	F0	54	B4	F1
36	94	E1	BF	61	B9	29	F4	32	C6	9D	29	07	49	F8	F3	3A
37	F5	AE	0E	99	64	37	2E	13	01	C9	04	E0	5A	B8	F7	27
38	29	50	07	5B	06	CB	FF	1C	9F	21	4F	22	39	65	22	3D
39	ED	64	82	17	05	28	D8	E1	2F	FD	B6	DF	7A	7E	49	0F
3A	43	2F	EA	1F	03	5E	B2	E3	98	28	51	3B				

The last two packets in Table 11.10 have PID 0x0233, which is the same as the first packet. The **continuity_counter** fields of these three packets can be seen to be increasing by 1. ■

11.8. USE OF MPEG-2 SYSTEMS IN DIGITAL TELEVISION

This section describes the specific use of MPEG-2 systems in the ATSC and DVB digital television standards. Both systems place a number of restrictions on the MPEG-2 systems.

11.8.1. Use of MPEG-2 Systems in ATSC

MPEG-2 systems provides the multiplex for ATSC digital television. The ATSC standard, however, places a number of restrictions on the MPEG-2 implementation of multiplexing and PSI, which are discussed in this section.

11.8.1.1. Implementation of Multiplexing in ATSC

The MPEG-2 transport stream is the only multiplex used; the program stream is not used.

Within the PES header

- scrambling is not used (this means that scrambling in ATSC is based on scrambling of transport stream packets);
- the ESCR is not transmitted;
- the optional CRC is not transmitted;
- private data is not carried; and
- fields relating to the program stream are coded as zero or not carried.

For video elementary streams, the **data_alignment_indicator** in the PES header always has value 1, and the PES_packet_length is specified as 0. This means that the length of PES packets is not bounded by the PES header and that alignment is guaranteed between the video and PES layers in every PES packet.

AC-3 audio is assigned a value of 0xBD for the **stream_id** in the PES header.

Still pictures are not supported.

11.8.1.2. Implementation of PSI in ATSC

MPEG-2 PSI is used by ATSC, with extensions to add tables and descriptors of its own, using values of **descriptor_tag** and **table_id** left for private definition by MPEG-2. These extensions are described fully in Chapter 12.

Table 11.11 Example for Program 51 (0x0033).

Name	PID value
base_PID	0x0330
PMT_PID	0x0330
Video_PID	0x0331
PCR_PID	0x0331
Audio_PID	0x0334

ATSC assigns the **stream_type** value 0x81 in the PMT to AC-3 audio.

Programs carried by a transport stream that comply with the requirements of ATSC are indicated by the placement of a registration descriptor in the program map table, with the **format_identifier** set to 0x4741 3934. AC-3 audio streams are carried as private data and indicated by associating a registration descriptor with **format_identifier** set to 0x4143 2D33 with the audio elementary stream in the program map table.

Using the MPEG-2 PSI, random access to a transport stream is limited to those points where a complete set of PSI is transmitted. Rather than limit random access to these points, ATSC defines program numbers and assigns fixed PIDs to the elementary streams making up each program. For each program number (assigned in the program association table) between 1 and 255, a base_pid is calculated by multiplying the program number by 16. The PMT for this program is carried in transport stream packets whose PID is PMT_PID = base_PID + 0x0000. The video elementary stream (Video_PID) and PCR (PCR_PID) are both carried with PID base_PID + 0x0001. The primary audio for the program is carried in Audio_PID = base_PID + 0x0004. An example of this calculation is shown in Table 11.11.

11.8.2. Use of MPEG-2 Systems in DVB

MPEG-2 systems provides the multiplex for DVB digital television. The DVB standards, however, place a number of restrictions on the MPEG-2 implementation of multiplexing and PSI, which are discussed in this section.

11.8.2.1. Implementation of Multiplexing in DVB

DVB does not use the following features of MPEG-2 systems [5]:

- the use of trick modes (such as fast forward, fast rewind, pause);
- ESCR;
- private data in PES headers; and
- the program stream, including all associated signaling.

A DVB decoder is required to be able to scan any valid MPEG-2 systems bit stream but may ignore any components associated with unused features.

11.8.3. Implementation of PSI in DVB

MPEG-2 PSI is used by DVB, along with additional tables and descriptors of its own, using values of **descriptor_tag** and **table_id** left for private definition by MPEG-2. These extensions are described fully in Chapter 12.

11.9. CONCLUSION

MPEG-2 systems provide a range of services for multiplexing elementary streams, timing and synchronization, buffer management, and control.

In following chapters, the following are of particular importance:

- The value of the **PID** field in the transport stream packet header is associated with one source of data, either an elementary stream or a PSI table.
- A particular transport stream can be uniquely identified by the **transport_stream_id** in the program association table.
- The type of an elementary stream can be identified from the **stream_id** field of the PES packet header and the **stream_type** field of the program map table.

PROBLEMS

11.1 Fill in the empty boxes in the following table of transport stream overheads in Table 11.12.

Table 11.12 Transport stream overheads for Q 11.1.

Frames per second	Slices per frame	Number of programs	Transport stream rate	Overhead
30	30	5	20 Mbit/s	
30	30	1		2%
30	1	5	20 Mbit/s	

11.2 For the bitstream in Table 11.13:

(a) Identify the locations at which transport stream packets start.

(b) For each transport stream packet, find its PID and ensure that all values of the continuity counter are valid.

(c) Find the packet(s) that contain the program map table and ensure that the correct value of **table_id** is present.

Table 11.13 Example transport stream bit stream for Q 11.2.

	0	1	2	3	4	5	6	7	8	9	A	B	C	D	E	F
00	43	2F	EA	1F	03	5E	B2	E3	98	28	51	3B	B2	E3	98	47
01	02	33	15	E4	C3	86	74	04	D2	71	9D	CA	EB	BC	2D	67
02	EF	EA	69	E4	0E	5A	D0	02	23	33	32	9A	45	32	03	BF
03	71	EE	77	6B	D8	86	33	AC	D6	05	AE	61	D4	80	B5	6D
04	4D	30	31	AE	4D	8A	26	B2	60	DC	DA	97	7F	E6	D2	A5
05	D1	A9	57	4A	57	88	BA	4F	D6	91	5E	B3	8B	71	B1	9F
06	CB	F4	85	E1	2C	FA	45	40	E0	BC	22	03	E4	32	4C	A9
07	48	78	10	FD	95	6C	83	55	6E	39	94	C2	87	A3	35	61
08	C8	AE	76	91	CB	0F	9A	0C	6A	4E	DF	03	C4	F8	FD	C9
09	70	7F	36	A4	51	F5	BA	69	BE	44	70	EE	AE	36	D6	A0
0A	22	35	9B	A1	5E	93	73	0B	06	50	03	62	AE	17	09	9C
0B	9B	04	04	30	96	0E	5E	A1	B7	B1	15	74	71	5A	27	AC
0C	B2	BA	7A	8E	1E	73	B7	E4	45	41	DD	47	40	00	15	05
0D	14	A4	30	D8	2C	00	3C	00	70	57	50	5D	64	97	1E	09
0E	75	DE	EF	43	29	DF	3C	A5	F7	AA	DE	02	23	D1	6E	E3
0F	BC	AF	58	2A	27	30	6C	DB	7D	D0	75	75	73	69	E6	01
10	4C	0C	B1	A6	FB	8D	66	32	A0	BB	60	02	6B	C0	CB	EB
11	D8	5E	9E	BB	31	E7	91	A1	3C	8C	EE	55	A7	64	A0	B2
12	65	69	A7	D6	5F	6C	98	90	B7	82	C6	7D	2F	B3	FB	CE
13	B4	7C	1D	AA	5D	23	91	D2	AC	FF	F6	0F	5C	8C	43	98
14	0C	92	B3	F6	C0	BD	6E	A2	CD	15	F2	EA	9A	40	DF	83
15	BB	6C	F6	12	8D	4A	DB	55	AE	0D	5B	7F	6F	8F	9D	1D
16	E5	C1	CA	D0	AB	33	45	A0	89	0F	16	45	68	79	E8	98
17	54	7A	98	29	D4	F4	98	07	CF	9C	B3	17	6C	60	2A	D5
18	D6	73	F4	25	DE	C4	71	47	02	33	16	30	7D	68	76	03
19	9C	12	50	9B	2C	9E	3E	96	81	76	8A	F1	57	66	4E	69
1A	49	64	80	B8	4E	1C	71	77	03	A9	B9	48	43	B5	C8	FC
1B	79	E7	73	CD	D4	2A	64	85	B7	91	75	71	16	71	5D	4D
1C	DA	C2	F3	8E	03	98	D0	FA	38	B4	85	EE	B6	3A	73	2C
1D	F8	5B	0C	C1	E5	49	40	EE	21	F0	B3	D9	35	74	14	D9
1E	8F	51	5F	DE	5F	12	33	0C	91	1F	85	1D	C5	60	D2	0B
1F	99	F2	49	E3	1A	10	3B	EE	10	43	FF	36	7F	4A	AC	F5
20	C4	AA	21	18	03	49	D1	FC	04	D1	9F	8F	3E	D2	43	C0
21	A8	36	9A	9A	A8	2E	A2	2B	8A	9F	AF	AD	E0	03	4F	C7
22	4E	ED	AD	13	12	03	3A	84	75	B4	95	82	13	31	61	46
23	C5	50	A3	FC	80	F2	D3	EA	1C	CF	E8	28	1F	C3	B8	A6
24	C1	A9	E2	00	00	16	D9	57	77	E9	3A	3A	3A	DC	A8	E4
25	7C	FE	5F	88	2E	80	6C	A9	AC	F5	31	1C	90	F8	06	DE
26	06	84	31	B7	40	EF	23	85	E5	F1	55	6F	78	26	22	88
27	B9	66	5B	49	DE	A0	3D	FA	A3	3A	AE	AA	22	05	43	1D
28	11	DA	2E	08	BB	89	46	5E	03	E3	DD	41	91	28	98	54
29	A8	DD	91	FB	CA	27	D5	31	A3	AB	C5	61	71	7B	9B	2D
2A	00	CA	83	36	1A	28	68	68	0D	F1	26	62	4F	2B	E5	52
2B	BB	69	66	81	2B	86	A4	04	D6	CD	B2	76	15	D2	31	72

Table 11.13 *(Continued)*

	0	1	2	3	4	5	6	7	8	9	A	B	C	D	E	F
2C	03	4F	E0	D5	55	E1	7A	8F	9D	A9	9D	AF	82	B6	83	9B
2D	F7	D2	51	96	21	41	CD	AA	03	8F	74	E7	48	10	7A	FB
2E	EC	8F	A6	C5	1B	00	8A	01	73	32	C9	9E	03	E4	C2	E8
2F	C2	61	54	81	90	C4	C7	7B	CD	78	33	94	AA	AD	F1	47
30	02	33	17	A1	CA	72	86	2B	21	98	38	1B	24	74	C9	47
31	39	E8	01	96	8A	A7	50	3B	6A	4C	AC	F0	57	90	1E	2B
32	47	8E	7C	F3	3B	7A	86	CA	31	E8	EC	03	C4	F2	D0	EC
33	32	AC	ED	58	98	9D	00	FB	E6	B1	70	B3	9C	4C	DB	1C
34	4A	18	65	55	F1	D6	42	0A	01	93	BE	CE	A3	40	24	A6
35	F2	D0	EE	4F	44	89	29	36	37	A6	0D	3A	AA	4F	4E	B8
36	F4	21	11	20	2A	E9	22	9D	44	38	B6	8C	F0	54	B4	F1
37	94	E1	BF	61	B9	29	F4	32	C6	9D	29	07	49	F8	F3	3A
38	F5	AE	0E	99	64	37	2E	13	01	C9	04	E0	5A	B8	F7	27
39	29	50	07	5B	06	CB	FF	1C	9F	21	4F	22	39	65	22	3D
3A	ED	64	82	17	05	28	D8	E1	2F	FD	B6	DF	7A	7E	49	0F

11.3 Construct transport stream headers (including offsets to the start of any PSI sections and start codes associated with PES packet headers) for the following data: **transport_error_indicator** = 0, **payload_unit_start_indicator** = 0, **transport_priority** = 0, PID = 563, **ts_scrambling_control** = 0, **ts_adaptation_control** = 1, **continuity_counter** = 6.

11.4 Construct transport stream headers (including offsets to the start of any PSI sections and start codes associated with PES packet headers) for the following data: **transport_error_indicator** = 0, **payload_unit_start_indicator** = 1, **transport_priority** = 0, PID = 563, **ts_scrambling_control** = 0, **ts_adaptation_control** = 1, **continuity_counter** = 6, for a transport stream packet containing PES packet data.

11.5 Construct transport stream headers (including offsets to the start of any PSI sections and start codes associated with PES packet headers) for the following data: **transport_error_indicator** = 0, **payload_unit_start_indicator** = 1, **transport_priority** = 0, PID = 63, **ts_scrambling_control** = 0, **ts_adaptation_control** = 1, **continuity_counter** = 6, for a transport stream carrying PSI data with offsets of 34 and 93 bytes to the starts of the PSI sections.

11.6 Construct a program association table for a transport stream carrying two programs whose program map tables have PID 0x0123 and 0x38F, with **transport_stream_id** = 0x083E. The network information table for this transport stream is carried in packets with PID 0x0562.

11.7 Construct a program map table for a program consisting of one video and one audio elementary stream, with the PCR carried in with the same PID as the video elementary stream. The video elementary stream is carried with PID 0x0437 and is the main profile at main level with 30-Hz frame rate and 4:2:0 chroma format. The audio elementary stream is constant bit-rate MPEG-2 Layer 2 audio with **free_format_flag** = 0, ID = 1 and is carried with PID 0x0346.

11.8 Use the data in Table 11.14, which describes the programs and elementary streams carried in an MPEG-2 transport stream, to construct a program association table and program map tables. The transport stream has a maximum bit rate of 20 Mbit/s, and **transport_stream_id** is 0x0034. In answering this question, it may be necessary to assign PIDs or other parameters. The network information table is carried in transport stream packets with PID 0x001F.

Table 11.14 Parameters of program and their constituent elementary streams for Q 11.6.

Program number	PID	Elementary stream type	Elementary stream description
1	0x0365	MPEG-2 video	Main profile at main level; 4:2:0; 25 Hz; aligned at slice layer
	0x366	MPEG-2 audio	English language; constant rate; Layer 2; ID = 1; **free_format_flag** = 0
	0x0365	PCR	Elementary stream carrying PCR for this program
2	0x0385	MPEG-2 video	Main profile at main level; 4:2:0; 30 Hz; aligned at picture layer
	0x386	MPEG-2 audio	English language; constant rate; Layer 2; ID = 1; **free_format_flag** = 0
	0x387	MPEG-2 audio	French language; constant rate; Layer 2; ID = 1; **free_format_flag** = 0
	0x0385	PCR	Elementary stream carrying PCR for this program

11.9 Calculate the overheads for the transport stream sublayer in the following cases:

(a) 30 slices per picture, 24 pictures per second, with each slice occupying one PES packet;

(b) 36 slices per picture, 25 pictures per second, with each slice occupying two PES packets; and

(c) 36 slices per picture, 25 pictures per second, with each slice occupying four PES packets.

11.10 Calculate the overheads due to the PES sublayer in the following cases:

(a) 30 slices per picture, 24 pictures per second, with each slice occupying one PES packet;

(b) 36 slices per picture, 25 pictures per second, with each slice occupying two PES packets; and

(c) 36 slices per picture, 25 pictures per second, with each slice occupying four PES packets.

11.11 Construct an MPEG-2 target background grid descriptor and MPEG-2 video window descriptor for a 720×576 pixel display with aspect ratio 4:3, containing a 352×288 window offset 50 pixels horizontally and 30 pixels vertically from the top left of the display;

11.12 Construct an MPEG-2 target background grid descriptor and MPEG-2 video window descriptor for a 720×576 pixel display with aspect ratio 4:3, containing a 352×288 window offset, 50 pixels horizontally and 30 pixels vertically from the bottom right of the display;

11.13 Construct an MPEG-2 IBP descriptor for an MPEG-2 video elementary stream in which each I-picture is immediately preceded by a GOP header, and each GOP contains exactly 12 pictures.

11.14 Construct an MPEG-2 video stream descriptor for a video elementary stream that carries an MPEG-1 constrained parameters bit stream with a frame rate of 30 Hz.

11.15 Construct an MPEG-2 ISO 639 language descriptor describing an audio elementary stream containing English language with no defined type and a French service for the hearing impaired.

11.16 Construct an MPEG-2 system clock descriptor for a system clock with accuracy 34 ppm.

11.17 Construct an MPEG-2 system clock descriptor for a system clock with accuracy 45 ppm.

11.18 Calculate the values of **CR_base** and **CR_ext** for the following values of the system time clock:

(a) 20,193,325;

(b) 5384.

11.19 Write down the closest time stamp values to the values of the system time clock in Q 11.18.

11.20 For the transport stream packet payload shown in Table 11.15, identify the **table_id** of each section appearing and the contents of each of these sections.

Table 11.15 Transport stream packet payload for Q 11.20.

	0	1	2	3	4	5	6	7	8	9	A	B	C	D	E	F
00	EF	EA	69	E4	0E	5A	D0	02	23	33	32	9A	45	32	03	BF
01	71	EE	77	6B	D8	86	33	AC	D6	05	AE	61	D4	80	B5	6D
02	2F	00	14	AE	4D	8A	26	B2	60	DC	DA	97	7F	E6	D2	A5
03	D1	A9	57	4A	57	88	BA	FF	FF	FF	FF	FF	FF	FF	FF	FF
04	FF	FF	FF	FF	FF	FF	FF	FF	FF	FF	FF	FF	FF	FF	FF	FF
05	FF	FF	FF	FF	FF	FF	FF	FF	FF	FF	FF	FF	FF	FF	FF	FF
06	FF	FF	FF	FF	FF	FF	FF	FF	FF	FF	FF	FF	FF	FF	FF	FF
07	FF	FF	FF	FF	FF	FF	FF	FF	FF	FF	FF	FF	FF	FF	FF	FF
08	FF	FF	FF	FF	FF	FF	FF	FF	FF	FF	FF	FF	FF	FF	FF	FF
09	FF	FF	FF	FF	FF	FF	FF	FF	FF	FF	FF	FF	FF	FF	FF	FF
0A	FF	FF	FF	FF	FF	FF	FF	FF								

REFERENCES

1. ISO/IEC 13818-1, *Information technology – Generic coding of moving pictures and associated audio information: Systems*, Geneva, Switzerland: International Standards Organisation, 1996.
2. While the structure shown in Figure 11.1 is consistent with the MPEG-2 standards, it is not mandated. There is no necessity, for example, that there exist a discrete entity that performs video encoding and

produces a video bit stream. What is required is that the encoder accept video and audio inputs and produce a valid systems bit stream.

3. As discussed above for the encoder, the MPEG-2 standards do not require that the decoder adopt this structure. What is required is that the decoder be capable of reconstructing video and audio signals from a received systems bit stream. Subject to this constraint, the internal architecture of the decoder may have any structure.
4. ISO 639, Code for the representation of names of languages.
5. Digital Video Broadcasting (DVB); Implementation guidelines for the use of MPEG-2 Systems, Video and Audio in satellite, cable and terrestrial broadcasting applications, ETR 154, Sophia Antipolis: ETSI, 1997.

Chapter 12

DVB Service Information and ATSC Program and System Information Protocol

12.1. INTRODUCTION

The systems part of MPEG-2 provides a range of functions needed to carry MPEG-2 video and audio elementary streams in a single multiplexed bit stream. The *Program-Specific Information* (*PSI*) in MPEG-2 systems (Chapter 11) allows limited information on the grouping of elementary streams to form programs to be delivered with the elementary streams. Both DVB and ATSC extend this concept, allowing much more side information on the services being delivered to be carried in the stream, each adding a number of its own new tables. In DVB, this additional information is known as *Service Information* (*SI*); in ATSC it is known as the *Program and System Information Protocol* (*PSIP*).

This chapter begins by examining the reasons why the information carried in the MPEG-2 PSI is not sufficient for a digital television service. This is followed by the operation of the DVB SI and ATSC PSIP. The aim of these sections is to present information on how the SI and the PSIP operate, rather than a complete examination of all the syntax involved. Readers interested in further information will find this in the relevant standards documents. It finishes by analyzing the constraints on the structure of transport streams that can enable embedding both DVB SI and ATSC PSIP in a single stream, allowing decoding by either ATSC or DVB decoders.

12.2. WHY SI AND PSIP?

The MPEG-2 PSI provides sufficient information on the structure of a transport stream for a decoder to identify packets carrying data for each elementary stream, to extract the elementary stream data, and to build complete programs from elementary streams. What is not provided, however, is any information that might be required to extract the transport stream from a distribution network. There is also no information

Digital Television, by John Arnold, Michael Frater and Mark Pickering.

that might be used to present an electronic program guide or other information on the available programs to the viewer.

The purpose of both the DVB SI and the ATSC PSIP is to extend the functionality of the MPEG-2 PSI in two areas: viewer information and system information. The viewer information may be used by a decoder to provide an electronic program guide. Examples include program titles; cast lists; and information on starting times and program classification, meeting regulatory requirements to indicate the presence of certain types of program content (such as violence).

The system information provides information used internally by a decoder, including the following:

- network information, describing one or more distribution networks;
- conditional access information, providing information to the decoder on any conditional access system in use;
- timing information, passing to the decoder the current time and date; and
- information on services not provided by MPEG-2, such as closed captioning/subtitling, teletext, and data services.

It is worth noting that, while both DVB and ATSC have been fit to define new tables and descriptors, neither has provided any performance requirements for decoders. Information provided in the profiles and levels for MPEG-2 video makes possible the guarantee that a compliant decoder can successfully decode a compliant bit stream. No such requirements exist for either SI or PSIP.

12.3. DVB-SI

The structure of DVB services is illustrated in Figure 12.1 [1]. A receiver or set-top box receives signals from one or more distribution systems, which are

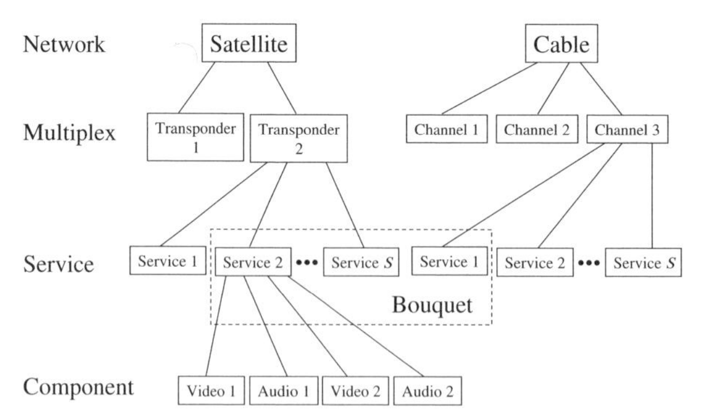

Figure 12.1 DVB service structure. © European Telecommunications Standards Institute 1997. © European Broadcasting Union 2006. Further use, modification, redistribution is strictly prohibited. ETSI standards are available from http://pda.etsi.org/pda/ and http:www.etsi.org/services_products/freestandard/home.htm.

known as networks in DVB. Possible networks are terrestrial broadcast, cable, satellite, and multipoint microwave distribution systems. Each of these networks delivers one or more multiplexes, each of which is an MPEG-2 system bit stream. Within each multiplexed stream, there are one or more services (corresponding to channels in analog television), each made up of a sequence of one or more programs. Each program consists of a sequence of one or more events. One or more components (corresponding to the MPEG-2 elementary stream) are combined to form each event.

Each network may contain the offerings of more than one service provider. The offerings of a service provider may also be spread across a number of networks. A *bouquet* is a group of services, usually corresponding to the services offered by a single provider. A bouquet may therefore be spread across a number of networks.

DVB SI consists of the four tables defined in the MPEG-2 PSI (program association table (PAT), program map table (PMT), network information table (NIT), conditional access table (CAT)) as well as a number of new tables, known as the *bouquet association table* (*BAT*), *service description table* (*SDT*), *event information table* (*EIT*), *running status table* (*RST*), *time and date table* (*TDT*), *time offset table* (*TOT*), and *stuffing table* (*ST*).

The **transport_stream_id** field in the program association table was intended by the designers of MPEG-2 to provide a unique identifier to each transport stream, potentially allowing one transport stream to carry data that refers to another identified transport stream. For ensuring that two network operators do not use the same value of **transport_stream_id**, fixed values can be assigned to each operator. The 16 bits allocated to this field allow 65,536 different streams to be distinguished, which is sufficient to distinguish one network's transport streams but insufficient to identify all transport streams on all networks or to allow permanent allocation of values to network operators. In order to solve this problem, DVB has introduced a new field, the **original_network_id**, carried in the network information table. A fixed assignment of values of the **original_network_id** to network operators is made. Each operator is at liberty to assign values of the **transport_stream_id**. The combination of **original_network_id** and **transport_stream_id** uniquely identifies a transport stream. When a transport stream crosses from one network to another, the value of the **original_network_id** is preserved. The relationship between **network_id**, **original_network_id**, and **transport_stream_id** is illustrated in Figure 12.2.

In addition to carrying SI tables describing itself, a DVB transport stream may also carry tables describing other transport streams. DVB uses the **transport_stream_id** field of the PAT to identify a transport stream described by a table. Tables describing data in the transport stream in which they are carried are said to refer to the *actual transport stream*, whereas the term *other transport stream* is used for tables describing data carried in a different transport stream. Tables referring to another transport stream identify the stream by the combination of its **original_network_id** and **transport_stream_id**. In the case of the network information table, information about the *actual* or *other* networks may also be carried.

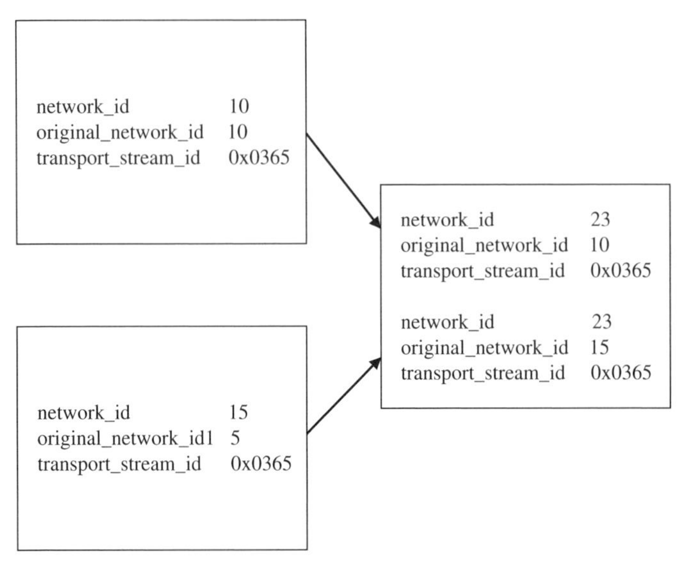

Figure 12.2 Use of network_id, original_network_id, and transport_stream_id to uniquely identify transport streams.

DVB tables are identified in the transport stream using the **table_id** field. DVB also assigns a fixed PID to each table (except for the PMT), which is not allowed to be used for other purposes. Some tables (such as the service description table and the bouquet association table) share PIDs, but have unique values of **table_id**. These PID and **table_id** values are shown in Table 12.1.

Where conditional access is used, tables are not scrambled except for the event information table carrying schedule information (**table_id** = 0x50–0x6F). The event information table may be scrambled because it is often considered that it carries proprietary information. Other tables do carry information that is effective in the public domain, so there is no protection for a network operator in scrambling them.

In DVB a new version of a section of a table becomes effective immediately after the last byte of the 32-bit CRC at the end of the section. Although the syntax supports sections with **current_next_indicator** equal to 0, these are never transmitted.

The decoders in consumer set-top boxes always use the SI data. Professional decoders (used by network operators) may use neither the SI nor the PSI, allowing streams with no SI or unreliable SI to be viewed and possibly debugged.

12.3.1. DVB Common Data Formats

A number of common elements appear in DVB tables and descriptors:

- The code used for text fields is based on the printable values of the ASCII code [2]. The length of text fields is sometimes specified explicitly

Table 12.1 DVB tables.

Table	Abbreviation	PID	table_id
Program association	PAT	0x0000	0x00
Conditional access	CAT	0x0001	0x01
Program map	PMT	Assigned in PAT	0x02
Network information—actual network	NIT	0x0010	0x40
Network information—other network	NIT	0x0010	0x41
Service description—actual TS	SDT	0x0011	0x42
Service description—other TS	SDT	0x0011	0x46
Bouquet association	BAT	0x0011	0x4A
Event information—actual TS present/following	EIT	0x0012	0x4E
Event information—other TS present/following	EIT	0x0012	0x4F
Event information—actual TS schedule	EIT	0x0012	0x50–0x5F
Event information—other TS schedule	EIT	0x0012	0x60–0x6F
Time and date	TDT	0x0014	0x70
Time offset	TOT	0x0014	0x73
Running status	RST	0x0013	0x71
Stuffing	ST	0x0010–0x0014	0x72

immediately prior to the beginning of the field. Otherwise, this length is derived from the value of **descriptor_length** (in a descriptor) or **section_length** (in a table).

- Countries are specified using the three-character (24-bit) codes specified by ISO 3166 [3]. For example, the code for the United Kingdom is "GBR."
- Languages are specified by the three-character (24-bit) codes specified by ISO 639 [4]. For example, the code for French is "FRE" and the code for English is "ENG."
- Time is coded as a 24-bit BCD number, specifying hours (*hh*), minutes (*mm*), and seconds (*ss*): *hhmmss*.
- Date is coded as a 16-bit modified Julian date (MJD), calculated as [5] $\text{MJD} = 14{,}956 + D + \lfloor (Y - L - 1900) \times 365.25 \rfloor + \lfloor (M + 1 + L \times 12) \times 30.6001 \rfloor$, where Y is the year, M is the month number (January = 1, February = 2, etc.), D is the day of the month (from 1 to 31), and $L = 0$ if $M = 1$ or $M = 2$ and $L = 1$ otherwise. $\lfloor x \rfloor$ means the integer part of x.

- Where the time and date are to be specified together, a 40-bit number is used in which the modified Julian date forms the most significant 16 bits and the time forms the remaining 24 bits.

12.3.2. DVB Descriptors

DVB defines a number of descriptors in addition to those defined by MPEG-2 (discussed in Section 11.6). The properties of these new descriptors, whose syntax follows the generic structure shown in Chapter 11, are shown in Table 12.2. The values of **descriptor_tag** used are all in the range assigned for private descriptors by MPEG-2 (0x40–0xFF). DVB claims values in the range 0x40–0x7F for its own use, leaving the remaining values for user definition.

Table 12.2 Descriptors defined by DVB in EN 300 468.

descriptor_tag	Descriptor	Purpose
0x40	Network name descriptor	Network name in text form.
0x41	Service list descriptor	Specifies a service type for a program number.
0x42	Stuffing descriptor	Allows overwriting of invalidated descriptors without changing size or timing of transport stream or inserting of dummy descriptors as place holders. May be carried in NIT, BAT, SDT, and EIT.
0x43	Satellite delivery system descriptor	Specifies the characteristics of a satellite delivery system.
0x44	Cable delivery system descriptor	Specifies the characteristics of a cable delivery system.
0x45–0x46		Reserved for future use.
0x47	Bouquet name descriptor	Text name of a bouquet.
0x48	Service descriptor	Specifies the name of the service provider and service in text form, along with the service type.
0x49	Country availability descriptor	Specifies countries in which the service is intended to be available and those in which the service is not intended to be available.
0x4A	Linkage descriptor	Specifies a service that can be presented at user request (e.g., additional information related to the service or electronic program guide) or when the selected service is not available (e.g., denied by CA).

Table 12.2 (*Continued*)

descriptor_tag	Descriptor	Purpose
0x4B	NVOD reference descriptor	Provides a list of services that form a near video-on-demand service, allowing a user to select near to the start of an event by choosing the appropriate service.
0x4C	Time-shifted service descriptor	Used in place of the service descriptor to indicate a time-shifted copy of a service.
0x4D	Short event descriptor	Provides the name of an event and a short text description in a specified language.
0x4E	Extended event descriptor	Detailed text description of an event, including a specification of the language used in the description. Up to 15 of these descriptors may be combined where insufficient space is available in one descriptor.
0x4F	Time-shifted event descriptor	Used in place of the short event descriptor to indicate a time-shifted copy of an event.
0x50	Component descriptor	Identifies the content of an elementary stream (video, audio, or DVB-defined data) and the type of the content (e.g., video 4:3 aspect ratio, audio single channel), and assigns a text name to it.
0x51	Mosaic descriptor	Specifies a mosaic of images for display.
0x52	Stream identifier descriptor	Tags a component to allow association with a component_descriptor.
0x53	CA identifier descriptor	Contains a 16-bit field that identifies the CA system in use. DVB document ETR 162 assigns the values of this field to CA vendors.
0x54	Content descriptor	Classifies content. See example below.
0x55	Parental rating descriptor	Specifies for individual countries the minimum suitable viewing age for service.
0x56	Teletext descriptor	Specifies the language used in the teletext and a magazine and page number.
0x57	Telephone descriptor	Specifies a telephone number for narrow-band dial-up interactive channel.
0x58	Local time offset descriptor	Provides a list of countries and regions with the associated time offsets from coordinated universal time (UTC), expressed in hours and minutes. Also allows for the next time offset and the date of the change to be specified (e.g., beginning or end of summer time) for each region.

(*continued*)

Table 12.2 (*Continued*)

descriptor_tag	Descriptor	Purpose
0x59	Subtitling descriptor	Identifies the type of subtitling and the language of the subtitles.
0x5A	Terrestrial delivery system descriptor	Specifies the characteristics of a terrestrial delivery system.
0x5B	Multilingual network name descriptor	Text description of network name in one or more languages.
0x5C	Multilingual bouquet name descriptor	Provides bouquet name in one or more languages.
0x5D	Multilingual service name descriptor	Provides text name of service provider and text description of service in one or more languages.
0x5E	Multilingual component descriptor	Provides a text description of a component in one or more languages.
0x5F	Private data specifier descriptor	Specifies the type of private data, using values listed in ETR 162.
0x60	Service move descriptor	Used to allow a service to be tracked by a receiver from one TS to another TS if it is moved.
0x61	Short smoothing buffer descriptor	A compact version of the MPEG-2 smoothing_buffer_descriptor.
0x62	Frequency list descriptor	Contains a list of center frequencies for a multiplex that is transmitted on more than one frequency.
0x63	Partial transport stream descriptor	Used with partial transport streams; usually associated with digital storage media.
0x64	Data broadcast descriptor	For data broadcast components, identifies the specification to which the broadcast adheres and the language of the data broadcast. An optional text description of the component is also available.
0x65	CA system descriptor	Reserved for definition by the Digital Audio Visual Council (DAVIC).
0x66	Data broadcast id descriptor	Short form of data broadcast descriptor containing only the type of the data broadcast (but without text or language).

Many of these descriptors may appear in more than one table. The scope of the descriptor (i.e., the objects to which it refers) is defined by the table in which it appears, as discussed in Section 12.3.3. The following sections provide a more detailed description of a number of these descriptors.

content_descriptor

*	content_nibble_level_1	{movie/drama,news/current affairs, sports, etc}
	content_nibble_level_2	{comedy, game show, football, etc}
	user_nibble	Unsigned integer
	user_nibble	Unsigned integer

Figure 12.3 Structure of the DVB content descriptor.

12.3.2.1. DVB Content Descriptor

The syntax of the DVB content_descriptor is shown in Figure 12.3, using the simplified syntax defined in Chapter 11. The placement of the bracket and "*" to the left of a field or group of fields indicates that the field or group is repeated zero or more times. This alternative representation is used throughout this chapter.

The body of the DVB content_descriptor consists of four 4-bit fields. The values of the first two of these (**content_nibble_1** and **content_nibble_2**) are standardized by ETSI EN 300 468 [1], allowing a high-level description of the content, such as "drama" or "comedy." The meanings associated with the remaining two fields (**user_nibble**) is not standardized, but is left for definition by the broadcaster. The mnemonic "uimsbf" means that the field is represented as an unsigned integer and the most significant bit is transmitted first. The purpose of the "for" loop is to allow multiple values to be transmitted for the body of the descriptor. "*N*" is the number of sets of values contained in the descriptor. This is not transmitted in the bit stream but is inferred from the value of the **descriptor_length** field.

EXAMPLE 12.1—Content Descriptor

The construction of the bit stream for an example content_descriptor is illustrated in Table 12.3. This descriptor specifies a movie/drama (**content_nibble_level_1** = 0x1), of type comedy (**content_nibble_level_2** = 0x4). The values assigned to the two user nibbles are 0x9 and 0xD. The bit stream generated is 0101 0100 0000 0010 0001 0100 1001 1101.

Table 12.3 Sample DVB content descriptor.

Bit stream	Syntax element
0101 0100	descriptor_tag = content_descriptor (0x54)
0000 0010	descriptor_length = 2 bytes
0001	content_nibble_level_1 = movie/drama (0x1)
0100	content_nibble_level_2 = comedy (0x4)
1001	user_nibble = 0x9
1101	user_nibble = 0xD

The alternative representation, avoiding the use of syntax, is shown in Figure 12.4.

content_descriptor	
content_nibble_level_1	movie / drama
content_nibble_level_2	comedy
user_nibble	0x9
user_nibble	0xD

Figure 12.4 Alternative representation of DVB content descriptor example. ■

12.3.2.2. DVB Delivery System Descriptors

DVB provides three descriptors that convey the properties of a particular delivery system: the cable deliver system descriptor, the satellite delivery system descriptor, and the terrestrial delivery system descriptor. All delivery system descriptors have length 13 bytes in order to facilitate swapping of descriptors when a transport stream moves between networks. Without this fixed length, retiming of the transport stream would often be required at network boundaries.

DVB Satellite Delivery System Descriptor The satellite delivery system descriptor carries information on satellite orbital position, modulation, and channel coding. The contents of this descriptor, shown in Figure 12.5, are the frequency (expressed in megahertz, with a resolution of 100 Hz[1]), the orbital position (in degrees east or west, with a resolution of 0.1°), the polarization of the transmitted signal (linear vertical/horizontal or circular left/right), the modulation (always QPSK), the symbol rate (in megasymbols (Msym) per second, with a resolution of 100 sym/s), and the coding rate for the inner convolutional coder (1/2, 2/3, 3/4, 5/6, or 7/8). (The operation of the convolutional coder is discussed in Chapter 13.)

satellite_delivery_system_descriptor	
frequency	frequency (in MHz)
orbital_position	degrees east or west
polarization	{linear (vertical/horizontal),circular (left/right)}
modulation	QPSK
symbol_rate	rate in Msym/s
FEC_inner	{1/2,2/3,3/4,5/6,7/8}

Figure 12.5 Simplified syntax for the DVB satellite delivery system descriptor.

EXAMPLE 12.2—Satellite Delivery System Descriptor

An example of the satellite delivery system descriptor, with a frequency of 500.5 MHz, orbital position 150.7° East, vertical polarization, a symbol rate of 10 Msym/s, and a 1/2 rate inner convolutional coder, is shown in Figure 12.6.

[1]This could also be expressed as an integer representing the frequency divided by 100 Hz. The description in the main text follows that used by the DVB standards.

satellite_delivery_system_descriptor	
frequency	500.5 MHz
orbital_position	150.6 degrees east
polarization	linear/vertical
modulation	QPSK
symbol_rate	10 Msym/s
FEC_inner	1/2 rate

Figure 12.6 Example satellite DVB delivery system descriptor. ■

cable_delivery_system_descriptor	
frequency	frequency in MHz
FEC_outer	{1/2,2/3/,3/4,4/5,5/6,7/8}
modulation	{16, 32, 64, 128, 256} QAM
symbol_rate	rate in Msym/s
FEC_inner	{1/2,2/3,3/4,5/6,7/8, none}

Figure 12.7 Structure of the DVB cable delivery system descriptor.

DVB Cable Delivery System Descriptor The cable delivery system descriptor is similar in structure to the satellite delivery system descriptor, as illustrated in Figure 12.7. It carries the frequency (in megahertz, with a resolution of 100 Hz), the coding rates for the inner and outer convolutional coders (1/2, 2/3, 3/4, 5/6, 7/8, or none), and the modulation type (16, 32, 64, 128, or 256 QAM). Modulation and the operation of convolutional coders are discussed in Chapter 13.

EXAMPLE 12.3—Cable Delivery System Descriptor

An example cable delivery system descriptor is shown in Figure 12.8, with frequency 10.5005 GHz, 7/8 rate outer convolution coder, 1/2 rate inner convolutional coder, 64-QAM modulation, and a symbol rate of 20 Msym/s.

cable_delivery_system_descriptor	
frequency	10,500.5 MHz
FEC_outer	7/8 rate
modulation	64-QAM
symbol_rate	20 Msym/s
FEC_inner	1/2 rate

Figure 12.8 Example DVB cable delivery system descriptor. ■

DVB Terrestrial Delivery System Descriptor The terrestrial delivery system descriptor, whose structure is shown in Figure 12.9, carries the center frequency of the channel (in megahertz, with a resolution of 10 Hz), the channel bandwidth (6, 7, or 8 MHz), the constellation (QPSK, 16 QAM, or 64 QAM), a flag indicating the presence or absence of the hierarchical mode, coding rates for the inner convolutional coder in the high- and low-priority streams, the guard interval (1/32, 1/16, 1/8, or 1/4), the transmission mode (2k or 8k), and a flag indicating whether or not other frequencies are in use. The modulation and channel coding used for DVB terrestrial broadcast delivery systems are described in Chapter 13.

terrestrial_delivery_system_descriptor	
center frequency	frequency (in MHz)
bandwidth	bandwidth in MHz
constellation	{QPSK,16-QAM,64-QAM}
hierarchy information	{Hierarchical,non-hierarchical}
code_rate-HP_stream	{1/2,2/3,3/4,5/6,7/8}
code_rate-LP_stream	{1/2,2/3,3/4,5/6,7/8}
guard_interval	{1/32,1/16,1/8,1/4}
transmission mode	{2k,8k}
other_frequency_flag	flag

Figure 12.9 Structure of the DVB terrestrial delivery system descriptor.

EXAMPLE 12.4—Terrestrial Delivery System Descriptor

Figure 12.10 shows the contents of a terrestrial delivery system descriptor, for a system transmitting at 500.5 MHz, with a bandwidth of 7 MHz using 64 QAM in the nonhierarchical mode with a 1/2 rate inner code, a 1/32 guard interval, operating in the 2k mode, and not using other frequencies.

terrestrial_delivery_system_descriptor	
center frequency	500.5 MHz
bandwidth	7 MHz
constellation	64-QAM
hierarchy information	Non hierarchical
code_rate	1/2
guard_interval	1/32
transmission mode	2k
other_frequency_flag	None

Figure 12.10 Example DVB terrestrial deliver system descriptor. ■

12.3.2.3. DVB Name Descriptors

DVB provides a number of descriptors whose function is to carry the name of an object, such as a network, a bouquet, a service, or a component.

DVB Network Name Descriptor The purpose of the network name descriptor is to carry a text name for a network, as illustrated in Figure 12.11. The network name is expressed as a single text string, whose length is limited by the maximum descriptor size of 256 bytes.

network_name_descriptor	
char	String

Figure 12.11 Structure of the DVB network name descriptor.

EXAMPLE 12.5—Network Name Descriptor

An example network name descriptor, carrying the text "News Network" is shown in Figure 12.12.

network_name_descriptor	
char	"News Network"

Figure 12.12 Example DVB network name descriptor. ■

bouquet_name_descriptor	
char	String

Figure 12.13 Structure of the DVB bouquet name descriptor.

DVB Bouquet Name Descriptor The bouquet name descriptor carries text for the name of a particular bouquet in an unspecified language, as illustrated in Figure 12.13. The only data-carrying field of the descriptor is the character string.

DVB Multilingual Network Name Descriptor The multilingual network name descriptor has the same purpose as the network name descriptor, that is, to carry the text name of a network. The multilingual version, however, is able to carry this name expressed in one or more languages, each of which is explicitly identified within the descriptor. The multilingual network name descriptor carries a language identifier (using the format of ISO 639) and a text string for each language in which the network name is specified, as shown in Figure 12.14. The network to which the descriptor refers is not specified in the descriptor, but is inferred from the table in which the descriptor appears.

multilingual_network_name_descriptor		
*	ISO_639_language_code	ISO 639 language code
	char	String

Figure 12.14 Structure of the DVB multilingual network name descriptor.

EXAMPLE 12.6—Multilingual Network Name Descriptor

A multilingual network name descriptor for a news service is shown in Figure 12.15. Two languages (English and German) are supported, each identified by an ISO 639 language code.

multilingual_network_name_descriptor	
ISO_639_language_code	English
char	"News Network"
ISO_639_language_code	German
char	"Nachrichten"

Figure 12.15 Example DVB multilingual network name descriptor. ■

multilingual_bouquet_name_descriptor

*	ISO_639_language_code	ISO 639 language code
	char	String

Figure 12.16 Structure of the DVB multilingual bouquet name descriptor.

multilingual_service_name_descriptor

*	ISO_639_language_code	ISO 639 language code
	service_provider_name	String
	service_name	String

Figure 12.17 Structure of the DVB multilingual service name descriptor.

DVB Multilingual Bouquet Name Descriptor Like the multilingual network name descriptor, the multilingual bouquet name descriptor carries its data in one or more specified languages, as shown in Figure 12.16. The purpose of this descriptor is to carry the name of a bouquet in one or more languages. The bouquet to which the descriptor refers is inferred from the table in which it is carried.

DVB Multilingual Service Name Descriptor The multilingual service name descriptor carries the name of a service provider and service name for one service in one or more specified languages, as shown in Figure 12.17. The service described is not explicitly identified in the descriptor, but is determined from the table in which the descriptor is carried.

DVB Multilingual Component Descriptor The multilingual component descriptor specifies a text name for a component in one or more specified languages, as illustrated in Figure 12.18. The component is identified explicitly in the descriptor by its unique **component_tag**, which appears before the text description. The value of **component_tag** is assigned to a particular elementary stream by placing a stream identifier descriptor in the program map table. The component tag is therefore guaranteed to be unique only within a transport stream.

multilingual_component_descriptor

	component_tag	Unsigned integer
*	ISO_639_language_code	ISO 639 language code
	char	String

Figure 12.18 Structure of the DVB multilingual component descriptor.

EXAMPLE 12.7—Multilingual Component Descriptor

An example of the multilingual component descriptor is shown in Figure 12.19, for which the component being named has **component_tag** equal to 0x45. The language of the string is English.

multilingual_component_descriptor

component_tag	0x45
ISO_639_language_code	English
char	"English audio"

Figure 12.19 Example DVB multilingual component descriptor. ■

component_descriptor

stream_content	{video, audio, teletext, subtitles}
component_type	{aspect ratio for video, subtitles & teletext, number of channels for audio}
component_tag	Unsigned integer
ISO_639_language_code	ISO 639 language code
text	String

Figure 12.20 Structure of the DVB component descriptor.

12.3.2.4. DVB Component Descriptor

The DVB component descriptor, whose structure is shown in Figure 12.20, describes one component (i.e., elementary stream) present in a transport stream. The **stream_content** tells a decoder that the stream is video, audio, subtitles, or teletext, indicating the appropriate elementary stream decoder to be used. The **component_type** further classifies the stream, specifying for video the aspect ratio and for audio the number of channels. The **component_tag** assigns a unique identifier to the component.

EXAMPLE 12.8—DVB Component Descriptor

An example DVB component descriptor is shown in Figure 12.21. The component described is a video elementary stream with 4:3 aspect ratio, component_tag equal to 0x23, English language, and name "Football-video."

component_descriptor

stream_content	video
component_type	4:3 aspect ratio
component_tag	0x23
ISO_639_language_code	ENG
text	"Football-video"

Figure 12.21 Example DVB component descriptor. ■

12.3.2.5. DVB Event Descriptors

DVB uses the term *event* to describe a portion of one program. Two descriptors are available to provide text descriptions of events: the short event descriptor and the extended event descriptor.

DVB Short Event Descriptor The short event descriptor (Fig. 12.22) carries the name of an event and a free-form text description of the event in one specified language. The event to which the descriptor refers is not specified, but is determined from the table in which the descriptor is carried.

short_event_descriptor	
ISO_639_language_code	ISO_639_language_code
event_name_char	String
text_char	String

Figure 12.22 Structure of the DVB short event descriptor.

EXAMPLE 12.9—Short Event Descriptor

An example short event descriptor is shown in Figure 12.23. This descriptor names the event as "Premier League Football," with a description that includes the names of the teams involved.

short_event_descriptor	
ISO_639_language_code	English
event_name_char	Premier League Football
text_char	Manchester United vs Arsenal

Figure 12.23 Example of the DVB short event descriptor. ■

DVB Extended Event Descriptor The extended event descriptor, which is intended for use in addition to the short event descriptor rather than as a replacement for it, is used to specify additional information about an event (Fig. 12.24). The descriptor begins by specifying the language used in its text, which is followed by one or more pairs of strings that are designed to be presented in a two-column format. The **item_description** carries the string for the left column; the **item_char** carries the data for the right column. This style is useful in an electronic program guide for presenting credits for the program. Further text, for presentation in a single column, can be carried in the **text_char**, which might carry an expanded version of the text in the short event descriptor and might be a general description of the event. The extended event descriptor is unique among descriptors, in that its syntax permits the size of the descriptor to be larger than 256 bytes.

Like the short event descriptor, the event to which the extended event descriptor refers is determined from the table in which it is carried.

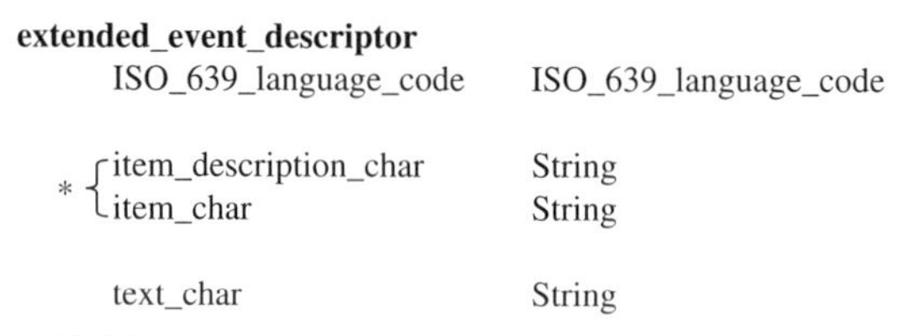

Figure 12.24 Simplified syntax of the DVB extended event descriptor.

EXAMPLE 12.10—Extended Event Descriptor

An example extended event descriptor is shown in Figure 12.25. The intended, two-column presentation of the **item_description_char** and **item_char** fields is shown in Table 12.4.

extended_event_descriptor

ISO_639_language_code	English
item_description_char	"Producer"
item_char	"John Smith"
item_description_char	"Director"
item_char	"Hugo Jones"
text_char	"Live from Manchester"

Figure 12.25 Example DVB extended event descriptor.

Table 12.4 Presentation of data from example extended event descriptor of Figure 12.25.

Producer	John Smith
Director	Hugo Jones

12.3.2.6. DVB Descriptors for Ancillary Services

DVB supports several ancillary services, including teletext and subtitles. Each of these ancillary services has an associated descriptor.

DVB Teletext Descriptor The teletext descriptor (Fig. 12.26) specifies for one or more teletext pages, the language, the type of data carried, the magazine number, and the page number. (The operation of ITU-T System B Teletext, which is used in DVB streams, is described in Chapter 14.) A teletext page may be specified as carrying subtitles, as an initial page or additional page, or a program schedule page. Without specification of the page type, the decoder treats a teletext stream in the same way as an analog receiver; use of these fields may support more sophisticated page navigation features. The teletext stream to which the descriptor refers is determined from the placement of the descriptor in the program map table.

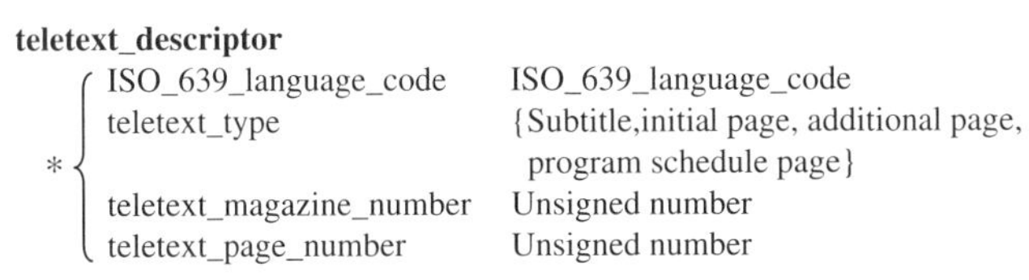

Figure 12.26 Structure of the DVB teletext descriptor.

EXAMPLE 12.11—Teletext Descriptor

An example teletext descriptor is shown in Figure 12.27. Identification of this page as a program schedule facilitates quick presentation, without a requirement for the viewer to explicitly enter the magazine and page number.

teletext_descriptor	
ISO_639_language_code	English
subtitling_type	Program schedule
teletext_magazine_number	5
teletext_page_number	12

Figure 12.27 Example DVB teletext descriptor. ■

subtitling_descriptor

*	ISO_639_language_code	ISO_639_language_code
	subtitling_type	{Normal/hearing impaired}
	composition_page_id	Unsigned number
	ancillary_page_id	Unsigned number

Figure 12.28 Structure of the DVB subtitling descriptor.

DVB Subtitling Descriptor The subtitling descriptor is used to identify the type of subtitles (Fig. 12.28). More than one subtitling service may be provided for one program, some supporting services for the hearing impaired while other providing text translation of dialog into other languages. This descriptor is usually carried in the program map table, from which the subtitling stream to which it refers is identified. The values of **composition_page_id** and **ancillary_page_id** refer to the location in the subtitling service where the subtitling data can be found.

DVB subtitling is described in Chapter 14.

EXAMPLE 12.12—Subtitling Descriptor

An example subtitling descriptor is shown in Figure 12.29, in which subtitling in two languages is specified. For English, hearing-impaired subtitles are provided (with a composition page id of 34 and ancillary page id 653), whereas for French, normal subtitles are provided (with a composition page id of 36 and ancillary page id 659). For each type of subtitle, the subtitle page is identified.

subtitling_descriptor	
ISO_639_language_code	English
subtitling_type	Hearing impaired
composition_page_id	34
ancillary_page_id	653
ISO_639_language_code	French
subtitling_type	Normal
composition_page_id	36
ancillary_page_id	659

Figure 12.29 Example of the DVB subtitling descriptor. ■

12.3.2.7. DVB Service Descriptor

The service descriptor (Fig. 12.30) specifies the type of a service, using one of the values shown in Table 12.5, the name of the service provider, and the name

service_descriptor

service_type	{digital television,digital radio,teletext, data broadcast, etc}
service_provider_name	String
service_name	String

Figure 12.30 Structure of the DVB service descriptor.

Table 12.5 Service types supported by the DVB service descriptor.

Service type
Digital television service
Digital radio sound service
Teletext service
NVOD reference service
NVOD time-shifted service
Mosaic service
PAL coded signal
SECAM coded signal
D/D2-MAC
FM radio
NTSC coded signal
Data broadcast service

of the service. The service to which these values refer is inferred from the service description table in which the descriptor is carried. "NVOD" is an abbreviation for near-video-on-demand, which refers to a group of services that are time-shifted from one another to provide a similar service to video-on-demand.

EXAMPLE 12.13—Service Descriptor

An example service descriptor for a news service provided by CNN is shown in Figure 12.31.

service_descriptor

service_type	digital television
service_provider_name	"CNN"
service_name	"News channel"

Figure 12.31 Example DVB service descriptor. ■

12.3.2.8. DVB Service List Descriptor

The service list descriptor, illustrated in Figure 12.32, provides a means for specifying the types (e.g., digital television, digital radio) of one or more services. It is typically carried in the network information table or bouquet association table. The

service_list_descriptor

*	service_id	Unsigned integer
	service_type	{digital television,digital radio,teletext,data broadcast, etc}

Figure 12.32 Structure of the DVB service list descriptor.

transport stream in which the service is carried must be inferred from the table in which the service list descriptor is carried. The format of the **service_type** field is the same as in the service descriptor.

EXAMPLE 12.14—Service List Descriptor

Figure 12.33 shows an example of the service list descriptor. The first service, with **service_id** 0x4535, is a digital television service, whereas the second service is a teletext service.

service_list_descriptor

service_id	0x4535
service_type	digital television
service_id	0x543D
service_type	teletext

Figure 12.33 Example DVB service list descriptor. ■

12.3.2.9. DVB Linkage Descriptor

The linkage descriptor specifies a service that provides extra information about a bouquet, service, or event and has the structure shown in Figure 12.34. The type of the associated information is specified by the **linkage_type** and may be an information service, an EPG service, a CA replacement service, (for use where the conditional access system does not grant access to the service), a service replacement service, or a data broadcast service. The service carrying the associated information is specified by the **transport_stream_id**, **original_network_id**, and **service_id** contained in the descriptor. The bouquet, service, or event to which the associated information is linked is inferred from the table in which the linkage descriptor is transmitted.

linkage_descriptor

transport_stream_id	Unsigned integer
original_network_id	Unsigned integer
service_id	Unsigned integer
linkage_type	{information,EPG, CA replacement, service replacement,data broadcast}

Figure 12.34 Structure of the DVB linkage descriptor.

EXAMPLE 12.15—Linkage Descriptor

An example linkage descriptor that specifies a service carrying an EPG associated with a service is shown in Figure 12.35.

linkage_descriptor

transport_stream_id	0xD234
original_network_id	0x5432
service_id	0x4545
linkage_type	EPG

Figure 12.35 Example DVB linkage descriptor. ■

parental_rating_descriptor

*	country_code	ISO-8859 country code
	rating	age

Figure 12.36 Structure of the DVB parental rating descriptor.

12.3.2.10. DVB Parental Rating Descriptor

The parental rating descriptor (Fig. 12.36) specifies the minimum age for which a particular content is suitable. A separate specification may be made for each country.

EXAMPLE 12.16—Parental Rating Descriptor

An example parental rating descriptor, specifying a minimum age of 15 for England and 10 for Sweden, is shown in Figure 12.37.

parental_rating_descriptor

country_code	England
rating	15
country_code	Sweden
rating	10

Figure 12.37 Example DVB parental rating descriptor. ■

12.3.2.11. DVB Local Time Offset Descriptor

The local time offset descriptor is used to specify one or more time offsets from *coordinated universal time* that are used in particular regions of one or more countries. UTC was previously known as *Greenwich mean time* (*GMT*). For each region and country, the current local time offset from UTC is specified in hours and minutes, with positive values indicating east and negative values indicating west. The date and time at which the next change in time offset is to occur is also specified (as a modified Julian date), accompanied by the value of this new offset (in hours and minutes). The most likely reason for this offset is a switch to or from summer time (Fig. 12.38).

local_time_offset_descriptor

*	country_code	ISO-3166 country code
	region_id	time zone within country
	local_time_offset	signed BCD number
	time_of_next_change	same format as UTC_time
	next_time_offset	signed BCD number

Figure 12.38 Structure of the DVB local time offset descriptor.

EXAMPLE 12.17—Local Time Offset Descriptor

An example local time offset descriptor is shown in Figure 12.39, specifying the time offset for the United Kingdom as plus 1 h (i.e., summer time). The descriptor also specifies a time and date for the next change in this time offset, at which the offset changes to zero.

local_time_offset_descriptor

country_code	United Kingdom
region_id	Only one time zone
local_time_offset	+0100
time_of_next_change	0xCBD7000000
next_time	+0000

Figure 12.39 Example DVB local time offset descriptor. ■

12.3.3. DVB Tables

MPEG-2 defines four tables: the program association table, the program map table, the conditional access table, and the network information table. These tables form a part of the DVB SI. In the case of the program association table, the program map table, and the conditional access table, the full syntax and semantics are defined by MPEG-2. For the network information table, only its existence is defined by MPEG-2; its syntax is defined by DVB. This section discusses special use made by DVB of the MPEG-2 PSI and the new tables introduced by DVB.

12.3.3.1. DVB Use of the Network Information Table

The syntax of the network information table is not defined by MPEG-2 but is defined by DVB, where it is always carried with a PID of 0x0010. Its purpose is to carry information on the physical organization of transport streams on a network and on the network itself. A network information table may describe the network on which it is transmitted (the *actual network*) or another network (*other network*). Transmission of the network information table for the actual network is mandatory.

One section of the network information table (Fig. 12.40) is required for each network to be described. The network information section consists of a unique network identifier (assigned in ETR 162 [6]), a group of descriptors describing this network, and a list of transport streams carried by the network, each of which may be accompanied by one or more descriptors describing that transport stream. Each

network_information_section-(actual/other)_network

	network_id	unsigned integer
*	<< descriptors>>	
* {	transport_stream_id	unsigned integer
	original_network_id	unsigned integer
	<< descriptors>>	

Figure 12.40 Simplified syntax of the DVB network information table.

of these transport streams is identified by its combination of **transport_stream_id** and **original_network_id**.

The network information table is not scrambled.

A number of descriptors have meaning when carried in the network information table, some describing the network as a whole and others describing one multiplex carried by the network. Descriptors that can be used to describe the network to which the table refers are the linkage descriptor, the network name descriptor, and the multilingual network name descriptor. The linkage descriptor is used to point to services associated with the network as a whole, such as electronic program guides or network information. The network name descriptor and multilingual network name descriptor carry a text form of the network's name, such as "BSkyB," or "ASTRA."

Descriptors carried by the network information table that describe individual multiplexes are the delivery system (cable, satellite, and terrestrial) descriptors, the service list descriptor, and the frequency list descriptor. The carriage of one delivery system descriptor describing each multiplex is required. The service list descriptor may contain a list of services carried by the multiplex, identified by their **service_id**, which is the same as the MPEG-2 **program_number**. The frequency list descriptor may carry a list of other frequencies used in transmission of the multiplex.

EXAMPLE 12.18—DVB Network Information Table

An example network information table is shown in Figure 12.41. This network has a network identifier of 0x0069 and a network name "Satellite Network." This network carries two

network_information_section-actual_network
network_id 0x0069
network_name_descriptor
char "Satellite Network"

transport_stream_id 0x0234
original_network_id 0x0001
satellite_delivery_system_descriptor
frequency 12.750 MHz
orbital_position 156.0° E
bandwidth 7 MHz
polarization linear/vertical
modulation QPSK
symbol_rate 14 Msym/s
FEC_inner 1/2

transport_stream_id 0x0239
original_network_id 0x0001
satellite_delivery_system_descriptor
frequency 12.720 MHz
orbital_position 156.0° E
bandwidth 10 MHz
polarization linear/vertical
modulation QPSK
symbol_rate 20 Msym/s
FEC_inner 3/4

Figure 12.41 Example DVB network information table.

multiplexes, each of which is identified by its unique combination of **transport_stream_id** and **original_network_id**. Each of these multiplexes is described by a satellite delivery system descriptor. ■

```
bouquet_association_section
        bouqet_id       unsigned integer
*       << descriptors>>

      { transport_stream_id          unsigned integer
*     { original_network_id          unsigned integer
      {  << descriptors>>
```

Figure 12.42 Simplified syntax for the DVB bouquet association table.

12.3.3.2. DVB Bouquet Association Table

The purpose of the bouquet association table is to provide information on bouquets, which are collections of services, possibly crossing network boundaries. The bouquet association table has the same structure as the network information table. Whereas the network information table provides information on the physical arrangement of multiplexes and services, the bouquet association table describes the logical grouping of services into bouquets.

Each section of the bouquet association table, whose syntax is illustrated in Figure 12.42, contains a **bouquet_id** that uniquely identifies the bouquet to which the table refers. These values are assigned in ETR 162. BSkyB, for example, is assigned values in the range 0x1000–0x101F. The **bouquet_id** is followed by a list of descriptors describing the bouquet as a whole. These descriptors may include the bouquet name descriptor and multilingual bouquet name descriptor, CA identifier descriptor, country availability descriptor, and linkage descriptor.

The bouquet name descriptor and the multilingual bouquet name descriptor may be used in the bouquet association table to provide a text name for the bouquet. The CA identifier descriptor identifies conditional access used by services within the bouquet. The country availability descriptor specifies countries in which the entire bouquet is to be available or not. The linkage descriptor is used to specify a service that carries information on the bouquet or an EPG.

For each transport stream carrying services belonging to a bouquet, the bouquet association table carries the **transport_stream_id**, **original_network_id**, and a service list descriptor that associates one or more services in that transport stream with the bouquet.

EXAMPLE 12.19—Bouquet Association Table

An example bouquet association table is shown in Figure 12.43. This bouquet is spread across two transport streams, encompassing two services on each (giving a total of four services).

bouquet_association_section
bouqet_id 0x1000
bouquet_name_descriptor
char "Satellite Bouquet"

transport_stream_id 0x0234
original_network_id 0x1056
service_list_descriptor
service_id 0x4520
service_type digital television

service_id 0x5123
service_type teletext

transport_stream_id 0x028F
original_network_id 0x1056
service_list_descriptor
service_id 0x4535
service_type digital television

service_id 0x543D
service_type digital radio

Figure 12.43 Example DVB bouquet association table. ■

12.3.3.3. DVB Service Description Table

The service description table defines the services carried by a network. A service is similar in concept to an analog television channel and to the MPEG-2 concept of a program.

Each section of the service description table refers to one transport stream, identified by its **transport_stream_id** and its **original_network_id**. For each service present in the transport stream, the section carries the **service_id**, flags to indicate the presence of event information tables for the schedule and the present and following events, and a flag to indicate the use of conditional access in one or more components of the service. The **service_id** uniquely identifies a service within a transport stream. A service can be uniquely identified among all services on all networks using its **service_id** with the **transport_stream_id** and **original_network_id** of its host transport stream. The **program_number** in the program map table has the same value as the **service_id**. It is recommended by the DVB standards that a fixed **service_id** (and therefore **program_number**) be allocated to each service so that a decoder can store a list of previously used services to allow the viewer to return easily (Fig. 12.44).

Each descriptor carried by the service description table refers to the one service with whose **service_id** it is grouped. Descriptors that are likely to be carried in this table include the service descriptor (specifying the name of the service and the service operator) and multilingual service descriptor, the bouquet name descriptor, the CA identifier descriptor, the country availability descriptor, the data broadcast descriptor, the linkage descriptor (used to specify services that contain additional information about this service, that carry an EPG for this service, or that is a replacement service for this service if it is unavailable), the mosaic descriptor, the

```
service_description_section-(actual/other)_transport_stream
          transport_stream_id   unsigned integer
          original_network_id   unsigned integer

      ┌   service_id                    unsigned integer
      │   EIT_schedule_flag             flag
      │   EIT_present_following_flag    flag
    * ┤   running_status                {running,pausing,...}
      │   free_CA_mode                  flag
      └   << descriptors>>
```

Figure 12.44 Simplified syntax for the DVB service description table.

NVOD reference descriptor, the telephone descriptor, and the time-shifted service descriptor.

The **EIT_schedule_flag** and **EIT_present_following_flag** indicate the presence (or absence) of the event information table for the schedule and the present/following events, respectively. The **running_status** is used to indicate whether a program is currently running, paused, or about to run. The **free_CA_mode** field is set to one for services in which conditional access is not used on any component.

EXAMPLE 12.20—Service Description Table

An example service description table is shown in Figure 12.45, carrying information on two digital television services carried in the same transport stream as the table. For each of these services, event information tables are present for both the schedule and the present/following events, and no conditional access is used. Both services are running.

```
service_description_section-actual_transport_stream
          transport_stream_id   0x0234
          original_network_id   0x1000

          service_id                    0x33
          EIT_schedule_flag             1
          EIT_present_following_flag    1
          running_status                running
          free_CA_mode                  1
          service_descriptor
                service_type                    digital television
                service_provider_name           "CNN"
                service_name                    "CNN Headline News"

          service_id                    0x35
          EIT_schedule_flag             1
          EIT_present_following_flag    1
          running_status                running
          free_CA_mode                  1
          service_descriptor
                service_type                    digital television
                service_provider_name           "NBC"
                service_name                    "NBC-Europe"
```

Figure 12.45 Example DVB service description table. ■

```
event_information_section
        service_id              unsigned integer
        transport_stream_id     unsigned integer
        original_network_id     unsigned integer

    ( event_id                  unsigned integer
    | start_time                time and date
    | duration                  time
  * < running_status            {running,pausing,...}
    | free_CA_mode              flag
    ( << descriptors >>
```

Figure 12.46 Simplified syntax for the DVB event information table.

12.3.3.4. DVB Event Information Table

The purpose of the event information table is to provide information in chronological order on the events within each service. Each section of the event information table contains information on one service, which is identified by its **service_id** and the **transport_stream_id** and **original_network_id** of the transport stream carrying the service (Fig. 12.46).

Unlike other tables, the event information table may be scrambled.

Each descriptor carried by the event information table refers to one event. Descriptors that are likely to be present include the component descriptor, the content descriptor, the data broadcast descriptor, the extended event descriptor, the linkage descriptor, the multilingual component descriptor, the parental rating descriptor, the short event descriptor, the telephone descriptor, and the time-shifted event descriptor.

EXAMPLE 12.21—Event Information Table

An example event information table is shown in Figure 12.47, showing information on two events (numbered 0x32 and 0x33) associated with the **service_id** 0x31. ■

12.3.3.5. DVB Running Status Table

The purpose of the running status table is to enable fast updating of the timing status of one or more events, which may be necessary when the starting time of an event is altered. This might occur when the preceding event finishes later or earlier than programmed. The running status of an event is one of the following values: "undefined," "not running," "starts in a few seconds," "pausing," or "running." The running status table is not scrambled. The running status table is not able to carry any descriptors.

EXAMPLE 12.22—Running Status Table

An example running status table is shown in Figure 12.48. This table contains entries for three events. The first two share the same transport stream (because they have the same value of **transport_stream_id** and **original_network_id**). The third entry refers to a dif-

event_information_section

service_id	0x31	
transport_stream_id	0x0300	
original_network_id	0x1020	
event_id	0x32	
start_time	0xCBD7000000	
duration	3600	
running_status	running	
free_CA_mode	1	
short_event_descriptor		
ISO_639_language_code		English
event_name_char		Premier League Football
text_char		Manchester United vs Arsenal
event_id	0x33	
start_time	0xCBD7000100	
duration	1800	
running_status	running	
free_CA_mode	1	
short_event_descriptor		
ISO_639_language_code		English
event_name_char		BBC News
text_char		BBC early edition news service
parental_rating_descriptor		
country_code		England
rating		5
country_code		Sweden
rating		3

Figure 12.47 Example DVB event information table.

ferent transport stream. The only means to determine if a reference in the running status table refers to a service in the same transport stream is to compare the transport stream's values of **transport_stream_id** and **original_network_id** to those in the running status table.

running_status_section

transport_stream_id	0x0300
original_network_id	0x1020
service_id	0x20
event_id	0x54
running_status	Pausing
transport_stream_id	0x0300
original_network_id	0x1020
service_id	0x31
event_id	0x65
running_status	Running
transport_stream_id	0x1000
original_network_id	0x17F1
service_id	0xF4
event_id	0x45
running_status	Starts in a few seconds

Figure 12.48 Example DVB running status table. ■

program time_and_date_section
UTC_time number

Figure 12.49 Simplified syntax of the DVB time and data table.

12.3.3.6. DVB Time and Date Table

The purpose of this table is to carry time and date information, with the syntax illustrated in Figure 12.49. Time is expressed as a six-digit number (hh:mm:ss), coded as BCD. The date is expressed as a modified Julian date and is coded as a 16-bit binary number. This table is always transmitted as a single section and is not scrambled. It is not able to carry any descriptors. The **UTC_time** field is 40 bits in length, with the most significant 16 bits corresponding to the least significant 16 bits of the MJD, and the remaining 24 bits carrying the six BCD digits for time.

EXAMPLE 12.23—Time and Date Table

The time 08:05:10 (i.e., 5 min and 10 s after 8 a.m.) is represented in BCD as 080510. The date "September 21, 2001" corresponds to $D = 21$, $M = 9$, $Y = 2001$, and $L = 1$, for which MJD = 52,173, or 0xCBCD. The contents of the time and date table for this example are shown in Figure 12.50.

time_and_date_section
UTC_time 0xCBCD080510

Figure 12.50 Example DVB time and date table. ■

12.3.3.7. DVB Time Offset Table

The purpose of the time offset table is to carry the current time and date information. Time and date information is carried in UTC. The table can also carry a list of countries and regions and their associated time offsets, allowing conversion of time and data to local values. This table is not scrambled and is always transmitted as a single section.

The header of the time offset table contains the current time (in UTC, coded in the same manner as that for time and date table) and date (expressed as a modified Julian date, coded as that for the time and date table). The local time offset descriptor is typically the only descriptor carried in this table (Fig. 12.51).

program time_offset_section
UTC_time number
<< descriptors >>

Figure 12.51 Simplified syntax for the DVB time offset table.

EXAMPLE 12.24—Time Offset Table

Figure 12.52 shows an example time offset table showing the same time and date as in the previous example of the time and date table. This time offset table also contains a local time offset descriptor, which specifies information for two countries: the United Kingdom and Australia. For the United Kingdom, the **region_id** shows that only one time zone is used. The current

local_time_offset is plus 1 h, which will change to zero at midnight on October 1, 2001. For Australia, the **region_id** specifies the eastern-most time zone, where the current time offset is plus 10 h, which will change to plus 11 h at 2 a.m. on October 7, 2001.

time_offset_section
UTC_time 0xCBCD080510

local_time_offset_descriptor

country_code	United Kingdom
region_id	Only one time zone
local_time_offset	+0100
time_of_next_change	0xCBD7000000
next_time	+0000
country_code	Australia
region_id	Eastern time zone
local_time_offset	+1000
time_of_next_change	0xCBDE020000
next_time	+1100

Figure 12.52 Example DVB time offset table. ■

12.3.3.8. *DVB Stuffing Table*

The purpose of the stuffing table is to allow a section becoming invalid as a transport stream crosses a network boundary to be overwritten with null data. This means that the data rate of the transport stream remains constant. A stuffing table may also be inserted in anticipation of a section being added when the transport stream crosses a network boundary. This table is not scrambled and contains no descriptors. All bytes after the header have the value 0xFF.

12.3.4. DVB Delivery Issues

12.3.4.1. *Size of Tables*

Although tables larger than 1024 bytes are supported by the DVB standard, the maximum size of each table section is limited to 1024 bytes. A table may be sufficiently long so that it requires two or more sections for transmission simply because an encoder places a large number of descriptors in the table. This may also occur because of the use of a number of free-text descriptors whose length is not constrained.

When a TS is remultiplexed, new descriptors may be added to one or more tables. Some descriptors may also be deleted. When descriptors are deleted, the system encoder may shorten the table. Alternatively, it may simply overwrite the deleted descriptor with a stuffing descriptor. This use of stuffing descriptors can lead to a growth in the length of tables, especially when a number of stages of remultiplexing occur. This growth in the length of tables is supported by the standard through the use of multisection tables. Many encoders, however, do not support multisection tables. Use of these encoders in the remultiplexing of streams that contain multisection tables can lead to corruption of SI data.

Table 12.6 Maximum intervals in seconds between retransmission of tables.

Table	Satellite/Cable	Terrestrial
NIT	10	10
BAT	10	10
SDT	Actual TS: 2 Other TS: 10	Actual TS: 2 Other TS: 10
TDT	30	30*
TOT	30	30*
EIT (Present/following)	Actual TS: 2 Other TS: 10	Actual TS: 2 Other TS: 10
EIT (Schedule Table)	First 8 days: 10 Later: 30	Actual TS First 8 days: 10* Later: 30* Other TS First 8 days: 60* Later: 300*

Note: entries marked "*" are recommended. All other entries are mandatory.

12.3.4.2. Table Entries

To be decoded, a service must contain one or more elementary streams (containing video, audio, or private data), an entry in the PAT for a PMT, and the PMT specified in the PAT. A service missing any one of these cannot be decoded. A DVB bit stream may, however, contain elementary streams that do not have corresponding PAT and PMT entries. There may also exist services that are defined by the SI but with no PSI or elementary streams, which are sometimes referred to as *blank programs*. Blank programs occur when SI for a service is carried in a transport stream separate from the one carrying the service itself.

12.3.4.3. Repetition Rates

The DVB standard defines the maximum intervals between transmission for each type of table, shown in Table 12.6 [7]. Regular retransmission of tables minimizes the impact of propagation errors on the decoded service and the time taken to begin decoding a transport stream. Different repetition rates are specified for satellite, cable, and terrestrial broadcast systems.

In a DVB system, approximately 5% of the network capacity is allocated to the transmission of SI tables and CA information. In a-40 Mbit/s satellite distribution system, the total SI and CA data is likely to be 2 Mbit/s.

12.4. ATSC PROGRAM AND SYSTEM INFORMATION PROTOCOL

ATSC's program and the PSIP provides a small collection of tables and associated descriptors that can be used to describe the contents of one or more transport streams. These

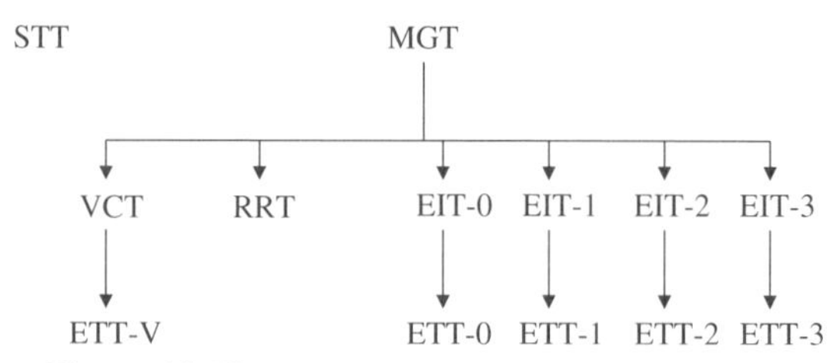

Figure 12.53 Logical structure of ATSC PSIP tables.

tables can also be used to describe the contents of analog transmissions. The purpose of ATSC PSIP [8] is similar to that of DVB SI, although it is much simpler in structure.

ATSC continues to use the program association table, program map table, and conditional access table defined by MPEG-2. In addition, ATSC defines

- a *system time table* (*STT*), used to specify the current UTC time;
- a *master guide table* (*MGT*), which provides an index to the other tables;
- a *virtual channel table* (*VCT*), which describes the virtual channels that are carried by one or more transport streams and also provides the facilities of the network information table;
- a *rating region* (*RRT*) table that conveys information on the program classification system in use in one or more regions;
- *event information tables*, which describe the upcoming programs on all virtual channel defined in the virtual channel table; and
- the *extended text table*, which provides text descriptions that can be used to augment the virtual channel table or event information table.

The logical structure of these tables is shown in Figure 12.53, where the master guide table sits at the top of the hierarchy, with the virtual channel table, rating region table, and event information tables below it. The virtual channel table and event information tables may each have a child extended text table. The system time table sits outside this hierarchy.

In this section, we begin with an overview of the common data formats used by the ATSC's PSIP. This is followed by a description of each of the descriptors and tables.

12.4.1. Common Data Formats

ATSC uses common data formats for the representation of strings and time and date.

12.4.1.1. String Representation

ATSC uses the multiple string structure, illustrated in Figure 12.54, as a common means for specifying strings in one or more identified languages. (The format for the structure is set out in Section 12.3.2.1.) The language is specified by an ISO 639 three-character code. Strings may be uncompressed, or compressed using one of the

multiple_string_structure

*		ISO_639_language_code	ISO 639 language code
	*	compression_type	{No compression,Huffman 1, Huffman 2}
		mode	Unsigned integer
		compressed_string	Compressed string

Figure 12.54 Structure of the ATSC multiple string structure.

two Huffman codes. The character set is based on an 8- or 16-bit code defined by ISO 10646, with the page number controlled by the **mode** field.

EXAMPLE 12.25—Multiple String Structure

Figure 12.55 shows an example multiple string structure carrying the string "Hello" in English and in French. In both cases the mode has value 0x00, the ISO 10646 value indicating the Latin-1 alphabet.

multiple_string_structure

ISO_639_language_code	English
compression_type	No compression
mode	0x00
compressed_string	"Hello"
ISO_639_language_code	French
compression_type	No compression
mode	0x00
compressed_string	"Bonjour"

Figure 12.55 Example ATSC multiple string structure. ■

12.4.1.2. ATSC Time and Date Formats

Time and date in ATSC tables is specified as a 32-bit unsigned integer representing the number of seconds since midnight, January 6, 1980, based on the GPS time reference. This GPS time is translated to UTC time by adding an offset carried by the system time table.

EXAMPLE 12.26—ATSC Time and Date

The GPS time at 12:00, August 7, 2001, is 681220800. ■

12.4.1.3. ATSC Virtual Channels

Each ATSC transport stream may carry several programs simultaneously. A means is required to identify uniquely each of these programs and possibly to identify programs in other transport streams. ATSC uses virtual channel numbers for this purpose, with each virtual channel number identifying a particular program that the viewer can select. Each virtual channel is uniquely identified by a major channel

Table 12.7 ATSC Descriptors.

descriptor_tag	Descriptor	Purpose
0x80	Stuffing descriptor	Acts as a place holder in a table.
0x81	AC-3 audio descriptor	Describes an AC-3 audio elementary stream.
0x86	Caption service descriptor	Carries characteristics of closed caption services associated with an event.
0x87	Content advisory descriptor	Conveys advisory indications of program classification for one or more regions.
0xA0	Extended channel name descriptor	Carries the long form of the name of a virtual channel.
0xA1	Service location descriptor	Specifies the stream types, PID, and language for each elementary stream.
0xA2	Time-shifted service descriptor	For one virtual channel, identifies other virtual channels that are time-shifted copies.
0xA3	Component name descriptor	Provides a text name for one component.
0xC0–0xFE	User private	Available for definition by users without interference with the operation of ATSC

number and a minor channel number. The major channel number is assigned based on the RF channel number on which the operator broadcasts. For broadcasters who already hold an NTSC license, the major channel number is the number of this channel. For new, digital-only broadcasters, the major channel number is the number of their assigned digital channel. This method of assignment means that, in a given geographic area, only one major channel number is used by each broadcaster, even if that broadcaster controls more than one RF channel. Minor channel numbers are used to identify individual virtual channels.

12.4.2. ATSC Descriptors

ATSC defines a number of descriptors in addition to those defined by MPEG-2 (Chapter 11), whose characteristics are summarized in Table 12.7, to complement those defined in MPEG-2 systems. The values of **descriptor_tag** assigned by ATSC are in the range 0x80–0xAF, which are values left unassigned by MPEG-2. ATSC restricts user-defined descriptor tags to the range 0xC0–0xFE.

12.4.2.1. ATSC AC-3 Audio Descriptor

The AC-3 audio descriptor [9] is used to provide information about an AC-3 audio elementary stream. It is usually transmitted in the program map table, with the structure shown in Figure 12.56.

The **bsid** and **bsmod** fields take the same values as the fields of the same name in the AC-3 audio elementary stream. The **bit_rate_code** specifies either the exact bit rate or the upper limit for the bit rate for values between 32 and 640 kbit/s. The **num_channels** field specifies either the number of audio channels encoded in the

AC3_audio_descriptor	
sample_rate_code	{48,44.1,32} kHz
bsid	bit string
bit_rate_code	{32-640} Kbps
surround_mode	{Not indicated, not Dolby surround, Dolby surround}
bsmod	bit string
num_channels	bit string
full_svc	flag
langcod	bit string
text	String

Figure 12.56 Structure of the ATSC AC-3 audio descriptor.

AC-3 elementary stream or an upper bound on the number of channels. The **langcod** field is set to the same value as the langcode field in the AC-3 elementary stream, and represents the language of the audio. A value of zero is used to indicate an unknown language. A value of zero for **full_svc** indicates that this audio elementary stream is not suitable for presentation on its own but should be combined with some other audio service. This might occur where an audio elementary stream contains only special effects and is designed to be combined with another containing dialog. The use of surround sound is indicated by the **surround_mode** field. The **text** field can be used to carry a text description of the audio elementary stream and can be encoded using either the 8-bit ASCII or the 16-bit Unicode formats. Unlike other text strings in ATSC descriptors, this field does not use the multiple string structure.

12.4.2.2. ATSC Caption Service Descriptor

The caption service descriptor (Fig. 12.57) conveys the characteristics of closed captioning that are associated with an event. The captioning systems used in ATSC are described in Chapter 14. Up to 16 different closed captioning services can be associated with one event, for each of which the language, its type (ATVCC or Line 21), the transmission field for Line 21 data, and the caption service number for ATVCC captions, and whether or not the captions are specially formatted for beginner readers or for a wide aspect ratio is specified.

caption_service_descriptor

```
  ( language                        ISO 639 language code
  | cc_type                         {ATVCC,Line 21} Captions
  | if cc_type == 0
  |         line21_field            {Field 1, Field 2}
* < else
  |         caption_service_number  Unsigned integer
  | easy_reader                     flag
  ( wide_aspect_ratio               flag
```

Figure 12.57 Structure of the ATSC caption service descriptor.

EXAMPLE 12.27—Caption Service Descriptor

An example caption service descriptor is shown in Figure 12.58, specifying two ATVCC closed captioning services. The first is in English, using the caption service number 4 and is aimed at beginner readers. The second service is in French and uses the caption service number 6.

caption_service_descriptor

language	English
cc_type	ATVCC Captions
caption_service_number	4
easy_reader	1
wide_aspect_ratio	0
language	French
cc_type	ATVCC Captions
caption_service_number	6
easy_reader	0
wide_aspect_ratio	0

Figure 12.58 Example ATSC caption service descriptor. ■

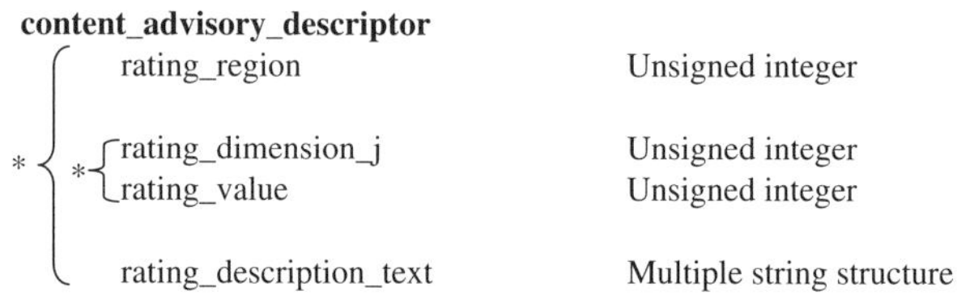

Figure 12.59 Structure of the ATSC content advisory descriptor.

12.4.2.3. ATSC Content Advisory Descriptor

The content advisory descriptor, carried in the rating region table, is used to indicate program classifications for up to eight regions for each event. The region rating table (Section 12.4.3.5) defines one or more dimensions of rating (e.g., language, sexual content) and one or more levels of content for each dimension. The content advisory descriptor presents a list of dimensions and the corresponding rating values for a particular region (Fig. 12.59).

12.4.2.4. ATSC Extended Channel Name Descriptor

The extended channel name descriptor provides the long channel name for a virtual channel. Its payload carries one multiple string structure (Fig. 12.60).

12.4.2.5. ATSC Service Location Descriptor

The service location descriptor specifies the stream types, PID, and language for each elementary stream. The Valid values for the **stream_type** field include MPEG-2 video and AC-3 audio. One service location descriptor is carried in the virtual channel table for each active channel and refers to the elementary streams associated with the current event on that channel (Fig. 12.61).

extended_channel_name_descriptor

long_channel_name_text	Multiple string structure

Figure 12.60 Structure of the ATSC extended channel name descriptor.

service_location_descriptor

	PCR_PID	
*	stream_type	{MPEG-2 video,AC-3 audio}
*	elementary_PID	unsigned integer
*	ISO_639_language_code	ISO 639 language code

Figure 12.61 Structure of the ATSC service location descriptor.

EXAMPLE 12.28—Service Location Descriptor

An example service location descriptor is shown in Figure 12.62 for an event with two associated elementary streams, one containing MPEG-2 video and the other containing AC-3 audio. For each elementary stream, the language is specified as English. The PCR is carried in the transport stream packets with the same PID as the video elementary stream.

service_location_descriptor

PCR_PID	0x0456
stream_type	MPEG-2 video
elementary_PID	0x0456
ISO_639_language_code	English
stream_type	AC-3 audio
elementary_PID	0x0765
ISO_639_language_code	English

Figure 12.62 Example ATSC service location descriptor. ■

12.4.2.6. ATSC Time-Shifted Service Descriptor

The time-shifted service descriptor, carried in the virtual channel table, links one virtual channel with one or more other virtual channels that carry the same content at a different time. NVOD services are the most likely application on this descriptor. As illustrated in Figure 12.63, for each linked virtual channel, this descriptor carries the time shift in minutes, and the major and minor channel numbers of the linked virtual channel. Up to 20 linked virtual channels may be specified.

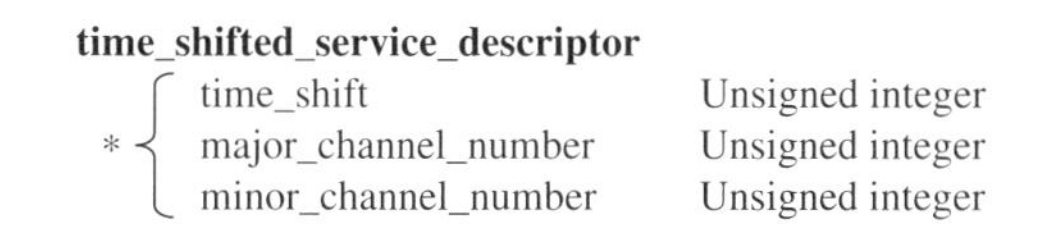

Figure 12.63 Structure of the ATSC time-shifted service descriptor.

EXAMPLE 12.29—Time-Shifted Service Descriptor

An example time-shifted service descriptor is shown in Figure 12.64, showing three linked virtual channels. The major channel number of each of the linked virtual channels is 12. The first linked virtual channel is time shifted by 60 min and has minor channel number 4, the second is time shifted by 30 min with a minor channel number of 5, and the third is time shifted by 90 min and has a minor channel number of 6.

time_shifted_service_descriptor	
time_shift	30
major_channel_number	12
minor_channel_number	4
time_shift	60
major_channel_number	12
minor_channel_number	5
time_shift	90
major_channel_number	12
minor_channel_number	6

Figure 12.64 Example ATSC time-shifted service descriptor. ■

12.4.2.7. ATSC Component Name Descriptor

The component name descriptor carries a single multiple string structure, which can be used as the text name of one component, such as a single video or audio elementary stream (Fig. 12.65).

12.4.2.8. ATSC Stuffing Descriptor

The stuffing descriptor may appear in any ATSC table and may be used as a place holder where it is known that it may be necessary to insert a new descriptor, for example, when the transport stream crosses a network boundary. The stuffing descriptor may also be used to replace a descriptor that is removed from a transport stream, avoiding the need for retiming the stream.

12.4.2.9. Descriptors for Inactive Channels

For virtual channels that are not currently active, no service location descriptor should be carried in the bit stream. Other descriptors for inactive virtual channels may, however, be carried by the transport stream.

12.4.3. ATSC Tables

The tables used by ATSC, including those defined by MPEG-2, are shown in Table 12.8. Of the new tables defined by ATSC, the event information table and the extended text table have PIDs assigned by the master guide table. All other ATSC-defined tables use PID 0x1FFB and are distinguished from one another only by the different values of **table_id** assigned.

12.4.3.1. ATSC Use of the Program Map Table

In an ATSC transport stream, the program map table may carry a number of ATSC-defined descriptors in addition to those defined by MPEG-2, including the AC-3

component_name_descriptor	
component_name_string	Multiple string structure

Figure 12.65 Structure of the ATSC component name descriptor.

Table 12.8. ATSC Tables.

Table	Abbreviation	PID	**table_id**
Program association	PAT	0x0000	0x00
Conditional access	CAT	0x0001	0x01
Program map	PMT	Assigned in PAT	0x02
Master guide	MGT	0x1FFB	0xC7
Virtual channel (terrestrial)	VCT	0x1FFB	0xC8
Virtual channel (cable)	VCT	0x1FFB	0xC9
Rating region	RRT	0x1FFB	0xCA
Event information	EIT	Assigned in MGT	0xCB
Extended text	ETT	Assigned in MGT	0xCC
System time	STT	0x1FFB	0xCD

audio descriptor (to describe audio elementary streams), the caption service descriptor, the content advisory descriptor, and the component name descriptor (to assign names to individual components).

12.4.3.2. ATSC System Time Table

The system time table specifies the current time and date and the status of daylight saving. The **protocol_version**, which currently must be set to zero, is included to facilitate future evolution of the protocol. The **system_time** is specified using the format described in Section 12.4.1.2. The **GPS_UTC_offset** (measured in seconds) can be subtracted from the system time, which is calibrated in GPS time, to obtain a UTC time.

The use of daylight savings in converting between UTC and local time is specified by the **DS_status**, **DS_day_of_month**, and **DS_hour** fields. **DS_status** takes the value 1 when daylight savings time is active, zero otherwise. In months where the status of daylight savings changes, **DS_day_of_month** indicates the date on which this is to occur, and **DS_hour** the time on that day. In other months, **DS_day_of_month** and **DS_hour** are set to zero.

Although not prohibited by the syntax, the system time table would not normally carry any descriptors (Fig. 12.66).

system_time_table_section

protocol_version	Unsigned integer
system_time	Time and date
GPS_UTC_offset	Unsigned integer
DS_status	flag
DS_day_of_month	Unsigned integer
DS_hour	Unsigned integer
<< descriptors >>	

Figure 12.66 Structure of the ATSC system time table.

EXAMPLE 12.30—System Time Table

The GPS time at 0800 UTC, August 7, 2001, with an offset of 5 s is 681206405. The system time table for this time, with daylight savings active and no change in daylight savings status during August, is shown in Figure 12.67.

system_time_table_section	
protocol_version	0
system_time	681206405
GPS_UTC_offset	5
DS_status	1
DS_day_of_month	0
DS_hour	0

Figure 12.67 Example ATSC system time table. ■

12.4.3.3. ATSC Master Guide Table

The master guide table carries information on all PSIP tables in a transport stream, except for the system time table, using the structure shown in Figure 12.68. Descriptors defined by MPEG-2 and ATSC are not usually carried in the master guide table, even though carriage of descriptors is supported by the syntax. The data of the master guide table begins with a **protocol_version** field, which is intended to enable future evolution of the protocol. For each PSIP table, the master guide table carries the table type, the PID of the transport stream packets in which this table is carried (**table_type_PID**), the current version number of the table (**table_type_version_number**), and the total number of bytes used by this table. Specification of the version number allows a decoder to determine from the master guide table when a new version of a table is transmitted, without having to monitor all PSIP data in the transport stream. Direct specification of the number of bytes used by a table is designed to assist a decoder to manage its memory. The master guide table is not scrambled.

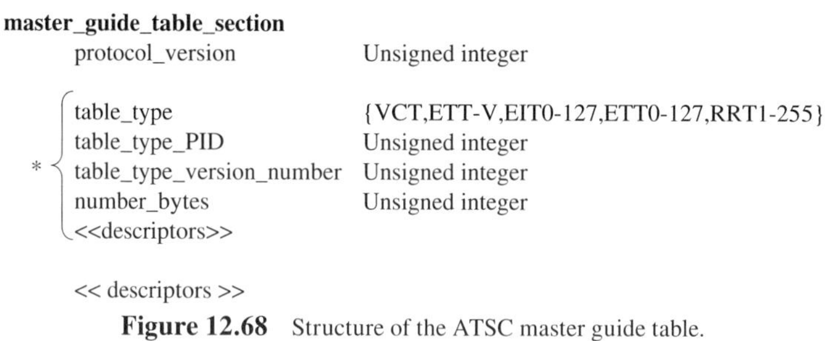

Figure 12.68 Structure of the ATSC master guide table.

EXAMPLE 12.31—System Time Table

A partial example of the ATSC master guide table is shown in Figure 12.69, for which the protocol version is 0. The virtual channel table is located in PID 0x1FFB (as required for this table), its current version number is 5 and length 354 bytes. The rating region table for region

1 is located in PID 0x1FFB (as required), with version number 3 and size 74 bytes. EIT-0 is located in PID 0x0543, with version number 45 and size 742 bytes, whereas EIT-1 is located in PID 0x523, with version number 47 and size 890 bytes.

master_guide_table_section

protocol_version	0
table_type	VCT
table_type_PID	0x1FFB
table_type_version_number	5
number_bytes	354
table_type	RRT-1
table_type_PID	0x1FFB
table_type_version_number	3
number_bytes	74
table_type	EIT-0
table_type_PID	0x0543
table_type_version_number	45
number_bytes	742
table_type	EIT-1
table_type_PID	0x523
table_type_version_number	47
number_bytes	890

Figure 12.69 Example ATSC master guide table. ■

12.4.3.4. ATSC Virtual Channel Table

ATSC specifies two versions of the virtual channel table, the first for terrestrial broadcast and the second for cable. Only the terrestrial virtual channel table is described here.

The virtual channel table uses the **protocol_version** field in the same way as the master guide table, allowing future evolution of the protocol. For each virtual channel described, the major channel number and minor channel number are specified, followed by the **modulation_mode**, which may be analog, ATSC (8 or 16 VSB), two modes designed for transmission over a cable network, or specified in private data. The carrier frequency is transmitted as a multiple of 10 Hz, followed by the MPEG-2 transport stream id and program number associated with the virtual channel. The presence of any extended text description is flagged by the **ETM_location**, whereas the use of conditional access on any component of a virtual channel is indicated by setting **access_controlled** to 1. A virtual channel may be hidden from viewing by entering its channel number using the **hidden** flag, or it may be hidden from an EPG using the **hide_guide** flag. This latter case may be useful for NVOD services, where the EPG may show a single virtual channel. The **service_type** specifies whether the virtual channel is an analog transmission, ATSC television, or audio or ATSC data broadcast service. The **source_id** indicates the programming source of the virtual channel and has provision for

virtual_channel_table_section

protocol_version	Unsigned integer
* { major_channel_number	Unsigned integer
minor_channel_number	Unsigned integer
modulation_mode	{Analog,ATSC(8VSB),ATSC(16VSB),Cable Mode 1,Cable Mode 2,Private}
carrier_frequency	Unsigned integer (MHz)
channel_TSID	Unsigned integer (MPEG-2 transport_stream_id)
program_number	Unsigned integer (MPEG-2 program number)
ETM_location	{No ETM,ETM in PSIP VC,ETM in channel_TDID}Q
access_controlled	flag
hidden	flag
hide_guide	flag
service_type	{analog, ATSC television,ATSC audio,ATSC data}
source_id	Unsigned integer
<<descriptors>> }	
<< additional descriptors >>	

Figure 12.70 Structure of the ATSC virtual channel table.

registration of values, enabling the unique identification of a content provider. No two virtual channels within any transport stream for which the virtual channel table carries data should have the same value of **source_id**, unless one virtual channel is a time-shifted version of the other. Descriptors are usually placed in the virtual channel table to describe one virtual channel, although there is provision in the syntax for additional descriptors that are not specific to one virtual channel (Fig. 12.70).

Descriptors typically carried in the virtual channel table are the extended channel name descriptor, the service location descriptor, and the time-shifted service descriptor. The virtual channel table is not scrambled.

EXAMPLE 12.32—Virtual Channel Table

An example virtual channel table is shown in Figure 12.71. Each of the three virtual channels has major channel number 12, has no extended text message (ETM), is not access controlled, is not hidden, and is not hidden in the EPG. The first and third virtual channels are ATSC digital television services, using 8-VSB modulation, with transport_stream_id 0x453D and transmission frequency 540.32 MHz. The MPEG-2 program number (which is used in the PMT) of the first virtual channel is 4 and that of the third virtual channel is 6. The analog virtual channel is assigned transport stream id and program number 0xFFFF.

12.4.3.5. ATSC Rating Region Table

The rating region table carries information on program rating for up to 255 distinct geographic regions. The purpose of the table is to define the ratings system used; the actual rating of a particular event is carried by the content rating descriptor in the event information table or in the program map table.

The region to which a section of the rating region table refers is determined by the value of the **rating_region** field. This value is also used to refer to this region

virtual_channel_table_section

protocol_version	0
major_channel_number	12
minor_channel_number	0
modulation_mode	ATSC(8VSB)
carrier_frequency	800.31 MHz
channel_TSID	0x453D
program_number	0x4
ETM_location	No ETM
access_controlled	0
hidden	0
hide_guide	0
service_type	ATSC television
source_id	0x1242
<<descriptors>>	
major_channel_number	12
minor_channel_number	1
modulation_mode	Analog
carrier_frequency	205.25 MHz
channel_TSID	0xFFFF
program_number	0xFFFF
ETM_location	No ETM
access_controlled	0
hidden	0
hide_guide	0
service_type	Analog
source_id	0x1242
<<descriptors>>	
major_channel_number	12
minor_channel_number	2
modulation_mode	ATSC(8VSB)
carrier_frequency	800.31 MHz
channel_TSID	0x453D
program_number	0x6
ETM_location	No ETM
access_controlled	0
hidden	0
hide_guide	0
service_type	ATSC television
source_id	0x1242
<<descriptors>>	

Figure 12.71 Example ATSC virtual channel table (without descriptors). ■

in other tables, such as the master guide table. The text name of the rating region is carried in the **rating_region_name_text** field. For each rating dimension (such as "violence," "nudity," or "language") to be defined, **dimension_name_text** specifies the name of the dimension. Dimensions for which a graduated scale of increasing content is to be defined are identified by setting **graduated_scale** to 1. For each value of the dimension, **abbrev_rating_value_text** contains an abbreviated version of the name of the rating value, whereas **rating_value_text** contains the full text of the name (Fig. 12.72).

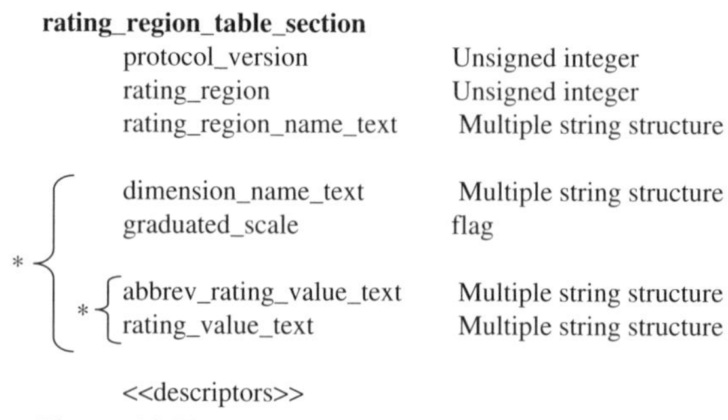

Figure 12.72 Structure of the ATSC rating region table.

The rating region table is not scrambled and does not usually carry ATSC-defined descriptors.

EXAMPLE 12.33—Rating Region Table

An example rating region table, which partially defines the television program classification system used in Australia, is shown in Figure 12.73. The **region_rating_name** shows the region as being Australia. Two dimensions are defined: "language" and "other." The language dimension is a graduated scale, using four levels: "some coarse language," "frequent coarse

rating_region_table_section	
protocol_version	0
rating_region	1
rating_region_name_text	"Australia"
dimension_name_text	"Language"
graduated_scale	1
abbrev_rating_value_text	"SCL"
rating_value_text	"Some coarse language"
abbrev_rating_value_text	"FCL"
rating_value_text	"Frequent coarse language"
abbrev_rating_value_text	"VCL"
rating_value_text	"Very coarse language"
abbrev_rating_value_text	"FVCL"
rating_value_text	"Frequent very coarse language"
dimension_name_text	"Other"
graduated_scale	0
abbrev_rating_value_text	"A"
rating_value_text	"Adult themes"
abbrev_rating_value_text	"N"
rating_value_text	"Nudity"

Figure 12.73 Example ATSC rating region table.

language," "very coarse language," and "frequent very coarse language." The other dimension is not a graduated scale, but instead indicates the presence of particular features, such as "nudity" or "adult themes." For each of these rating values, both the full text and an abbreviation are included. ■

12.4.3.6. ATSC Event Information Table

The event information table is used to carry information on the contents of virtual channels. Each section of the event information table carries information on one virtual channel, which is identified by its **source_id** (which can be used to link to the corresponding information in the virtual channel table).

The structure of the event information table is shown in Figure 12.74. Following the **source_id** and **protocol_version** fields, one or more events on the virtual channel are described. Each event is assigned a unique identifier (**event_id**), and the start time, length, title, and the location of any ETM event are specified.

Each event information table describes events in a 3-h window. Each 3-h window begins at 0000, 0300, 0600, 0900, 1200, 1500, 1800, or 2100 UTC. EIT-0 always specifies current events, EIT-1 specifies events in the next 3-h window, and the following event information tables specify events in future 3-h windows. For example, if the current time is 0500, EIT-0 contains information on events in the period 0300–0600, EIT-1 on 0600–0900, EIT-2 on 0900–1200, and so on. A minimum of four event information tables must be carried, describing events in the current 3-h period and the following three 3-h periods.

ATSC descriptors commonly carried in the event information table are the AC-3 audio descriptor, the caption service descriptor, and the content advisory descriptor. The event information table is not scrambled.

EXAMPLE 12.34—Event Information Table

An example event information table for EIT-0 valid at 0930 UTC, January 26, 2000, is shown in Figure 12.75.

The first event, whose title is "Who wants to be a millionaire," has **event_id** 0x0425, starts at 0830 (coded as 0x25B97308), and is 1-h long. The second event, whose **event_id** is 0x0625, starts at 0930 (0x25B98118), is 90 min long, has extended text in the virtual channel carrying the event and has the title "Friends." The third event (0x0429) starts at 1100 (0x25B99630), is 60 min long, and has the title "The Late Show."

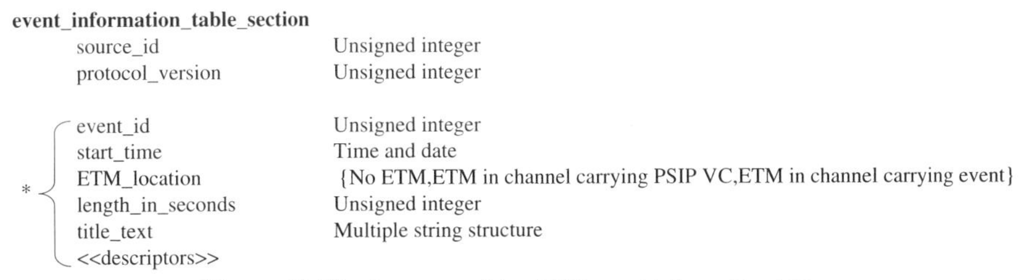

Figure 12.74 Structure of the ATSC event information table.

event_information_table_section

source_id	0x4352
protocol_version	0
event_id	0x0425
start_time	0x25B97308
ETM_location	No ETM
length_in_seconds	3600
title_text	"Who wants to be a millionaire"
<<descriptors>>	
event_id	0x0625
start_time	0x25B98118
ETM_location	ETM in channel carrying event
length_in_seconds	5400
title_text	"Friends"
<<descriptors>>	
event_id	0x0429
start_time	0x25B99630
ETM_location	No ETM
length_in_seconds	3600
title_text	"The Late Show"
<<descriptors>>	

Figure 12.75 Example ATSC event information table. ■

12.4.3.7. ATSC Extended Text Table

The extended text table is used to carry longer text descriptions of events and virtual channels, and contains one multiple string structure (Fig. 12.76). For an event, the **ETM_id** (32 bits) combines the **source_id** (16 bits) of the virtual channel and the **event_id** (16 bits) of the event. For a virtual channel, the **source_id** is combined with 16 zero bits.

The extended text table is not scrambled.

12.5. DVB SI AND ATSC PSIP INTEROPERABILITY

It may sometimes be desirable to create a transport stream that can be decoded by either a DVB or an ATSC decoder. This is most likely to occur when a transport stream is to be distributed into a number of countries.

In practice, interoperability in this respect does not mean that the DVB receiver decodes ATSC PSIP or that the ATSC receiver decodes DVB SI. What is meant is that the transport stream carries both DVB SI and ATSC PSIP in a manner such that no conflicts between the two systems arise, which imposes a number of restrictions

extended_text_table_section

protocol_version	Unsigned integer
ETM_id	Unsigned integer
extended_text_message	Multiple string structure

Figure 12.76 Structure of the extended text table.

on the structure of the transport stream to ensure that conflicts between DVB and ATSC do not arise [10].

The restrictions imposed by this requirement are in the areas of allocation of PIDs to transport stream packets, and the use of values of **table_id** or **descriptor_tag** assigned by either DVB or ATSC.

12.5.1. PIDs

PIDs in the range 0x1FFB–0x1FFD are reserved for use by ATSC (e.g., 0x1FFB is used by PSIP) and should not be used by DVB. Similarly, PIDs in the range 0x0010 through 0x0014 are reserved by DVB and should not be used by ATSC. This latter restriction means that ATSC program number 1 should not be used.

12.5.2. Use of table_id

The ranges of values reserved by DVB (0x40 through 0x7F) and ATSC (0xC0 through 0xFE) should not be used for user-private data.

12.5.3. Use of descriptor_tag

Descriptor tag values from 0x80 to 0xAF are used or reserved for use in the ATSC specification, whereas 0x40 through 0x7F are reserved in the DVB specification. User-private descriptors should use descriptor tag values outside this range. The range 0xB0 through 0xFE can be safely used for user-private descriptors, the use of which is controlled by the MPEG-2 registration descriptor.

12.6. CONCLUSION

DVB SI and ATSC PSIP extend the basic functionality of the MPEG-2 PSI, providing a range of additional information about the programs carried in one or more transport streams. This additional information supports applications such as electronic program guides, assists the viewer in navigating between different programs, and allows the appearance of service offerings to be made independent of the physical structure of MPEG-2 transport streams.

PROBLEMS

12.1 Using the format of Figure 12.40, construct a network information table for a satellite distribution system with two channels on one satellite with orbital position 130.5°E, **original_network_id** 0x1010, and network identifier 0x0432, with the following characteristics:

- Channel 1 is transmitted with **transport_stream_id** 0x0786, at a frequency of 3587 MHz, with a channel bandwidth of 10 MHz, vertical polarization, a symbol rate of 10 Msym/s, and an inner coding rate of 2/3.
- Channel 2 is transmitted with **transport_stream_id** 0x0765, at a frequency of 3456 MHz, with a symbol rate of 20 Msym/s, a bandwidth of 20 MHz, left circular polarization, and an inner coding rate of 1/2.

12.2 Answer the following questions, using the information in the example network information table shown in Figure 12.77.

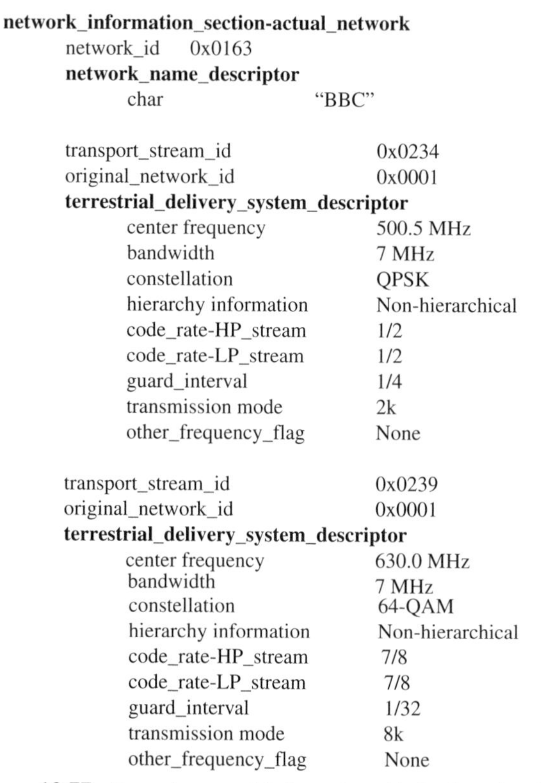

network_information_section-actual_network
network_id 0x0163
network_name_descriptor
char "BBC"

transport_stream_id	0x0234
original_network_id	0x0001
terrestrial_delivery_system_descriptor	
center frequency	500.5 MHz
bandwidth	7 MHz
constellation	QPSK
hierarchy information	Non-hierarchical
code_rate-HP_stream	1/2
code_rate-LP_stream	1/2
guard_interval	1/4
transmission mode	2k
other_frequency_flag	None

transport_stream_id	0x0239
original_network_id	0x0001
terrestrial_delivery_system_descriptor	
center frequency	630.0 MHz
bandwidth	7 MHz
constellation	64-QAM
hierarchy information	Non-hierarchical
code_rate-HP_stream	7/8
code_rate-LP_stream	7/8
guard_interval	1/32
transmission mode	8k
other_frequency_flag	None

Figure 12.77 Example network information table for Question 12.2.

- How many transport streams are carried by this network?
- Which fields uniquely identify each transport stream?
- For each descriptor, whether it refers to one transport stream or to the whole network?

12.3 A transport stream carries five programs with no conditional access. If each table consists of exactly one section, and each section occupies one transport stream packet, calculate the overhead in bytes per second in the following cases:

(a) DVB SI (consisting of program association table, program map table, network information table, bouquet association table, network information table, service description table, event information table (present/following), and time offset table) is broadcast twice per second.

(b) ATSC PSIP (consisting of program association table, program map table, system time table, master guide table, virtual channel table, event information table, and rating region table) is broadcast twice per second.

(c) Both DVB SI and ATSC PSIP are carried.

12.4 Fill in the blank entries in Table 12.9.

Table 12.9 Corresponding values of date and MJD for Question 12.4.

Date	MJD
January 5, 2001	
September 9, 1999	
	52035
	51106

12.5 Fill in the blank entries in Table 12.10. The number of days since the reference date can be found using the difference between their values of MJD.

Table 12.10 Corresponding entries of GPS time and date and time for Q 12.5.

Time and date	GPS time
8:45 a.m., August 8, 1995	
3:40 p.m., February 29, 1996	
	673275360
	545217240

12.6 Use the data in Table 12.11, which describes the programs and elementary streams carried in a DVB transport stream, to construct a program association table and program map tables. The transport stream has a maximum bit rate of 20 Mbit/s, **original_network_id** is 0x0100, and **transport_stream_id** is 0x0034. In answering this question, it may be necessary to assign PIDs or other parameters. No conditional access is used in this transport stream.

12.7 Construct a bouquet association table based on the data provided in Tables 12.11–12.13.

12.8 Construct a service description table based on the data provided in Tables 12.12 and 12.13. All services are running and event information tables are carried in the transport stream for present/following events but not for the schedule.

12.9 Construct an event information table for Service 1, using the data in Tables 12.12–12.14.

12.10 Is the DVB transport stream defined in Tables 12.11–12.14 compatible with ATSC, in the sense that ATSC PSIP could be added to the transport stream to allow decoding by an ATSC decoder?

Table 12.11 Elementary stream data for Questions 12.6–12.9.

Program number	PID	Elementary stream type	Component name	Elementary stream description
—	0x0056	EPG		EPG for Bouquet 1
—	0x0057	EPG		EPG for Bouquet 2
1	0x0365	MPEG-2 video	BBC1—video	Main profile at main level; 4:2:0; 25 Hz; aligned at slice layer
	0x366	MPEG-2 audio	BBC1—audio	English language; constant rate; Layer 2; ID = 1; **free_format_flag** = 0
	0x375	DVB subtitle	BBC1—subtitles	English (hearing impaired, **composition_page_id** = 1, **ancillary_page_id** = 10) and French language (normal, **composition_page_id** = 11, **ancillary_page_id** = 43)
	0x377	DVB teletext	BBC1—teletext	English (initial page in magazine 0, page 10) and French language (initial page in magazine 1, page 10)
	0x0365	PCR		Elementary stream carrying PCR for this program
2	0x0385	MPEG-2 video	BBC2—video	Main profile at main level; 4:2:0; 30 Hz; aligned at picture layer
	0x386	MPEG-2 audio	BBC2—English audio	English language; constant rate; Layer 2; ID = 1; **free_format_flag** = 0
	0x387	MPEG-2 audio	BBC2—French audio	French language; constant rate; Layer 2; ID = 1; **free_format_flag** = 0
	0x0385	PCR		Elementary stream carrying PCR for this program
3	0x0435	MPEG-2 video	Sky—video	Main profile at main level; 4:2:0; 30 Hz; aligned at picture layer
	0x436	MPEG-2 audio	Sky—English audio	English language; constant rate; Layer 2; ID = 1; **free_format_flag** = 0

Table 12.11 *(Continued)*

Program number	PID	Elementary stream type	Component name	Elementary stream description
	0x437	MPEG-2 audio	Sky—German audio	German language; constant rate; Layer 2; ID = 1; **free_format_flag** = 0
	0x0435	PCR		Elementary stream carrying PCR for this program

Table 12.12 Service description information associated with components defined in Table 12.11.

Service identifier	Service name	Service provider
1	BBC-1	British Broadcasting Corporation
2	BBC-2	British Broadcasting Corporation
3	Sky Sports	Sky Television

Table 12.13 Bouquet information associated with data in Tables 12.12 and 12.13.

Bouquet identifier	Bouquet name	Services	Country availability	EPG PID
1	BBC	1, 2	UK, France	0x0056
2	News	1, 3	UK, Germany	0x0057

Table 12.14 Data for event information table in Questions 12.9 and 12.14.

Event identifier	Start time	Duration (min)	Running status	Event name	Description
0x23	7:00 p.m., January 1, 2000	90	Running	Premier league football	Manchester United versus Arsenal
0x6B	8:30 p.m., January 1, 2000	30	About to start	Nightly news	National and international news from the BBC

12.11 Use the data in Table 12.15, which describes the programs and elementary streams carried in an ATSC transport stream, to construct a program association table and program map tables. The transport stream has a maximum bit rate of 20 Mbit/s, and **transport_stream_id** is 0x0034. In answering this question, follow the ATSC program number convention. No conditional access is used in this transport stream.

12.12 Using the data in Tables 12.15 and 12.16, construct an ATSC virtual channel table for a station using Channel 7 for its preexisting NTSC analog service and Channel 14 for its digital service. There is no extended text for these programs.

12.13 Construct an ATSC master guide table for the services in Questions 12.9–12.11. Where necessary, assign PIDs for tables that have not yet been defined. All tables have version number 7 and size 875 bytes.

12.14 Use the data in Table 12.14 to construct an ATSC event information table for the transport stream defined in Tables 12.15 and 12.16.

Table 12.15 Elementary stream data for Questions 12.11–12.14.

Program number	PID	Elementary stream type	Component name	Elementary stream description
2	0x0021	MPEG-2 video	NBC-video	Main profile at main level; 4:2:0; 30 Hz; aligned at slice layer, with English (Line 21) closed captioning in Field 1, designed for beginner readers.
	0x0024	AC-3 audio	NBC-audio	English language; 44.1 kHz, 320 kbit/s, two channels.
3	0x0031	MPEG-2 video	CBS-audio	Main profile at main level; 4:2:0; 30 Hz; aligned at picture layer; with German (ATVCC) closed captioning using caption service number 3.
	0x0034	AC-3 audio	CBS-English audio	English language; 48 kHz, 320 kbit/s, two channels.
	0x0038	AC-3 audio	CBS-French audio	French language; 44.1 kHz, 320 kbit/s, two channels.
4	0x0041	MPEG-2 video	ABC-video	Main profile at main level; 4:2:0; 30 Hz; aligned at picture layer; with English (Line 21) closed captioning in Field 2.
	0x0044	AC-3 audio	ABC-English audio	English language; 44.1 kHz, 320 kbit/s, two channels.
	0x0047	AC-3 audio	ABC-German audio	German language; 44.1 kHz, 640 kbit/s, two channels.

Table 12.16 Data for virtual channel table for Question 12.12.

Minor channel number	Program number	Service type	Carrier frequency (MHz)	Source identifier
1	—	Analog	175.25	0x5345
2	2	ATSC television	470.310	0x5345
3	3	ATSC television	470.310	0x5345
4	4	ATSC television	470.310	0x5345

12.15 Is the transport stream described by Question 12.10 capable of carrying DVB-SI and thereby being decodable by a DVB decoder?

12.16 Construct an ATSC rating region table from the data in Table 12.17.

Table 12.17 Rating region data for Question 12.15.

Dimension name	Graduated scale	Rating value text	Abbreviated rating value text
Violence	1	Some violence	SV
		Frequent violence	FV
		Strong violence	StV
Sex	1	Sexual reference	SR
		Sex scenes	SS
		Strong sex scenes	StS
Other	0	Medical procedures	M
		Horror	H

12.17 Construct an ATSC system time table specifying the time as 0020 UTC on 4th March, 1999. The offset between UTC and GPS is 14 s, and daylight savings is inactive. The next change of daylight savings status is to occur on October 1, 1999, at 2 a.m.

12.18 Construct a DVB time offset table for New Zealand (which has only one time zone), specifying that the current time is 8 p.m., January 3, 1998 (local time), with an offset of +13 h from UTC, and also specifying that this offset will change to +12 h at 2 a.m. (local time), April 1, 1998.

MATLAB EXERCISE 12.1

ATSC and DVB make use of different time and date representations.

The aim of this exercise is to implement MATLAB functions to calculate the modified Julian date and GPS time, and to use these to calculate the days of week.

1. Using the formula provided in Section 12.3.1, implement a MATLAB function to calculate the modified Julian date, given the year, month, and day of month.

2. Write a MATLAB function to translate an array in the format of the output of the MATLAB clock() function into the GPS time used by ATSC. (Hint: The number of days since January 6, 1980, can be easily found using the modified Julian date function from Part 1 of this exercise.)
3. Implement a function to calculate the days of week (i.e., Sunday, Monday, etc.) from the modified Julian date. (Hint: Start by picking a reference date for which the day of week is known. Find the modified Julian date of this reference date and the date for which the day of week is required. The required day of week can be calculated from the difference in the modified Julian dates.

REFERENCES

1. DVB specifications relating to SI are
   (a) Digital Video Broadcasting (DVB); *Specification for Service Information (SI) in DVB Systems*, EN 300 468, Sophia Antipolis: ETSI, 1998.
   (b) Digital Video Broadcasting (DVB); *Guidelines on Implementation and Usage of Service Information (SI)*, ETR 211, Sophia Antipolis: ETSI, 1997.
   (c) Digital Video Broadcasting (DVB); *Allocation of Service Information (SI) Codes for DVB Systems*, ETR 162, Sophia Antipolis: ETSI, 1995.
2. The formal definitions for character sets used in DVB-SI are contained in
   (a) ISO 6937, *Information technology—coded graphic character set for text communication—Latin alphabet.*
   (b) ISO 8859, *Information processing—8-bit single-byte coded graphic character sets, Latin alphabets.*
3. ISO 3166, *Codes for the representations of names of countries.*
4. ISO 639, *Code for the representation of names of languages.*
5. Digital Video Broadcasting (DVB); *Specification for Service Information (SI) in DVB Systems*, EN 300 468, Sophia Antipolis: ETSI, 1998, Annex C, pp. 71–72.
6. Digital broadcasting systems for television, sound and data services; *Allocation of Service Information (SI) Codes for Digital Video Broadcasting (DVB) Systems*, ETR 162, Sophia Antipolis: ETSI, 1995.
7. Digital Video Broadcasting (DVB); *Guidelines on implementation and usage of Service Information (SI)*, ETR 211, Sophia Antipolis: ETSI, 1997, Section 4.4, pp. 27–28.
8. ATSC Standard A/65A, *Program and system information protocol for terrestrial broadcast and cable*, Advanced Television Systems Committee, May 2000.
9. This descriptor is specified in ATSC Standard A/58, *Digital audio compression standard (AC-3)*, Advanced Television Systems Committee, December 1995.
10. This issue is discussed further in
    (a) Digital Video Broadcasting (DVB); *Guidelines on implementation and usage of Service Information (SI)*, ETR 211, Sophia Antipolis: ETSI, 1997, Annex A.
    (b) ATSC Standard A/58, *Harmonization with DVB SI in the use of the ATSC Digital Television Standard*, Advanced Television Systems Committee, September 1996.
    (c) AS 4599-1999, *Digital television—terrestrial broadcasting—characteristics of digital terrestrial television transmissions*, Homebush NSW: Standards Australia, 1999, Annex A to Section 7.

Chapter 13

Digital Television Channel Coding and Modulation

13.1. INTRODUCTION

Although the MPEG-2 standards specify source coding and multiplexing for digital television, they do not specify the channel coding and modulation to be used to transmit the systems bit stream over any particular channel. This chapter describes the channel coding and modulation provided for terrestrial broadcast by the DVB and ATSC digital television standards.[1] Generic concepts of channel coding and modulation, common to both standards are described first, followed by more detailed description of the methods employed in DVB [1] and ATSC. The sections relating to DVB and ATSC are written so that each can be read independent of the other.

The aim of this chapter is to provide an overview of the modulation techniques used in digital television systems. Some familiarity with digital modulation is therefore assumed. A comprehensive coverage of digital modulation and channel coding would require a book in its own right.

13.2. GENERIC CONCEPTS

Digital signals passing through transmission channels are subject to errors introduced in the transmission process. Although all channels introduce errors, some channels, especially radio channels, tend to introduce high error rates. The types of impairments introduced by an imperfect transmission channel include additive noise and channel perturbations. Additive noise may take the form of Gaussian noise with stationary statistics, impulsive noise that is not always stationary or easy to characterize, or even deliberate jamming. Channel perturbations may occur due to fading in radio channels, synchronization slip in digital channels, and breaks in transmission.

[1]We do not discuss satellite or cable delivery systems in this chapter. Information on these systems can be found in the relevant standards.

Digital Television, by John Arnold, Michael Frater and Mark Pickering.

The likely effects of these channel impairments on a digital signal are one of *uniformly random errors*—errors occurring individually and independently, with approximately uniform probability density, primarily due to noise (often just called *random noise*); *burst errors*—errors grouped in clusters, mainly the result of a combination of noise and channel perturbations; and *erasures*—irregular intervals when it is known that no reliable signal can be detected because of severe channel perturbation.

Minimization of the impact of transmission errors involves a combination of channel coding and modulation, each of which are addressed in the following sections.

13.2.1. Channel Characteristics and Intersymbol Interference

Transfer of information across a communications channel is limited by the bandwidth of the channel and the combination of its bandwidth and signal-to-noise ratio.

For a channel that introduces *additive, white, Gaussian noise* (*AWGN*), the precise relationship between channel capacity, bandwidth, and signal-to-noise ratio is given by Shannon's criterion:

$$C = B\log_2\left(1+\frac{S}{N}\right)$$

where C is the capacity of the channel in bits per second, B is the bandwidth of the channel in hertz, S is the signal power at the receiver, and N is the noise power at the receiver [2]. In practice, the throughput of a channel is usually significantly less than the maximum indicated by the value of C. The major reason for this is limitations in the performance of channel coding systems. Shannon's criterion tells us that an increase in channel capacity can be achieved either by increasing its bandwidth or increasing its signal-to-noise ratio. It also tells us that channel bandwidth has a much greater impact than its signal-to-noise ratio.

Limited channel bandwidth can blur transitions where changes of state occur in a bit stream. This effect, known as intersymbol interference, is illustrated in Figure 13.1. An equalizer is a digital filter that has been designed to correct for

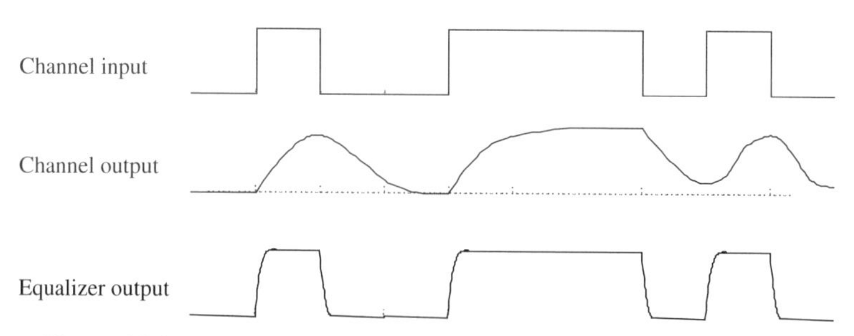

Figure 13.1 Intersymbol interference and equalization for a binary waveform.

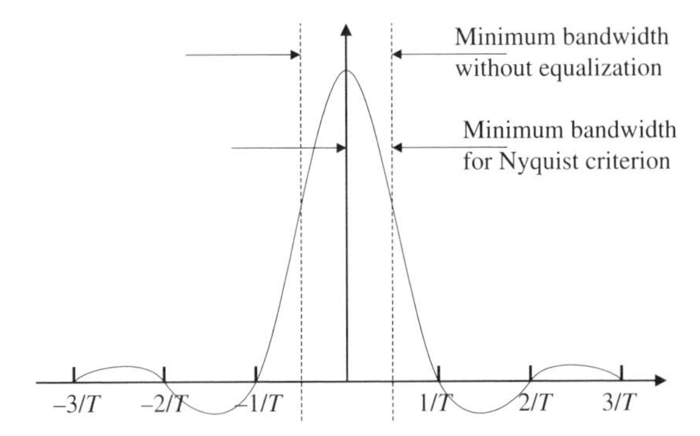

Figure 13.2 Minimum channel bandwidths required to meet the Nyquist criterion and to avoid the need for equalization.

the intersymbol interference of the channel. The use of an appropriately designed *equalizer* can allow the original signal to be recovered so long as the channel bandwidth meets the Nyquist criterion, which is discussed below.

Independent of any limitation on information-carrying capacity implied by Shannon's criterion, the bandwidth of a channel limits the symbol rate that can be carried. The Nyquist criterion says that the minimum bandwidth required to carry a given symbol rate is equal to half the symbol rate [3]. Figure 13.2 illustrates the minimum bandwidth requirements with and without equalization for a signal with symbol period T.

EXAMPLE 13.1—Nyquist Criterion

A bit stream is transmitted at a rate of 120,000 bit/s, with each bit carried in one symbol. What is the minimum channel bandwidth required to meet the Nyquist criterion? What is the impact if each symbol carries two bits of information?

The minimum channel bandwidth to meet the Nyquist criterion is 60 kHz. If the same bit stream is transmitted with two bits in each transmitted symbol (that is, at 60,000 sym/s), then the minimum bandwidth to meet the Nyquist criterion is 30 kHz. ■

As the data rate rises or channel bandwidth falls, the impact of intersymbol interference increases. If significant intersymbol interference occurs, a receiver must incorporate an equalizer to recover the original bit stream.

In this section the following notation is used:

- $c(t)$—the impulse response of channel;
- $C(f)$—the frequency response of channel, which is the Fourier transform of the impulse response;
- W—the bandwidth of bandlimited channel;
- $\theta(f)$—the phase component of $C(f)$; and
- $\tau(f)$—the delay response $= -(1/2\pi)\, d\theta(f)/df$.

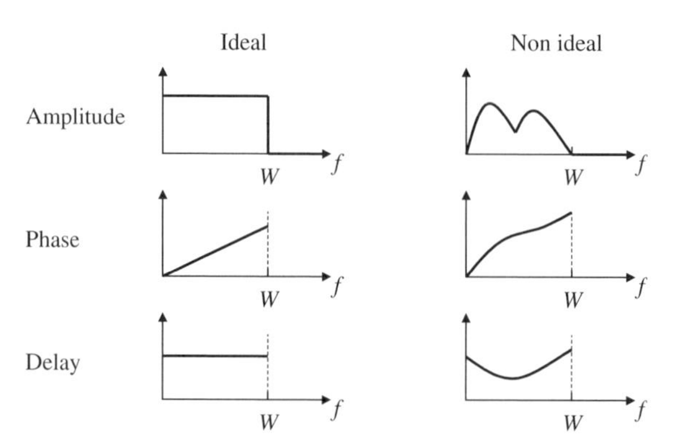

Figure 13.3 Ideal versus nonideal channels.

The characteristics of ideal and nonideal channels are illustrated in Figure 13.3. An ideal channel is one with the following characteristics:

- *Constant in-band amplitude response.* $|C(f)|$ is constant for $|f| \leq W$. Violation $\Rightarrow$ "amplitude distortion."
- *Zero out-of-band amplitude response.* $C(f) = 0$ for $|f| > W$.
- *Phase linearity.* $\theta(f)$ is a linear function of f for $|f| \leq W$. Violation $\Rightarrow$ "phase distortion."
- *Constant delay.* $\tau(f)$ is constant for $|f| \leq W$. Violation $\Rightarrow$ "delay distortion."

13.2.2. Modulation

A digital bit stream is a representation of data in terms of the binary values 0 and 1. Inside a video coder or decoder, these values are often represented as voltage levels. They cannot normally be transmitted over a communications system directly in this form. The purpose of modulation is to convert the binary values into a form that is suitable for transmission. Currently used transmission systems usually represent the information content of the bit stream in the amplitude, the frequency, or the phase of the transmitted signal. For digital signals, these forms of modulation are referred to as *amplitude-shift keying* (*ASK*), *frequency-shift keying* (*FSK*), and *phase-shift keying* (*PSK*).

The forms of modulation used for digital television are PSK or combinations of ASK and PSK. The simplest form of modulation used is binary PSK (BPSK). The waveforms used by BPSK to represent the binary values 0 and 1 are illustrated on the left side of Figure 13.4. For each bit entering the modulator, one symbol is produced at the output. Each BPSK symbol, therefore, carries one bit of data. These waveforms are simply +/− cosine functions. An alternative representation, in terms of the phase of waveforms, is shown on the right side of Figure 13.4. BPSK offers a high level of robustness in the presence of additive channel noise (significantly higher than, e.g., ASK) because the information is contained in the phase of the

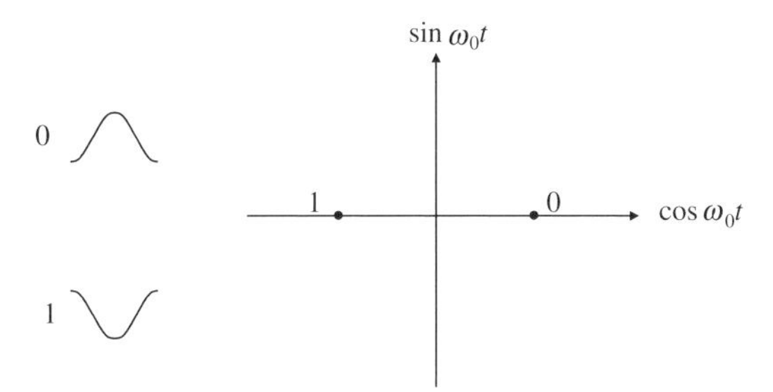

Figure 13.4 Constellation diagram for BPSK.

transmitted signal, which is not directly affected by the additive noise, rather than in the amplitude, which is directly affected by additive noise.

Differential binary PSK, often denoted by DPSK, represents a binary 1 by a change of phase of π radians (one half cycle) and a binary 0 by no change of phase.

In quadrature PSK (QPSK), the number of symbols is increased from two to four, producing the constellation diagram shown in Figure 13.5. One QPSK symbol is produced at the output of the modulator for each pair of bits at its input. Each QPSK symbol, therefore, carries two bits of data. QPSK transports data at twice the rate as BPSK in the same channel bandwidth and offers identical error performance to BPSK with the same energy per bit on the transmission channel. This is possible because the QPSK waveform is made up of two orthogonal components: the in-phase (I) componenent, shown as the horizontal (real) axis in Figure 13.5, and the quadrature (Q) component, shown as the vertical (imaginary) axis. The cost of increase in channel capacity depends on a small increase in encoder and decoder complexity. For encoders and decoders implemented within modern integrated circuits, this additional cost is negligible.

The data carrying capacity of a channel can be increased beyond that available using QPSK without increasing the channel bandwidth, by increasing the number of bits of data carried by each modulation symbol. This requires increasing the number of points in the constellation diagram, which implies a reduction in the spacing between these points for a fixed transmission power. For digital television applications, this increase is achieved by making use of both the amplitude and phase of the

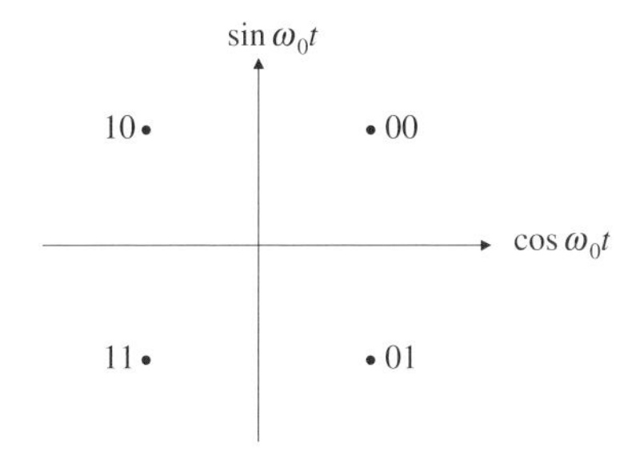

Figure 13.5 Constellation diagram for QPSK.

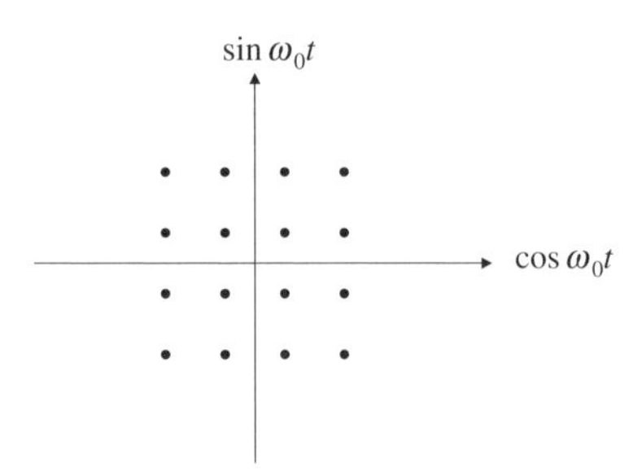

Figure 13.6 Constellation diagram for 16-QAM.

transmitted signal to carry information. The cost of this increase in channel capacity depends both on the complexity of encoders and decoders (which is acceptable with modern integrated circuits) and on a reduced robustness to additive noise introduced in the channel, resulting in an increased error rate.

The use of this combination of amplitude and phase modulation to carry four bits per symbol results in the 16-symbol constellation diagram of 16-quadrature amplitude modulation (16-QAM) as shown in Figure 13.6. Further increasing the number of bits per symbol to 6 leads to the constellation diagram of 64-QAM (Fig. 13.7). The further increase in data rate going from 16-QAM to 64-QAM comes with a corresponding reduction in robustness to channel noise. Using 16-QAM, a channel bandwidth of at least 3 kHz is required to carry a bit stream with rate 12 kbit/s without equalization, which reduces to 2 kHz for 64-QAM.

The amplitude spectrum of QPSK, 16-QAM, and 64-QAM is shown in Figure 13.8. The width of the central lobe is $2/T$, where T is the symbol period.

An alternative strategy is the combination of ASK with QPSK, which is used to combine ASK with BPSK. An example of this approach is the 8-level ASK modulation scheme illustrated in Figure 13.9. There are eight possible symbols, each of which carries three bits of data. These are illustrated on the left side of Figure 13.9, with the corresponding constellation diagram on the right side. The spectrum of this form of modulation is the same as illustrated in Figure 13.8, with T being the symbol

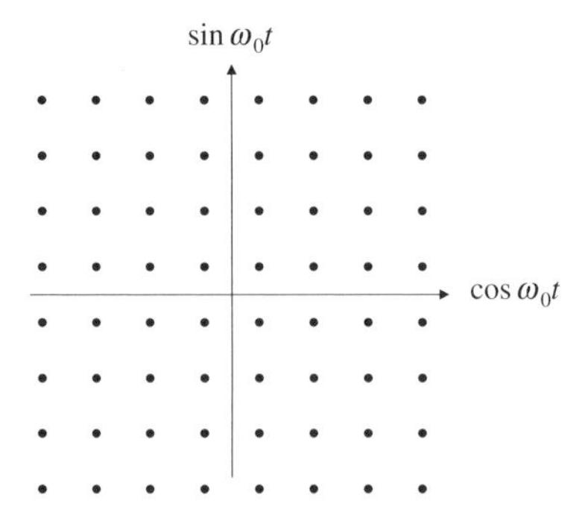

Figure 13.7 Constellation diagram for 64-QAM.

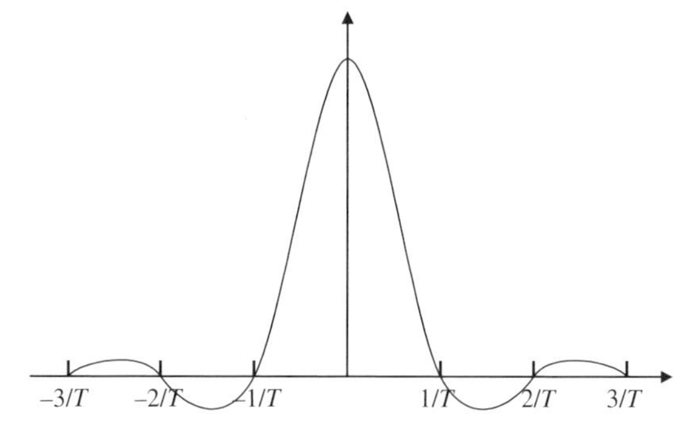

Figure 13.8 Spectrum of QPSK, 16-QAM, and 64-QAM.

period. Using this 8-level ASK, a channel bandwidth of at least 4 kHz is required to carry a bit stream with rate 12 kbit/s without significant intersymbol interference.

This 8-level ASK waveform may be filtered to reduce the transmission bandwidth. Theoretically, the minimum bandwidth that can be achieved without loss of information is the Nyquist bandwidth. This can be achieved by retaining the upper sideband and removing the lower sideband by filtering. In practice, a small amount of the lower sideband is usually retained, giving vestigial sideband modulation.

BPSK and QPSK have identical error performance because QPSK is formed from an orthogonal combination of two BPSK signals. In the same way, 64-QAM can be formed by the combination of two of the 8-level ASK modulation illustrated in Figure 13.9. The 8-level ASK modulation scheme should, therefore, have the same error performance as 64-QAM.

The bandwidth efficiency of this 8-level ASK scheme can be maximized by filtering, as illustrated in Figure 13.10, to limit the total bandwidth to the Nyquist

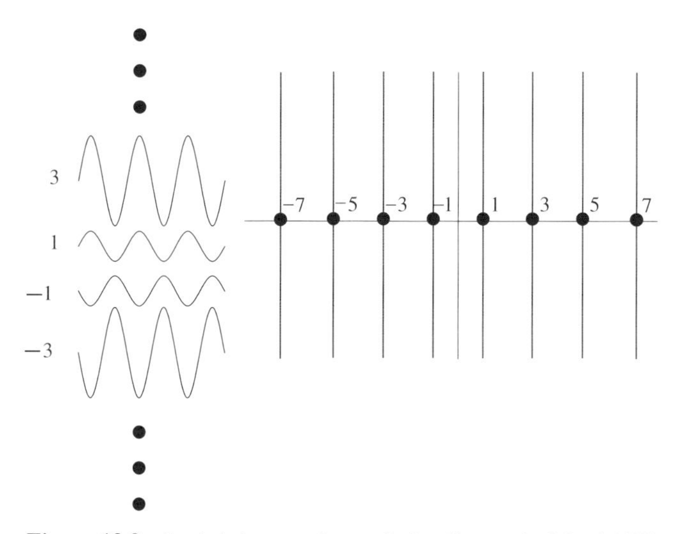

Figure 13.9 Symbol shapes and constellation diagram for 8-level ASK.

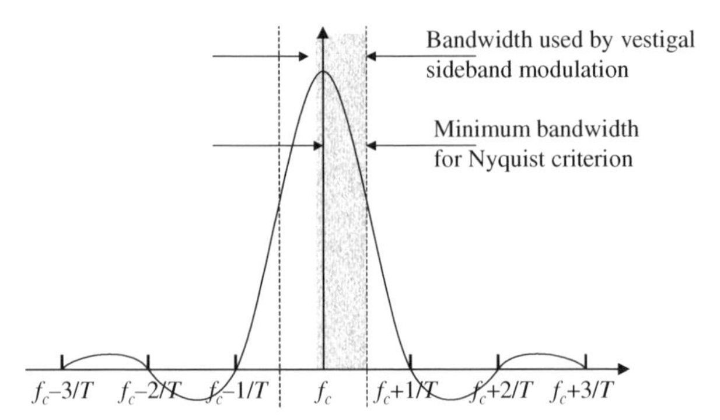

Figure 13.10 Frequency components used in 8-level vestigial sideband (8-VSB) modulation with carrier frequency f_c.

bandwidth. T is the period of a single 8-level symbol. Because the power density of the signal at the boundary between upper and lower sidebands is nonzero, the use of upper-sideband or lower-sideband transmission is not possible. Vestigial sideband modulation, in which the upper sideband and a small proportion of the lower sideband are retained, is preferred.

13.2.3. Equalization

Usually, an equalizer is an integral part of the demodulator. In the general case, modeling of the impact of channel and equalizer is quite difficult. Given the restrictions, however, that the channel noise is additive, white Gaussian and the channel is time invariant, a linear transversal model can be used for the combined impact of modulator, channel, and demodulator, that is, their combined effect can be modeled as an IIR filter [4]. This is illustrated in Figure 13.11.

13.2.3.1. Fixed-Tap Equalizer

In situations where the channel response is known in advance and remains constant, an optimal equalizer can be designed off-line. The equalizer consists of two parts:

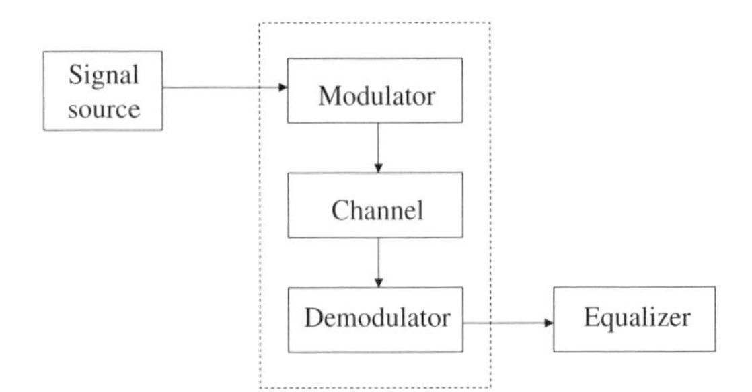

Figure 13.11 Linear-transversal model for channel.

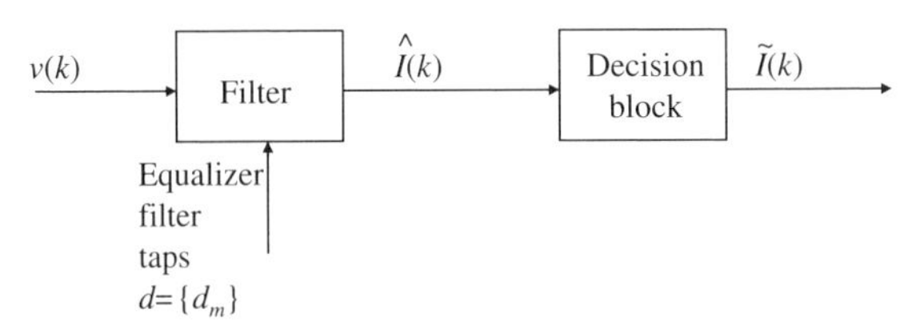

Figure 13.12 Fixed-tap equalizer.

an FIR filter and a decision block (Fig. 13.12). The filter approximately reverses the intersymbol interference introduced by the channel. The filter block is usually an FIR linear transversal filter, with taps denoted by the M-element vector $\{d_m\}$. The decision block takes the output of the filter and selects the most appropriate symbol as the output of the equalizer. The channel input at time k is denoted by $I(k)$, whereas the channel output at time k is denoted by $v(k)$. The equalizer output is an estimate of the channel input, so the equalizer output at time k is denoted by $I(k)$, so

$$\hat{I}(k) = \sum_{m=0}^{M-1} d_m \cdot v(k-m)$$

13.2.3.2. Trained Adaptive Equalizer

In a trained adaptive equalizer, the optimal values for the filter taps are estimated using knowledge of the original value of the incoming bit stream. This information is available in systems, such as ATSC, where a known training sequence is transmitted at regular intervals.

The operation of the filter and decision blocks is the same as that for a fixed equalizer (Fig. 13.13).

The adaptation block calculates updates to the equalizer filter coefficients based on the equalizer outputs and original symbol values. Because the optimal value of the equalizer filter taps is now estimated rather than known *a priori*, it is usual to denote the values of the taps by $\hat{d}$ (k), with the mth tap of the equalizer denoted by $\hat{d}_m$ (k).

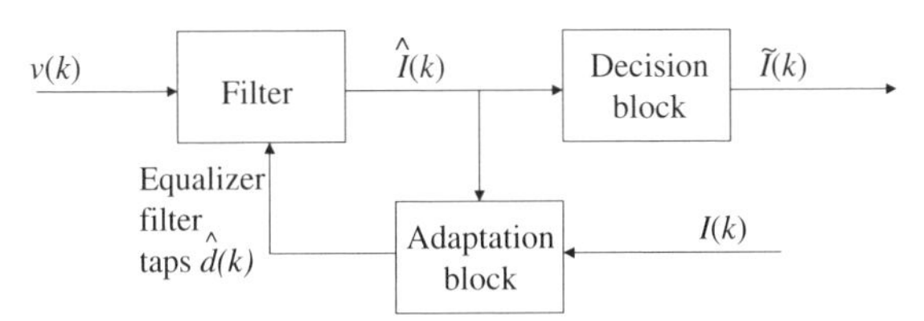

Figure 13.13 Trained adaptive equalizer.

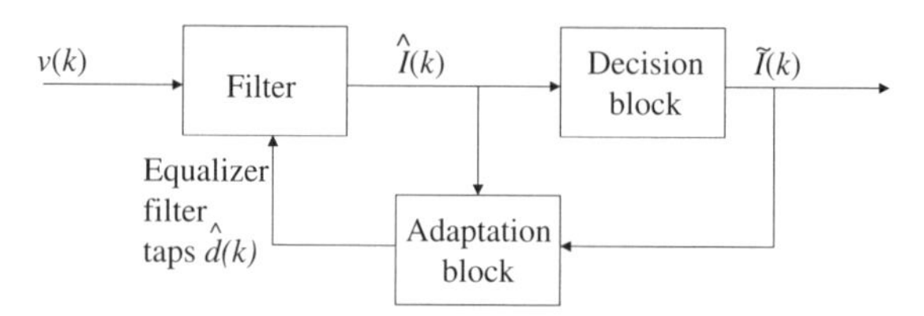

Figure 13.14 Blind adaptive equalizer.

13.2.3.3. *Blind Adaptive Equalizer*

In a blind-adaptive equalizer, the optimal values for the filter taps are estimated without access to the original value of the incoming bit stream. Instead, estimates of these symbol values based on equalizer output are used. Blind-adaptive equalizers do not require training sequences or explicit knowledge of channel characteristics.

The operation of the filter and decision blocks is identical to that discussed above for trained equalizers. The adaptation block calculates updates to the equalizer filter coefficients based on the equalizer outputs and decision block outputs (Fig. 13.14).

13.2.3.4. *The LMS Algorithm*

The least mean square (LMS) algorithm is one of the most fundamental techniques used in adaptive equalization and adaptive control. The LMS algorithm updates equalizer taps $d(k)$ in accordance with

$$\hat{d}_m(k) = \hat{d}_m(k-1) + \Delta\varepsilon(k)v^*(k-m)$$

where Δ is known as the step size, and where the error is given by

$$\varepsilon(k) = I(k) - \hat{I}(k)$$

LMS is a gradient descent algorithm. This means that the equalizer uses its measurement of the error to determine the magnitude and direction of the update at each step using a notational performance surface in which the best performance lies at the bottom of a well.

Convergence of the LMS algorithm is guaranteed if Δ is small enough. Increasing Δ has two effects:

- the rate of convergence increases and
- the long-term error is increased.

13.2.3.5. *Practical Blind Adaptive Equalizers*

Blind LMS In cases where a training sequence is not available, the LMS algorithm can be used for blind adaptation by replacing the training sequence input to the adaptation process by the equalizer's output, using the same notation as in the previous section

$$\hat{d}_m(k) = \hat{d}_m(k-1) + \Delta\varepsilon(k)v^*(k-m)$$

where

$$\varepsilon(k) = \tilde{I}(k) - \hat{I}(k)$$

This algorithm has many of the same properties as the LMS algorithm. Its convergence, however, cannot be guaranteed. It is only suitable for situations where a reasonable approximation to the channel response is known *a priori*. This approach might be used to continue adaptation throughout a long transmission where a training sequence is used to provide the initial adaptation at the beginning of the transmission.

Godard's Algorithm Another method for blind equalization is that proposed by Godard [5]. In this algorithm, equalizer taps are updated as follows:

$$\hat{d}_{mk}(k) = d_m(k-1) + \Delta v^*(k-m)\hat{I}(k)\left(R_2 - \left|\hat{I}(k)\right|^2\right)$$

where

$$R_2 = \frac{E\left(|I(k)|^4\right)}{E\left(|I(k)|^2\right)}$$

For modulation techniques whose alphabets include symbols with nonzero imaginary parts, all calculations in the equalizer are carried out on complex numbers (Table 13.1).

Convergence of the Godard algorithm can be guaranteed if the symbol stream is pseudorandom, an equalizer of infinite length is used, and the equalizer is initialized with all its taps set to zero, except for the center tap that is set to one.

13.2.4. Randomization

Before the demodulator in a receiver can recover a bit stream from the received signal, it must extract timing information. As a minimum, this timing information is used to identify the boundaries between symbols. A demodulator often relies on changes in the value of received symbols to identify these boundaries.

Figure 13.15 illustrates a BPSK received signal with and without transitions in the symbol value. In the former case, the demodulator has no means of identifying

Table 13.1 Alphabets and R_2 values for various modulation techniques using the Godard equalizer.

Modulation	Alphabet	R_2
BPSK	{1, −1}	1
QPSK	{1+*j*, 1−*j*, −1+*j*, −1−*j*}	2
16-QAM	{3+3*j*, 3+*j*, 1+3*j*, 1+*j*, −3+3*j*, −3+*j*, −1+3*j*, −1+*j*, 3−3*j*, 3−*j*, 1−3*j*, 1−*j*, −3−3*j*, −3−*j*, −1−3*j*, −1−*j*}	13.2
64-QAM	{7+7*j*, 7+5*j*, 5+7*j*, 5+5*j*, 7+*j*, 7+3*j*, 5+*j*, 5+3*j*, 1+7*j*, 1+5*j*, 3+5*j*, 3+7*j*, 1+*j*, 3+*j*, 1+3*j*, 3+3*j*, −7+7*j*, −7+5*j*, −5+7*j*, −5+5*j*, −7+*j*, −7+3*j*, −5+*j*, −5+3*j*, −1+7*j*, −1+5*j*, −3+5*j*, −3+7*j*, −1+*j*, −3+*j*, −1+3*j*, −3+3*j*,7−7*j*, 7−5*j*, 5−7*j*, 5−5*j*, 7−*j*, 7−3*j*, 5−*j*, 5−3*j*, 1−7*j*, 1−5*j*, 3−5*j*, 3−7*j*, 1−*j*, 3−*j*, 1−3*j*, 3−3*j*, −7−7*j*, −7−5*j*, −5−7*j*, −5−5*j*, −7−*j*, −7−3*j*, −5−*j*, −5−3*j*, −1−7*j*, −1−5*j*, −3−5*j*, −3−7*j*, −1−*j*, −3−*j*, −1−3*j*, −3−3*j*}	58
8-VSB	{7, 5, 3, 1, −1, −3, −5, −7}	37

where boundaries between symbols lie. In the latter case, the presence of transitions between different symbol values makes this possible.

Long runs of identical symbols may often occur in digital television. Examples include start codes used in video and systems layers and transport stream null packets. In order to prevent these runs of identical symbols, a process of *transport stream randomization* is applied by both DVB and ATSC to the transport stream prior to channel coding. This is illustrated in Figure 13.16. The encoder generates a pseudorandom sequence. The exclusive-or of each bit of this pseudorandom sequence with the transport stream generated by the MPEG-2 systems encoder is passed through to the channel coding.

The decoder generates the same pseudorandom sequence as the encoder. The exclusive-or of this sequence with the received stream recovers the original bit

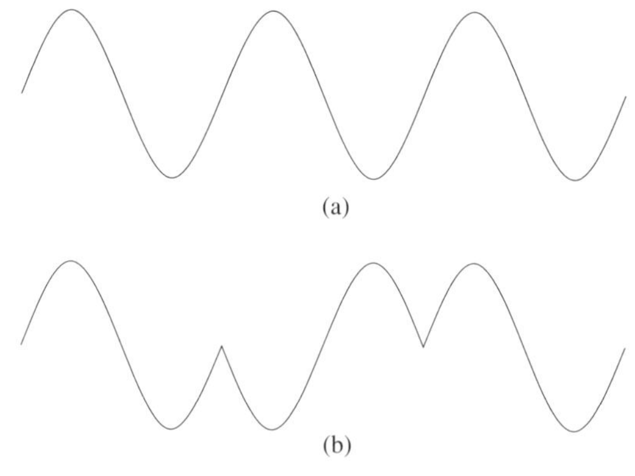

Figure 13.15 Received BPSK signal (a) without transitions and (b) with transitions in symbol value.

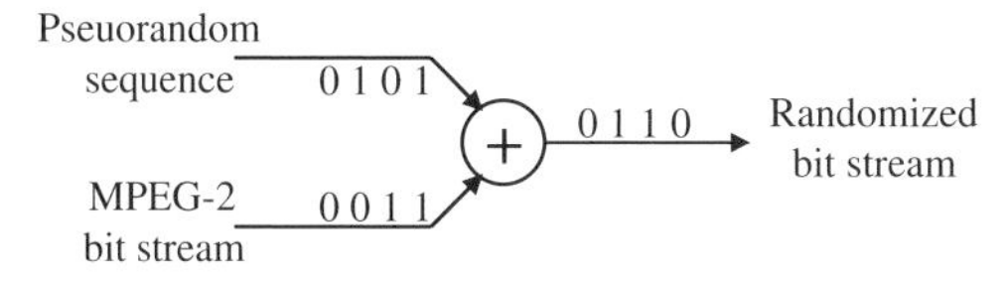

Figure 13.16 Transport stream randomization.

stream. To facilitate decoder operation, both DVB and ATSC reset the state of the pseudorandom sequence generator periodically.

Transport stream randomization does not guarantee that long runs of identical symbols cannot occur; rather it makes these events very unlikely. Although a similar effect could be achieved by the use of a periodic sequence, long streams of identical symbols would then be converted into long periodic streams. These long periodic streams may also upset the operation of a receiver.

13.2.5. Channel Coding Technology

Channel coding is used to correct errors caused by channel impairments through the introduction of controlled redundancy, enabling messages corrupted in transmission to be corrected before further processing [6]. With this controlled redundancy, only a subset of all possible transmitted messages (bit sequences) contains valid messages. This subset is called a code and the valid messages are called *code words* or *code vectors.* A good code is one in which code words are so separated that the likelihood of errors corrupting one into another is kept small.

Error detection is simplified to answering this question: Is the received message a code word or not? If it is a code word, one assumes that no errors have occurred. The probability of an undetected error getting through is then the probability of sufficient errors occurring to transform the real transmitted code word into another, apparently correct but in reality a false one.

If an error is detected, it can be corrected (at least in principle) by *automatic repeat request* (*ARQ*) or *forward error correction* (*FEC*). In ARQ systems, the receiver checks received blocks of data for errors. If an error is detected, the receiver sends a request for retransmission to the transmitter. For broadcast services, such as digital television, ARQ is not feasible because it is not possible for a receiver to request retransmission. FEC is, therefore, used to protect transmitted data.

In FEC, the recipient corrects the errors by finding the valid code word 'nearest' to the received message, on the assumption that the nearest is the most likely because few corrupting errors are more likely than many. There are two types of FEC: *block coding* and *convolutional coding.*

In block coding, source data is partitioned into blocks of k bits, converted by the encoder into blocks of n ($>k$) bits with enough checks to enable the decoder to correct errors of the more probable kinds. The most common types of block codes [7] are Golay Codes, Bose–Chadhuri–Hocquenghem (BCH) Codes, and Reed–Solomon (RS) Codes. Block codes are used when information is naturally structured in blocks, when channel capacity is relatively low and we do not want to waste it further with unnecessarily low code rates, and when quick efficient decoding is required because of limited processing time available.

For a convolutional code, the encoder operates not on disjoint blocks but on a running block of bits held in a shift register, generating a sequence of higher rate. The correcting capabilities of convolutional codes are not so clear-cut as with block codes. Probabilistic decoding, approximating maximum likelihood, is generally used. When long streams of relatively unstructured data are transmitted on high-capacity channels (such as terrestrial broadcast channels with capacities of approximately 20 Mbit/s) and when the complexity of the decoder represents a relatively small proportion of the total cost of the receiving equipment (such as a digital television set-top box), then convolutional codes can offer the best error-correcting solutions.

13.2.5.1. Block Codes

FEC allows transmission errors to be corrected by adding redundancy to a bit stream. This may seem to contradict the aim of source coding, which is to minimize the number of bits to be transmitted. The key difference between the original source redundancy and that added for FEC is that FEC adds (usually small amounts of) controlled redundancy in such a way as to maximize error correction.

A very simple example of a block code would be to represent each bit of data with a 5-bit symbol: 1→ 11111 and 0 → 00000. Using this code, two errors in each code word can be detected and corrected. If, for example, the transmitted code word is 11111 and the received code word 01110, the receiver converts this code word to a 1 because its value is closer to 11111 than to 00000. The obvious drawback of this very simple approach is that it has a very high overhead (400%).

Practical block codes group symbols into blocks and provide parity symbols for each block. Each symbol may consist of one or more bits, with the number of bits per symbol fixed for all symbols in the block. This greatly reduces the overhead, although at the cost of some reduction in performance. This is illustrated in Figure 13.17. This code with k information symbols and n-k parity symbols is referred to as an (n, k) code.

If all valid code words differ in at least $d_{\min}$ symbols, the number of symbol errors t that the code is capable of correcting is

$$t = \left\lfloor \frac{d_{\min} - 1}{2} \right\rfloor$$

where $\lfloor x \rfloor$ denotes the largest integer whose value is not greater than x.

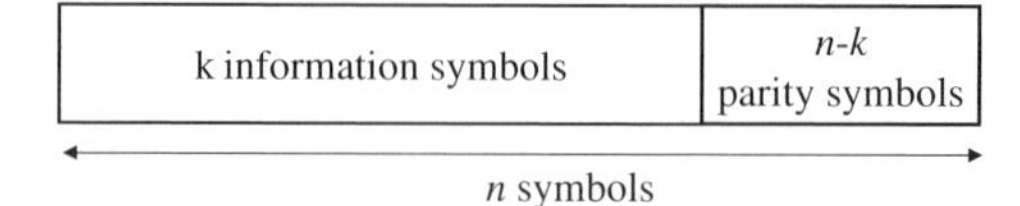

Figure 13.17 Structure of a practical block code.

1	0	0	1	1	1	1
0	0	1	0	1	0	0
1	0	0	0	0	1	0

Five data symbols Two redundancy symbols

Figure 13.18 Example of block code using (7,5) RS code.

In the simple example above, each bit is one symbol, $d_{\min}$ is 5, so $t = 2$, meaning that two bit errors can be corrected.

The block codes used in digital television transmission are Reed–Solomon (RS) codes. RS codes use a nonbinary alphabet, that is, each symbol consists of more than one bit. One code word of an (7,5) RS code is illustrated in Figure 13.18. Each symbol contains three bits. Seven symbols make up one code word. The first five of these symbols contain the original data; the other two provide the redundancy for error correction.

This code can correct one symbol error, where a symbol error is defined as any number of bit errors in a single symbol. Reed–Solomon codes are therefore good for correcting burst errors, because there is no additional cost for correcting two errors in one symbol compared to a single error. In this example, the code could correct up to three bit errors provided they are all in the same symbol.

For k-bit symbols, the maximum code word length (n) for a Reed–Solomon code is $2k-1$. Typically 8-bit symbols are used, which means that each code word can contain up to 255 bytes.

A Reed–Solomon code that can correct t symbol errors requires $2t$ parity symbols. A Reed–Solomon code that can correct t symbol errors can also correct $2t$ erasures (i.e., where position of errors are known). In general, correction is possible if

$$2s + r < 2t$$

where s is the number of symbol errors and r is the number of erasures.

Reed–Solomon codes can be shortened by setting a number of symbols to zero at the encoder, transmitting the code word without these zero symbols and then reinserting these zero symbols at the decoder. For example, a (255,223) code can correct 16 symbol errors. A (200,168) code could operate by adding 55 zero bytes to create a (255,223) code and then transmitting only 168 data bytes plus 32 parity bytes. The notation (200,168,8) is often used to represent an RS (200,168) coder that operates on symbols of 8 bits.

The performance of RS codes is illustrated in Figure 13.19. The error rate at the input to the encoder is expressed as the symbol error rate on the horizontal axis. The residual error rate at the output of the coder (expressed as a bit error rate) is shown on the vertical axis.

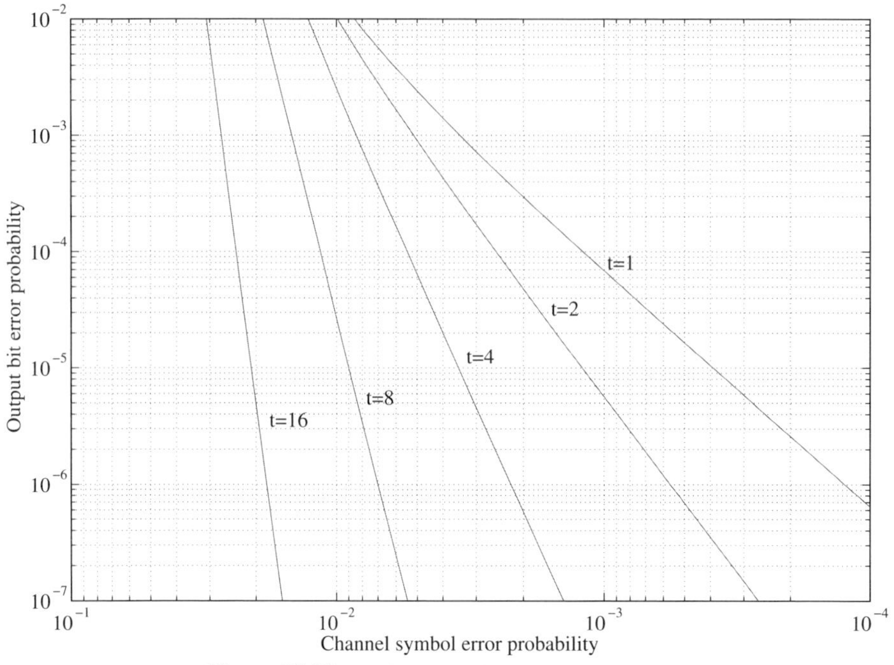

Figure 13.19 Performance of Reed–Solomon codes.

One disadvantage of block codes is that their performance depends on the location of errors in the bit stream. In the top bit stream illustrated in Figure 13.20, a particular pattern of three errors is split across the boundary of two blocks, with two errors occurring in the first block and one in the second. In the bottom bit stream, all three errors occur in the same block. The use of a block code capable of correcting two errors leads to an error-free output bit stream in the top case, but not in the bottom case.

13.2.5.2. Convolutional Codes

Convolutional codes operate on continuous streams of data and are commonly used on high-capacity wireless communications channels such as terrestrial broadcast

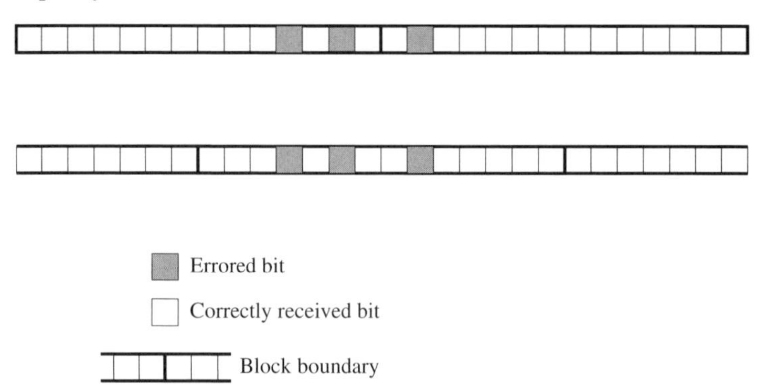

Figure 13.20 Impact of location of block boundaries on the performance of block codes.

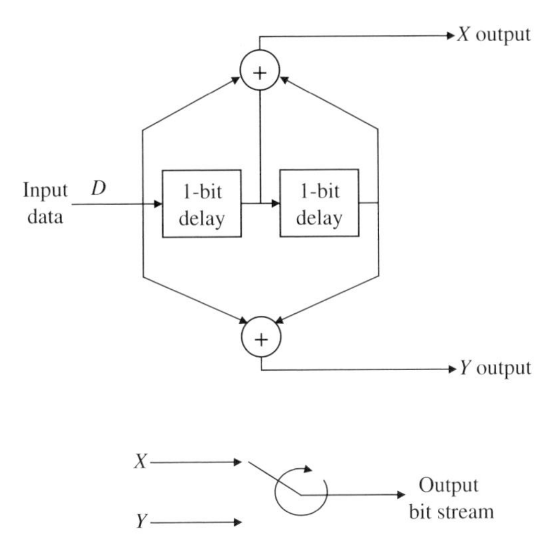

Figure 13.21 Example of 1/2 rate convolutional encoder with constraint length 3.

digital television. The basic structure of a simple convolutional encoder is shown in Figure 13.21. Data arrives at the left of the figure and passes to a shift register, which is clocked once for each arriving bit. For each input bit *D*, two output bits are generated by calculating the modulo-2 sum of nominated bits in the shift register. Each output bit uses a different combination of these bits. In this example, the output bits are labeled *X* and *Y* and are multiplexed into a single output stream X_1 Y_1 X_2 Y_2 X_3 Y_3 X_4 Y_4 X_5 Y_5 X_6 Y_6 …, whose rate is twice that of the input data stream. Since two output bits are generated for each input bit, this is an example of a 1/2 rate encoder.

Unlike block coders, convolutional coders do not usually have the input bit stream appearing as a subset of the output bits. Even when no errors are introduced into a bit stream, some logic is required in the receiver to recover the original bits. In the case of the simple example above, values of *D* can be recovered from correctly received *X* and *Y* using

$$D_i = X_{i+1} \oplus Y_{i+1}$$

In the more usual case where the decoder must be able to recover from errors, decoding techniques are based on reconstructing possible internal states of the encoder. The Viterbi algorithm provides an optimal method [8].

The major factors that control the performance of a convolutional code are the length of the shift register (usually referred to as the constraint length) and the coding rate. The example shown above can be easily extended to 1/3 rate codes by the addition of another set of taps and exclusive-or function. The use of puncturing, where some bits are deleted before multiplexing at the output of the convolutional encoder, allows lower coding overheads at the cost of some loss of error correction performance. For the example coder, a 2/3 rate code might be produced by deleting every second bit on the *X* path, leaving a stream: X_1 Y_1 Y_2 X_3 Y_3 Y_4 X_5 Y_5 Y_6 …

One of the important advantages of convolutional codes is that their performance is less dependent on the location of errors in the bit stream than for a block code. Because there are no block boundaries for convolutional codes, the effects illustrated in Figure 13.20 cannot occur.

The error correcting capability of a convolutional code is not as easily quantifiable as for a block code. Although it is given by

$$t = \left\lfloor \frac{d_f - 1}{2} \right\rfloor$$

where d_f is the minimum free distance of the convolutional code and t is the number of errors that can be corrected within an interval of a few constraint lengths, it is not possible to quantify "a few" without describing the distribution of errors. Generally, "a few" is between three and five.

13.2.5.3. Interleaving

Error correcting codes can be used to detect and correct random bit errors. The codes are effective so long as the number of errors close together remains small. In many types of channel, especially radio channels, however, the channel errors occur in bursts of many errors followed by long periods with almost no errors.

The problem of bursty channel errors can be overcome by interleaving the transmitted data. This is achieved by rearranging the coded data at the transmitter in a predefined pseudorandom order. This means that a burst of errors will be randomized at the receiver when the bits are placed back in their original order.

The simplest form of interleaver is a cyclic interleaver, which operates on fixed-length blocks of data. Each block of input data is read into the register. The output is formed by reading this data out with a different (known) start position. A cyclic interleaver with a start point of zero simply reproduces its input stream.

EXAMPLE 13.2—Cyclic Interleaver

What is the output of a 7-register cyclic interleaver with start position 3?

A 7-register cyclic interleaver with start position 3 and an input sequence 1, 2, 3, 4, . . . , 14 produces the output 4, 5, 6, 7, 1, 2, 3, 11, 12, 13, 14, 8, 9, 10. ■

A block interleaver consists of an $M \times N$ block of memory into which arriving bits are scanned in horizontally and scanned out vertically as illustrated in Figures 13.22 and 13.23.

EXAMPLE 13.3—Block Interleaver

What is the output of the 8 × 3 block interleaver shown in Figure 13.22, in which the data arrives at the interleaver in the order 1 2 3 4 5 6 7 ... 24?

The corresponding scan out is shown in Figure 13.23, showing that data leave the interleaver in the order 1 9 17 2 10 18 3 11 19 4 12 20 5 13 21 . . . 8 16 24.

1	2	3	4	5	6	7	8
9	10	11	12	13	14	15	16
17	18	19	20	21	22	23	24

Figure 13.22 Example 8 × 3 block interleaver — scan in.

1	2	3	4	5	6	7	8
9	10	11	12	13	14	15	16
17	18	19	20	21	22	23	24

Figure 13.23 Example 8 × 3 block interleaver — scan out. ■

The convolutional interleaver is more commonly used in digital television systems, and this is illustrated in Figure 13.24.

With $n = 1$ and a 24-element input block with input order 1 2 3 4 5 6 7 8 9 10 ..., the convolutional interleaver illustrated in Figure 13.24 will provide data with output order 1 X X X X X 7 2 X X X X 13 8 3 X X X 19 14 9 4 X X 25 20 15 10 5 X 31 26 21 16 11 6 37 32 27 22 17 12 43 38 33 28 23 18 49 44 39 34 29 24, where X denotes data loaded into the interleaver prior to '1.' The *period* of the convolutional interleaver is the amount of data that is loaded into the shift registers during one cycle of the input data distributor, and it is the product of the number of levels and the size of the registers, denoted by D in Figure 13.24. For the convolutional interleaver shown in Figure 13.24, the period is $6D$.

The performance of a convolutional interleaver is similar to a block interleaver. The primary advantages of the convolutional interleaver are that it causes half the delay and requires only half the memory of a corresponding block interleaver.

The primary purpose of an interleaver is to protect the decoder of a block code from burst errors introduced during transmission. The aim is, therefore, usually to

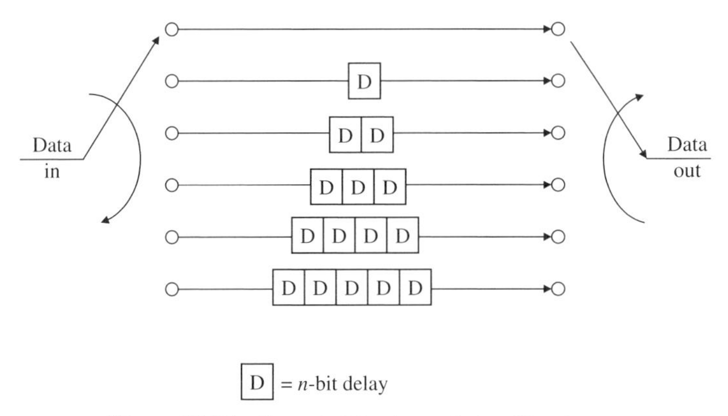

Figure 13.24 Example 6-level convolutional interleaver.

distribute each code word of the block code over several code word transmission times. For example, a 12-level convolutional interleaver with a register size of 17 bytes might be used to protect an (204,188) RS code. Although there is no necessary relationship between the period of the input to the interleaver and the length of the block code word, they often take the same value, which can be used to ensure that the first byte of a code word always passes through the zero-delay path of the interleaver. In this case, the period of the input to the interleaver is 204, which is the same as the code word length. Each 204-byte RS code word is distributed over 12 times its normal transmission interval.

13.2.5.4. Concatenated Codes

Concatenated codes use two levels of coding as illustrated in Figure 13.25. In the encoder, the input data passes through the outer encoder, after which its order is modified by an interleaver before passing through the inner coder. In the decoder, the order of these processes is reversed. The data received from the channel is passed through the inner decoder, which corrects most of the errors introduced during transmission. Where the inner coder fails to correct errors, it is likely to leave bursts of errors at its output. The deinterleaver rearranges the order of the data, spreading these bursts of errors before passing the data to the outer decoder.

Residual errors at the output of the inner decoder tend to be bursty. This would lead to poor performance by the outer decoder, so an interleaver is used to change the order of received data, spreading out bursts of errors to maximize the correcting capability of the outer decoder. A further interleaver may also be used between the inner encoder and the modulator to minimize the likelihood of bursts of errors at the input to the decoder for the inner code.

One of the most popular systems uses a convolutional inner code and an RS outer code. The RS coder is chosen because it can operate on symbols that consist of a number of bits. Like other FEC, it operates best on isolated symbol errors. Because the symbols may consist of a number of bits, the RS coder is quite effective in correcting bursts of bit errors. The interleaver in such systems is likely to be based on the same symbol size as the RS code; there is no value in reordering errors within one symbol of the RS code. Both DVB and ATSC employ such a concatenated coder.

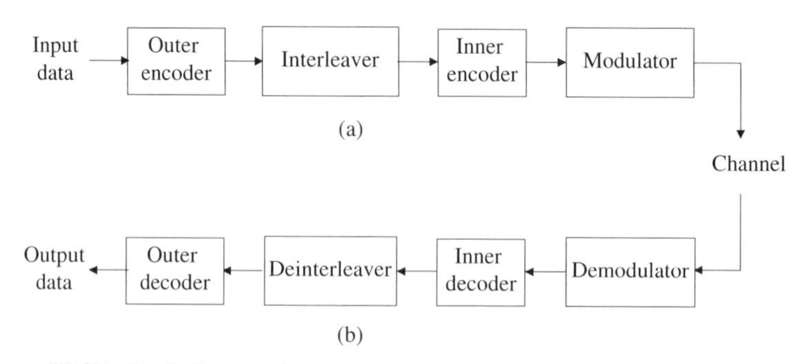

Figure 13.25 Block diagram of (a) a concatenated coder and (b) the corresponding decoder.

13.3. CHANNEL CODING AND MODULATION FOR ATSC

ATSC supports a payload data rate for terrestrial broadcast of 19.28 Mbit/s in one 6 MHz channel [9]. A high data-rate mode is also available that supports a data rate of 38.57 Mbit/s, but this is not intended for use in terrestrial broadcast and is, therefore, not considered further here.

The ATSC coder consists of the following blocks as illustrated in Figure 13.26:

- a data stream randomizer (Section 13.3.3.4);
- a concatenated channel coder (Section 13.3.3), based on:
  - * an RS outer coder,
  - * an outer convolutional interleaver, and
  - * a trellis inner coder (that incorporates an inner block interleaver);
- a framing module in which one segment carries one TS packet, 313 data segments form one data field, and two data fields form one data frame (Section 13.3.2); and
- a modulator, employing 8-VSB modulation (Section 13.3.1).

13.3.1. ATSC 8-VSB Modulation

All data transmitted by an ATSC transmitter is modulated using an 8-VSB modulator at a rate of 10.762 Msym/s, with the eight symbol values denoted by $\{-7,-5,-3,-1,1,3,5,7\}$. The output of the modulator is passed through a filter to reduce the bandwidth of the transmitted signal to 6 MHz.

The Nyquist criterion tells us that the maximum rate at which symbols can be transmitted over a bandwidth-limited channel and detected by a receiver is equal to twice the channel bandwidth (Section 13.2.1). For a 6 MHz digital television channel, the Nyquist criterion limits the symbol rate to 12 Msym/s, which is higher than the 10.76 Msym/s used by ATSC. As we saw in Section 13.2.1, satisfying the Nyquist criterion does not mean that intersymbol interference will not occur; generally a channel bandwidth at least equal to the symbol rate (that is, twice the Nyqist limit) is required to prevent intersymbol interference. A channel equalizer is, therefore, always required in an ATSC terrestrial receiver. This adaptive equalizer is typically based on the LMS algorithm and employs an FIR filter with approximately 64 taps to equalize received data, with a further 192-tap filter being used in the adaptation process that adjusts the tap values of the 64-tap filter.

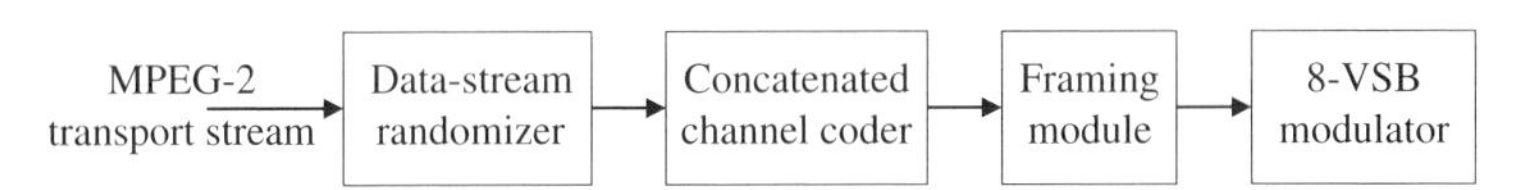

Figure 13.26 Outline structure of ATSC encoder.

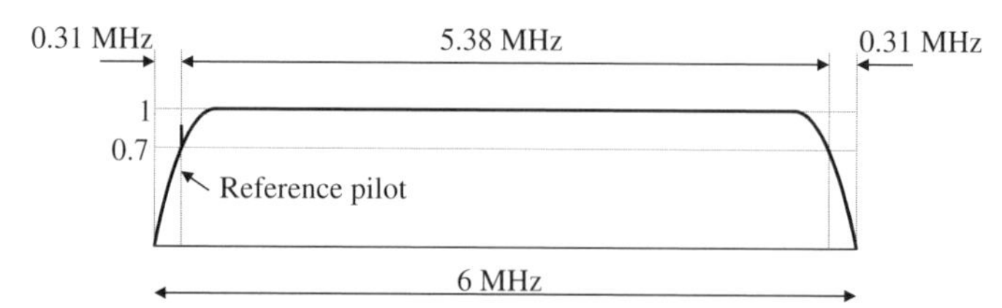

Figure 13.27 Nominal channel occupancy of 8-VSB ATSC transmitted signal. © Advanced Television Standards Committee Inc 2001. A copy of this standard is available at http://www.atsc.org.

Transmission of the 10.762 Msym/s, 8-VSB signal in a 6 MHz channel requires filtering at the output of the modulator. The nominal channel occupancy of the filtered signal is shown in Figure 13.27. The half-power bandwidth of the transmitted signal is 5.38 MHz, lying in the center of the 6 MHz channel. A reference pilot signal is transmitted 0.31 MHz from the bottom of the channel, which is the frequency of the suppressed carrier for the VSB signal. A linear-phase, raised-cosine filter is used to produce this band-limited signal.

The primary advantages of 8-VSB are its relatively high bandwidth efficiency and the simplicity created by using only the I component (that is, the value of the symbol does not depend on its Q component). The use of only the I component means that only one analog-to-digital converter is required, and equalization can be carried out using real multiplications and additions.

13.3.2. ATSC Data Framing

The 8-VSB data symbols are grouped into data segments, each consisting of 832 symbols, as illustrated in Figure 13.28. Each data segment is transmitted over 77.3 μs. The first four symbols of the segment are used for synchronization and always have

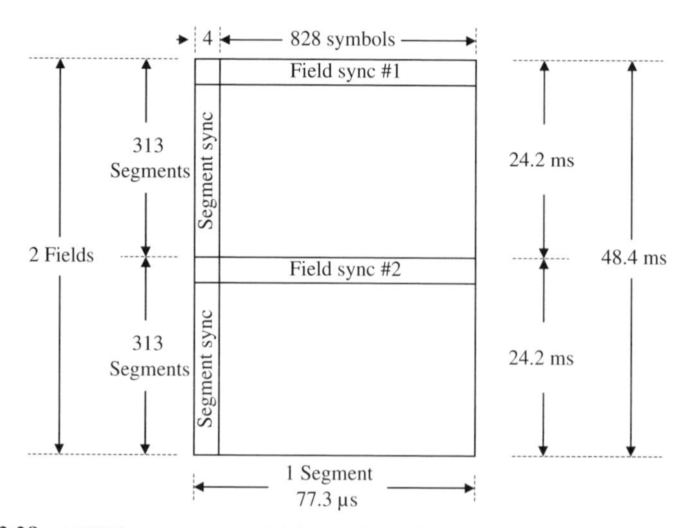

Figure 13.28 ATSC frame structure. © Advanced Television Standards Committee Inc 2001. A copy of this standard is available at http://www.atsc.org.

the values {5,−5,−5,5}. The remaining 828 symbols (carrying 2484 bits) form the payload of the segment and have exactly the capacity required to carry one MPEG-2 TS packet, complete with all overhead for error correction.

Data segments are grouped into data fields, each consisting of 313 data segments, transmitted over approximately 24 ms. The payload of the first segment of each field contains only a special synchronization sequence, that is, it carries no audio–visual data. This synchronization sequence is used by the decoder to identify the start of a data field and as a training sequence for the channel equalizer. This first segment carries the same 4-symbol synchronization header as all other segments.

A data frame is made up of two data fields and is, therefore, transmitted over approximately 48 ms. The first and second data fields of a data frame are distinguished by the different synchronization sequences carried in their first segments.

13.3.3. ATSC Concatenated Channel Coder

ATSC employs a concatenated coder. The outer coder is an RS coder, which is followed by a convolutional interleaver. The inner coder is a trellis coder that incorporates a stage of inner interleaving.

13.3.3.1. ATSC RS Coder

An (207,187,8) RS code is used to protect each MPEG-2 transport stream packet, allowing up to 10 symbol errors to be corrected in each code word. On entering the RS coder, the length of each transport stream packet is 187 bytes because the sync byte was removed during randomization. The purpose of the outer code is to allow removal of residual errors remaining after decoding of the inner code.

13.3.3.2. ATSC Interleaver

ATSC employs a byte-oriented, 52-level convolutional outer interleaver as shown in Figure 13.29.

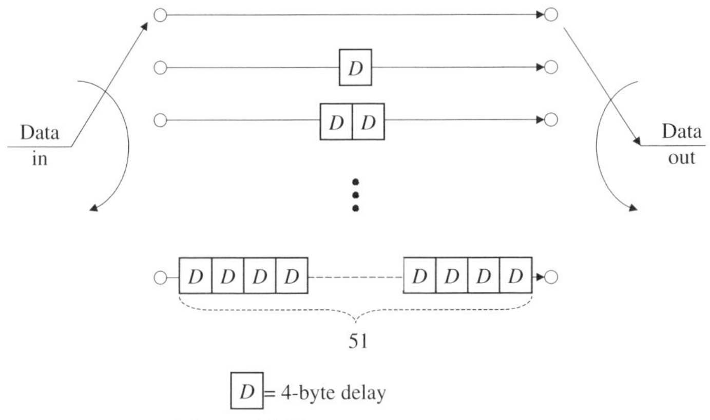

Figure 13.29 ATSC outer interleaver.

This interleaver has a period of 208 bytes, which corresponds to the combined size of the MPEG-2 transport stream sync byte and the outer RS code word. Each input code word is distributed over 52 TS-packet periods, corresponding to approximately 4.0 ms with a data rate of 19.28 Mbit/s or 1/6 of the data field size. The MPEG-2 sync byte is replaced by the 4-symbol data-segment synchronization signal, and always passes through the zero-delay path at the encoder.

In each 207-byte code word, the RS outer coder can correct 10 symbol errors (that is, bytes that contain one or more errors) or 20 erasures. The outer interleaver distributes the 208-byte protected MPEG-2 transport stream packets across 52 TS-packet transmission times in 4-byte groups. The RS coder can, therefore, always correct errors occurring in two of these 4-byte groups within one RS code word, and sometimes correct errors occurring in three or more.

13.3.3.3. ATSC Inner Coder and Interleaver

The ATSC inner coder is a 2/3-rate trellis coder whose input symbols are formed from pairs of output bits from the outer interleaver and whose output consists of 3-bit symbols for the VSB modulator.

At the input to the inner coder, data is formed in bytes. These bytes are then demultiplexed across 12 identical trellis coders (Figure 13.30), whose output is re-multiplexed to form the input symbol stream for the 8-VSB modulator. This demultiplexing and remultiplexing is a form of interleaving, whose effect is to spread burst errors on the channel across a number of independent trellis decoders.

Each of the twelve trellis coders has the structure shown in Figure 13.31. Each input byte is broken into four 2-bit symbols, with bits 7, 5, 3, and 1 or the bytes being the more significant bit of their symbol and bits 6, 4, 2, and 0 the less significant bit. The more significant bit (X_2) of each symbol passes through an interference filter precoder. The trellis coder passes the less significant bit (X_1) of each symbol through a 1/2 rate convolutional encoder but leaves the more significant bit uncoded. The outputs of the precoder (Z_2) and convolutional encoder (Z_1 and Z_0) then pass to the symbol mapper that converts the 3-bit symbol inputs into one of the eight levels at its output. This mapping, in which the distance Z_2 is more significant than Z_1 and Z_0, explains why it is reasonable that redundancy is added to X_1 but not to X_2.

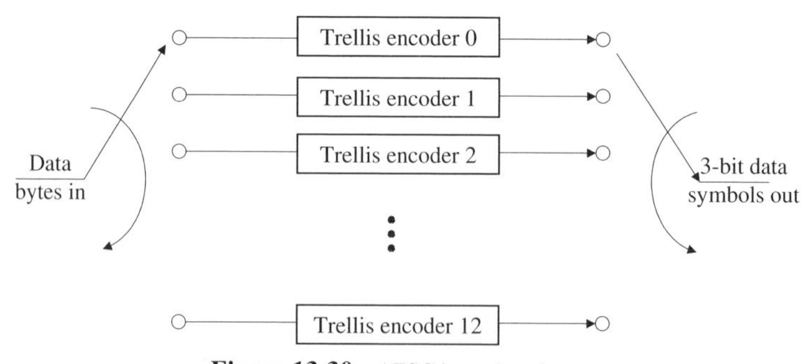

Figure 13.30 ATSC inner interleaver.

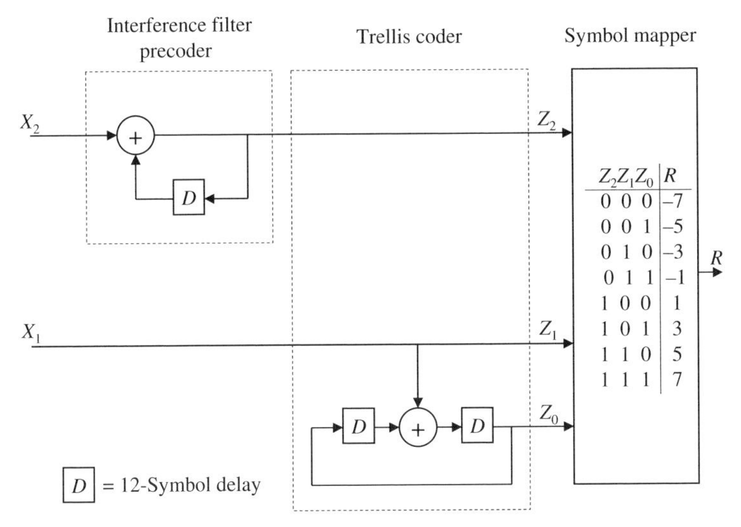

Figure 13.31 VSB trellis coder with interference filter precoder and symbol mapper. © Advanced Television Standards Committee Inc 2001. A copy of this standard is available at http://www.atsc.org.

13.3.3.4. ATSC Randomization

Data stream randomization in ATSC is applied only to the 187-byte payloads of MPEG-2 TS packets. The sync byte of MPEG-2 TS packets is discarded and replaced by the data segment sync signal. The randomization is carried out by taking the exclusive-or of the payload with the output of a pseudorandom binary sequence generator, which is initialized at the beginning of every data field. This initialization, combined with the unique data field sync signal that occupies the first data segment of the data field, makes random access possible at the beginning of any data field, in other words, random access is possible once in every 24 ms.

The pseudorandom sequence generator is a 16-bit linear shift register with eight feedback taps and eight outputs as shown in Figure 13.32. Each element in the shift register is labeled with its index. At each clock cycle, one output byte $[D_7 \ldots D_0]$ is produced. This output byte is bit-wise exclusive-ored with one byte of the MPEG-2 TS packet. The shift register is initialized to the value 0xF180 at the beginning of every data field (that is, registers 16, 15, 14, 13, 9, and 8 are loaded with 1, the other registers with 0).

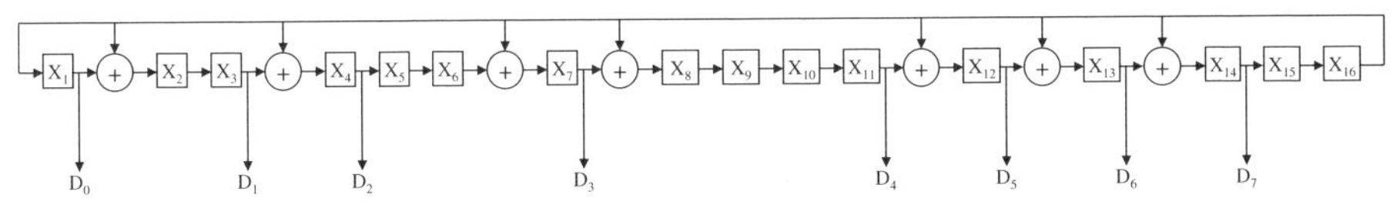

Figure 13.32 Detailed structure of ATSC PN generator. © Advanced Television Standards Committee Inc 2001. A copy of this standard is available at http://www.atsc.org.

13.3.4. ATSC Channel Capacity

The capacity of the ATSC channel can be derived as follows. The raw symbol rate on the channel is 10.762 Msym/s. Each symbol carries 3 bits of data. The trellis coder is a rate 2/3 coder, so that two out of every three bits transmitted are information (the other is overhead). This rate is reduced by the overheads for transmission of field sync segments (312/313) and the outer RS code (188/208). The useful transmission capacity is, therefore,

$$U = 10.762 \times 3 \times \frac{2}{3} \times \frac{312}{313} \times \frac{188}{208} = 19.392\,\text{Mbit/s}$$

This value for U includes the MPEG-2 transport stream sync bytes, which are mapped to the segment sync headers before transmission. This method of calculation differs from that used in the ATSC standard, but it provides a value that is directly useable in assessing the channel capacity in terms relevant to the MPEG-2 transport stream. ATSC documents often quote a useful data rate of 19.29 Mbit/s, which excludes the MPEG-2 transport stream sync bytes.

13.4. CHANNEL CODING AND MODULATION FOR DVB

The DVB channel coder takes an MPEG-2 transport stream as input and passes it through the following processes prior to modulation [10] as illustrated in Figure 13.33:

- transport adaptation and randomization,
- outer coding using Reed–Solomon code,
- outer interleaving,
- inner coding using a punctured convolutional code, and
- inner interleaving.

The output of the channel coder is passed to the modulator for modulation using QPSK, 16-QAM, or 64-QAM and orthogonal frequency-division multiplexing. Each of these processes is described in the following sections.

13.4.1. DVB Modulation

Having completed the processes of outer coding and interleaving and inner coding and interleaving, a DVB encoder proceeds to modulation. Data is modulated using QPSK, 16-QAM, or 64-QAM. The first step of the modulation process is to group

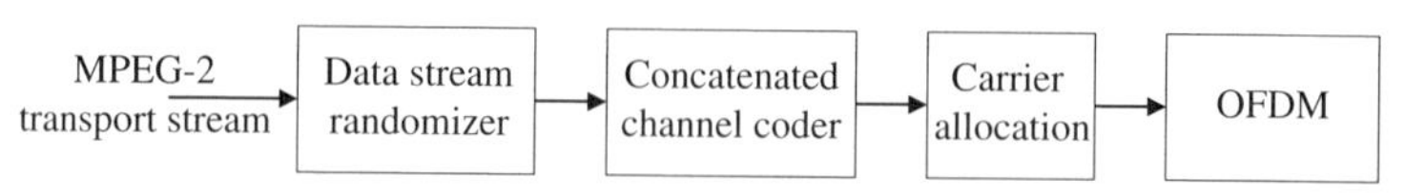

Figure 13.33 Outline structure of DVB channel coding and modulation.

the data bits to form complex-valued modulation symbols, denoted by z, based on the constellation diagrams in Section 13.2.1. For QPSK the constellation diagram in Figure 13.5 is used. Each QPSK symbol is made of 2 bits. The symbol value obtained from Figure 13.5 is then normalized so that the average symbol energy is 1. For QPSK, this requires dividing the symbol value by $\sqrt{2}$, giving a normalized symbol value c. For example, the symbol value z for the bits 10 is $z = -1 + j$, where j denotes the square root of -1, and $c = -0.7071 + 0.7071j$.

Figure 13.6 shows the corresponding constellation diagram for 16-QAM, for which each symbol consists of 4 bits. Normalization for 16-QAM requires dividing each symbol value by $\sqrt{10}$. Figure 13.7 shows the constellation diagram for the 6-bit symbols used for 64-QAM, for which normalization requires division of the symbol value by $\sqrt{42}$.

The normalized complex symbols are then converted to a sequence of sample values using orthogonal frequency-division multiplexing (OFDM), during which pilot signals and signaling information are incorporated into the signal.

13.4.1.1. DVB Orthogonal Frequency-Division Multiplexing

The basic idea of orthogonal frequency-division multiplexing (OFDM) is to split the data to be carried on a channel among several subchannels, and to frequency-division multiplex these subchannels within the bandwidth of the original channel. A very simple implementation, using four subchannels, is illustrated in Figure 13.34.

Figure 13.34(a) shows a block diagram of the four subchannel OFDM modulator. Figure 13.34(b) shows the approximate power spectrum for a conventional signal, whereas Figure 13.34(c) shows the power spectrum of the OFDM signal. The OFDM signal provides a more even distribution of power across the bandwidth of the channel and has sharper drop off in power at the edge of the channel, which simplifies the design of transmitter output filters.

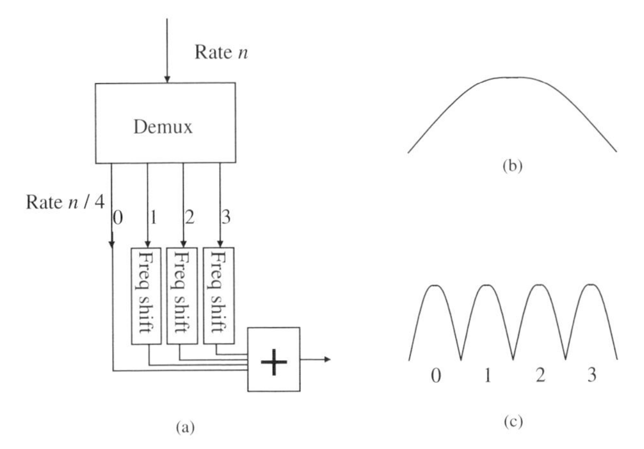

Figure 13.34 Simple OFDM with four carriers.

The major requirements for an effective implementation of OFDM are that

- the subchannels must be synchronized so that remultiplexing is possible at the decoder to reconstruct exactly the original bit stream, and
- the modulation waveforms for the individual subchannels should be orthogonal to provide high spectral efficiency.

Under these circumstances, the principal advantages of OFDM are

- *spectral efficiency*—signal is more evenly spread over channel than for a single data stream;
- *immunity to fading and interference*—fading or interference associated with one modulated carrier may not affect others; when data is protected by FEC, this provides a powerful means for overcoming narrowband fading and interference; and
- *interference with other signals*—minimization of signal level outside channel is easily achieved without complicated filtering.

The DVB OFDM system has two modes of operation: the 2k mode and the 8k mode. In the 2k mode, there are 2048 notional carriers spread across the whole channel bandwidth, with constant intercarrier spacing. Only a subset of the carriers (1705) at the center of the channel is actually used for transmission. The remaining notional carriers are not generated and, therefore, help with suppression of out-of-channel emissions. The signals transmitted on the 2048 carriers are each orthogonal to every other signal (i.e., their cross correlation is zero), enabling high spectral efficiency.

For the 8k mode, there are 8192 notional carriers spread evenly across the channel bandwidth. Of these carriers, the 6817 lying in the middle of the channel bandwidth are used for transmission. Once again the signals transmitted on the carriers are orthogonal, enabling high-spectral efficiency.

Each carrier may be modulated by QPSK, 16-QAM, or 64-QAM. The group of symbols transmitted simultaneously on all carriers is called an OFDM symbol. OFDM symbols are grouped together to form OFDM frames, each of which contains 68 OFDM symbols. Frames are grouped to form super frames, each of which contains four frames.

More formally, the output s of the OFDM modulator is given by

$$s(t') = \mathrm{Re}\left\{ e^{j2\pi f_c t} \sum_{k=k_{\min}}^{k=k_{\max}} c_{m,l-1,k} e^{j2\pi k'(t'-\Delta)/T_U} \right\}$$

where t' is the time relative to the beginning of the symbol transmission period; T is the elementary period, which is the reciprocal of the clock period; k denotes the carrier number; l denotes the OFDM symbol number; m denotes the transmission frame number; K is the number of transmitted carriers (1705 for the 2k mode and 6817 for the 8k mode); T_U is the inverse of the spacing between OFDM carriers; ($T_U = 2048 \times T$ for the 2k mode and $T_U = 8192 \times T$ for the 8k mode); Δ is the duration of the guard interval; $T_S = \Delta + T_U$ is the symbol duration; f_c is the center frequency of the RF signal; $K_{\min}$ is the index of the lowest frequency carrier (0 for both 2k and 8k modes); $K_{\max}$ is the index of the highest frequency carrier (1704 for 2k mode, 6816 for 8k mode); $k' = k - (K_{\max} - K_{\min})/2$ is carrier index relative to the channel center frequency f_c; $c_{m,l-1,k}$ is the normalized complex symbol for carrier k of the data symbol number l in frame m.

For a channel bandwidth of 7 MHz, the clock frequency is 8 MHz and the elementary period $T = 1/8$ µs. DVB also supports operation using 6 MHz and 8 MHz channel bandwidths. The only changes required to encoder and decoder operation are in the clock frequency (and therefore the elementary period T). For a 6 MHz channel, the clock frequency is 48/7 MHz and the elementary period $T = 7/48$ µs. For an 8 MHz channel, the clock frequency is 64/7 MHz and the elementary period $T = 7/64$ µs. A 6 MHz channel, therefore, operates at 6/7 the symbol rate of a 7 MHz channel and has 6/7 the useful bit rate. An 8 MHz channel operates at 8/7 the symbol rate of a 7 MHz channel and has 8/7 the useful bit rate.

In converting from the OFDM symbol to a time sequence s to be transmitted, the sampling rate must be at least twice the bandwidth of the symbol to ensure that aliasing does not occur; in other words, the sampling rate for s must be greater than or equal to $2/T$. One common implementation of this conversion is by means of an inverse Fourier transform. The choice of the number of notional carriers as a power of two in both the 2k and 8k modes facilitates this transform, making possible the use of the inverse fast Fourier transform (IFFT). The principle advantage of the IFFT over other implementations of the transform is that its computational complexity is of the order of $(n \log n)$, rather than of the order of (n^2), where n is the number of carriers.

In the 2k mode, DVB uses 1705 carriers in each OFDM symbol, of which 193 are used for pilots and signaling (described in Sections 13.4.1.4 and 13.4.1.5 below), leaving 1512 useful carriers. In the 8k mode, there are 6817 carriers in each OFDM symbol, with 769 used for pilots and signaling, leaving 6048 useful carriers.

13.4.1.2. DVB Guard Interval

The choice of the intercarrier spacing (which is effectively the channel bandwidth for the stream of signals transmitted on one carrier) to be the reciprocal of the OFDM symbol duration T_U means that essentially all the intersymbol interference in a DVB system is caused by the channel. The purpose of the guard interval is to reduce the impact of intersymbol interference by increasing the spacing between symbols and allowing time for the impulse response of the channel to decay between the point where the encoder changes symbol value and the point where the decoder uses the received signal in demodulation. The DVB guard interval is created by making a copy of that portion of symbol to be transmitted last and transmitting this immediately in front of the symbol as shown in Figure 13.35. The size of the DVB guard interval determines what fraction of the symbol is copied in this way. In DVB, the guard interval may be chosen to be 1/4, 1/8, 1/16, or 1/32 of the OFDM symbol period, and it would usually be chosen to ensure that the impulse response of the channel has decayed before the beginning of the symbol to be decoded. This is especially

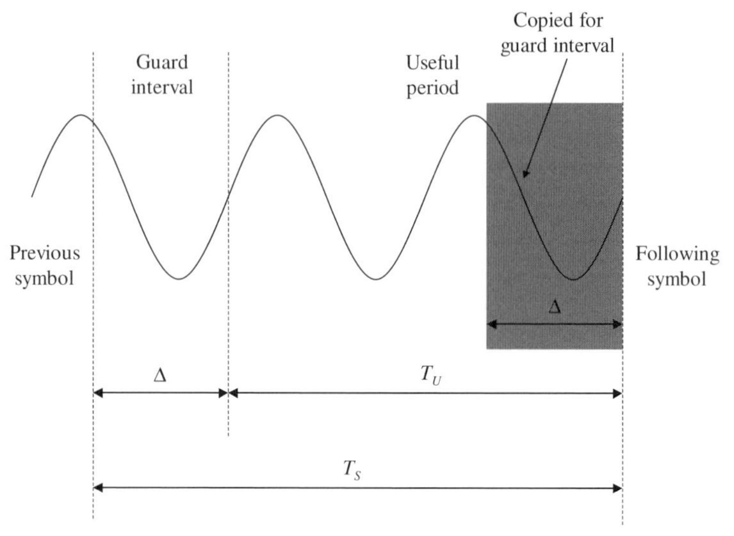

Figure 13.35 Construction of DVB guard interval.

important for channels in which the signal travels between transmitter and receiver over multiple, simultaneous paths (i.e., multipath channels). Without this guard interval, multipath propagation leads to significant intersymbol interference. Because the OFDM symbol period is longer for the 8k mode than that for the 2k mode, the 8k mode is more suitable for channels with longer impulse responses than that for the 2k mode. One of the major advantages of this use of the guard interval is that the requirement for channel equalization is reduced to one complex multiplication for each OFDM carrier if the decoder in signal processing is based on a Fourier transform (i.e., reversing the inverse Fourier transform at the encoder).

The choice of guard interval length directly affects the useful channel capacity available. Table 13.2 shows how the number of OFDM symbols per second varies

Table 13.2 Impact of guard interval on number of OFDM symbols per second.

Mode	Δ/T_U	ODFM symbols per second	Loss of capacity due to guard interval (%)
2k	1/4	3125	20
	1/8	3472	11
	1/16	3677	6
	1/32	3788	3
	0[a]	3906	0
8k	1/4	781	20
	1/8	868	11
	1/16	919	6
	1/32	947	3
	0[a]	977	0

[a]Not supported by DVB. The number of OFDM symbols per second is $1/T_U$ in this case.

with the mode (2k or 8k) and size of guard interval (Δ/T_U) for a 7 MHz channel. Changing between modes does not affect the relative loss of capacity for fixed Δ/T_U. It is likely, however, that a smaller value of Δ/T_U would be used for a given channel in the 8k mode than in the 2k mode, so that the length of the guard interval Δ is the same. This means that for a fixed transmission channel, less loss of capacity is likely to be required for guard intervals in the 8k mode than in the 2k mode. Alternatively, the 8k mode allows operation with channels whose impulse response takes longer to decay than is possible in the 2k mode.

13.4.1.3. Single-Frequency Network

In conventional television broadcast, black spots of extremely poor reception occur due to terrain. Additional, local repeaters are sometimes used to provide an acceptable signal quality in such areas. Because of their potential to cause interference with other signals, it is usual that such repeaters are allocated a separate channel from the main transmitter. Sometimes, it is even necessary to allocate a separate channel to each repeater in an area. This is a very inefficient use of the limited bandwidth available for terrestrial-broadcast television. The use of OFDM in combination with channel coding (known as *coded OFDM*) in DVB offers the possibility of operating a large number of repeaters on one channel. Interference will occur on a small proportion of the carriers, causing fading and very high bit error rates. The combination of inner and outer coding, spread across a large number of carriers by interleaving, overcomes this interference, leading to a high-quality bit stream at the input to the MPEG-2 systems decoder. This is known as a single-frequency network.

A single-frequency network is illustrated in Figure 13.36. Two transmitters ("A" and "B") are used. Clearly, the use of two transmitters in close proximity, operating on the same frequency, can lead to interference between their signals. The difference between the propagation delays from the two transmitters to the receiver is the major potential cause of degradation in the received signal.

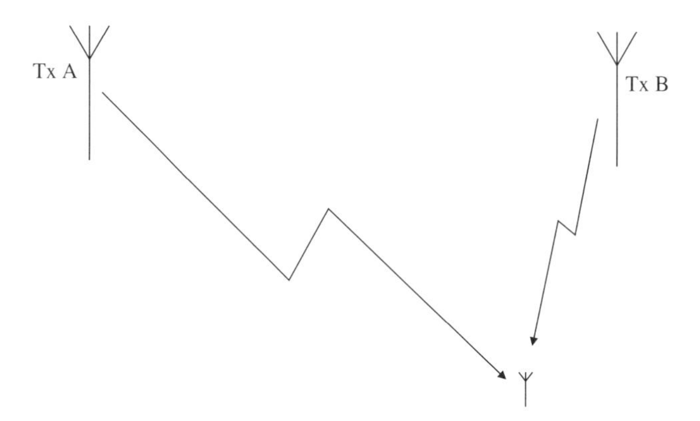

Figure 13.36 A simple single-frequency network.

The first impact of the difference in propagation delay is that it can cause intersymbol interference. So long as the difference in propagation delays from the two transmitters is less than the DVB guard interval, the contribution of the single-frequency network to intersymbol interference should be negligible.

The second effect of receiving two signals from different sources is that the received power depends on the relative phase of the two signals, which changes with frequency. If the phase difference between the two signals is one half cycle, destructive interference causes the received power to be zero. In analyzing the impact of this interference, we assume the worst case, that is, that the signal strengths from A and B are the same. In this case, nulls occur at frequencies where the difference in path lengths between the transmitters and the receiver takes particular values satisfying

$$d_{null} = \frac{\lambda(2n+1)}{2}$$

for integer values of n. The frequency spacing between these nulls is equal to the reciprocal of the difference in propagation delay from A to the receiver and B to the receiver.

If the signals received from transmitters A and B have the same electric field strength, we can write

$$E_A = \sin(2\pi ft + \phi/2)$$

and

$$E_B = \sin(2\pi ft - \phi/2)$$

where ϕ is the phase shift caused by the difference in propagation delays between A and the receiver and B and the receiver. The total electric field strength of the received signal is therefore

$$E_{A+B} = 2\cos(\phi/2)\sin(2\pi ft)$$

In other words, the amplitude of the received signal is $2\cos(\phi/2)$. The implications of this calculation are that

- the peak received power is four times higher than that for one transmitter,
- the received power is higher than that would be received from a single transmitter for 2/3 of the OFDM carriers, and
- the received power is higher than half that of what would be received from a single transmitter for 77 % of the OFDM carriers.

In a DVB system, the OFDM carriers for which the received power is small tend to be clustered in groups. It is for this reason that DVB does not assign symbols to OFDM carriers simply in order of increasing frequency, but assigns symbols to carriers in a pseudorandom order.

Table 13.3 Relationship between received power and signal cancellation in a single-frequency network.

Received power (dB)					
A	**B**	Peak power (dB)	Carriers with power ≥ 0 dB	Carriers with ≥ −3 dB	Carriers with ≥ −6 dB
0	0	6.0	66 %	77 %	84 %
0	−3	4.6	61 %	75 %	85 %
0	−6	3.5	58 %	77 %	100 %
0	−12	1.9	53 %	100 %	100 %

EXAMPLE 13.4—Single-Frequency Network

Calculate the difference in propagation delay experienced by two signals in a single-frequency network where one transmitter is 3 km closer to the receiver than the other. Calculate also the spacing between frequencies for which the received signal strength is zero if the received power from each of the two transmitters is the same.

A difference in path length of 3 km is equivalent to a 10 μs difference in propagation delay. The frequency spacing between nulls in the received power is 100 kHz. ■

In the more general case where the strengths of the signals received from the two transmitters differ, the peak power is less than that indicated by the above analysis; but the received power never drops to zero. The proportion of the OFDM carriers for which the power is less than half the stronger of the received powers is also reduced. This is illustrated in Table 13.3 and Figure 13.37.

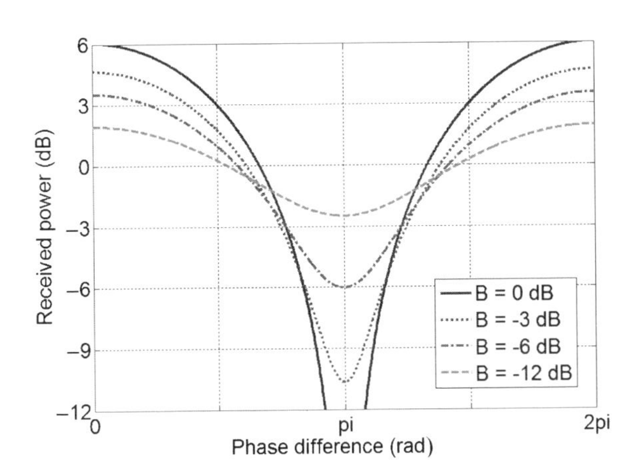

Figure 13.37 Total received power versus phase difference for 0 dB received from A and various powers from B.

Table 13.4 Carrier indices for TPS.

2k Mode	8k Mode
34 50 209 346 413 569 595 688 790 901 1073 1219 1262 1286 1469 1594 1687	34 50 209 346 413 569 595 688 790 901 1073 1219 1262 1286 1469 1594 1687 1738 1754 1913 2050 2117 2273 2299 2392 2494 2605 2777 2923 2966 2990 3173 3298 3391 3442 3458 3617 3754 3821 3977 4003 4096 4198 4309 4481 4627 4670 4694 4877 5002 5095 5146 5162 5321 5458 5525 5681 5707 5800 5902 6013 6185 6331 6374 6398 6581 6706 6799

13.4.1.4. DVB Transmission Parameter Signaling

A DVB decoder is required to be able to identify the parameters of a received signal and begin decoding it without user intervention. This means that the decoder must identify features, such as the type of modulation, the inner coding rate, and guard interval, which cannot be easily inferred from the encoded data itself. *Transmission parameter signaling* (*TPS*) is used to provide this type of information to the decoder in an easily accessible way.

TPS data is carried on a number of dedicated OFDM carriers, 17 for the 2k mode and 68 for the 8k mode, each of which carries an identical data stream. The indices of these carriers are shown in Table 13.4. These carriers are never used for any other purpose. All TPS data is modulated using differential BPSK (DBPSK), so that a decoder does not need to identify the modulation type before decoding this data.

Each OFDM symbol carries one TPS bit. Sixty eight consecutive OFDM symbols form an OFDM frame. The 68 bits of TPS data is carried by each OFDM frame form one TPS block. Each TPS block contains one initialization bit, followed by 16 synchronization bits, 37 information bits, and 14 redundancy bits for error protection. The contents of each field are described in Table 13.5.

13.4.1.5. DVB Reference Signals

Of the 1705 transmitted carriers in the 2k mode, 1512 are available for carrying data and the remaining 193 are used to carry reference signals. (Similarly, in the 8k mode, there are 6048 useful carriers and 769 are used for reference signals.) In addition to TPS, DVB uses two other types of reference signals: continual pilots and scattered pilots. Both continual and scattered pilots transmit a pseudorandom sequence, which is modulated using BPSK.

Scattered pilots are always transmitted on the lowest and highest frequency carriers. They are also transmitted on every twelfth carrier, starting at $K_{\min}$ for symbols 0, 4, 8, ..., 64 in a frame; $K_{\min}+3$ for symbols 1, 5, 9, ..., 65; $K_{\min}+6$ for symbols 2, 6, 10, ..., 66; and $K_{\min}+9$ for symbols 3, 7, 11, ..., 67. Continual pilots are always transmitted on the same carriers, which are listed in Table 13.6.

Table 13.5 Usage of fields in each TPS block.

Field purpose	Length	Description
Initialization bit	1	Initialization bit for DBPSK.
Synchronization bits	16	Alternating 0011 0101 1110 1110 and1100 1010 0001 0001.
Length indicator	6	010 111 if cell identifier not supported, 011 111 otherwise.
Frame number	2	Frame number in superframe.
Constellation	2	Identifies QPSK, 16-QAM, or 64-QAM as the modulation used for data.
Hierarchy information	3	Selects hierarchical/nonhierarchical mode.
Code rate, HP stream	3	Selects inner code rate from 1/2, 2/3, 3/4, 5/6, and 7/8 for nonhierarchical or high-priority stream of hierarchical mode.
Code rate, LP stream	3	Selects inner code rate from 1/2, 2/3, 3/4, 5/6, and 7/8 low-priority stream of hierarchical mode.
Guard interval	2	Identifies guard interval from 1/32, 1/16, 1/8, and 1/4.
Transmission mode	2	Selects between 2k mode and 8k mode. This field must be guessed first by the receiver, but transmission in TPS can be used to confirm this guess.
Cell identifier	8	Used to identify the cell from which the signal comes.
Reserved for future use	6	Set all to zero.
Error Protection	14	BCH code protecting the TPS block.

Table 13.6 Carrier indices for continual pilots.

2k Mode	8k Mode
0 48 54 87 141 156 192 201 255 279 282 333 432 450 483 525 531 618 636 714 759 765 780 804 873 888 918 939 942 969 984 1050 1101 1107 1110 1137 1140 1146 1206 1269 1323 1377 1491 1683 1704	0 48 54 87 141 156 192 201 255 279 282 333 432 450 483 525 531 618 636 714 759 765 780 804 873 888 918 939 942 969 984 1050 1101 1107 1110 1137 1140 1146 1206 1269 1323 1377 1491 1683 1704 1752 1758 1791 1845 1860 1896 1905 1959 1983 1986 2037 2136 2154 2187 2229 2235 2322 2340 2418 2463 2469 2484 2508 2577 2592 2622 2643 2646 2673 2688 2754 2805 2811 2814 2841 2844 2850 2910 2973 3027 3081 3195 3387 3408 3456 3462 3495 3549 3564 3600 3609 3663 3687 3690 3741 3840 3858 3891 3933 3939 4026 4044 4122 4167 4173 4188 4212 4281 4296 4326 4347 4350 4377 4392 4458 4509 4515 4518 4545 4548 4554 4614 4677 4731 4785 4899 5091 5112 5160 5166 5199 5253 5268 5304 5313 5367 5391 5394 5445 5544 5562 5595 5637 5643 5730 5748 5826 5871 5877 5892 5916 5985 6000 6030 6051 6054 6081 6096 6162 6213 6219 6222 6249 6252 6258 6318 6381 6435 6489 6603 6795 6816

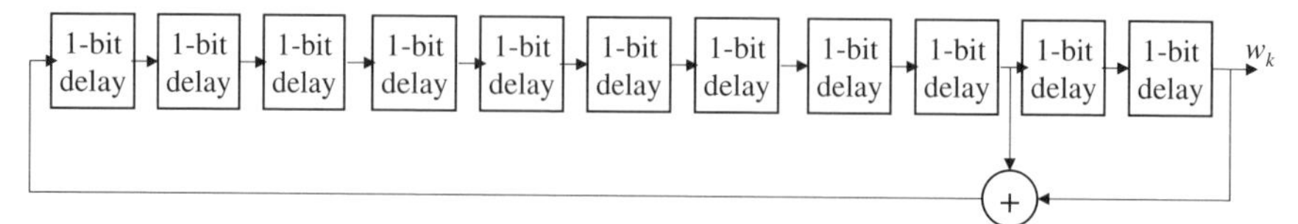

Figure 13.38 Pseudo random sequence generator for pilot signals.

The outline structure of the pseudorandom sequence carried by both continual and scattered pilots is shown in Figure 13.38. The generator is reset at the beginning of every OFDM symbol by setting all elements of the shift register to 1. One bit w_k is generated for each carrier of the OFDM symbol, regardless of whether this bit carries data or a pilot signal; that is, if carrier k carries a pilot signal, the value of the pilot is w_k.

13.4.1.6. DVB Spectrum Characteristics

DVB supports channel bandwidths 6, 7, and 8 MHz. One important advantage of the use of OFDM is that changing from one channel bandwidth to another requires only scaling the elementary period T in proportion to the new channel bandwidth required.

Figure 13.39 shows the power spectrum of a DVB transmission signal for an 8 MHz channel in the 2k mode with a guard interval $\Delta = T_U/4$. Frequency is shown relative to the center frequency f_c of the channel. The power spectral density is simply the sum of the power spectral densities of the individual modulated carriers. In this mode of operation, the power spectral density of the transmitted signal is attenuated by more than 30 dB outside the allocated channel. The frequency of the highest frequency carrier is approximately 3.83 MHz above the channel center frequency, and the frequency of the lowest frequency carrier is approximately 3.83 MHz below the center frequency.

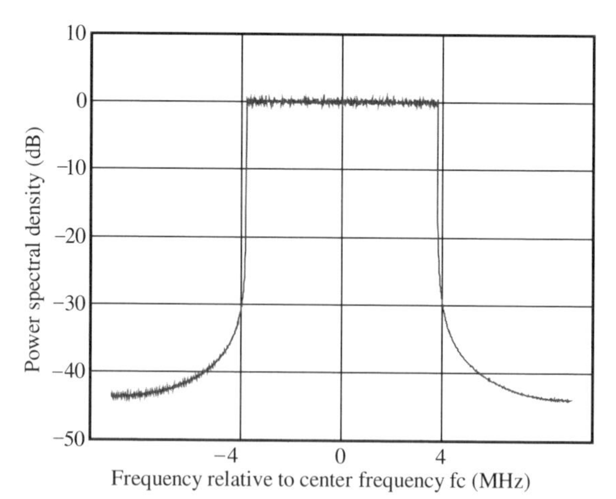

Figure 13.39 DVB transmission signal spectrum for 8MHz channel operating in 2k mode with guard interval $\Delta = T_U/4$.

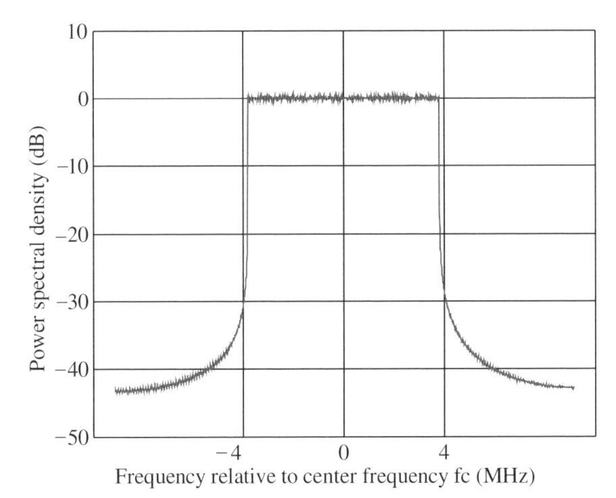

Figure 13.40 DVB transmission signal spectrum for 8 MHz channel operating in 2k mode with guard interval $\Delta = T_U/32$.

Alteration of the guard interval to $\Delta = T_U/32$ produces the power spectral density shown in Figure 13.40. There are no significant differences between this spectrum and that shown previously for $\Delta = T_U/4$. This is because the change in guard interval makes no difference to the center frequencies of OFDM carriers. A change in the line width around individual carriers does occur, because a change in guard interval changes the number of OFDM symbols per second (Table 13.2). The line width around individual carriers, therefore, may vary by up to 20%, but the overall impact on the bandwidth of the channel is negligible.

Changing from the 2k mode to the 8k mode increases the attenuation at the edge of the channel's allocated bandwidth by approximately 6–36 dB as shown in Figure 13.41.

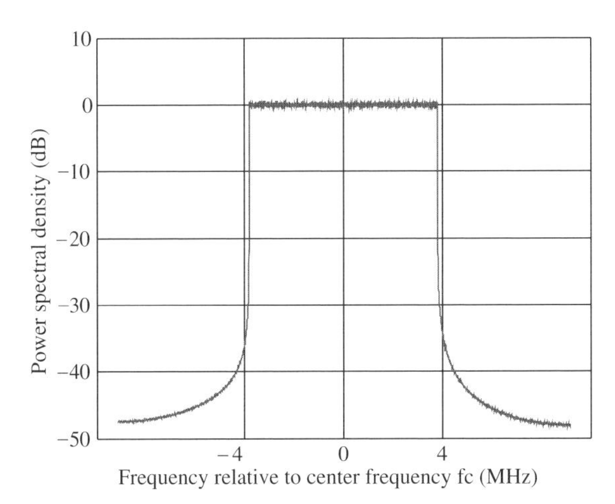

Figure 13.41 DVB transmission signal spectrum for 8 MHz channel operating in 8k mode with guard interval $\Delta = T_U/4$.

This occurs because the OFDM symbol rate in the 8k mode is one quarter of that in the 2k mode, causing the bandwidth of individual modulated carriers in the 8k mode to be one quarter of that in the 2k mode leading to a reduction in their contribution to out-of-band power.

Other properties of the DVB transmitted signal, such as modulation type and inner code rate, do not significantly affect the transmitted power spectral density.

13.4.2. DVB Channel Coding

The DVB channel coder employs data randomization, an RS outer coder, an outer convolutional interleaver, an inner convolutional coder, and an inner block interleaver.

13.4.2.1. DVB Randomization

Transport-stream randomization in DVB is performed using a pseudorandom sequence generated by a 15-stage shift register shown in Figure 13.42. The pseudorandom sequence generator is initialized with the value 100101010000000 and used to generate 1503 bytes of data. These bytes are bit-wise exclusive ored with eight transport-stream packets, starting immediately after the sync byte of the first transport-stream packet. Sync bytes of transport-stream packets are not randomized, but 1 byte of the sequence is skipped for the sync byte of all subsequent transport-stream packets. The first transport-stream packet in the group of eight is identified in the transmitted bit stream by having its sync byte inverted (that is, 0x47 is transmitted as 0xB8).

13.4.2.2. DVB Outer Coding

DVB's outer coder uses an RS (204, 188, 8) code, with the information part of each code word representing one MPEG-2 transport stream packet. This code can correct up to eight symbol errors in each received code word. The construction of the code word is illustrated in Figure 13.43.

Each byte of the data to be protected is treated as one symbol. The first step is to prefix the 188-symbol transport stream packet with 51 symbols containing all zeros, giving a 239-symbol block. The parity symbols for this block are then calculated as

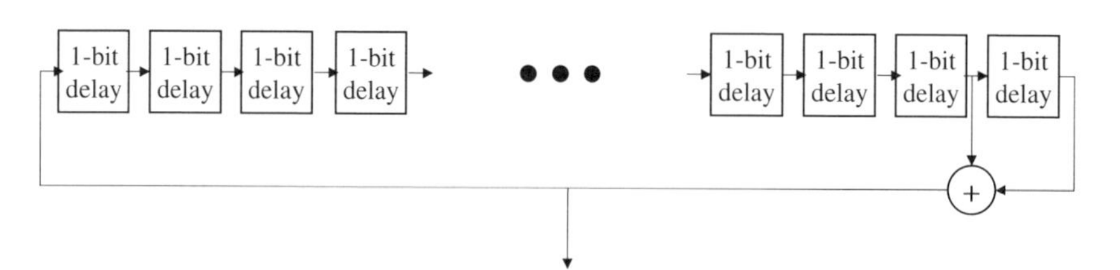

Figure 13.42 Structure of the pseudorandom sequence generator for DVB transport randomization.

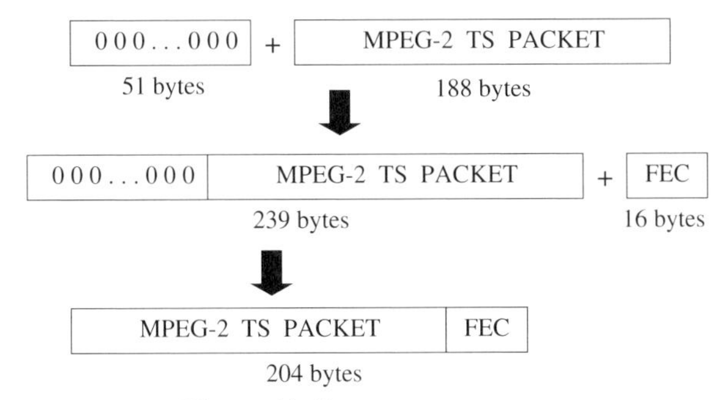

Figure 13.43 DVB outer coding.

for an RS (255,239) code, generating a 16-symbol parity block. The 51 zero symbols are then discarded, leaving a 204-symbol code word, made up of the 188-byte transport stream packet and 16 parity bytes.

13.4.2.3. DVB Outer Interleaving

The DVB outer interleaver is a 12-level, byte-oriented convolutional interleaver as shown in Figure 13.44. Each delay element D corresponds to a delay of 17 bytes, leading to a period of 204 bytes. The sync byte of the first MPEG-2 TS packet is always routed through the zero-delay path, causing the sync bytes of all future TS packets to also be routed through the zero-delay path (Table 13.7).

This interleaver is byte oriented because the RS coder it precedes in the decoder treats each byte of data as a single symbol. Where burst errors occur, it is therefore desirable to keep these errors within the symbols in which they occur, rather than spread them to other symbols and thereby increase the symbol error rate.

In each 204-byte code word, the RS outer coder can correct eight symbol errors (i.e., bytes that contain one or more errors) or 16 erasures. The outer interleaver distributes the 204-byte RS code words across 12 TS-packet transmission times in 17-byte groups.

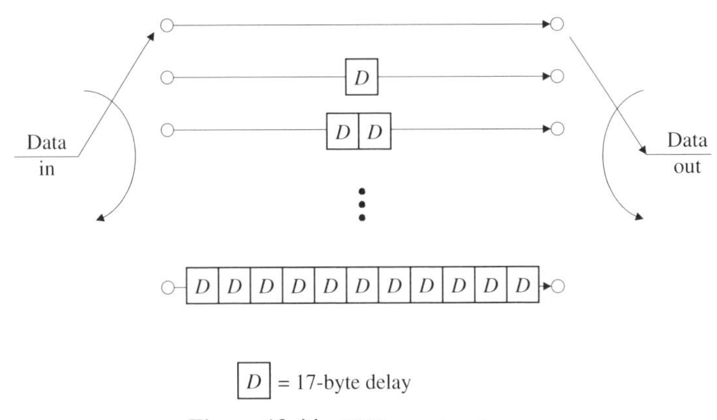

Figure 13.44 DVB outer interleaver.

Table 13.7 Temporal spread of an interleaved TS packet in a DVB transmission signal.

Useful bit rate (Mbit/s)	Interleaving spread (ms)
5	3.6
10	1.8
15	1.2
20	0.9
25	0.7
30	0.6

13.4.2.4. DVB Inner Coding

The DVB inner coder is based on the 1/2 rate convolutional coder with 64 states (i.e., a constraint length of seven) as shown in Figure 13.45. The data arrives at the left side of the figure and passes through the 6-bit shift register. For each input bit, the contents of the shift register are shifted one position to the right side and two output bits, X and Y, are generated. These two output bits are then multiplexed to provide two bits for the single-coded output bit stream. This particular code offers the best performance available from a 1/2 rate, constrain-length seven convolutional code [11]. This code has a free distance d_f equal to 10, meaning that it can correct bursts of up to four errors, following the discussion of Section 13.2.5.2.

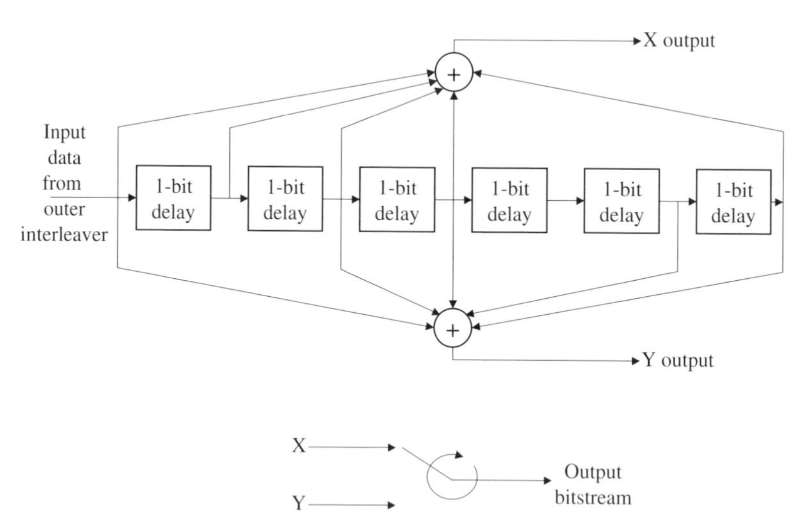

Figure 13.45 The 1/2 rate convolutional coder used for DVB's inner coder. © European Telecommunications Standards Institute 1997. © European Broadcasting Union 2004. Further use, modification, redistribution is strictly prohibited. ETSI standards are available from http://pda.etsi.org/pda/ and http://www.etsi.org/services_products/freestandard/home.htm.

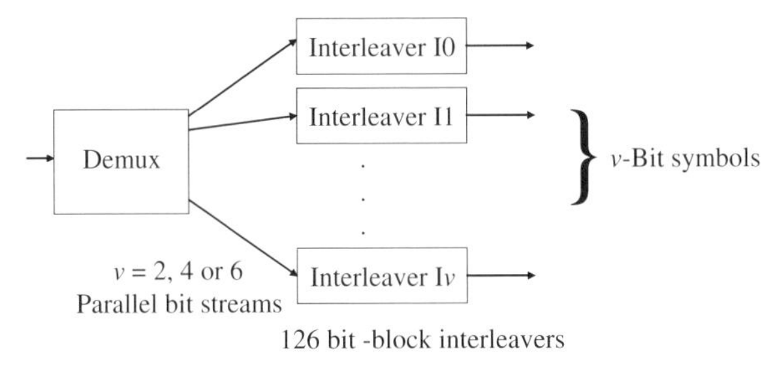

Figure 13.46 DVB inner interleaver.

For operation at lower coding rates, puncturing is used, allowing coding rates of 2/3, 3/4, 5/6, and 7/8 to be achieved. Puncturing is commonly used with convolutional codes to allow a variety of different levels of error protection to be achieved, without requiring more than one decoder to be implemented. Changing the error protection overhead with block codes, in contrast, requires the use of a different code and decoder for each level of protection.

13.4.2.5. DVB Inner Interleaving

The purpose of inner interleaving is to randomize the order of errors being passed into the inner (convolutional) decoder. The first stage of inner interleaving is to break the bit stream into v-bit symbols for modulation. For QPSK, $v=2$; for 16-QAM, $v=4$; and for 64-QAM, $v=6$. Each bit of the v-bit symbols is passed through a bit-wise interleaver, after which the symbols are reconstructed and assigned to OFDM carriers by the symbol interleaver.

The outline structure of the inner interleaver is shown in Figure 13.46. Up to six different block interleavers are used depending on the modulation (QPSK, 16-QAM, or 64-QAM). Each of the bit-wise interleavers I0 … I5 operates on a 126-bit block and performs a cyclic shift of this block. The starting points for the cyclic shift are shown in Table 13.8.

The symbol interleaver is responsible for assigning interleaved data symbols to OFDM carriers. Each OFDM symbol carries 12 126-symbol blocks in the 2k mode and 48 126-symbol blocks in the 8k mode.

Table 13.8 Start points for 126-bit cyclic interleavers.

Interleaver	Start point
I0	0
I1	63
I2	105
I3	42
I4	21
I5	84

13.4.3. DVB Channel Capacity

The useful transmission capacity of a DVB channel depends on the elementary period, the modulation (QPSK, 16-QAM, or 64-QAM), the length of the guard interval, and the coding rate used for the inner coder. The useful transmission capacity U is given by

$$U = \frac{v}{T}\frac{1512}{2048}\frac{1}{(1+\Delta/T_U)}\frac{188}{204}C$$

where v is the number of bits per symbol (2 for QPSK, 4 for 16-QAM, and 6 for 64-QAM); T is the elementary period (1/8 μs for a 7 MHz channel, 7/48 μs for a 6 MHz channel, and 7/64 μs for an 8 MHz channel); Δ/T_U is the length of the guard interval; and C is the rate of the inner convolutional coder.

The value of U can be constructed as follows. The basic data rate of the DVB system is v/T. This is reduced by the fact that not all carriers are used to carry data (1512/2048); some transmission capacity is used for the guard intervals $1/(1+\Delta/T_U)$, the overhead for the outer code (188/204), and the overhead for the inner code (C).

EXAMPLE 13.5—DVB Channel Capacity

Find the capacity of a 7-MHz DVB channel using QPSK modulation with a guard interval of one quarter of the symbol period and a 7/8-rate inner code.

A QPSK (v = 2), 7 MHz (T = 1/8 μs) channel with guard interval Δ/T_U = 1/4 and a 7/8-rate convolutional coder has a capacity of 7.62 Mbit/s. ■

13.5. CONCLUSION

Channel coding and modulation are an integral part of any digital television system. Although DVB and ATSC have developed different systems, both are based on a concatenated coder, employing a convolutional inner coder, a convolutional interleaver, and a Reed–Solomon outer coder. This is no coincidence, representing a very effective trade-off between performance and complexity, using the technology available at the time of development.

PROBLEMS

13.1 Fill in the blank cells in Table 13.9, relating the DVB channel capacity to the various parameters of the channel.

Table 13.9 DVB channel capacities for Q 13.1.

U (Mbit/s)	v	Channel BW (MHz)	Δ/T_U	C
4.3544	2	7	1/4	1/2
15.834		7	1/32	3/4
	4	6	1/16	2/3
21.772	6	7		5/6
14.929	6		1/4	1/2

13.2 Complete the blank entries in Table 13.10. The notation is that used in Section 13.2.5.1.

Table 13.10 Reed–Solomon characteristics for Q 13.2.

n	k	Max(s)	Max(r)
255	251	4	
255	239		8
255	223	16	
	15	4	8
255		4	8

13.3 Using the data provided in Tables 13.4 and 13.6 and Section 13.4.1.5, confirm that each OFDM symbol uses 193 carriers for reference signals in the 2k mode and 769 carriers for reference signals in the 8k mode.

13.4 If a convolutional encoder of the type shown in Figure 13.21 is used to encode a data stream D into two streams X and Y, write down a logic equation that can be used to extract the stream D from

(a) the current value of X and the previous values of D and

(b) the current value of Y and the previous values of D.

13.5 Write down a logic equation to decode the output of the DVB inner convolutional coder and recover its input data in the error free case using

(a) the current value of X and previous values of D and

(b) the current value of Y and previous values of D.

13.6 A 5-register cyclic interleaver has input stream 10, 3, 4, 2, 8, 9, 3, 11, 19, 0, 5, 7, 2, 1, 12. Write down the output stream for start positions zero and four.

13.7 A 5×2 block interleaver has input stream 10, 3, 4, 2, 8, 9, 3, 11, 19, 0, 5, 7, 2, 1, 12, 14, 7, 5, 2, 1. Write down the output stream.

13.8 A 3-level convolutional interleaver with $n = 1$ has input stream 7, 2, 1, 7, 4, 2, 1, 12, 14, 10, 3, 4, 2, 8, 9, 3, 11, 19, 60, 5, followed by a continuous stream of value 17. All cells of the interleaver have the value zero prior to loading the first element of the input stream. Write down the output stream beginning from the time that the first element of the input stream appears at the output and ending when the element with value 5 is output.

Table 13.11 Relationship between input bit-stream rate, code rate, and output bit-stream rate for Q 13.9.

Input bit-stream rate	Code rate	Output bit-stream rate
100 kbit/s	1/2	200 kbit/s
	1/3	3 Mbit/s
300 kbit/s		450 kbit/s
150 bit/s	5/6	
	3/4	160 Mbit/s

13.9 Complete the missing entries in Table 13.11, relating the input data rate, code rate, and output bit-stream rate for a convolutional coder.

13.10 Figure 13.31 shows the VSB trellis encoder used for ATSC. Draw the block diagram of a decoder to reproduce X_1 and X_2 from R, assuming that no transmission errors occur.

13.11 Calculate the total bit rate of the DVB transmission parameter signaling in the 2k and 8k modes.

13.12 A DVB system, using the 8k mode, employs two transmitters on a single-frequency network. What is the maximum permissible difference in path length between a receiver and each of the transmitters before the difference in arrival times of the two transmissions exceeds the length of the guard interval? Perform the calculation for $\Delta/T_U = 1/4$ and 1/32.

13.13 Calculate the average number of DVB transmission parameter signaling blocks transmitted each second on the carrier with index 34 in 2k and 8k modes.

13.14 Draw the block diagram of a decoder to recover the values of X_1 and X_2 from the sequence R generated by the encoder shown in Figure 13.31. This decoder need is only able to function correctly where all values of R are correctly received.

13.15 Calculate the number of data frames, data fields, data segments, and MPEG-2 transport packets per second carried by an ATSC digital television channel.

13.16 Calculate the proportions of the ATSC channel capacity used for carriage of

(a) field sync,

(b) segment sync,

(c) trellis coder overhead,

(d) Reed–Solomon coder overhead, and

(e) MPEG-2 transport stream packets.

13.17 Calculate the proportions of the DVB channel capacity used for carriage of

(a) TPS,

(b) scattered and continual pilots,

(c) convolutional coder overhead,

(d) Reed–Solomon coder overhead, and

(e) MPEG-2 transport stream packets.

13.18 Calculate the minimum bandwidth required to transmit a 64 kbit/s signal without significant intersymbol interference using

(a) BPSK,

(b) QPSK,

(c) 16-QAM, and
(d) 64-QAM.

13.19 Calculate the minimum bandwidth (based on the Nyquist criterion) required to transmit a 128 kbit/s signal using

(a) BPSK,
(b) QPSK,
(c) 16-QAM, and
(d) 64-QAM.

13.20 Calculate the number of data bits carried by one DVB OFDM symbol in 2k and 8k modes for each of the three modulation schemes: QPSK, 16-QAM, and 64-QAM.

MATLAB EXERCISE 13.1

The aim of this exercise is to implement the DVB 8k mode using the inverse Fourier transform. The sequence of operations is to pack bits into modulation symbols, to interleave these modulation symbols across OFDM symbols, to convert this array of OFDM symbols to a sequence of samples to be sent to the transmitter, and finally to add the guard intervals.

1. Write a MATLAB function to convert an input array of bits into an output array of symbols for QPSK, 16-QAM, or 64-QAM.
2. Write a MATLAB function that takes as its input an array of QPSK, 16-QAM, or 64-QAM symbols, which it interleaves across 8k OFDM symbols using the parameters set out in Section 13.4.1.1, and returns an array of 8k OFDM symbols.
3. Implement an 8k OFDM modulator in accordance with the description in Section 13.4.1.1.
4. Add the guard interval to the beginning of each OFDM symbol.

MATLAB EXERCISE 13.2

The aim of this exercise is to evaluate the performance of the channel coding systems used in DVB and ATSC.

1. Write a MATLAB function that generates a random bit stream and convolutionally encodes this bit stream using the DVB convolutional encoder.
2. Write a MATLAB function that decodes this random bit stream using hard-decision Viterbi decoding.
3. Write a function that introduces random errors into the coded bit stream at a nominated error rate.
4. Grouping the output bits from the Viterbi decoder into 8-bit symbols, plot the output symbol error rate of the Viterbi decoder against its input bit error rate. This should be done for input bit error rates in the range 10^{-1}–10^{-6}. You

may need to run these simulations a number of times and average the results to obtain useful plots.

5. Assuming that the input symbol errors to the DVB RS decoder are randomly spaced, use the information in the graph in Figure 13.19 to plot the output bit error rate of the DVB concatenated coder versus its input bit error rate.

Hint: the convenc and vitdec functions in the MATLAB communications toolbox provide convolutional encoding and decoding.

MATLAB EXERCISE 13.3

The aim of this exercise is to implement a convolutional interleaver in MATLAB.

1. Implement a 6-level convolutional interleaver using a MATLAB function. The argument to this function is an array of data to be interleaved. The delay D is one array element.
2. Extend the interleaver so that the number of levels can be specified as a second argument to the function.
3. Further extend the interleaver function so that the register size (expressed in array elements of the input data) can be specified as a third argument to the function.

REFERENCES

1. While DVB also supports satellite and cable delivery, the channel coding and modulation used for these channels differs from that used for terrestrial broadcast, and is not described here.
2. C. E. Shannon, A mathematical theory of communication, *Bell Syst. Tech. J.* **27**, 1948, 379–423 and 623–656. Alternatively, see B. Sklar, *Digital Communications*, Englewood Cliffs, NJ: Prentice-Hall, 1988.
3. H. Nyquist, Certain topics on telegraph transmission theory, *Trans. Am. Inst. Electr. Eng.*, **47**, 1928, 617–644.
4. The discussion here follows that of Proakis (J. G. Proakis, *Digital Communications*, 2nd edn., New York: McGraw-Hill, 1989.)
5. D. N. Godard, Self-recovering equalization and carrier tracking in two-dimensional data communication systems, *IEEE Trans. Commun.* **COM-28**, 1980, 1867–1875.
6. See, for example, B. Sklar, *Digital Communications*, Englewood Cliffs, NJ: Prentice-Hall, 1988; J. G. Proakis, *Digital Communications*, 2nd edn., New York: McGraw-Hill, 1989.
7. See, for example, B. Sklar, *Digital Communications*, Englewood Cliffs, NJ: Prentice-Hall, 1988; J. G. Proakis, *Digital Communications*, 2nd edn., New York : McGraw-Hill, 1989.
8. A. J. Viterbi, Error bounds for convolutional codes and an asymptotically optimal decoding algorithm, *IEEE Trans. Inf. Theo.* **IT13**, 1967, 260–269.
9. Channel coding and modulation for ATSC are described in: ATSC Standard A/53A, *ATSC Digital Television Standard*, Advanced Television Systems Committee, 2001, pp. 24–26; ATSC Document A/54, *Guide to the Use of the ATSC Digital Television Standard*, Advanced Television Systems Committee, 1995, p. 60.
10. Framing structure, channel coding and modulation for digital terrestrial television, *Digital Video Broadcasting (DVB)*, EN 300 744, Sophia Antipolis: ETSI, 2001.
11. J. P. Odenwalder, *Error Control Coding Handbook*, San Diego, CA: Linkabit Corp., 1976.

Chapter 14

Closed Captioning, Subtitling, and Teletext

14.1. INTRODUCTION

One of the major advantages offered by digital television is the ability to augment programs with a variety of text, graphics, and even subsidiary audio–visual information. The simplest of these ancillary services are closed captioning, also known as subtitling and teletext.

Subtitling and teletext services in digital television extend those provided in analog television, which make use of the otherwise unused transmission capacity in the vertical blanking interval (VBI) to transmit text and simple graphics. Very low coding efficiencies are obtained when this information in the VBI is coded using the video coding tools of MPEG-2. Therefore, both DVB and ATSC carry this ancillary text information in special-purpose data channels.

14.2. DVB SUBTITLES AND TELETEXT

The ancillary services provided by DVB are as follows:

- *subtitling*, which provides a similar closed-captioning service to that provided by analog television but with significantly greater flexibility and
- *teletext*, which replicates the analog television teletext service based on System-B teletext and is primarily intended to provide backward compatibility with previous systems and will eventually be superseded by datacasting.

This section describes both of these services.

Both subtitling and teletext data in a DVB stream are carried in PES packets. Table 14.1 shows the information used by a decoder to identify subtitling and teletext data so that it can be passed to the appropriate decoder. The **stream_type** field of the program map table entry pointing to these PES packets identifies the stream as PES packets containing private data. The headers of these PES packets also identify the payload as private data (by setting the **stream_id** field to the value 0xBD). DVB

Digital Television, by John Arnold, Michael Frater and Mark Pickering.

Table 14.1 Identification of subtitling and teletext data in DVB.

Field	Value for subtitle	Value for teletext
stream_type in PMT	0x06	0x06
stream_id in PES header	0xBD	0xBD
data_identifier in PES payload	0x20	0x10 – 0x1F

has a convention that the first payload byte of a PES packet containing private data carries the **data_identifier** field, which is used to identify uniquely the type of data in the payload of the packet.

Further information about subtitling and teletext streams is provided to the decoder in the component, subtitling, and teletext descriptors described in Chapter 12.

14.2.1. Subtitles

The subtitling function of DVB is designed for carrying the simple text and graphics required for closed captioning. These services are used traditionally to provide text for the hearing impaired and in lieu of dubbing for foreign language content.

Subtitling in DVB is defined in ETS 300 743 [1], which specifies the transmission and coding of graphical elements. Data is transmitted in the form of pixel structures, which define objects that can be displayed. These pixel structures are reusable but are not suitable for downloading character sets because there is no special error protection provided for pixel structures, which leads to error expansion when an incorrectly received pixel structure is reused. DVB does not provide any specific character set or font for subtitles. A means is provided to utilize ROM-based pixel structures, however, although no ROM-based pixel structure is defined by the standard. A network operator may use ROM-based pixel structures for character generation.

In DVB subtitling, a *region* is composed of one or more objects present in the decoder's memory. A group of regions that are displayed simultaneously forms a *page*.

Color information is coded with a resolution of 2, 4, or 8 bits, requiring color lookup tables with 4, 16, or 256 colors, respectively. Pixel data is compressed by applying run-length coding to consecutive pixels of the same color.

14.2.1.1. Timing

DVB subtitles are delivered in the MPEG-2 stream as private data, carried in PES packets. The MPEG-2 presentation time stamp defines the display time for a subtitle. This allows subtitle display to be synchronized with other elements of the displayed program including video and audio.

14.2.1.2. Example Subtitle Operation

A simple example of the operation of the DVB subtitling system is shown in the figures below. Figure 14.1 shows the initial (empty) display. The subtitling systems'

Figure 14.1 Initial display.

internal region list is represented on the left side and the output on the display on the right side.

Figure 14.2 shows the contents of the region list and display after the introduction of regions. Placing region A in the page specifies that this region is to be displayed. Regions outside the page (B and C) are not displayed. In this example, region A will be used to display a logo, whereas buffers B and C will be used to display text. Although the appearance of this logo is similar to the watermarks used in analog television, DVB subtitling does not provide a satisfactory means for conveying watermarks. A key characteristic of a watermark is that it is difficult to remove from the watermarked image. Because it is transmitted separately from the video, a DVB subtitle is easily removed.

Once the regions have been defined, the pixel structures defining graphics to be displayed can be downloaded as illustrated in Figure 14.3. Because region A is contained within the current version of the displayed page, the logo is displayed immediately after it is downloaded.

Figure 14.4 shows the delivery of the first text. This text is placed in the region B, but is not displayed immediately, because region B is not defined to lie within the page.

When the time for display of the first text arrives (as indicated by its associated presentation time stamp), a new version of the page description is activated in which region B is defined to lie at a particular location within the display. Although this text is displayed (Fig. 14.5), the next text to be displayed can be downloaded into region C.

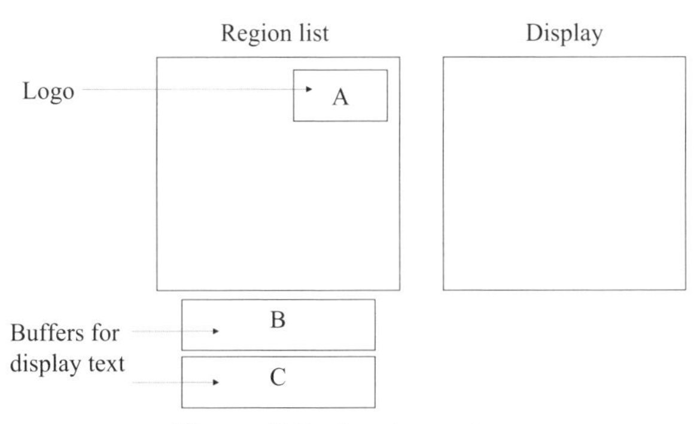

Figure 14.2 Introduce regions.

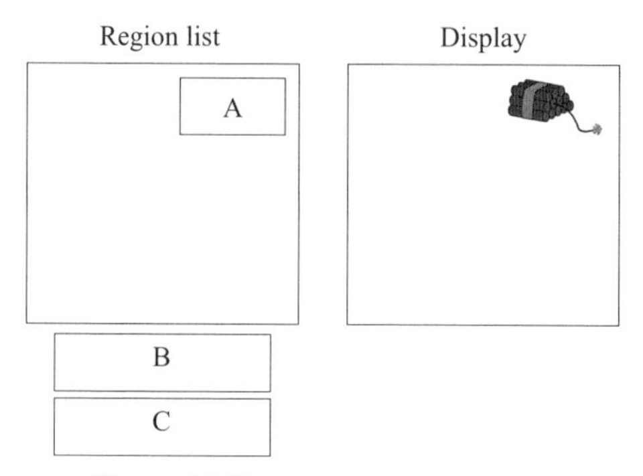

Figure 14.3 Deliver and display logo.

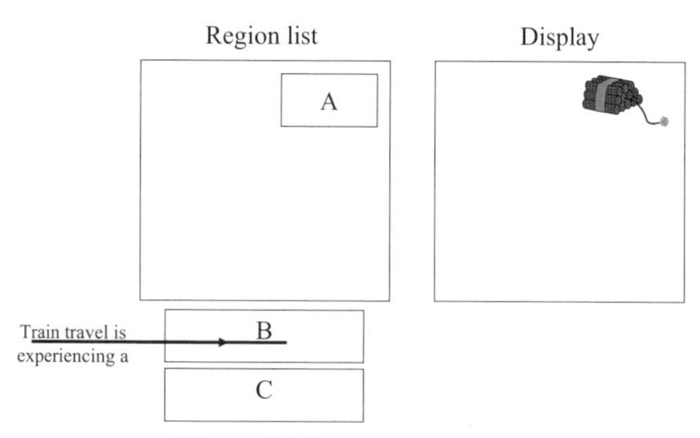

Figure 14.4 Deliver first text.

The second text is displayed by the use of a new page definition, in which region B is outside the display, but region C is now inside the display. Once the second text is displayed (Fig. 14.6), the third text can be downloaded into region B.

Finally as shown in Figure 14.7, the third text can be displayed.

This example illustrates the use of regions for immediate display (A) and the use of regions (B and C) for deferred display and simultaneous offscreen download of updates.

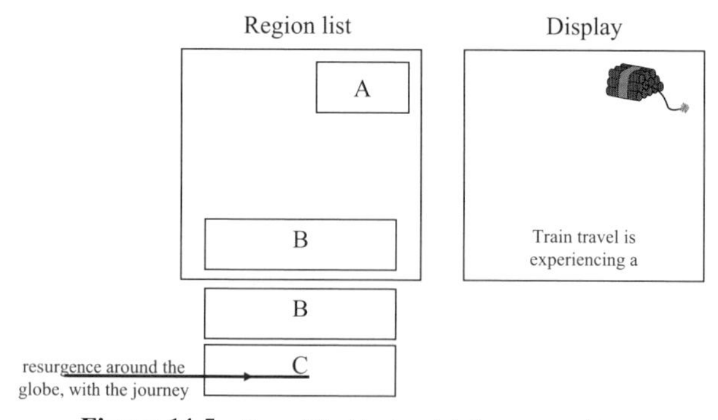

Figure 14.5 Reveal first text and deliver second text.

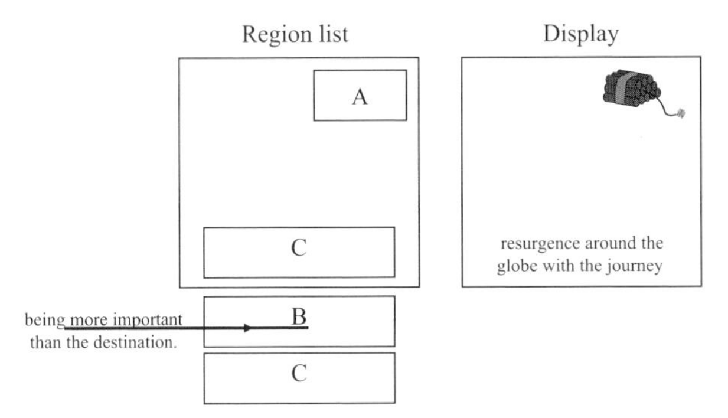

Figure 14.6 Reveal second text and deliver third text.

14.2.1.3. Coding for DVB Subtitling Data

Four new syntactic structures are defined to carry the DVB subtitling data. The *page composition segment* (PCS) defines the locations within the picture of those regions that are to be displayed as part of the page. Each region is defined by a *region composition segment*, which specifies the objects that make up that region. The structure of each object is defined by an *object data segment*, which specifies either the pixel structure for the object or one or more characters from a character set that make up the object. The *color lookup table* (*CLUT*) defines the 2, 4, or 8-bit colormap to be used for a particular page.

Page Composition Segment The PCS defines regions within the displayed picture into which data is subsequently loaded. Each region is specified to lie within a specific page. The PCS defines the following for one page:

- its **page_id**, which uniquely identifies the page;
- the **page_time_out**, which specifies a time in seconds after which the page is to be erased, protecting against the loss of a subsequent PCS that would cause this erasure;

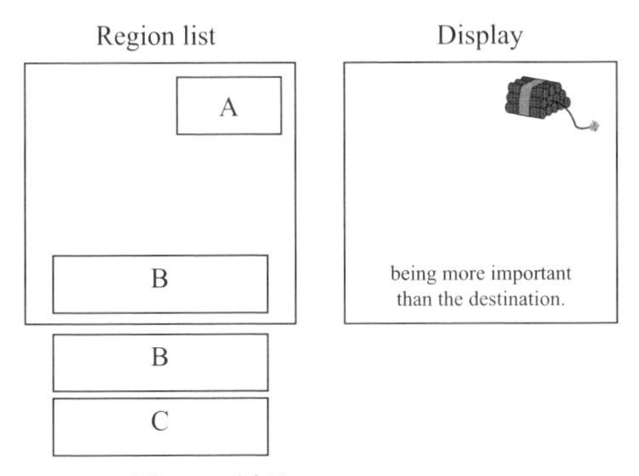

Figure 14.7 Reveal third text.

- the **page_version_number**, which is increased by 1 (mod 16) each time the page is modified;
- the **page_state**, which is used to specify where sufficient information is provided to permit random access and those points at which the memory plan of the decoder is to be changed;
- a list of regions (specified by their **region_id**) and their horizontal and vertical addresses in the display.

Region Composition Segment The region composition table defines for a particular region

- the **region_id**, which is used to uniquely identify the region;
- the **region_version_number**, which is increased by 1 (mod 16) each time the contents of the region are changed;
- the **region_fill_flag** and **region_fill_n-bit_color**, which are used to allow a region to be filled with a particular color before composition of objects begins;
- the **region_width** and **region_height**, specifying the horizontal and vertical size of the region, respectively;
- the **region_depth**, specifying the maximum pixel depth to be used in rendering the region and the **region_level_of_compatibility**, specifying the minimum CLUT size to be used for rendering the region;
- a list of objects from which the region is composed, each of which is identified by its unique **object_id** and their locations within the region.

Object Data Segment The object data segment is used to download data into a specified region in a particular page. Where a predefined character set is used, object data takes the form of indices in this character set and is carried directly in the object data segment. Where pixel-based subtitling is to occur, pixel data is carried within a number of pixel-data subblocks. The syntax of the object data segment is shown in Table 14.2.

The **sync_byte** has value "0000 1111," and is used to ensure that correct synchronization is maintained during decoding. The value of **segment_type** for the object data segment is 0x13. The **page_id** field identifies the page in which this object appears, whereas the **segment_length** specifies the length in bytes of the remainder of the object data segment.

The **object_id** uniquely identifies this object, with the **object_version_number** used to distinguish between different versions of the same object. The **object_version_number** is increased by 1 each time the content of an object is modified.

The **object_coding_method** specifies whether the object is coded as a bitmap ("00") or string of characters ("01"). The **non_modifying_color_flag** is set to 1 if the CLUT entry 1 does not overwrite other objects.

Table 14.2 Syntax for object data segment.

Syntax	Number of bits	Mnemonic
object_data_segment() {		
sync_byte	8	bslbf
segment_type	8	bslbf
page_id	16	bslbf
segment_length	16	uimsbf
object_id	16	bslbf
object_version_number	4	uimsbf
object_coding_method	2	bslbf
non_modifying_color_flag	1	bslbf
reserved	1	bslbf
if (object_coding_method=='00') {		
top_field_block_length	16	uimsbf
bottom_field_block_length	16	uimsbf
while(processed_length<top_field_block_length)		
pixel-data_sub-block()		
while(processed_length<bottom_field_block_length)		
pixel-data_sub-block()		
if (!wordaligned())		
"0000 0000"	8	bslbf
}		
if (object_coding_method=='01'){		
number_of_codes	8	uimsbf
for (*i*=1, *i*<=number_of_codes, *i*++)		
character_code	16	bslbf
}		
}		

Where the object is coded as a bitmap, that is, pixel data is to be transmitted explicitly, the **object_coding_method** is transmitted as "00." Pixel data for the two fields of the interlaced display is transmitted separately. The number of bytes of data present for the top field is specified by **top_field_block_length**, whereas the number of bytes of data for the bottom field is specified by the value of **bottom_field_block_length**.

Specifying the value "01" for the **object_coding_method** field tells the decoder that the object makes use of a predefined character set. In this case, the number of characters in the object is determined by the value of **number_of_codes** and the values of the characters by the contents of the **character_code** fields. Because character sets are not standardized, but are agreed privately between network operators and equipment manufacturers, these characters are not restricted to any particular code.

Each call to pixel-data_subblock() returns pixel values for one line of pixel data in the object.

EXAMPLE 14.1—Object Data Segment

Table 14.3 shows the makeup of an object data segment for an object with **page_id** equal to 0x10, **object_id** equal to 0x23, **object_version_number 2**, and coded as characters. The contents of the object are the character string "HELLO WORLD," represented as ASCII characters.

Table 14.3 Example object data segment.

Field name	Field value (binary)	Meaning
sync_byte	0000 1111	Sync byte is always 0000 1111.
segment_type	0001 0011	0x13 means that this segment is an object data segment.
page_id	0000 0000 0001 0000	Page id is 0x10.
segment_length	0000 0000 0001 1011	27 data bytes follow in this object data segment.
object_id	0000 0000 0010 0011	Object id is 0×23.
object_version_number	0010	Object version number is 2.
object_coding_method	01	Object coded as characters.
non_modifying_color_flag	0	CLUT entry 1 modifies underlying objects.
Reserved	0	Always 0.
number_of_codes	0000 0000 0000 1011	Eleven characters follow.
character_code	0000 0000 0100 1000	"H"
character_code	0000 0000 0100 0101	"E"
character_code	0000 0000 0100 1100	"L"
character_code	0000 0000 0100 1100	"L"
character_code	0000 0000 0100 1111	"O"
character_code	0000 0000 0010 0000	" "
character_code	0000 0000 0101 0111	"W"
character_code	0000 0000 0100 1111	"O"
character_code	0000 0000 0101 0010	"R"
character_code	0000 0000 0100 1100	"L"
character_code	0000 0000 0100 0100	"D"

The bit stream representing this segment is therefore 0000 1111 0001 0011 0000 0000 0001 0000 0000 0000 0001 1011 0000 0000 0010 0011 0010 0100 0000 0000 0000 1011 0000 0000 0100 1000 0000 0000 0100 0101 0000 0000 0100 1100 0000 0000 0100 1100 0000 0000 0100 1111 0000 0000 0010 0000 0000 0000 0101 0111 0000 0000 0100 1111 0000 0000 0101 0010 0000 0000 0100 1100 0000 0000 0100 0100. ■

14.2.1.4. Pixel-Data Subblock

The pixel-data subblock carries pixel-based subtitling data, that is, all subtitling data not based on a predefined character set. The subset of the syntax of the pixel-data_sub-block required to transmit pixel data for an 8-bit color lookup table is shown in Table 14.4.

The **sync_byte, segment_type, page_id**, and **segment_length** fields have the same meaning as for the object data segment. The **data_type** field specifies the pixel depth of the color lookup table using the value 0x12 for an 8-bit table.

The **8-bit/pixel_code_string** field uses a run-length code to specify a run of pixels of a particular value in an 8-bit color lookup table. The possible values of the variable length code used for the **8-bit/pixel_code_string** field and their meanings are shown in Table 14.5. The length of each of these possible values is 16 or 24 bits, both of which are a multiple of 8, ensuring that whole bytes of data are always generated. This avoids the need for padding of incomplete bytes at the end of a line of pixels and simplifies the implementation of the decoder.

Using the values shown in Table 14.5, 1 pixel whose index in the color lookup table is 12 would be represented by the 8-bit code 0000 1100. A string of 17 consecutive pixels with color index 34 would be represented by the 24-bit code 0000 0000 1001 0001 0010 0010.

Table 14.4 Syntax for pixel-data_sub-block.

Syntax	Number of bits	Mnemonic
pixel-data_sub-block() {		
sync_byte	8	bslbf
segment_type	8	bslbf
page_id	16	bslbf
segment_length	16	uimsbf
data_type	8	bslbf
...		
if (data_type=='0x12') {		
repeat {		
8-bit/pixel_code_string		vlclbf
} until (next_bits == '0000 0000 0000 0000')		
0000 0000 0000 0000	16	bslbf
}		
...		
}		

Table 14.5 Eight bits per **pixel_code_string**.

Value	Meaning
0000 0001 – 1111 1111	One pixel in color 1 to one pixel in color 255.
0000 0000 0LLL LLLL	L (1–127) pixels in color 0.
0000 0000 1LLL LLLL CCCC CCCC	L (3–127) pixels in color C.

EXAMPLE 14.2—Pixel-Data Subblock

Table 14.6 shows the fields making up a pixel-data subblock to carry a line of 15 pixels whose indices in the current color lookup table are 0 0 0 0 0 5 10 10 10 10 10 10 10 0 5 10.

Table 14.6 Example pixel-data subblock.

Field name	Field value (binary)	Meaning
sync_byte	0000 1111	Sync byte is always 0000 1111.
segment_type	0001 0011	0x13 means that this segment is an object data segment.
page_id	0000 0000 0001 0000	Page id is 0x10.
segment_length	0000 0000 0000 1101	Thirteen data bytes follow.
data_type	0001 0010	0x12 means 8-bit/pixel color lookup table.
8-bit/pixel_code_string	0000 0000 0000 0101	Five pixels of color 0.
8-bit/pixel_code_string	0000 0101	One pixel of color 5.
8-bit/pixel_code_string	0000 0000 1000 0110 0000 1010	Six pixels of color 10.
8-bit/pixel_code_string	0000 0000 0000 0001	One pixel of color 0.
8-bit/pixel_code_string	0000 0101	One pixel of color 5.
8-bit/pixel_code_string	0000 1010	One pixel of color 10.
	0000 0000 0000 0000	End of pixel data.

The bit stream produced to carry this pixel-data subblock is therefore 0000 1111 0001 0011 0000 0000 0001 0000 0000 0000 0000 1101 0001 0010 0000 0000 0000 0101 0000 0101 0000 0000 1000 0110 0000 1010 0000 0000 0000 0001 0000 0101 0000 1010 0000 0000 0000 0000. ■

14.2.1.5. Color Lookup Table

The color lookup table specifies the color to be associated with each index in the color table. For each index, the luminance, two chrominance components, and transparency are defined.

14.2.2. Teletext

The DVB teletext service is intended to carry teletext coded in accordance with the ITU-R System B Teletext recommendation [2] and is defined by EN 300 472 [3]. The major purpose of this part of the standard is to provide backward compatibility with existing equipment, belonging to both the network operator and their customers. In this section, the characteristics of ITU-R System B Teletext are described, followed by a discussion of the carriage of this data in a DVB stream.

14.2.2.1. ITU-R System B Teletext

Teletext provides a text-based service that allows users to browse pages of information. Each page consists notionally of 25 lines, each of which contains 40 characters. Up to 99 pages may exist in one magazine. Up to eight magazines are supported. Teletext is based on a data carousel, which means that data is transmitted continuously, with new copies of all data transmitted regularly. A new copy of a page usually completely replaces its previous contents. By regularly retransmitting all pages, a receiver need to only have a very limited storage capability, certainly not sufficient to hold the whole data carousel.

All teletext data is transmitted in packets of 360 bits organized as 45 bytes, transmitted LSB first. In analog television, all teletext data is transmitted in the VBI. Up to 16 lines per field (32 lines per picture) are allocated for this purpose, with each teletext packet occupying one VBI line. The maximum average raw data rate is therefore 11.52 kbit per picture or 288 kbit/s. There are three types of packet: the page header packet, normal packets intended for direct display, and nondisplayable packets.

ITU-R System B Teletext Packet Header Packets begin with a 2-byte clock run-in sequence (0x5555) and a 1-byte framing code (0x27). Each packet header specifies a 3-bit magazine number and 5-bit line number to which the content of the packet is applicable. Page header packets also identify a page number within the magazine, a page subcode, and control information. All subsequent packets with the same magazine number refer to this page until another page header packet with the same magazine number is received.

For packet numbers between 0 and 24, the packet number indicates the line of the display on which the packet's data should be displayed. Packet numbers larger than 24 are used to transmit nondisplayed control data and are described fully in ETS 300 706.

Teletext data is always transmitted least-significant-bit-first.

The magazine and packet number fields are protected against errors using a Hamming (8,4) code. The Hamming (8,4) code is generated as a Hamming 7/4 code with an additional parity bit to ensure that the byte has odd parity. The 4 parity bits produced are as follows:

$$P_1 = 1 \oplus D_1 \oplus D_3 \oplus D_4$$
$$P_2 = 1 \oplus D_1 \oplus D_2 \oplus D_4$$
$$P_3 = 1 \oplus D_1 \oplus D_2 \oplus D_3$$
$$P_4 = 1 \oplus D_1 \oplus D_2 \oplus D_4 \oplus P_1 \oplus P_2 \oplus P_3$$

A 4-bit input word $D_4\ D_3\ D_2\ D_1$ is transmitted as the 8-bit code word $D_4\ P_4\ D_3\ P_3\ D_2\ P_2\ D_1\ P_1$.

EXAMPLE 14.3—(8,4) Hamming Code

The Hamming (8,4) code word for the input value 0x5 is 0xDC, while the code word for 0x0 is 0x15. ■

Protection of the magazine number and packet number is achieved by forming an 8-bit word in which the magazine number forms the least significant 3 bits and the packet number forms the most significant 5 bits. Each nibble of this 8-bit word is coded using the Hamming (8,4) code, giving a total size for these fields of 16 bits.

ITU-R System B Teletext Page Header The page header (i.e., Packet 0 of each page) also indicates a page subcode, which allows one page to contain a number of subpages, each of which is designed to be displayed individually. There are four subcode components, denoted as S1, S2, S3, and S4, whose sizes are 4, 3, 4, and 2 bits, respectively.

A number of control bits are also transmitted in the page header, whose meanings are shown in Table 14.7.

Table 14.7 Allocation of control bits in Packet 0.

Control bit	Function
C4	Erase page—packets from a previous transmission of the page should be erased from decoder's memory before decoding and storing new transmission.
C5	Indicates a news flash that should be displayed boxed inset into normal video.
C6	Subtitle—should be displayed inset into normal picture when subtitles are selected.
C7	Suppress header—data transmitted in line 0 should be ignored.
C8	Update indicator—data in page has changed since previous transmission.
C9	Interrupted sequence—the associated page is not in numerical order of page sequence.
C10	Inhibit display—data from page should not be displayed.
C11	Magazine serial—set when all pages from one magazine are transmitted before any pages from another magazine.
C12, C13, C14	National character option subset—used to control national character sets.

Table 14.8 Contents of page subcode and control bits.

Byte 1	Byte 2	Byte 3	Byte 4	Byte 5	Byte 6
S1	S2 C4	S3	S4 C5 C6	C7 – C10	C11 – C14

Six bytes are used to transmit the page subcodes and control bits, with each byte protected by the Hamming 8/4 code. The first byte carries S1, the second S2 and C4, the third S3, the fourth S4, C5, and C6, with C7-C10 in byte 5 and C11-C14 in byte 6 as shown in Table 14.8. In each case, leftmost bits are the least significant, and bytes are transmitted least-significant-bit first.

ITU-R System B Teletext Character Set The basic teletext character set (known as the G0 character set), which is based on the ASCII standard, supports minor variations for language-specific characters. The English variant is shown in Table 14.9. The precise appearance of characters is a decoder option, although the teletext standard provides a sample character set based on a 10×12 pixel grid.

ITU-R System B Teletext Attributes The presentation of characters may be modified by the selection of other character sets or by changing the attributes of the basic G0 set using *spacing attributes*, a selection of which is shown in Table 14.10.

Attributes labeled "set-after" apply to subsequent character positions; those labeled "set-at" apply from the current character position. Each position occupied

Table 14.9 Basic teletext character set.

D6 D5 D4 D3	0100	0101	0110	0111	1000	1001	1010	1011	1100	1101	1110	1111
D2 D1 D0												
000		(	0	8	@	H	P	X	—	h	p	x
001	!	)	1	9	A	I	Q	Y	a	I	q	y
010	"	*	2	:	B	J	R	Z	b	j	r	z
011	£	+	3	;	C	K	S	←	c	k	s	¼
100	$	,	4	<	D	L	T	½	d	1	t	‖
101	%	-	5	=	E	M	U	→	e	m	u	¾
110	&	.	6	>	F	N	V	↑	f	n	v	÷
111	'	/	7	?	G	O	W	#	g	o	w	■

Table 14.10 Spacing attributes for ITU-R System B teletext.

Code	Function
0x00	Alpha black ("set-after")—sets the text color to black.
0x01	Alpha red ("set-after")—sets the text color to red.
0x02	Alpha green ("set-after")—sets the text color to green.
0x03	Alpha yellow ("set-after")—sets the text color to yellow.
0x04	Alpha blue ("set-after")—sets the text color to blue.
0x05	Alpha magenta ("set-after")—sets the text color to magenta.
0x06	Alpha cyan ("set-after")—sets the text color to cyan.
0x07	Alpha white ("set-after")—sets the text color to white—this is the default at the beginning of a packet.
0x08	Flash ("set-after")—causes foreground pixels subsequent characters to flash between foreground and background colors.
0x09	Steady ("set-at")—cancels a previous flash attribute for subsequent characters.
0x0A	End box ("set-after")—terminates a box beginning with the "Start box" attribute.
0x0B	Start box ("set-after")—Used with subtitles to specify a part of the picture that is to be boxed on top of the normal picture. Characters outside this area are not displayed.
0x0C	Normal size ("set-at")—cancels a previous double height, double width, or double size code for subsequent characters.
0x0D	Double height ("set-after")—characters have double height and are stretched into the next row down. Any characters transmitted in these positions are ignored.
0x0E	Double width ("set-after")—subsequent characters are displayed at double their normal width.
0x0F	Double size ("set-after")—subsequent characters are displayed at double their normal height and width.
0x18	Conceal ("set-at")—subsequent characters are to be displayed as spaces until a color code is received. A decoder or user action may reveal these characters.
0x1C	Black background ("set-at")—sets background color for subsequent characters to black—default at beginning of row.
0x1D	New background ("set-at")—the current foreground color is adopted as the background color. The foreground color must be changed before subsequently transmitted characters become visible.

by a spacing attribute is usually displayed as a space. The action of an attribute persists until the end of a row or explicit modification by the transmission of another attribute.

ITU-R System B Teletext Syntax An outline of the syntax used by the teletext system is shown in Table 14.11. Each packet begins with the magazine and packet

Table 14.11 Syntax of ITU-R System B Teletext.

Syntax	Number of bits	Mnemonic
teletext_packet() {		
magazine_packet_number	16	bslbf
if (packet == 0) {		
page_number	8	bslbf
subcode_and_control	48	bslbf
}		
if (packet_number <= 24)		
for ($k = n$; $k < 45$; k++)		
char_or_attribute_byte	8	bslbf
else {		
...		
}		
}		

numbers encoded together as two 8/4 Hamming code words. A packet number of zero indicates a new page, for which the page number is transmitted as an 8/4 Hamming code word, followed by the 6-byte subcode and control field (Table 11.8), of which each byte is an 8/4 Hamming code word. For packets with packet number less than or equal to 24, the remaining bytes of each packet are filled with attribute bytes (Table 11.9) or text characters (Table 11.10).

EXAMPLE 14.4—ITU-R System B Teletext

Figure 14.8 shows an example of teletext page, with a black (default color) background and white (default color) text. The first line of text is double height, double width. The second line is double height, whereas the third line is double width, and the last line is normal height and width. The construction of this simple page in page 10 of magazine 4 with all subcode and control bits set to 0 is shown in Table 14.12.

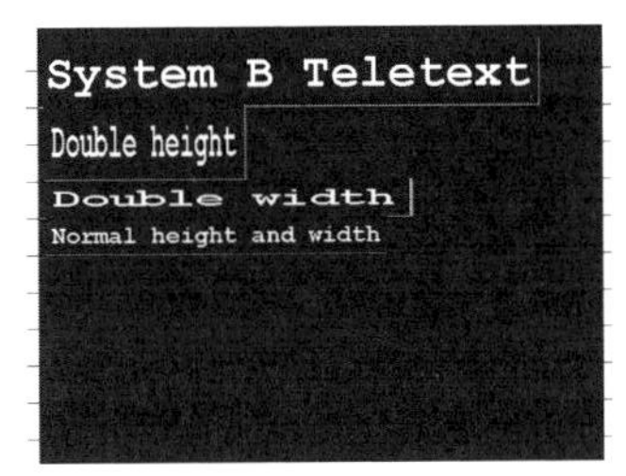

Figure 14.8 Example of teletext page.

Table 14.12 Construction of sample teletext page in Figure 11.8.

Command/data	Transmitted data (expressed as 7-bit hexadecimal words, excluding parity)
Clock run-in and framing	55 55 27
Magazine 4, Packet 0	D0 15
Page number	02 15
Page subcode and control	15 15 15 15 15 15
Set double size attribute	0F
Text "System B Teletext"	53 79 73 74 65 6D 20 42 20 54 65 6C 65 74 65 78 74
Fill remainder of packet	20 20 20 20 20 20 20 20 20 20 20 20 20 20
Clock run-in and framing	55 55 27
Magazine 4, Packet 2	D0 49
Set double height attribute	0D
Text "Double height"	44 6F 75 62 6C 65 20 68 65 69 67 68 74
Fill remainder of packet	20 20 20 20 20 20 20 20 20 20 20 20 20 20 20 20 20 20 20 20 20 20 20 20 20 20
Clock run-in and framing	55 55 27
Magazine 4, Packet 4	D0 25
Set double width attribute	0E
Text "Double width"	44 6F 75 62 6C 65 20 77 69 64 74 78
Fill remainder of packet	20 20 20 20 20 20 20 20 20 20 20 20 20 20 20 20 20 20 20 20 20 20 20 20 20 20 20
Clock run-in and framing	55 55 27
Magazine 4, Packet 5	D0 33
Text "Normal height and width"	4E 6F 72 6D 61 6C 20 68 65 69 67 68 74 20 61 6E 64 20 77 69 64 74 78
Fill remainder of packet	20 20 20 20 20 20 20 20 20 20 20 20 20 20 20 20 20

In this example, because the first two lines are double height, any data transmitted in Packets 1 or 3 is ignored. The second line of text is therefore transmitted in Packet 2 and the third line in Packet 4 and the fourth line in Packet 5. ■

14.2.2.2. DVB Teletext

DVB allows teletext information to be carried as private data in PES packets of an MPEG-2 stream. The PID of the PES packets used to carry teletext data is defined in the program map table for that service. DVB teletext may be used to provide either the teletext service or as an alternative means of subtitling; this usage is identified within the teletext and also by DVB's teletext descriptor that is carried in the program map table (see Section 11.6.2.3). Multiple streams of teletext may be present in one PID. These can be distinguished by different values of **data_identifier** between 0x10 and 0x1F in the PES packet header. Each teletext stream should use a single value of the **data_identifier**.

The data carried in these packets is decoded by the set-top box, which may provide directly for the display of this data or may simply transcode this data into the VBI of an output analog television signal. In the former case, the set-top box stores the teletext data and provides a user interface to switching between teletext and video and the selection of a particular page of teletext. In the latter case, a teletext decoder in the analog television receiver is responsible for decoding and display of the transcoded teletext data. By restricting the amount of data in each teletext stream to 16 lines per field, DVB guarantees that transcoding of this teletext data in to an analog television signal is always possible.

14.3. ATSC CLOSED CAPTIONING

There are two closed-captioning (i.e., subtitling) systems specified by the ATSC standard, both developed by the Electronic Industry Association. The first is the "Line 21 Data Service" [4], which provides closed captioning for NTSC analog transmissions. The second, specification for advanced television closed captioning (ATVCC), is specifically designed for digital television [5]. Support for closed captioning is mandated by the FCC for all digital television receivers in the United States.

All closed-captioning data in an ATSC transmission is carried as user data in the MPEG-2 video picture header [6], which is intended to ease the task of transcoding between digital and analog signals. Data is carried in quanta of 16 bits, with each picture header able to carry between 0 and 31 quanta, that is, up to 496 bits. It is required that an average data rate of 9600 bit/s be maintained for closed-caption data (if present), which equates to an average of 320 bits per picture with a transmission rate of 30 pictures per second.

The type of closed-captioning data (i.e., Line-21 data service or ATVCC) is indicated within the picture-header user data. ATSC's caption service descriptor (described in Chapter 12), which may be transmitted in the program map table or event information table, also tells a decoder which subtitling system is used by a particular program.

This section describes both the line-21 data service and ATVCC.

14.3.1. Line 21 Data Service

The Line 21 data service provides a text-based closed-captioning system. Its name comes from the fact that the data for captions in analog television is transmitted in line 21 of each frame (and optionally in line 248), which is part of the VBI. Each field carries two 7-bit characters of data, each of which is protected by 1 parity bit (using odd parity). Seventeen bits are transmitted in each field, beginning with 1 start bit, which is followed by the first 7-bit character (LSB first) and its parity bit, followed by the second 7-bit character (LSB first) and its parity bit. On a standard 4:3 interlaced display, the Line 21 data service can display 15 lines (numbered 1–15), each containing up to 32 characters. In ATSC digital television, the 2 bytes of the Line 21 data service are carried in the MPEG-2 video bit stream in the **user_data** field of the picture header.

The Line 21 data service supports a number of operating modes, including the following:

- *Text mode.* The text mode displays text in 7–15 rows of the display and displays text immediately after it is decoded. The cursor is initially at the top left position and moves to the right side as text is displayed and down as carriage-return commands are received. When the display is full, it scrolls up one line and places new text on the bottom line. If more than 32 characters are transmitted on one line, the 33rd and subsequent characters overwrite the 32nd character. Text mode is entered on receiving one of the commands "resume text display" or "text restart."
- *Roll-up captioning.* The bottom 2, 3, or 4 rows of the display contain the caption. New text is always displayed on the bottom line (15), and the display scrolls when this line becomes full or a carriage-return command is received, while the cursor returns to the beginning of line 15. Roll-up captioning mode is entered on receiving one of the commands "roll-up captions, 2-rows," "roll-up captions, 3-rows," or "roll-up captions, 4-rows."
- *Pop-on captioning.* Two text buffers are provided, one for the caption currently being displayed and the other for the caption currently being received. The received caption text is buffered without changing the displayed caption. When an "end of caption" command is received, the buffers swap roles. Four rows of text (either rows 1–4 or 12–15) are supported. This mode is entered when the "resume caption loading" command is received. The "erase nondisplayed memory" is used to remove the contents of the receiver buffer before transmitting its new contents.
- *Paint-on captioning.* One text buffer is provided, into which caption text is both loaded on decoding and simultaneously displayed. This mode is entered on receipt of the "resume direct captioning" command.

All transmitted data is either a control code, a text character, or a null character. All control codes have the form 001*H*XXX XXXXXXX, which allows them to be distinguished from text characters, all of which have the form 1XXXXXX or 01XXXXX. Two channels of caption data are supported, distinguished by the 1-bit value of "*H*" in the control code. A null character is used as a filler where no data is to be transmitted.

Control codes consist of 2 bytes of data, which are transmitted twice. There are three types of control codes: preamble address codes, midrow codes, and miscellaneous control codes. Prototypes of the three types of control codes are shown in Table 14.13. *H* is used to distinguish commands associated with the two

Table 14.13 Prototypes of control codes.

Code	Prototype
Preamble address code	$0\ 0\ 1\ H\ L_3\ L_2\ L_1\ 1\ L_0\ N\ C_2\ C_1\ C_0\ U$
Midrow code	$0\ 0\ 1\ H\ 0\ 0\ 1\ 0\ 1\ 0\ C_2\ C_1\ C_0\ U$
Miscellaneous control code	$0\ 0\ 1\ H\ 1\ 0\ F\ 0\ 1\ 0\ M_3\ M_2\ M_1\ M_0$

captioning channels supported. U is set to one if the caption is to be underlined. If N is zero, $C_2\ C_1\ C_0$ determines the color in which text is displayed in accordance with Table 14.14. If N is one, text is displayed in white and indented by $4n$ spaces, where n is the 3-bit, unsigned binary number represented by $C_2\ C_1\ C_0$ spaces. $L_3\ L_2\ L_1\ L_0$ determines the row number to which text is addressed for the preamble control code, following the assignment set out in Table 14.15. The commands associated with the miscellaneous control codes are represented by the value of $M_3\ M_2\ M_1\ M_0$ and are defined in Table 14.16. F is zero if the command is transmitted in line 21, one if the command is transmitted in line 248.

Miscellaneous control codes are used to set the operating mode of the closed-captioning decoder and sometimes to control its operation within this mode (e.g., to flip memories in the pop-on captioning mode).

Preamble address codes are always used at the beginning of a line of text, except in the caption-roll-up and text modes. In these latter cases, the line on which text is displayed is determined automatically by the mode. In other modes, the specification of the line number in the preamble address code is used by the decoder to position text in the display. Color attributes are automatically reset at the beginning of each line.

Midrow codes are used within a line of text to alter its attributes and may also be used at the beginning of a line of text in caption-roll-up and text modes. Changing the color automatically turns off italics; changing the italics has no effect on the color.

Table 14.14 Color codes.

$C_2\ C_1\ C_0$	Color
000	White
001	Green
010	Blue
011	Cyan
100	Red
101	Yellow
110	Magenta
111	*Italic*

Table 14.15 Codes for row addresses.

$L_3\, L_2\, L_1\, L_0$	Row number
0000	11
0001	Not used.
0010	1
0011	2
0100	3
0101	4
0110	12
0111	13
1000	14
1001	15
1010	5
1011	6
1100	7
1101	8
1110	9
1111	10

Table 14.16 Commands for miscellaneous control codes.

$M_3\, M_2\, M_1\, M_0$	Command
0000	Resume caption loading
0001	Backspace
0010	Reserved
0011	Reserved
0100	Delete to end of row
0101	Roll-up captions, 2 rows
0110	Roll-up captions, 3 rows
0111	Roll-up captions, 4 rows
1000	Flash on
1001	Resume direct captioning
1010	Text restart
1011	Resume text display
1100	Erase displayed message
1101	Carriage return
1110	Erase nondisplayed memory
1111	End of caption (flip memories)

Table 14.17 Basic character set of EIA-608 closed captioning.

D6 D5 D4 D3	0100	0101	0110	0111	1000	1001	1010	1011	1100	1101	1110	1111
D2 D1 D0												
000		(	0	8	@	H	P	X	ú	h	p	x
001	!	)	1	9	A	I	Q	Y	a	i	q	y
010	"	á	2	:	B	J	R	Z	b	j	r	z
011	#	+	3	;	C	K	S	[	c	k	s	ç
100	$	,	4	<	D	L	T	é	d	l	t	÷
101	%	-	5	=	E	M	U	]	e	m	u	Ñ
110	&	.	6	>	F	N	V	í	f	n	v	ñ
111	'	/	7	?	G	O	W	ó	g	o	w	■

Characters for display may follow a preamble address code or a midrow control code. The basic character set for captions is shown in Table 14.17. These codes mostly follow the ASCII standard.

EXAMPLE 14.5—Line-21 Data Service

What data is transmitted by the Line-21 data service to display the text:

Hello world
Line-21 data standard is published
by the EIA

with the first and last lines colored green and the middle line colored blue in the "roll-up captions, 2 rows" mode.

Table 14.18 shows the sequence of transmitted data required in order to transmit the text. The transmission begins with the miscellaneous command to set the transmission mode. This has the binary form $001H10F010M_3M_2M_1M_0$. For this command, $M_3M_2M_1M_0 = 0101$, so if $H = 0$ and $F = 0$, the binary form of the command is 0010100 00100101, which is equivalent to 14 25 in hexadecimal.

Once the transmission mode is set, a midrow command is used to set the attributes of the row of text to green. After the line of text is transmitted, a carriage-return command moves the display to the next line of text, and two midrow commands are required to set the text to blue italics. Following some text, a midrow command setting the color is used to turn off italics, after which more text is transmitted on the same line. This line is followed by another carriage-return command, a midrow command to begin the next line and the text of the line.

Table 14.18 Sequence of transmitted data for Line-21 example.

Command/data	Transmitted data (expressed as 7-bit hexadecimal words, excluding parity)
Roll-up captions, 2-rows	14 25
Midrow command, green	11 22
"Hello world"	48 65 6C 6C 6F 20 77 6F 72 6C 64
Carriage return	14 2D
Midrow command, blue	11 24
Midrow command, italics	11 2E
"Line-21 Data Standard "	4C 69 6D 65 2D 32 31 20 44 61 74 61 20 43 74 61 6D 64 61 72 64 20
Midrow command, blue	11 24
"is published by"	69 72 20 70 75 62 6C 69 73 68 65 64 20 62 79
Carriage return	14 2D
Midrow command, green	11 22
"the EIA"	74 68 65 20 45 49 41 2E
Carriage return	14 2D

The complete stream of characters to be transmitted is therefore 14 25 11 22 48 65 6C 6C 6F 20 77 6F 72 6C 64 14 2D 11 24 11 2E 4C 69 6D 65 2D 32 31 20 44 61 74 61 20 43 74 61 6D 64 61 72 64 20 11 24 69 72 20 70 75 62 6C 69 73 68 65 64 20 62 79 14 2D 11 22 74 68 65 20 45 49 41 2E 14 2D. ■

14.3.2. Advanced Television Closed Captioning

The closed captioning used for all ATSC services is defined by EIA-708-B [7] and is known as ATVCC. This caption system supports text captions that are displayed in windows on the screen. The structure of the caption services associated with a video stream is defined by the ATVCC service directory carried in the caption service descriptor in the program map table and/or event information table.

All ATSC television services carry ATVCC data. Each second of video carries a nominal 9600 bits, which is normally allocated as 20 bytes per frame with a frame rate of 60 Hz. This constant-bit-rate approach means that insertion of closed-captioning data into a precoded video stream can be accomplished without the need for retiming of the bitstream (with its accompanying difficulties of recoding the clock references in the transport stream).

All ATVCC data is transmitted in packets, each of which may overlap frame boundaries. Data is inserted into the video bit stream in "frame order."

The ATVCC specification assumes that the underlying ATSC system provides a nominally error-free channel. Consequently, no channel coding is provided by ATVCC.

ATVCC allows several different captioning services to be multiplexed together, allowing the viewer to select the most appropriate. This would allow, for example, subtitles for a movie to be carried in a number of different languages, while at the same time providing a separate captioning service for hearing-impaired viewers.

Data from each captioning service is displayed in windows, which can be positioned on the video display. The ATVCC bit stream includes information that describes the attributes of these windows (such as size and color) and directs text into a particular window for display.

14.3.2.1. Caption Display Model

In the ATVCC channel, data is assigned to one of 63 caption services. Caption service 1 is the primary caption service, whereas caption service 2 is known as the secondary language service. It is intended that a viewer selects one particular caption service to be decoded. Support for multiple caption services means that simultaneous support can be provided for several different secondary languages, as well as providing captions for the hearing impaired.

All caption data are displayed as text in windows, very much like the windows provided by a personal computer. Each window is displayed over the top of the program's video material. A caption service may define up to eight windows.

Each caption window is assigned a number of attributes, including its size and background color. Each window has an associated pen, which also has attributes that control the properties of characters as they are displayed.

This section presents the ATVCC models for the screen, window and pen, and the method used for specifying color.

Color Representation Color in ATVCC is specified using a palette of red, green, and blue. Each color is specified with a resolution of 2 bits, making possible a total of 64 distinct colors. For example, white is specified by the (r, g, b) values (3,3,3), black by (0,0,0), and bright green by (0,3,0).

In some circumstances, opacity can be specified. Opacity is defined by a 2-bit number, with the meanings shown in Table 14.19. "Solid" means that pixels lying behind the object are not visible; "translucent" means that pixels lying behind the object are partially visible, and transparent means that pixels lying behind the object are fully visible (and the object itself is not visible). "Flashing" means that the object alternates between solid and transparent.

Caption Screen Coordinate System ATVCC defines a rectangular grid of cells that sits over the top of the displayed picture. Normally, this grid does not cover the whole of the displayed picture, leaving margins at the left, right, top, and bottom

Table 14.19 Opacity values associated with ATVCC colors.

Opacity	Meaning
0	Solid
1	Flashing
2	Translucent
3	Transparent

of the screen. For a 16:9 aspect ratio display, the grid is 210 horizontal cells by 75 vertical cells. For a 4:3 aspect ratio, the grid is 160 cells horizontally and 75 cells vertically.

Grid coordinates are specified as pairs of integers, with the top left cell being (0, 0). The first number specifies the horizontal location (column) of the cell; the second specifies the vertical location (row).

Caption Windows Each service may define up to eight caption windows. The position of a window on the screen may be specified using either relative or absolute positioning.

All caption data are displayed in windows, which occupy a part of the display screen. A window must be defined before caption data can be displayed in it. All windows are rectangular in shape. For each window, the following parameters are defined:

- *Priority.* The relative priority of two windows determines which one appears on top if they overlap. Eight different priorities are supported. A lower value of the priority indicates a higher priority.
- *Anchor identifier.* The anchor of a window is the point in the window that is used to specify its location. The anchor has two components: one horizontal and the other vertical. The horizontal component may refer to the left, middle, or right side of the window. The vertical component may refer to the top, middle, or bottom portion of the window. In total, there are therefore nine different possible types of anchor.
- *Anchor location.* The anchor location specifies a horizontal and vertical value for the location of the window's anchor. Both horizontal and vertical values are specified in terms of the screen's grid of cells.
- *Window size.* The size of the window is specified as the number of rows of characters contained in the window vertically and the number of columns horizontally. The actual displayed size of the window therefore depends on the font size.
- *Row lock and column lock.* These parameters are used to specify whether or not the display is allowed to change the number of rows and columns in a window on the fly.
- *Visibility.* The visibility of a window may be set to 1 (displayed) or 0 (hidden).
- *Row and column count.* The row and column count specify the width and height of the window, respectively.
- *Window style id.* Allows window attributes to be selected from one of the seven predefined styles when the window is created.
- *Pen style id.* The attributes associated with the pen in the current window may be specified by the pen style identifier.
- *Justification.* Displayed text may be left, right, center, or full justified.
- *Print direction.* The direction in which characters are written may be left-to-right, right-to-left, top-to-bottom, or bottom-to-top.

- *Scroll direction.* When carriage returns occur, the scroll direction controls the direction in which scrolling occurs. The options for scroll direction are the same as for print direction. Horizontal scrolling is permitted only for vertical print directions. Likewise, vertical scrolling is only allowed for horizontal print directions.
- *Word wrap.* Word wrap specifies whether the decoder is allowed to force words onto a new line where a transmitted line is too long to be displayed in the window.
- *Display effect.* The display effect controls the use of snap, wipe, and fade when a window is displayed or hidden.
- *Effect direction.* Where an effect such as wipe is used in displaying or hiding a window, the direction of this effect can be specified.
- *Effect speed.* The effect speed determines the time for a window to transition fully between the hidden and displayed states.
- *Fill color.* The fill color is the color of the window background, using the color format defined in section "Color Representation."
- *Fill opacity.* The fill opacity is the opacity of the window fill as specified in Table 14.19.
- *Border type.* Window borders may be used to surround windows. These borders may be raised, depressed, uniform, or drop shadowed.
- *Border color.* The border color is the color of the window border, using the color format defined in section "Color Representation."

Caption Pen The caption pen is a notional pen that draws new text to be displayed onto a window. The pen has a number of attributes, which control the properties of characters that are moved to the display. Changing the pen attributes only affects future characters displayed. Characters already in the display remain unchanged.

The attributes associated with the caption pen are described in section "ATVCC Pen Commands."

14.3.2.2. Caption Packet Format

Each caption packet consists of a 1-byte header and between 1 and 127 bytes of caption data. The packet header contains a 2-bit sequence number, used by the receiver to detect lost packets, and a 6-bit packet size code. The total length of the packet (including header) is twice the value specified in the packet size code.

14.3.2.3. Service Block Payload

Within each packet, caption data is carried in service blocks. Each service block carries data for one service and cannot overlap a caption packet boundary. A packet carries one or more service blocks. Each service block consists of a 1 or 2 byte header and a payload, which carries between 0 and 31 bytes of data for one service.

Table 14.20 Assignment of codes in ATVCC service block payload.

Most significant bits of byte	Use
0	Subset of ASCII control codes.
1	
2	Displayed characters, using a slightly modified form of the standard ASCII codes.
3	
4	
5	
6	
7	
8	Caption control codes (see Section 14.3.2.4).
9	
A	Additional characters following the ISO 8859-1 Latin character set and room for future expansion.
B	
C	
D	
E	
F	

For services 1–7, the service block has a 1-byte header, consisting of a 3-bit service number and a 5-bit payload length. For services 8–63, the first header byte specifies a service number of zero, and a second header byte carries a 6-bit service number field. The remaining 2 bits of this second byte are set to 0.

Each payload byte of a service block carries one text character or forms part of a control code. The decoder uses the most significant 4 bits of a byte to determine whether it is a character to be displayed or a control code as set out in Table 14.20.

14.3.2.4. ATVCC Commands

ATVCC commands are used to control the operation of the closed-captioning system in the receiver. They are grouped into three classes:

- *Windows commands.* Windows commands are used to create, delete, clear, display, and hide closed-captioning windows. They can also be used to set the attributes of windows, including their color and position.
- *Pen commands.* ATVCC pen commands are used to control the attributes, color, and location of the pen.
- *Synchronization commands.* The ATVCC synchronization commands are used to program delays between when caption data is received and displayed.

All of the commands in each of these three classes are discussed in the remainder of this section.

Table 14.21 Structure of the ATVCC SetCurrentWindow command.

	Bit number							
Byte	7	6	5	4	3	2	1	0
0	1	0	0	0	0	*id*		

ATVCC Windows Commands The ATVCC windows commands control the operation of windows, allowing the current window to be set, new windows to be defined, the attributes of windows to be changed, the display status (displayed or hidden) of windows to be altered, and windows to be deleted.

SetCurrentWindow The SetCurrentWindow command is a 1-byte command that selects one of the eight windows to be the current window. This window must have already been defined using the DefineWindow command.

The structure of the ATVCC SetCurrentWindow command, including its argument, is shown in Table 14.21. The number of the window (between 0 and 7) chosen to be the new current window is specified as the 3-bit *id* value.

DefineWindow The ATVCC DefineWindow command is a 7-byte command that is used to create a new window. The structure of the ATVCC DefineWindow command, including its arguments, is shown in Table 14.22. The type (*ap*) and horizontal and vertical coordinates (*ah* and *av*, respectively) of the anchor point for the window are specified, along with its priority (*p*), relative positioning (*rp*), row count (*rc*), column count (*cc*), row lock (*rl*), column lock (*cl*), visibility (*v*), window style id (*ws*), and pen style id (*ps*). The available values for each of these parameters are set out in previous sections.

ClearWindows The structure of the ATVCC ClearWindows command is a 2-byte command that specifies a list of windows whose contents are to be cleared. This command can be used to clear simultaneously between one and eight windows.

Table 14.22 Structure of the ATVCC DefineWindow command.

	Bit number							
Byte	7	6	5	4	3	2	1	0
0	1	0	0	1	1		*id*	
1	0	0	*v*	*rl*	*cl*		*p*	
2	*rp*				*av*			
3				*ah*				
4		*ap*				*rc*		
5	0	0			*cc*			
6	0	0		*ws*			*ps*	

Table 14.23 Structure of the ATVCC ClearWindows command.

	Bit number							
Byte	7	6	5	4	3	2	1	0
0	1	0	0	0	1	0	0	0
1	$w_7 \ldots w_0$							

The structure of the ATVCC ClearWindows command, including its arguments, is shown in Table 14.23. The second byte of the command is a binary mask that specifies which windows are to be cleared. For example, if window 0 is to be cleared, bit 0 is set in this byte; if window 5 is to be cleared, bit 5 is set.

DeleteWindows The structure of the ATVCC DeleteWindows command, including its arguments, is shown in Table 14.24. The bit mask for specifying the windows to be cleared has the same structure as for the ClearWindows command.

DisplayWindows The ATVCC DisplayWindows command is a 2-byte command, which specifies which of the eight windows are to be visible on the screen.

The structure of the ATVCC DisplayWindows command, including its arguments, is shown in Table 14.25. The bit mask for specifying the windows to be cleared has the same structure as for the ClearWindows command. For each bit set to 1, the corresponding window is made visible. For each bit set to 0, the display status of the corresponding window is not changed.

Table 14.24 Structure of the ATVCC DeleteWindows command.

	Bit number							
Byte	7	6	5	4	3	2	1	0
0	1	0	0	0	1	1	0	0
1	$w_7 \ldots w_0$							

Table 14.25 Structure of the ATVCC DisplayWindows command.

	Bit number							
Byte	7	6	5	4	3	2	1	0
0	1	0	0	0	1	0	0	1
1	$w_7 \ldots w_0$							

Table 14.26 Structure of the ATVCC HideWindows command.

	Bit number							
Byte	7	6	5	4	3	2	1	0
0	1	0	0	0	1	0	1	0
1				$w_7 \ldots w_0$				

HideWindows The ATVCC HideWindows command specifies windows that are to be hidden. The structure of the ATVCC HideWindows command, including its arguments, is shown in Table 14.26. The bit mask for specifying the windows to be cleared has the same structure as for the ClearWindows command. For each bit set to 1, the corresponding window is hidden. For each bit set to 0, the display status of the corresponding window is not changed.

ToggleWindows The ATVCC ToggleWindows command is a 2-byte command. Its purpose is to toggle the visibility of nominated windows, that is, hidden windows are displayed and displayed windows are hidden.

The structure of the ATVCC ToggleWindows command, including its arguments, is shown in Table 14.27. The bit mask for specifying the windows to be cleared has the same structure as for the ClearWindows command. For each bit set to 1, the visibility of the corresponding window is toggled. For each bit set to 0, the visibility of the corresponding window is not changed. A mask value of 0xFF causes the visibility of all windows to be toggled.

SetWindowAttributes The ATVCC SetWindowAttributes command is a 5-byte command, used to set attributes for a window that has already been defined using the DefineWindow command.

The structure of the ATVCC SetWindowAttributes command, including its arguments, is shown in Table 14.28. The attributes set are the fill color and opacity (*fr*, *fg*, *fb*, and *fo*), the border color (*br*, *bg*, and *bb*), the border type (*bt*), the word wrap (*ww*), print direction (*pd*), scroll direction (*sd*), justification (*j*), effect speed (*es*), effect direction (*ed*), and display effect (*de*). The available values for each of these parameters are set out in previous sections.

Table 14.27 Structure of the ATVCC ToggleWindows command.

	Bit number							
Byte	7	6	5	4	3	2	1	0
0	1	0	0	0	1	0	1	1
1				$w_7 \ldots w_0$				

Table 14.28 Structure of the ATVCC Set Window Attributes command.

	Bit number							
Byte	7	6	5	4	3	2	1	0
0	1	0	0	1	0	1	1	1
1	*fo*		*fr*		*fg*		*fb*	
2	*bt*		*br*		*bg*		*bb*	
3		*ww*	*pd*		*sd*		*j*	
4		*es*			*ed*		*de*	

ATVCC Pen Commands. The three ATVCC pen commands are used to set the attributes, color, and location of the pen.

SetPenAttributes The ATVCC SetPenAttributes command is used to set the attributes associated with the pen for the current window. These attributes include the pen size, font style, text tag, and control of italics and underlining. Pen attributes for a window can be changed at any time and remain in effect until either the next SetPenAttributes command or the window is closed.

The structure of the ATVCC SetPenAttributes command, including its arguments, is shown in Table 14.29. The arguments to the SetPenAttributes command are as follows:

- *Text tag (tt)*. Sixteen caption function tags are defined. They are used to indicate that the text displayed is dialog, source or speaker identification, electronically synthesized voice, dialog in a language different to the primary language, voiceover, and a range of other effects.
- *Offset (o)*. Displayed text may be one of superscript, subscript, or normal.
- *Pen size (s)*. Specifies whether the pen is small, medium, or large.
- *Italics (i)*. A value of 1 indicates the use of italics.
- *Underline (u)*. A value of 1 indicates that text is to be underlined.
- *Edge type (et)*. Edge types for the text include raised, depressed, and uniform.
- *Font style (fs)*. Although a specific font cannot be specified, font styles such as the use of serifs, mono versus proportional spacing, and the use of small capitals can be specified.

Table 14.29 Structure of the ATVCC SetPenAttributes command.

	Bit number							
Byte	7	6	5	4	3	2	1	0
0	1	0	0	1	0	0	0	0
1			*tt*		*o*		*s*	
2	*i*	*u*		*et*		*fs*		

Table 14.30 Structure of the ATVCC SetPenColor command.

	Bit number							
Byte	7	6	5	4	3	2	1	0
0	1	0	0	1	0	0	0	1
1	*fo*		*fr*		*fg*		*fb*	
2	*bo*		*br*		*bg*		*bb*	
3	0	0	*er*		*eg*		*eb*	

SetPenColor The ATVCC SetPenColor command is used to set the color that is used for characters that are written to the window. Three colors are specified: the foreground color, the background color, and the edge color, each specified using the format described in section "Color Representation."

The structure of the ATVCC SetPenColor command, including its arguments is shown in Table 14.30. The red, green, and blue values for the text foreground color are defined by *fr*, *fg*, and *fb*, respectively, with the foreground opacity defined by *fo*. Similarly, the red, green, and blue values for the text background color are defined by *br*, *bg*, and *bb* and the background opacity by *bo*. The color for the edge of the text is specified by *er*, *eg*, and *eb*. The opacity of the text edge is the same as the foreground.

SetPenLocation The ATVCC SetPenLocation command sets the location at which the next text is written to the current caption window (known as the *pen location*). This command takes two arguments, which are the row address (*r*) and the column address (*c*) to which the pen is to be moved. The structure of the ATVCC SetPenLocation command, including its arguments, is shown in Table 14.31.

A conflict exists between the window justification and the location specified by the SetPenLocation command if the justification is not left. The current window justification is not left:

- if the print direction is left-right or right-left, the column value *c* is ignored, or
- if the print direction is top-bottom or bottom-top, the row value *r* is ignored.

Table 14.31 Structure of the ATVCC SetPenLocation command.

	Bit number							
Byte	7	6	5	4	3	2	1	0
0	1	0	0	1	0	0	1	0
1	0	0	0	0	*r*			
2	0	0	*c*					

Table 14.32 Structure of the ATVCC Delay command.

	Bit number							
Byte	7	6	5	4	3	2	1	0
0	1	0	0	0	1	1	0	1
1	t							

ATVCC Synchronization Commands The ATVCC synchronization commands are used to control the decoding of caption services. They can be used to either define a delay between receipt of data and its decoding or to reset the decoder.

Delay The ATVCC Delay command is a 2-byte command, which suspends interpretation of the current caption service's command input buffer for t tenths of a second. Interpretation recommences when one of the following occurs:

- the specified time delay expires,
- a DelayCancel command is received,
- the service's input buffer becomes full, or
- a Reset command is received for the service.

The structure of the ATVCC Delay command, including its argument, is shown in Table 14.32.

DelayCancel The ATVCC DelayCancel command is a 1-byte command, which cancels all active delays in the caption decoder. It takes no arguments. The structure of the ATVCC DelayCancel command is shown in Table 14.33.

Reset. The ATVCC Reset command is a 1-byte command that reinitializes the caption service for which it is received. It takes no arguments. The structure of the ATVCC Reset command is shown in Table 14.34. The reset command causes all of the service's windows to be deleted and removed from the display, all attributes of these windows and their pens to be removed and the service input buffer to be flushed.

Table 14.33 Structure of the ATVCC DelayCancel command.

	Bit number							
Byte	7	6	5	4	3	2	1	0
0	1	0	0	0	1	1	1	0

Table 14.34 Structure of the ATVCC Reset command.

	Bit number							
Byte	7	6	5	4	3	2	1	0
0	1	0	0	0	1	1	1	1

14.4. CONCLUSION

Both ATSC and DVB provide a range of ancillary text services. For ATSC, these are aimed at text-only closed captioning, whereas DVB supports both subtitling based on graphic elements and text-only teletext. In both cases, the ability to provide backward compatibility with systems used in existing analog television has been a major design driver.

Ancillary text services provide a simple means for carrying basic data in a way that is compatible with the analog television systems that might be used as displays for digital television decoders. Although the life of a text-only teletext service is probably limited by the impending availability of a broad range of competing multimedia services, subtitling (or closed captioning) will continue to have a roll to provide text for language translation and access for the hearing impaired for the foreseeable future.

PROBLEMS

14.1 Using the syntax of Section 14.2.1.4, encode the following pixel sequence into DVB subtitling pixel-data subblocks. Each number represents the index in an 8-bit color lookup table of 1 pixel: 0 0 0 5 5 5 5 5.

14.2 Using the syntax of Section 14.2.1.4, encode the following pixel sequence into DVB subtitling pixel-data subblocks. Each number represents the index in an 8-bit color lookup table of 1 pixel: 1 2 3 0 0 0 0 0 0 0 0 68 248 248 248.

14.3 Decode the following DVB subtitling pixel-data subblock: 0000 1111 0001 0011 0000 0000 0000 0101 0000 0000 0000 0111 0001 0010 0000 0000 1000 1000 1010 0111 0000 0000 0100 0000 0001 0001.

14.4 Using the syntax of section "Object Data Segment," encode the string "Television Broadcast" into a DVB subtitling object data segment, with page id equal to 2, object id 5, object version number 1.

14.5 Decode the following DVB subtitling object data segment, using the syntax of section "Object Data Segment": 0000 1111 0001 0011 0001 1010 0000 0000 0001 1100 0000 0101 0001 0100 0000 0000 0000 1100 0100 0111 0110 1111 0110 1111 0110 0100 0010 0000 0100 1101 0110 1111 0111 0010 0110 1110 0110 1001 0110 1110 0110 0111.

14.6 Decode the following ITU-R System B Teletext packet (all values are in hexadecimal): 55 55 27 CB 15 0F 02 64 65 63 6F 64 65 20 20 20 20 20 20 20 20 20 20 20 20 20 20 20 20 20 20 20 20 20 20 20 20 20 20 20 20 20 20 20 20 20.

14.7 Decode the following ITU-R System B Teletext packet (all values are in hexadecimal): 55 55 27 38 15 0E 15 15 15 15 15 15 0D 65 6E 63 6F 64 65 20 20 20 20 20 20 20 20 20 20 20 20 20 20 20 20 20 20 20 20 20 20 20 20 20 20 20.

14.8 Decode the following ITU-R System B Teletext packet (all values are in hexadecimal): 55 55 27 E6 19 04 6F 6E 65 01 74 77 6F 20 20 20 20 20 20 20 20 20 20 20 20 20 20 20 20 20 20 20 20 20 20 20 20 20 20 20 20 20 20 20 20 20.

14.9 Decode the following ITU-R System B Teletext packet (all values are in hexadecimal): 55 55 27 34 15 C7 15 15 15 15 15 15 15 74 68 72 65 65 20 20 20 20 20 20 20 20 20 20 20 20 20 20 20 20 20 20 20 20 20 20 20 20 20 20 20 20.

14.10 Encode the following set of information into ITU-R System B Teletext packets, expressing the encoded data as hexadecimal numbers: Magazine 3, packet 0, page 4, page subcode and control 0x15 0x15 0x15 0x15 0x15 0x15, double size, blue, "This is a test."

14.11 Encode the following set of information into ITU-R System B Teletext packets, expressing the encoded data as hexadecimal numbers: Magazine 2, packet 1, double height, green, "of manual," blue, "encoding."

14.12 Encode the following set of information into ITU-R System B Teletext packets, expressing the encoded data as hexadecimal numbers: Magazine 1, packet 20, double width, white, "which is," blue, flash on, "quite painful."

14.13 Encode the following data using the Line-21 format (with parity bit set to 0), expressing the encoded data as hexadecimal numbers: Roll-up captions (2 rows) mode, blue, "This is a simple closed," carriage return, blue, "that can," italic, "be hand-encoded."

14.14 Encode the following data using the Line-21 format (with parity bit set to 0), expressing the encoded data as hexadecimal numbers: Text mode, row 5, yellow, "One," row 10, blue, italic, "two."

14.15 Decode the following of Line-21 hexadecimal data stream: 14 2B 53 61 6D 70 6C 65 11 26 74 65 78 74.

14.16 Decode the following of Line-21 hexadecimal data stream: 14 27 11 2E 4D 6F 72 65 14 2D 74 65 78 74.

REFERENCES

1. *Digital Video Broadcasting (DVB); DVB subtitling system, ETS 300 743*, Sophia Antipolis: ETSI, 1997.
2. The complete specification for System B teletext can be found in:
   ITU-R Recommendation 653, *System B, 625/50 Television Systems; or*
   Enhanced Teletext Specification, ETS 300 706, Sophia Antipolis: ETSI, 1997.
3. *Digital Video Broadcasting (DVB); Specification for conveying ITU-R System B Teletext in DVB Bitstreams, EN 300 472*, Sophia Antipolis: ETSI, 1997.
4. EIA-608, *Recommended Practice for Line 21 Data Service*, Electronic Industry Association.
5. EIA-708B, *Specification for Advanced Television Closed Captioning (ATVCC)*, Electronic Industry Association, December 1999.
6. See: ATSC Standard A/53A, ATSC Digital Television Standard, Advanced Television Systems Committee, April 2001, Annex D, pp. 46–57; ATSC Document A/54, *Guide to the Use of the ATSC Digital Television Standard*, Advanced Television Systems Committee, October 1995, pp. 96–135.
7. EIA-708B, *Specification for Advanced Television Closed Captioning (ATVCC)*, Electronic Industry Association, December 1999.

Appendix

MPEG Tables

Table A.1 Frequencies, critical band rates, and absolute threshold for Layer I at sampling rates of 32, 44.1, and 48 kHz.

Index	Frequency (Hz)			Critical band rate (z)			Absolute threshold (dB)		
i	32	44.1	48	32	44.1	48	32	44.1	48
1	93.75	96.13	62.50	0,925	0,850	0,617	24.17	25.87	33.44
2	187.50	172.27	125.00	1,842	1,694	1,232	13.87	14.85	19.20
3	281.25	258.40	187.50	2,742	2,525	1,842	10.01	10.72	13.87
4	375.00	344.53	250.00	3,618	3,337	2,445	7.94	8.50	11.01
5	468.75	430.66	312.50	4,463	4,124	3,037	6.62	7.10	9.20
6	562.50	516.80	375.00	5,272	4,882	3,619	5.70	6.11	7.94
7	656.25	602.93	437.50	6,041	5,608	4,185	5.00	5.37	7.00
8	750.00	689.06	500.00	6,770	6,301	4,736	4.45	4.79	6.28
9	843.75	775.20	562.50	7,457	6,959	5,272	4.00	4.32	5.70
10	937.50	861.33	625.00	8,103	7,581	5,789	3.61	3.92	5.21
11	1,031.25	9,479.46	697.50	8,708	8,169	6,289	3.26	3.57	4.80
12	1,125.00	1,033.59	750.00	9,275	8,723	6,770	2.93	3.25	4.45
13	1,218.75	1,119.73	812.50	9,805	9,244	7,233	2.63	2.95	4.14
14	1,312.50	1,205.86	875.00	10,301	9,734	7,677	2.32	2.67	3.96
15	1,406.25	1,291.99	937.50	10,765	10,195	8,103	2.02	2.39	3.61
16	1,500.00	1,378.13	1,000.00	11,199	10,629	8,511	1.71	2.11	3.37
17	1,593.75	1,464.26	1,062.50	11,606	11,037	8,901	1.38	1.83	3.15
18	1,687.50	1,550.39	1,125.00	11,988	11,421	9,275	1.04	1.53	2.93
19	1,781.25	1,636.52	1,187.50	12,347	11,783	9,632	0.67	1.23	2.73
20	1,875.00	1,722.66	1,250.00	12,684	12,125	9,974	0.29	0.90	2.53
21	1,968.75	1,808.79	1,312.50	13,002	12,448	10,301	−0.11	0.56	2.32
22	2,062.50	1,894.92	1,375.00	13,302	12,753	10,614	−0.54	0.21	2.12
23	2,156.25	1,981.05	1,437.50	13,586	13,042	10,913	−0.97	−0.17	1.92
24	2,250.00	2,067.19	1,500.00	13,855	13,317	11,199	−1.43	−0.56	1.71
25	2,343.75	2,153.32	1,562.50	14,111	13,578	11,474	−1.89	−0.96	1.49
26	2,437.50	2,239.45	1,625.00	14,354	13,826	11,736	−2.34	−1.38	1.27
27	2,531.25	2,325.59	1,687.50	14,585	14,062	11,988	−2.79	−1.79	1.04

(continued)

Table A.1 (*Continued*)

Index	Frequency (Hz)			Critical band rate (*z*)			Absolute threshold (dB)		
i	32	44.1	48	32	44.1	48	32	44.1	48
28	2,625.00	2,411.72	1,750.00	14,807	14,288	12,230	−3.22	−2.21	0.80
29	2,718.75	2,497.85	1,912.50	15,018	14,504	12,461	−3.62	−2.63	0.55
30	2,812.50	2,583.98	1,875.00	15,221	14,711	12,684	−3.98	−3.03	0.29
31	2,906.25	2,670.12	1,937.50	15,415	14,909	12,898	−4.30	−3.41	0.02
32	3,000.00	2,756.25	2,000.00	15,602	15,100	13,104	−4.57	−3.77	−0.25
33	3,093.75	2,842.38	2,062.50	15,783	15,284	13,302	−4.77	−4.09	−0.54
34	3,187.50	2,928.52	2,125.00	15,956	15,460	13,493	−4.91	−4.37	−0.83
35	3,281.25	3,014.65	2,187.50	16,124	15,631	13,678	−4.98	−4.60	−1.12
36	3,375.00	3,100.78	2,250.00	16,287	15,796	13,855	−4.97	−4.78	−1.43
37	3,468.75	3,186.91	2,312.50	16,445	15,955	14,027	−4.90	−4.91	−1.73
38	3,562.50	3,273.05	2,375.00	16,598	16,110	14,193	−4.76	−4.97	−2.04
39	3,656.25	3,359.18	2,437.50	16,746	16,260	14,354	−4.55	−4.98	−2.34
40	3,750.00	3,445.31	2,500.00	16,891	16,406	14,509	−4.29	−4.92	−2.64
41	3,843.75	3,531.45	2,562.50	17,032	16,547	14,660	−3.99	−4.81	−2.93
42	3,937.50	3,617.58	2,625.00	17,169	16,685	14,807	−3.64	−4.65	−3.22
43	4,031.25	3,703.71	2,687.50	17,303	16,820	14,949	−3.26	−4.43	−3.49
44	4,125.00	3,789.84	2,750.00	17,434	16,951	15,087	−2.86	−4.17	−3.74
45	4,218.75	3,875.98	2,812.50	17,563	17,079	15,221	−2.45	−3.87	−3.98
46	4,312.50	3,962.11	2,875.00	17,688	17,205	15,351	−2.04	−3.54	−4.20
47	4,406.25	4,048.24	2,937.50	17,811	17,327	15,478	−1.63	−3.19	−4.40
48	4,500.00	4,134.38	3,000.00	17,932	17,447	15,602	−1.24	−2.82	−4.57
49	4,687.50	4,306.64	3,125.00	18,166	17,680	15,841	−0.51	−2.06	−4.82
50	4,875.00	4,478.91	3,250.00	18.392	17,905	16,069	0.12	−1.32	−4.96
51	5,062.50	4,651.17	3,375.00	18,611	18,121	16,287	0.64	−0.64	−4.97
52	5,250.00	4,823.44	3,500.00	18,823	18,331	16,496	1.06	−0.04	−4.86
53	5,437.50	4,995.70	3,625.00	19,028	18,534	16,697	1.39	0.47	−4.63
54	5,625.00	5,167.97	3,750.00	19,226	18,731	16,891	1.66	0.89	−4.29
55	5,812.50	5,340.23	3,875.00	19,419	18,922	17,078	1.88	1.23	−3.87
56	6,000.00	5,512.50	4,000.00	19,606	19,108	17,259	2.08	1.51	−3.39
57	6,187.50	5,684.77	4,125.00	19,788	19,289	17,434	2.27	1.74	−2.86
58	6,375.00	5,857.03	4,250.00	19,964	19,464	17,605	2.46	1.93	−2.31
59	6,562.50	6,029.30	4,375.00	20,135	19,635	17,770	2.65	2.11	−1.77
60	6,750.00	6,201.56	4,500.00	20,300	19,801	17,932	2.86	2.28	−1.24
61	6,937.50	6,373.83	4,625.00	20,461	19,963	18,089	3.09	2.46	−0.74
62	7,125.00	6,546.09	4,750.00	20,616	20,120	18,242	3.33	2.63	−0.29
63	7,312.50	6,718.36	4,975.00	20,766	20,273	18,392	3.60	2.82	0.12
64	7,500.00	6,890.63	5,000.00	20,912	20,421	18,539	3.89	3.03	0.48
65	7,687.50	7,062.89	5,125.00	21,052	20,565	18,682	4.20	3.25	0.79
66	7,875.00	7,235.16	5,250.00	21,188	20,705	18,823	4.54	3.49	1.06
67	8,062.50	7,407.42	5,375.00	21,318	20,840	18,960	4.91	3.74	1.29
68	8,250.00	7,579.69	5,500.00	21,445	20,972	19,095	5.31	4.02	1.49
69	8,437.50	7,751.95	5,625.00	21,567	21,099	19,226	5.73	4.32	1.66
70	8,625.00	7,924.22	5,750.00	21,684	21,222	19,356	6.18	4.64	1.81

Table A.1 (*Continued*)

Index i	Frequency (Hz) 32	44.1	48	Critical band rate (z) 32	44.1	48	Absolute threshold (dB) 32	44.1	48
71	8,812.50	8,096.48	5,875.00	21,797	21,342	19,482	6.67	4.98	1.95
72	9,000.00	8,268.75	6,000.00	21,906	21,457	19,606	7.19	5.35	2.08
73	9,375.00	8,613.28	6,250.00	22,113	21,677	19,847	8.33	6.15	2.33
74	9,750.00	8,957.81	6,500.00	22,304	21,882	20,079	9.63	7.07	2.59
75	10,125.00	9,302.34	6,750.00	22,482	22,074	20,300	11.08	8.10	2.86
76	10,500.00	9,646.88	7,000.00	22,646	22,253	20,513	12.71	9.25	3.17
77	10,875.00	9,991.41	7,250.00	22,799	22,420	20,717	14.53	10.54	3.51
78	11,250.00	10,335.94	7,500.00	22,941	22,576	20,912	16.54	11.97	3.89
79	11,625.00	10,680.47	7,750.00	23,072	22,721	21,098	18.77	13.56	4.31
80	12,000.00	11,025.00	8,000.00	23,195	22,857	21,275	21.23	15.31	4.79
81	12,375.00	11,369.53	8,250.00	23,309	22,984	21,445	23.94	17.23	5.31
82	12,750.00	11,714.06	8,500.00	23,415	23,102	21,606	26.90	19.34	5.89
83	13,125.00	12,058.59	8,750.00	23,515	23,213	21,760	30.14	21.64	6.50
84	13,500.00	12,403.13	9,000.00	23,607	23,317	21,906	33.67	24.15	7.19
85	13,875.00	12,747.66	9,250.00	23,694	23,415	22,046	37.51	26.89	7.93
86	14,250.00	13,092.19	9,500.00	23,775	23,506	22,178	41.67	29.84	8.75
87	14,625.00	13,436.72	9,750.00	23,852	23,592	22,304	46.17	33.05	9.63
88	15,000.00	13,781.25	10,000.00	23,923	23,673	22,424	51.04	36.52	10.58
89	15,375.00	14,125.78	10,250.00	23,991	23,749	22,538	56.29	40.25	11.60
90	15,750.00	14,470.31	10,500.00	24,054	23,821	22,646	61.94	44.27	12.71
91	16,125.00	14,814.84	10,750.00	24,114	23,888	22,749	68.00	48.59	13.90
92	16,500.00	15,159.38	11,000.00	24,171	23,952	22,847	68.00	53.22	15.18
93	16,875.00	15,503.91	11,250.00	24,224	24,013	22,941	68.00	58.18	16.54
94	17,250.00	15,848.44	11,500.00	24,275	24,070	23,030	68.00	63.49	18.01
95	17,625.00	16,192.97	11,750.00	24,322	24,125	23,114	68.00	68.00	19.57
96	18,000.00	16,537.50	12,000.00	24,368	24,176	23,195	68.00	68.00	21.23
97	18,375.00	16,882.03	12,250.00	24,411	24,225	23,272	68.00	68.00	23.01
98	18,750.00	17,226.56	12,500.00	24,452	24,271	23,345	68.00	68.00	24.90
99	19,125.00	17,571.09	12,750.00	24,491	24,316	23,415	68.00	68.00	26.90
100	19,500.00	17,915.63	13,000.00	24,528	24,358	23,482	68.00	68.00	29.03
101	19,875.00	18,260.16	13,250.00	24,564	24,398	23,546	68.00	68.00	31.28
102	20,250.00	18,604.69	13,500.00	24,597	24,436	23,607	68.00	68.00	33.67
103		18,949.22	13,750.00		24,473	23,666		68.00	36.19
104		19,293.75	14,000.00		24,508	23,722		68.00	38.86
105		19,638.28	14,250.00		24,542	23,775		68.00	41.67
106		19,982.81	14,500.00		24,574	23,827		68.00	44.63
107			14,750.00			23,876			47.76
108			15,000.00			23,923			51.04

Table A.2 Frequencies, critical band rates, and absolute threshold for Layer II at sampling rates of 32, 44.1, and 48 kHz.

Index	Frequency (Hz)			Critical band rate (z)			Absolute threshold (dB)		
i	32	44.1	48	32	44.1	48	32	44.1	48
1	31.25	43.07	46.88	0.309	0.425	0.463	58.23	45.05	42.10
2	62.50	86.13	93.75	0.617	0.850	0.925	33.44	25.87	24.17
3	93.75	129.20	140.63	0.925	1,273	1,385	24.17	18.70	17.47
4	125,00	172,27	187.50	1,232	1,694	1,842	19.20	14.85	13.87
5	156.25	215.33	234.38	1,538	2,112	2.295	16.05	12.41	11.60
6	187.50	258.40	281.25	1,842	2,525	2,742	13.87	10.72	10.01
7	218.75	301.46	328.13	2,145	2,934	3,184	12.26	9.47	8.84
8	250.00	344.53	375.00	2,445	3,337	3,618	11.01	8.50	7.94
9	281.25	387.60	421.88	2,742	3,733	4,045	10.01	7.73	7.22
10	312.50	430.66	468.75	3,037	4,124	4,463	9.20	7.10	6.62
11	343.75	473.73	515.63	3,329	4,507	4,872	8.52	6.56	6.12
12	375.00	516.80	562.50	3,618	4,982	5,272	7.94	6.11	5.70
13	406.25	559.86	609.38	3,903	5,249	5,661	7.44	5.72	5.33
14	437.50	602.93	656.25	4,185	5,608	6,041	7.00	5.37	5.00
15	468.75	646.00	703.13	4,463	5,959	6,411	6.62	5.07	4.71
16	500.00	689.06	750.00	4,736	6,301	6,770	6.28	4.79	4.45
17	531.25	732.13	796.88	5,006	6,634	7,119	5.97	4.55	4.21
18	562.50	775.20	843.75	5,272	6,959	7,457	5.70	4.32	4.00
19	593.75	819.26	890.63	5,533	7,274	7,785	5.44	4.11	3.79
20	625.00	861.33	937.50	5,789	7,581	8,103	5.21	3.92	3.61
21	656.25	904.39	984.38	6,041	7,879	8,410	5.00	3.74	3.43
22	687.50	947.46	1,031.25	6,289	8,169	8,708	4.80	3.57	3.26
23	718.75	990.53	1,078.13	6,532	8,450	8,996	4.62	3.40	3.09
24	750.00	1,033.59	1,125.00	6,770	8,723	9,275	4.45	3.25	2.93
25	781.25	1,076.66	1,171.88	7,004	8,987	9,544	4.29	3.10	2.78
26	812.50	1,119.73	1,218.75	7,233	9,244	9,805	4.14	2.95	2.63
27	843.75	1,162.79	1,265.63	7,457	9,493	10,057	4.00	2.81	2.47
28	875.00	1,205.86	1,312.50	7,677	9,734	10,301	3.86	2.67	2.32
29	906.25	12,489.93	13,599.38	7,892	9,968	10,537	3.73	2.53	2.17
30	937.50	1,291.99	1,406.25	8,103	10,195	10,765	3.61	2.39	2.02
31	968.75	1,335.06	1,453.13	8,309	10,416	10,986	3.49	2.25	1.86
32	1,000.00	1,378.13	1,500.00	8,511	10,629	11,199	3.37	2.11	1.71
33	1,031.25	1,421.19	1,546.88	8,708	1,09,836	11,406	3.26	1.97	1.55
34	1,062.50	1,464.26	1,593.75	8,901	11,037	11,606	3.15	1.83	1.38
35	1,093.75	1,507.32	1,640.63	9,090	11,232	1,19,800	3.04	1.68	1921
36	1,125.00	1,550.39	1,687.50	9,275	11,421	11,988	2.93	1.53	1.04
37	1,156.25	1,593.46	1,734.38	9,456	11,605	12,170	2.83	1.38	0.86
38	1,187.50	1,636.52	1,781.25	9,632	11,783	12,347	2.73	1.23	0.67
39	1,218.75	1,679.59	1,828.13	9,805	11,957	12,518	2.63	1.07	0.49
40	1,250.00	1,722.66	1,875.00	9,974	12,125	12,684	2.53	0.90	0.29
41	1,281.25	1,765.72	1,921.88	10,139	12,289	12,845	2.42	0.74	0909
42	1,312.50	1,808.79	1,968.75	10,301	12,448	13,002	2.32	0.56	−0.11

Table A.2 (*Continued*)

Index	Frequency (Hz)			Critical band rate (z)			Absolute threshold (dB)		
i	32	44.1	48	32	44.1	48	32	44.1	48
43	1,343.75	1,851.86	2,015.63	10,459	12,603	13,154	2.22	0.39	−0.32
44	1,375.00	1,894.92	2,062.50	10,614	12,753	13,302	2.12	0.21	−0.54
45	1,406.25	1,937.99	2,109.38	10,765	12,900	13,446	2.02	0.02	−0.75
46	1,437.50	1,981.05	2,156.25	10,913	13,042	13,586	1.92	−0.17	−0.97
47	1,468.75	2,024.12	2,203.13	11,058	13,181	13,723	1.81	−0.36	−1.20
48	1,500.00	2,067.19	2,250.00	11,199	13,317	13,855	1.71	−0.56	−1.43
49	1,562.50	2,153.32	2,343.75	11,474	13,578	14,111	1.49	−0.96	−1.88
50	1,625.00	2,239.45	2,437.50	11,736	13,826	14,354	1.27	−1.38	−2.34
51	1,687.50	2,325.59	2,531.25	11,988	14,062	14,585	1.04	−1.79	.2.79
52	1,750.00	2,411.72	2,625.00	12,230	14,288	1,49,807	0.80	−2.21	−3.22
53	1,812.50	2,497.85	2,718.75	12,461	14,504	15,018	0.55	−2.63	−3.62
54	1,875.00	2,583.98	2,812.50	12,684	14,711	15,221	0.29	−3.03	−3.98
55	1,937.50	2,670.12	2,906.25	12,898	14,909	15,415	0.02	−3.41	−4.30
56	2,000.00	2,756.25	3,000.00	13,104	15,100	15,602	−0.25	−3.77	−4.57
57	2,062.50	2,842.38	3,093.75	13,302	15,284	15,783	−0.54	−4.09	−4.77
58	2,125.00	2,928.52	3,187.50	13,493	15,460	15,956	−0.83	−4.37	−4.91
59	2,187.50	3,014.65	3,281.25	13,678	15,631	16,124	−1.12	−4.60	−4.98
60	2,250.00	3,100.78	3,375.00	13,855	15,796	16,287	−1.43	−4.78	−4.97
61	2,312.50	3,186.91	3,468.75	14,027	15,955	16,445	−1.73	−4.91	−4990
62	2,375.00	3,273.05	3,562.50	14,193	16,110	16,598	−2.04	−4.97	−4.76
63	2,437.50	33,599.18	3,656.25	14,354	16,260	16,746	−2.34	−4.98	−4955
64	2,500.00	3,445.31	3,750.00	14,509	16,406	16,891	−2.64	−4.92	−4.29
65	2,562.50	3,531.45	3,843.75	14,660	16,547	17,032	−2.93	−4.81	−3.99
66	2,625.00	3,617.58	3,937.50	14,807	16,685	17,169	−3.22	−4.65	−3.64
67	2,687.50	3,703.71	4,031.25	14,949	16,820	17,303	−3.49	−4.43	−3.26
68	2,750.00	3,789.84	41,259.00	15,087	16,951	17,434	−3.74	−4.17	−2.86
69	2,812.50	3,875.98	4,218.75	15,221	17,079	17,563	−3.98	−3.87	−2.45
70	2,875.00	3,962.11	4,312.50	15,351	17,205	17,688	−4.20	−3.54	−2.04
71	2,937.50	4,048.24	4,406.25	15,478	17,327	17,811	−4.40	−3.19	−1.63
72	3,000.00	4,134.38	4,500.00	15,602	17,447	17,932	−4.57	−2.82	−1.24
73	3,125.00	4,306.64	4,687.50	15,841	17,680	18,166	−4.82	−2.06	−0.51
74	3,250.00	4,478.91	4,875.00	16,069	17,905	18,392	−4.96	−1.32	0.12
75	3,375.00	4,651.17	5,062.50	16,287	18,121	18,611	−4.97	−0.64	0.64
76	3,500.00	4,823.44	5,250.00	16,496	18,331	18,823	−4.86	−0.04	1.06
77	3,625.00	4,995.70	5,437.50	16,697	18,534	19,028	−4.63	0.47	1.39
78	3,750.00	5,167.97	5,625.00	16,891	18,731	19,226	−4.29	0.89	1.66
79	3,875.00	5,340.23	5,812.50	17,078	18,922	19,419	−3.87	1.23	1.88
80	4,000.00	5,512.50	6,000.00	17,259	19,108	19,606	−3.39	1.51	2.08
81	4,125.00	5,684.77	6,187.50	17,434	19,289	19,798	−2.86	1.74	2.27
82	4,250.00	5,957.03	6,375.00	17,605	19,464	19,964	−2.31	1.93	2.46
83	4,375.00	6,029.30	6,562.50	17,770	19,635	20,135	−1.77	2.11	2.65
84	4,500.00	6,201.56	6,750.00	17,932	19,801	20,300	−1.24	2.28	2.86
85	4,625.00	6,373.83	6,937.50	18,089	19,963	20,461	−0.74	2.46	3.09

(*continued*)

Table A.2 (*Continued*)

Index	Frequency (Hz)			Critical band rate (z)			Absolute threshold (dB)		
i	32	44.1	48	32	44.1	48	32	44.1	48
86	4,750.00	6,546.09	7,125.00	18,242	20,120	20,616	−0.29	2.63	3.33
87	4,875.00	6,718.36	7,312.50	18,392	20,273	20,766	0.12	2.82	3.60
88	5,000.00	6,890.63	7,500.00	18,539	20,421	20,912	0.48	3.03	3.89
89	5,125.00	7,062.89	7,687.50	18,682	20,565	21,052	0.79	3.25	4.20
90	5,250.00	7,235.16	7,875.00	18,823	20,705	21,188	1.06	3.49	4t54
91	5,375.00	7,407.42	8,062.50	18,960	20,840	21,318	1.29	3.74	4.91
92	5,500.00	7,579.69	8,250.00	19,095	20,972	21,445	1.49	4.02	5.31
93	5,625.00	7,751.95	8,437.50	19,226	21,099	21,567	1.66	4.32	5.73
94	5,750.00	7,924.22	8,625.00	19,356	21,222	21,684	1.81	4.64	6.18
96	6,000.00	8,268.75	9,000.00	19,606	21,457	21,906	2.08	5.35	7.19
97	6,250.00	8,613.28	9,375.00	19,847	21,677	22,113	2.33	6.15	8.33
98	6,500.00	8,957.81	9,750.00	20,079	21,882	22,304	2.59	7.07	9.63
99	6,750.00	9,302.34	10,125.00	20,300	22,074	22,482	2.86	8.10	11.08
100	7,000.00	9,646.88	10,500.00	20,513	22,253	22,646	3.17	9.25	12.71
101	7,250.00	9,991.41	10,875.00	20,717	22,420	22,799	3.51	10.54	14.53
102	7,500.00	10,335.94	11,250.00	20,912	22,576	22,941	3.89	11.97	16.54
103	7,750.00	10,680.47	11,625.00	21,098	22,721	23,072	4.31	13.56	18.77
104	8,000.00	11,025.00	12,000.00	21,275	22,857	23,195	4.79	15.31	21.23
105	8,250.00	11,369.53	12,375.00	21,445	22,984	23,309	5.31	17o23	23.94
106	8,500.00	11,714.06	12,750.00	21,606	23,102	23,415	5.88	19.34	26.90
107	8,750.00	12,058.59	13,125.00	21,760	23,213	23,515	6.50	21.64	30.14
108	9,000.00	12,403.13	13,500.00	21,906	23,317	23,607	7.19	24.15	33.67
109	9,250.00	12,747.66	13,875.00	22,046	23,415	23,694	7.93	26.88	37.51
110	9,500.00	13,092.19	14,250.00	22,178	23,506	23,775	8.75	29.84	41.67
111	9,750.00	13,436.72	14,625.00	22,304	23,592	23,852	9.63	33.05	46.17
112	10,000.00	13,781.25	15,000.00	22,424	23,673	23,923	10.58	36.52	51.04
113	10,250.00	14,125.78	15,375.00	22,538	23,749	23,991	11.60	40o25	56.29
114	10,500.00	14,470.31	15,750.00	22,646	23,821	24,054	12.71	44.27	61.94
115	10,750.00	14,814.84	16,125.00	22,749	23,888	24,114	13.90	48.59	68.00
116	11,000.00	15,159.38	16,500.00	22,847	23,952	24,171	15.18	53.22	68.00
117	11,250.00	15,503.91	16,875.00	22,941	24,013	24,224	16.54	58.18	68.00
118	11,500.00	15,848.44	17,250.00	23,030	24,070	24,275	18.01	63.49	68.00
119	11,750.00	16,192.97	17,625.00	23,114	24,125	24,322	19.57	68.00	68.00
120	12,000.00	16,537.50	18,000.00	23,195	24,176	24,368	21.23	68.00	68.00
121	12,250.00	16,882.03	18,375.00	23,272	24,225	24,411	23.01	68.00	68.00
122	12,500.00	17,226.56	18,750.00	23,345	24,271	24,452	24.90	68.00	68.00
123	12,750.00	17,571.09	19,125.00	23,415	24,316	24,491	26.90	68.00	68.00
124	13,000.00	17,915.63	19,500.00	23,482	24,358	24,528	29.03	68.00	68.00
125	13,250.00	18,260.16	19,875.00	23,546	24,398	24,564	31.28	68.00	68.00
126	13,500.00	18,604.69	20,250.00	23,607	24,436	24,597	33.67	68.00	68.00
127	13,750.00	18,949.22		23,666	24,473		36.19	68.00	
128	14,000.00	19,293.75		23,722	24,508		38.86	68.00	

Table A.2 (*Continued*)

Index i	Frequency (Hz) 32	44.1	48	Critical band rate (z) 32	44.1	48	Absolute threshold (dB) 32	44.1	48
129	14,250.00	19,638.28		23,775	24,542		41.67	68.00	
130	14,500.00	19,982.81		23,827	24,574		44.63	68.00	
131	14,750.00			23,876			47.76		
132	15,000.00			23,923			51.04		

Table A.3 The FFT index of the lower, geometric mean, and upper frequencies in each critical band for Layer I at sampling rates of 32, 44.1, and 48 kHz.

crit_ band	Lower 32	44.1	48	geom_mean 32	44.1	48	Upper 32	44.1	48
0	1	1	1	1	1	1	1	1	1
1	2	2	2	2	2	2	3	2	2
2	4	3	3	4	3	3	5	3	3
3	6	4	4	6	4	4	7	5	4
4	8	6	5	8	6	5	9	6	5
5	10	7	6	10	7	6	11	8	6
6	12	9	7	12	9	7	13	9	7
7	14	10	8	14	10	8	15	11	9
8	16	12	10	17	12	10	18	13	10
9	19	14	11	20	14	11	21	15	12
10	22	16	13	23	16	13	24	17	14
11	25	18	15	26	19	15	27	20	16
12	28	21	17	30	22	18	32	23	19
13	33	24	20	35	25	20	37	27	21
14	38	28	22	41	30	23	44	32	25
15	45	33	26	48	35	27	52	37	29
16	53	38	30	57	41	32	62	45	35
17	63	46	36	68	49	38	74	52	41
18	75	53	42	81	57	46	88	62	50
19	89	63	51	96	68	54	104	74	58
20	105	75	59	114	81	63	124	88	68
21	125	89	69	136	98	75	148	108	82
22	149	109	83	166	120	91	184	132	100
23	185	133	101	212	156	112	240	180	124
24		181	125		206	144		232	164
25			165			190			216

Table A.4 The FFT index of the lower, geometric mean, and upper frequencies in each critical band for Layer II at sampling rates of 32, 44.1, and 48 kHz.

crit_ band	Lower			geom_mean			Upper		
	32	44.1	48	32	44.1	48	32	44.1	48
0	1	1	1	1	1	1	1	1	1
1	2	2	2	2	2	2	3	2	2
2	4	3	3	5	3	3	6	3	3
3	7	4	4	8	4	4	10	5	5
4	11	6	6	12	6	6	13	7	7
5	14	8	8	15	9	8	17	10	9
6	18	11	10	19	12	11	21	13	12
7	22	14	13	23	15	13	25	16	14
8	26	17	15	28	18	16	30	19	17
9	31	20	18	33	21	19	35	22	20
10	36	23	21	38	24	22	41	26	24
11	42	27	25	44	28	26	47	30	27
12	48	31	28	51	33	30	54	35	32
13	55	36	33	59	38	35	64	40	37
14	65	41	38	69	43	40	74	46	42
15	75	47	43	81	50	46	88	54	50
16	89	55	51	96	59	54	104	64	58
17	105	65	59	114	70	64	124	76	70
18	125	77	71	136	83	76	148	90	82
19	149	91	83	162	97	91	176	104	100
20	177	105	101	192	114	108	208	124	116
21	209	125	117	228	136	126	248	148	136
22	249	149	137	272	162	150	296	176	164
23	297	177	165	332	196	182	368	216	200
24	369	217	201	423	240	224	480	264	248
25		265	249		311	288		360	328
26		361	329		411	379		464	432

Table A.5 Possible number of bits per subband for the following input sampling rate and output bit rate combinations: Fs = 32 kHz and bit rates per channel = 56, 64, and 80 kbits/s; Fs = 44.1 kHz and bit rates per channel = 56, 64, and 80 kbits/s; Fs = 48 kHz and bit rates per channel = 56, 64, 80, 96, 112, 128, 160, 192, and kbits/s and free format.

		Allocation index															
sb	*nbal*	0	1	2	3	4	5	6	7	8	9	10	11	12	13	14	15
0	4		5	9	12	15	18	21	24	27	30	33	36	39	42	45	48
1	4		5	9	12	15	18	21	24	27	30	33	36	39	42	45	48
2	4		5	9	12	15	18	21	24	27	30	33	36	39	42	45	48
3	4		5	7	9	10	12	15	18	21	24	27	30	33	36	39	48
4	4		5	7	9	10	12	15	18	21	24	27	30	33	36	39	48
5	4		5	7	9	10	12	15	18	21	24	27	30	33	36	39	48
6	4		5	7	9	10	12	15	18	21	24	27	30	33	36	39	48
7	4		5	7	9	10	12	15	18	21	24	27	30	33	36	39	48
8	4		5	7	9	10	12	15	18	21	24	27	30	33	36	39	48
9	4		5	7	9	10	12	15	18	21	24	27	30	33	36	39	48
10	4		5	7	9	10	12	15	18	21	24	27	30	33	36	39	48
11	3		5	7	9	10	12	15	48								
12	3		5	7	9	10	12	15	48								
13	3		5	7	9	10	12	15	48								
14	3		5	7	9	10	12	15	48								
15	3		5	7	9	10	12	15	48								
16	3		5	7	9	10	12	15	48								
17	3		5	7	9	10	12	15	48								
18	3		5	7	9	10	12	15	48								
19	3		5	7	9	10	12	15	48								
20	3		5	7	9	10	12	15	48								
21	3		5	7	9	10	12	15	48								
22	3		5	7	9	10	12	15	48								
23	2		5	7	48												
24	2		5	7	48												
25	2		5	7	48												
26	2		5	7	48												
27	2		5	7	48												
28	2		5	7	48												
29	2		5	7	48												
30																	
31																	

Table A.6 Possible number of bits per subband for the following input sampling rate and output bit rate combinations: Fs = 32 kHz and bit rates per channel = 96, 112, 128, 160, 192, and kbits/s and free format; Fs = 44.1 kHz and bit rates per channel = 96, 112, 128, 160, 192, and kbits/s and free format.

		Allocation index															
sb	*nbal*	0	1	2	3	4	5	6	7	8	9	10	11	12	13	14	15
0	4		5	9	12	15	18	21	24	27	30	33	36	39	42	45	48
1	4		5	9	12	15	18	21	24	27	30	33	36	39	42	45	48
2	4		5	9	12	15	18	21	24	27	30	33	36	39	42	45	48
3	4		5	7	9	10	12	15	18	21	24	27	30	33	36	39	48
4	4		5	7	9	10	12	15	18	21	24	27	30	33	36	39	48
5	4		5	7	9	10	12	15	18	21	24	27	30	33	36	39	48
6	4		5	7	9	10	12	15	18	21	24	27	30	33	36	39	48
7	4		5	7	9	10	12	15	18	21	24	27	30	33	36	39	48
8	4		5	7	9	10	12	15	18	21	24	27	30	33	36	39	48
9	4		5	7	9	10	12	15	18	21	24	27	30	33	36	39	48
10	4		5	7	9	10	12	15	18	21	24	27	30	33	36	39	48
11	3		5	7	9	10	12	15	48								
12	3		5	7	9	10	12	15	48								
13	3		5	7	9	10	12	15	48								
14	3		5	7	9	10	12	15	48								
15	3		5	7	9	10	12	15	48								
16	3		5	7	9	10	12	15	48								
17	3		5	7	9	10	12	15	48								
18	3		5	7	9	10	12	15	48								
19	3		5	7	9	10	12	15	48								
20	3		5	7	9	10	12	15	48								
21	3		5	7	9	10	12	15	48								
22	3		5	7	9	10	12	15	48								
23	2		5	7	48												
24	2		5	7	48												
25	2		5	7	48												
26	2		5	7	48												
27																	
28																	
29																	
30																	
31																	

Table A.7 Possible number of bits per subband for the following input sampling rate and output bit rate combinations: Fs = 44.1 kHz and bit rates per channel = 32 and 48 kbits/s; Fs = 48 kHz and bit rates per channel = 32 and 48 kbits/s.

		Allocation index															
sb	*nbal*	0	1	2	3	4	5	6	7	8	9	10	11	12	13	14	15
0	4		5	7	10	12	15	18	21	24	27	30	33	36	39	42	45
1	4		5	7	10	12	15	18	21	24	27	30	33	36	39	42	45
2	3		5	7	10	12	15	18	21								
3	3		5	7	10	12	15	18	21								
4	3		5	7	10	12	15	18	21								
5	3		5	7	10	12	15	18	21								
6	3		5	7	10	12	15	18	21								
7	3		5	7	10	12	15	18	21								
8	3		5	7	10	12	15	18	21								
9	3		5	7	10	12	15	18	21								
10	3		5	7	10	12	15	18	21								
11	3		5	7	10	12	15	18	21								

Table A.8 Possible number of bits per subband for the following input sampling rate and output bit rate combination: Fs = 32 kHz and bit rates per channel = 32 and 48 kbits/s.

		Allocation index															
sb	*nbal*	0	1	2	3	4	5	6	7	8	9	10	11	12	13	14	15
0	4		5	7	10	12	15	18	21	24	27	30	33	36	39	42	45
1	4		5	7	10	12	15	18	21	24	27	30	33	36	39	42	45
2	3		5	7	10	12	15	18	21								
3	3		5	7	10	12	15	18	21								
4	3		5	7	10	12	15	18	21								
5	3		5	7	10	12	15	18	21								
6	3		5	7	10	12	15	18	21								
7	3		5	7	10	12	15	18	21								
8																	
9																	
10																	
11																	

Index

Digital Television, by John Arnold, Michael Frater and Mark Pickering.